$9.26 ✓

D1314898

WATER QUALITY

Water Quality: Management of a Natural Resource

James Perry
Department of Forest Resources
College of Natural Resources
University of Minnesota
St. Paul, Minnesota

Elizabeth Vanderklein
Science Writer
College of Natural Resources
University of Minnesota
St. Paul, Minnesota

Consulting Editor:
John Lemons
Department of Life Sciences
University of New England
Biddeford, Maine

**Blackwell
Science**

Blackwell Science

Editorial offices:
238 Main Street, Cambridge, Massachusetts 02142,
    USA
Osney Mead, Oxford OX2 0El, England
25 John Street, London WC1N 2BL, England
23 Ainslie Place, Edinburgh EH3 6AJ, Scotland
54 University Street, Carlton, Victoria 3053,
    Australia

Other Editorial Offices:
Arnette Blackwell SA, 224, Boulevard Saint Ger-
    main, 75007 Paris, France
Blackwell Wissenschafts-Verlag GmbH Kurfürsten-
    damm 57, 10707 Berlin, Germany
Zehetnergasse 6, A-1140 Vienna, Austria

Distributors:
USA
    Blackwell Science, Inc.
    238 Main Street
    Cambridge, Massachusetts 02142
    (Telephone orders: 800-215-1000 or 617-876-
7000; Fax orders: 617-492-5263)
Canada
    Copp Clark, Ltd.
    2775 Matheson Blvd. East
    Mississauga, Ontario
    Canada, L4W 4P7
    (Telephone orders: 800-263-4374 or 905-238-
6074)
Australia
    Blackwell Science Pty., Ltd.
    54 University Street
    Carlton, Victoria 3053
    (Telephone orders: 03-9347-0300; fax orders
03-9349-3016)
Outside North America and Australia
    Blackwell Science, Ltd.
    c/o Marston Book Services, Ltd.
    P.O. Box 269
    Abingdon
    Oxon OX14 4YN
    England
    (Telephone orders: 44-01235-465500; fax or-
ders 44-01235-465555)

Acquisitions: Jane Humphreys
Development: Kathleen Broderick
Production: Irene Herlihy

Manufacturing: Lisa Flanagan
Typeset by BookMasters, Inc.
Printed and bound by Capital City Press
©1996 by Blackwell Science, Inc.

Printed in the United States of America
    97 98 99 5 4 3 2

Library of Congress Cataloging-in-Publication Data

Perry, James A. (James Alfred), 1945–
    Water quality : management of a natural
resource / James Perry and Elizabeth Vanderklein.
        p.   cm.
    Includes bibliographical references and index.
    ISBN 0-86542-469-1
    1. Water quality management—Government
policy.   2. Water quality management—Social
aspects.   3. Water rights.   I. Vanderklein,
Elizabeth.   II. Title.
HC79.W32P44   1996
333.91—dc20   96-8574
                                              CIP

Contents

Preface

The Aral Sea has been a pivotal feature of the Central Asian landscape for millennia, moderating the regional climate sufficiently to enable productive agriculture and civilization. Massive evaporation from the sea's 64,000 km^2 surface was lifted by the cold north winds and carried thousands of kilometers to the Pamir and Tyan-Shan mountains, where it was deposited as snow. From there, melting snows carried life-giving water throughout the Central Asian plains.

In the 1960s, Russian scientists and policy makers set out to drain the Aral Sea and reclaim the drained basin for cotton production. Scientists claimed that drying the Aral Sea would be far more advantageous than preserving it, that productivity of the reclaimed land would more than make up for lost fisheries and transportation industries, and that disappearance of the sea would not affect the region's landscape.

The result: the Aral Sea will disappear by the year 2010, leaving behind an ecological and social desert. Massive irrigation projects in the region have reduced the Aral Sea to less than 40% of its original volume and more than tripled its salinity. Ground levels have subsided (in some areas by as much as 8 m) due to groundwater pumping. More than 80% of the animals found in the region have disappeared. Increasing wind erosion has covered agricultural land with salt deposits from the newly exposed sea bed, and both daily and annual temperature ranges are increasing significantly. As a final injustice, draining the Aral Sea has changed the regional climate sufficiently so that it can no longer support the vital cotton crop for which the sea was originally sacrificed. The amazing story of the Aral Sea is not unique to Russia: worldwide, scientists and decision makers alike have a long way to go to understand the wide-ranging implications of managing water quality in a landscape.

The Need for a New Text: New Paradigms, New Tools, New Perspectives

Water quality management was once a purely technical subdiscipline of hydrology; its goal was accurate determination of water chemistry. While the current field of water quality management would not be possible without these technical roots, its scope has now expanded far beyond the laboratory. It is now a social and political discipline whose concerns range from ensuring adequate health standards to preserving biological diversity and ecosystem integrity. Both the evolution of the field and its elevation in societal importance have taken place rapidly. Consequently, the literature offers today's water quality managers and policy makers only technical manuals that provide little guidance or support for the everyday decisions they must make. Complementing these technical manuals is an increasing array of specialty publications that address water quality as a subset of water availability or environmental degradation. None of these addresses the field in its current broad context. This information gap constitutes the first reason for this book: *the technical aspects of the field are well understood, but the social, biophysical, land use, and policy considerations that are now part of the field are rarely addressed.*

The Need for an Integrated Approach

Historically, water quality has been studied by a loose association of specialists working in related fields: hydrologists, chemists, biologists, engineers, and policy makers all carved out pieces of the water quality puzzle. This reductionist approach has led to a long list of unidimensionally successful and multidimensionally tragic water projects. In the example of the Aral Sea, by focusing on only agricultural productivity, cotton harvests in Soviet Central Asia were increased temporarily but the Aral Sea has been reduced to a salt marsh and regional ecology transformed. In another example, through a desire to improve marketability of Lake Victoria fish, an introduced exotic provided a temporary increase in fish harvests but over the long term has resulted in a permanently altered and impoverished food chain of the lake and its surrounding riparian zones.

The fact that water quality management is not successful when divided among sectors constitutes the second need for this book: it is increasingly clear that dissecting water quality among specialties in this way yields an incomplete picture and results in more cases of management failure than success. Instead of taking the specialist's view and focusing on problems narrowly, it is now evident that understanding water quality requires an integrative and large-scale view: looking at whole landscapes and whole regions to see how hydrology, chemistry, biology, geology, land use, demographics, public attitude, policy, and an expanding world view all interact to determine the quality of our water resources. Consequently, *our second rationale for this book is to communicate a synthetic, landscape-scale perspective that will increase our ability to take appropriate remedial action in response to a given problem.* More importantly, it enables us to be proactive—to anticipate how ecosystems will respond to societal and natural pressures on water resources so that we may plan accordingly.

Management and Ecology on the Verge of Change

The third reason for this book is that significant developments in the field of ecology have changed the way in which management must be approached. Management reflects changes in both society and scientific understanding. Thus, the future of water quality management lies in the social and scientific paradigms of today, just as the management of today is based on the paradigms of 10 years ago. Today's science presents a unique opportunity, however. For the first time in the history of natural resource management, enough understanding exists in different fields that management might no longer have to be reactive. It is possible that management can begin to be, and remain, a proactive field that looks at anticipated needs, anticipated effects, and social values to plan management strategies and practices. *Our third rationale is to develop and present a proactive view, requiring an awareness of the social contexts within which management occurs, an awareness of new ecological theory with respect to ecosystem responses to stress, and an awareness of how policy is implemented in different situations and different countries.*

A Human Problem—A Global View

Finally, this book reflects changes in the way in which management decisions are made worldwide. International comparisons in natural resource management were once just interesting asides, but in today's globalized and internationalized world, it is critical to understand how decisions are made across boundaries. U.S. and European Community guidelines for water quality management each grew out of unique social and biophysical conditions, yet these standards increasingly are being applied in other countries where they often prove inappropriate. Additionally, an expanding dimension of water quality management deals with cross-boundary pollution. Both of these conditions require understanding the policy framework within which decisions are made and understanding which factors can and cannot be generalized. Another international issue of which managers must now be aware is the role of water quality in international and regional conflicts (e.g., it has been used both as a weapon, as in the case of the Persian Gulf War in 1991, and as a tangible factor contributing to ethnic and religious battles such as those between Israelis and Palestinians, or between North and South Koreans). *Thus, our fourth rationale is to "internationalize" the perspectives of scholars and students interested in water quality science and management.*

Consequently, this book offers four critical new dimensions to the existing literature on water quality management: (1) a *social dimension* and an understanding of how water quality management has evolved to its present state; (2) an *integrated dimension* in which the interactions among chemical, physical, biological, and social aspects are addressed; (3) an *ecological dimension* that uses new concepts of the hierarchical structure of ecosystems, new understanding of the response of ecosystems to stress, and a new frame of reference to ecological problems (e.g., regionalization) to yield a proactive management strategy; and (4) an *international dimension* that demonstrates the care required in applying water quality management across borders and regions as well as demonstrating similarities in different societies' approaches to water quality management.

The Structure of the Book

The specific goals societies have for their water resources stem from cultural traditions and perceptions of resource availability. From these goals emerge each society's water quality management needs and approaches. Consequently, a study of water quality must begin with a discussion of the ways societies use water and the regional, cultural, and historical factors that determine a society's valuation of its natural resources. We focus on these issues in the first section of the book.

Section two begins with a chapter that links the social dimension with the biophysical one: the importance of scale issues in ecology and in our perception of ecological problems. The section then presents background information on the biophysical environment. The approach used here is to highlight the characteristics of the hydrological cycle, lakes, rivers, coastal zones, and wetlands that are most important from a water quality perspective. This information on the basic water resources and their components provides vital background for understanding impacts of different uses on water quality and the management implications of those impacts.

Water quality is inherently an ecological science. Consequently, section three reviews important new developments in the fields of ecology—in particular, how ecosystems are structured as vertical and horizontal hierarchies and the fact that many ecosystems have similar responses to stress. This section also includes a chapter on the move toward management based on ecoregional characteristics of water quality and land use. Chapters in this section focus on our concern that water quality management must be implemented at a landscape or regional level. They are critical background chapters in which we give readers the conceptual tools with which to approach problems and dilemmas in water quality management.

In section four, we introduce a series of water quality case studies involving different land uses and their impacts on water quality. This section begins with an overview of water quality concerns and the ways in which water quality is affected by land use activities. This overview is followed by more detailed chapters on water quality effects of forestry, agriculture, urbanization, and special issues such as exotic species, global warming, acidification, and eutrophication. In each chapter, we discuss how social contexts, biophysical constraints, and issues of scale affect responses to these problems in different parts of the world. Chapters in this section are divided among sectors for convenience, but each chapter is designed to demonstrate the ways in which the different areas of human management are integrated and inseparable. Each chapter is also structured to give an overview of the dimensions of the issue (e.g., land use), the ways in which it may affect water quality, and actual impacts and significance of those changes and solutions (i.e., physical remediation or policy actions). Each chapter concludes with management case studies that illustrate the complex contexts within which water quality decisions are made.

In section five, we return to the social context of management, looking at water quality policy issues and decision making around the globe. Chapter 25 discusses the cultural dimensions of policy making such as the influences of religion and philosophy, historical experiences with pollution, political structures, and socioeconomic status. Chapter 26 discusses the elements of integrated water quality management, analyzing both the enormous potential for integrated management as a new management paradigm and some of the serious constraints to its successful implementation. Finally, in chapter 27 we present a series of case studies that illustrate the principles outlined throughout the book: principles of biological appropriateness in management, of the importance of understanding user values, and of the nature of overarching political and social concerns in the application of water quality management.

Management—A Multidimensional Problem

Throughout this book, we present the perspective that water quality management is multidimensional, operating across spatial and temporal scales, across political and social boundaries, and across land uses. We stress that water quality impacts may result from an array of sources as varied as the choices of personal household items such as soaps and paints, to complex effluents from factories owned by multinational conglomerates, to the choices of crop and tilling patterns on agricultural lands. Thus, the biophysical changes in water quality are inseparable from the cultural practices that caused them and the social and cultural dimensions must be included in any effective management plan. We also stress that water quality changes cause both ecological and social chain reactions. The social chain reaction begins with ecological impacts and progresses through changes in land use practices brought about because of those impacts. This water quality chain reaction can affect every dimension of a society: changing its dominant economic activities such as fisheries or agriculture, changing recreation patterns, altering health conditions, and altering the distribution and success of industrial activities. Unequal impacts of poor water quality on different sectors of a society make social justice as much a part of water quality decisions as effluent standards for different industries. The concept of an ecological chain reaction is more familiar and few would be surprised to read that water quality changes can affect the long-term viability of ecosystems. Because of the range of areas affected by water quality, managers must be skilled technically, but must also be astute social scientists, able to observe and stave off inevitable conflicts among users and affected human and nonhuman populations. Societies must engage in productive discussions about "acceptable" versus "unacceptable" changes in water quality and to mediate disputes about costs and benefits of management choices.

Goals and Intended Audience

This book is intended to serve two audiences: third- to fifth-year students in water quality management, and applied problem solvers and decision makers interested in a broad and integrated approach to water quality management. We assume that the reader has some background in science (e.g., basic physics, chemistry, and biology at the college level) but not necessarily any previous exposure to coursework in aquatic sciences or natural resource management.

Anticipating that this material will be used in classes led us to choose a format of numerous short chapters rather than a few long ones. Each chapter treats a well-defined topic and contains appropriate reading material for a discrete section of a class. Each begins with a chapter overview and concludes with a list of review papers, texts, and primary journal articles that are suggested for further reading. A more extensive literature list, including all material referenced in the chapters, is presented at the end of the book.

The measure of a text such as this one is its utility: a text for managers must reflect the needs and experiences of its users. We urge our readers to write to us so that we may improve the book.

Acknowledgments

Many people have contributed immeasurably to this effort. We are especially grateful to our respective families (Peggy, Kendra, Leslie, and Ben; Dirk, Kim, and Ian) for their support and patience. Mary Kay Corazalla provided valuable assistance with figures and literature sources as well as ideas and editorial assistance. Many colleagues and students offered review comments, including Pam Davis, Gray Henderson, Bill McDowell, Brian Shelley, and Nels Troelstrup. John Hall designed the cover. Jane Humphreys and Debra Lance, both of Blackwell Science, provided valuable guidance. In addition, this project was supported in part by the Minnesota Experiment Station through Project 42-25 of the McIntirre Stennis Cooperative Forestry Act.

On a personal note, Leigh wishes to thank the Center for the Environment at Cornell University, and in particular the Center's former director, Jim Lassoie, for offering me a visiting position while I was away from Minnesota. Access to Cornell's libraries, internet system, lectures, and astounding gardens has made this possible. On an even more personal and fundamental level, I would like to thank Jim Perry. I owe him professional thanks for all we have written and talked about over the years, and personal thanks for his unending patience, support, and encouragement. He has been a true mentor and friend. Lastly, I would like to thank Kim and Ian for sharing their mom with the computer and for helping me keep my priorities straight; Dirk for being my partner; and my small, but increasingly intimate circle of family for banding behind me. I hope this can contribute in some small way to changes for Kim, Ian, and Marshall's generation.

1

Water Quality Management: An Evolving Field for Changing Values

Overview

The ways in which a society manages water quality is a telling reflection of political, cultural, and economic processes within that society. In many cases, the same approaches are used by different nations to achieve notably different goals. In other cases, differences in administrative traditions may cause countries with virtually identical water quality value statements to adopt radically different approaches toward implementing their goals. The variations in both goals and approaches are reflected in the earliest stages of water quality management: assigning use-oriented values to different water resources and planning for implementation of goals based on those values.

All management decisions stem from the context of use-oriented goals. Consequently, in societies dominated by development goals such as improving riverways for transportation, constructing irrigation systems, and using water for waste disposal, water quality management is generally an afterthought in the planning process. Many societies around the globe, however, are no longer driven by unchallenged development pressures. Most now recognize a broader range of values, such as those involving public health, amenity, recreation, and ecological integrity (e.g., "environmental" values). As these sometimes compatible and often competing uses are balanced, water quality management is moving into a more primary position at the beginning of planning processes rather than as a tail-end consequence.

Water quality management is, as a result, a very fluid phenomenon: as society and societal values change, new values must be reflected in management directives and strategies. At any given place and time, water quality management goals will reflect a different balance between social pressures such as development status and economic priorities, perceptions of resource scarcity or abundance, environmental values, traditions of public incentives, or even experiences with pollution crises. In all cases, water quality management is a multidisciplinary field, whose practice requires knowledge of hydrology and chemistry, biology, sociology, and politics.

Coincident with the response of water quality management to value shifts and scientific knowledge, a larger-scale shift in water quality management is also occurring. Globally, water quality policies are undergoing a shift from linear planning in which goal statements define specific policies and tactics to a more integrated approach that incorporates ecological reality, demands for multiple-use planning, monitoring, and feedback. These elements are all critical in today's world, and their omission in the past has been at the heart of most failures of water quality management to date. New developments in ecological theory, coupled with new global awareness in society, necessitates development of a new management model for water quality. This model integrates multiple spatial, temporal, and political scales (hierarchies) and makes broad use of feedback loops to improve management actions continually.

Introduction: Why Manage Water Quality?

Water is the most basic natural resource. People subsist where there are no forests, where there are no fish, where soil cannot be utilized, and where wildlife cannot be found. Yet even in a desert, human society centers around the oasis, just as in the rain forest it centers around the rivers. For plants and animals alike, water is essential for life, making up as much as 65% of the human body (90% of an infant's body). It also provides homes for fish and animals, refuge, food, navigation, electricity and mechanical power, coolant, a waste stream, and opportunities for recreation. No body of water, however, can support all of these different uses without suffering some degree of degradation: under poor management, degradation may result from even one primary use.

Water quality management serves largely as an intermediary or interpreter between any water body and its users, balancing the biophysical capabilities of the water resource against the multitude of uses that may affect it. Notably, different societies, cultures, and regions all have different priorities based on their water-related needs so the water quality manager in each situation must balance a different set of considerations. For example, if a community's needs center around meeting minimum domestic uses, less concern will arise over specific quality considerations. For societies in which availability is not a critical limitation, however, valuation of water bodies will reflect a broader range of human uses. An important reality in water quality management is that no body of water can be all things to all people: uses frequently conflict, and valuation of water quality depends on the social, political, and cultural contexts of those uses.

Conflicts over water use are as old as civilization, and water laws mandating water use rights and outlining quality concerns related to this peculiarly "common" resource are among the most ancient of laws. While legal systems still form the framework for water use decisions and are the traditional form through which conflicts are resolved, changes in use patterns, population increases, industrial stress on ecosystems, and increasing pressure on water resources have created a revolution in water quality management. Water quality awareness and concerns have prompted managers and decision makers to consider water resources and water quality as part of a more integrated framework. Rather than the single-focus, deterministic model assumed by a long history of water development management, emerging models of water quality management recognize the contextual nature of water management decisions, the integrated nature of social and ecological effects, and considerations over a larger temporal and spatial scale.

This chapter stresses the human context of all water quality management decisions, and outlines both the traditional, deterministic approach to management and the evolving integrated approach. With this background, subsequent introductory chapters will consider global patterns of water availability and use in an historical context (chapter 2), how societal attitudes are reflected in goals and management strategies (chapter 3), and how goals and strategies are translated into designated uses (chapter 4). The final chapters of this introductory section are devoted to a discussion of environmental toxicology, the traditional science behind water quality policy statements (chapter 5), and the evolving use of bioindicators and other environmental assessments for policy and classification purposes (chapter 6).

From Unidimensional to Multidimensional Approaches

Traditionally, water resource management has been unidimensional, with actions designed to address single-purpose needs such as hydropower, irrigation, or navigation. If fisheries, biodiversity, agricultural, or other uses of the water resource were negatively impacted through construction of dams or navigation channels, it was generally seen as a necessary sacrifice to "progress."

Traditional, unidimensional water resource management is still practiced in areas throughout the world where multiple uses of water or ecological consequences of water development receive minimal, if any, overt valuation. Ribesame (1992), however, echoes a growing trend in water quality management: that the dominant feature of recent water resources planning around the world is no longer a technical or financial issue, but a "growing intolerance of environmental and social impacts." Consequently, changing global values increasingly demand that water quality integrate environmental, public health, recreation, and amenity values with economic ones. Consequently, present-day water quality management must be multidimensional, as managers must contend with the multiple ways in which land use practices affect water quality, and water quality's subsequent effect on land use, health, and economic and biological viability.

A corollary concern that distinguishes traditional and emerging water quality paradigms is the concern with ecological integrity of water bodies. More than a concern over specific ecological effects, ecological integrity implies the ability of a system to be self-perpetuating and sustainable. One important lesson learned throughout the history of water resource management has been that multiple uses cannot be ensured unless integrity of the entire system is maintained. Severely degraded water bodies can provide neither utilitarian nor aesthetic nor ecological benefits. Consequently, although emphasis on (or even awareness of) integrity is quite new from an historical context, it lies at the heart of how water quality management today is distinguished from traditional water resource management approaches.

The Role of Societal Values

Management is a peculiarly human practice and the starting point of management is, almost by definition, a designated (human) use. A more useful frame of reference, however, is provided by recognizing that *uses* originate from the different *values* humans place on a given resource. Thus, the job of management is to ensure that the chemical, physical, and biological integrity of a water resource is maintained within the bounds required for a suite of human values. The value context explains why water quality management must draw heavily on social and managerial sciences in addition to biological, chemical, and physical sciences. It also explains some of the forces behind changes in the field of water resource and water quality management.

Because management occurs within the context of human valuation, each society develops unique management systems and unique management goals. Individual and societal perceptions of natural resources reflect biophysical realities (e.g., scarcity and abundance of a resource) but also reflect cultural values, historical experiences, and political realities (see chapters 2 and 25). In turn, water quality questions and answers change from society to society, and from region to region. Thus, the water quality manager asks the same question in each society and each instance: is quality (and quantity) sufficient to support a given mix of uses? The answers to that question change depending on the audience. Even within a single state, perceptions of water quality vary with the expectations of the users. For example, recreational users expect significantly higher clarity of lake water in the northern areas of Minnesota than in the South; awareness of those regional expectations is critical to the success of management actions.

Different value systems may also reflect economic realities. On a large scale, economic forces are often "averaged out" to a country- or region-wide set of accepted standards. Economic conditions of nations, as a whole, may create vastly different national standards. National standards are only one part of water quality decision making, however. In practice, water quality decisions are carried out at local levels where local economies and local resources may create vast discrepancies between the national "consensus" values and the true local valuation of a resource. For example, local-scale concern over long-term water quality degradation is often less important to people if their short-term survival is dependent on the polluting industry or land use practice. For example, high lead threatens health and productivity of whole communities in Poland, yet the factories continue to run because workers need tomorrow's meal; likewise, in the 1980s California farm workers judged the ecological and health threat of selenium poisoning—caused by farm irrigation—less important than the continued operations of the farms on which they worked. Equally common, however, is the scenario in which local communities become concerned over a quality issue and force more stringent controls into effect than would be expected under national level value systems.

In addition to broad societal differences in valuation of resources, each different use is associated with its own nested set of values and expectations. Water quality impacts on a body of water will be judged according to these unique value sets. For example, although fishing and swimming in a lake usually are compatible uses, each has its own set of standards and water quality may affect each use differently. Specifically, acid rain results in many changes in a lake, often including significant increases in water clarity. It also often leads to dramatic reductions in fish populations. To recreational swimmers or boaters, water clarity has a high value and the fish population may be largely irrelevant. Consequently, for these uses acidification may be perceived as an improvement in water quality; if swimmers constituted the only users of a water body, then the political and social forces to mediate effects of acidification might be low. While fishermen might consider clarity to be pleasant aesthetically, their judgment of water quality is more closely tied to the health and abundance of fish populations: to a fishing-dependent community, the lack of fish would certainly constitute low water quality. Because fishing values factor in with concerns over ecological integrity and possible health effects from acidified waters, remediating the effects of acidification is usually a high local priority.

While uses and users help define the values ascribed to different water bodies and water quality threats, water quality values are also a result of bias filters used by the decision makers. As described by Gerlach (1993):

> Many technocrats recognize that they also interpret reality through the cultural filters of their respective groups. They are not aloof bystanders to the decisions that follow from their measurements and calculations; rather, they hold stakes in such decisions—stakes of objective interest and subjective identity. Jobs and careers are at stake, as is pride in working to help the world. But officials and specialists are also biased by their group culture. Agreements and disagreements among specialists and between specialists and the public are a function of many factors, only some of which are rooted in the accuracy and plenitude of scientific research [p. 283]

Gerlach (1993) continues to describe how during a 1980s drought in the Midwest, public health officials, engineers, politicians, and natural resource managers all had vastly different interpretations of the same data, with regard to whether public supplies were low enough to constitute a threat and whether extra water should be released from reservoirs.

The Traditional Approach to Management Planning

Given value conflicts and the perceptual filters of different users, and decision makers, how does society decide which values to ascribe to a water body? That is, how does a community decide whether water clarity, as in the acidification example above, is significant or not? The planning process begins when these diverse values and uses are summed up in a societally determined goal statement. In the traditional approach to resource planning, translating these goals into actions was (and in many places still is) a linear process in which goals were used to define policies, policies were used to define objectives, objectives defined strategies, and strategies defined implementable tactics. Goals, therefore, constitute the top level in a multitiered planning process, representing the

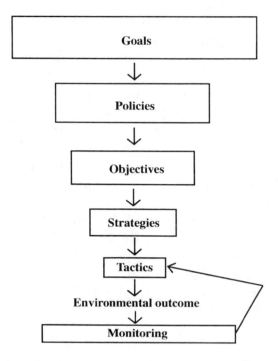

Fig. 1.1a. Traditional model of water quality planning. This approach follows a linear path in which feedback is limited and public input is minimal. Scientific advances and changing societal values, however, are forcing a change to hierarchical, iterative models (see Fig. 1.1b.).

broadest level of thought about a resource. From that point, the manager's task has always been to define increasingly focused answers to how to implement the goals (Fig. 1.1a). These tiers may be described as follows:

- *Goals* (e.g., portable water, fishable waters, electricity generation) are broadly defined and represent directions society wishes management to take. Goals are used to introduce laws and policies and to set the tone for major documents. For example, the U.S. Clean Water Act sets "fishable, swimmable" waters as its broad goal. In other cases, a goal may be expressed as "provide acceptable drinking water to 100% of households by the year 2000" or "ensure that fish inhabiting waterways in all major metropolitan areas are safe to eat."
- *Policies* are more specific statements that describe how an organization intends to accomplish its goals. They represent the current and currently obtainable practices. A policy under the "fishable, swimmable" goal might be "all municipal wastewaters shall receive at least secondary treatment." A policy under the drinking water goal might be "all citizens shall have

reasonable access to safe drinking water." In some societies (e.g., the United States), policies are legally binding and policy makers can be sued in civil court for failure to comply. In parliamentary governmental states (e.g., the United Kingdom), policies do not have the force of law.

- *Objectives* are more quantitative, shorter-term, and more specific; they are intended to represent steps to be accomplished in complying with policies. In other words, a policy constitutes a general directive to be accomplished, and objectives break down that directive into the components necessary to realize that policy. Generally, an organization will establish several specific, perhaps relatively quantitative, objectives that must be accomplished to comply with a policy. Exemplary objectives might be "establish a testing program for all city water by year's end" or "establish 1000 new hookups to new residences every 6 months for the next 10 years."
- *Strategies* are planning approaches that will be followed to achieve objectives. Each objective may have several attendant strategies, such as "ensure adequate staffing and facilities for water quality testing objective."
- *Tactics* are specific actions used to achieve a management strategy. For example, tactics might include "hire and train 15 laboratory technicians to implement testing procedures," "convert unused lab space into a water quality testing lab," or "raise city taxes to pay for new facilities." They may include both physical actions (e.g., a management practice) and administrative actions (e.g., meetings, discussions, reports).

Where a goal starts and a policy or objective begins is not, in practice, a clearly drawn line. The exercise of planning from the general toward the specific is, however, important in clarifying the goals toward which we are managing and in determining how we will conduct our management actions. Management is a problem-solving process; a clear statement of the problem and its attendant possible solutions is a prerequisite to good management. Yet it is not a guarantee that management will be implemented. A stepwise planning process has been adopted in most countries and regions of the world, but all too frequently plans remain on the shelves and legislation is rubber-stamped until some crisis provokes partial and ineffective implementation.

Even more importantly, this linear planning model does not leave sufficient room for incorporating biophysical realities, scientific input, or feedback and reassessment. It serves a valuable function in that it focuses questions and answers about societal needs. Yet, a manager might follow this linear model perfectly and still arrive at a management plan that is ecologically and socially disastrous. This fallacy is highlighted in the Aral Sea example at the end of this chapter, in which highly linear and sectoralized planning trivialized the ecological ramifications of decision making. Managers in today's world require a more complex, less linear model with which to make informed and sustainable decisions.

A New Conceptual Framework: Hierarchies and Issues of Scale

As natural resource management, including water quality management, evolves from unidimensional to multidimensional concerns, decision making must become less linear than in the past. In part, this change comes about because of multiple-use demands and the implicit necessity to arrive at management compromises. It also stems from recognizing the need to incorporate biophysical realities into management plans. Long ignored in natural resource management, the biophysical potential and limitations of a resource are at last earning equal footing with cultural values. Ideally, resource management actions represent a compromise between human goals and the physical and biological potential of a given water resource.

To achieve the goals of balancing use with ecological integrity, management must pursue planning along several lines: (1) an understanding of the current resource condition and its potential to deliver goods and services; (2) goals of society or the decision maker; and (3) the appropriate physical and institutional mechanisms to accomplish those goals. Throughout implementation of a management strategy, room must routinely be made for feedback and modification of goals or policies and for incorporation of new scientific evidence about the effects of management decisions.

The importance of explicit goals—*why* we wish to manage the water resource—is as important in multidimensional planning as in traditional linear planning. An important difference between the two models is that an accurate understanding of the characteristics of the water resource itself ranks on equal footing with goal definition in multidimensional planning. The failure of many management plans is traceable to a poor definition of goals; failure of the remaining plans is usually attributable to ignoring geologic, hydrologic, and biological realities of the water resource. Combining explicit and realistic management goals with a realistic assessment of biophysical conditions and potentials gives us the vital understanding of "where we are" and "where we want to be" in our management strategy. This knowledge, in turn, provides the framework for deciding "how to get there" (i.e., the appropriate institutional and biophysical mechanisms for management).

Another important distinction between traditional linear and hierarchical (multidimensional) planning involves the latter's use of feedback loops at each stage. These loops enable managers to monitor and evaluate not only program success but also the appropriateness of the original goals and the accuracy of scientific understanding. Imagine, for example, a popular trout stream in a national forest. The trout require clear, cool water and management of the forest requires revenue from fishing for part of its political support and managerial actions. Consequently, fishing becomes a protected or beneficial use. The qualities of water clarity and temperature upon which the fishing depends dictate management practices such as forest harvesting; harvests must be scheduled in such a way they do not violate temperature or sediment standards for the stream reach. In practice, management actions are conducted and results are monitored to ensure that there is actual compliance with applicable standards and that the standards applied to the water resource are effective in maintaining desired quality.

In addition to incorporating biophysical realities and feedback loops, another critical feature distinguishing linear from "integrated" planning is involvement of the local populace in the decision-making process. Decades of natural resource management failures and some notable successes clearly illustrate that unless the affected population has some "ownership" of a management plan, the plan will be disregarded and ineffective. When the affected community is fully involved in the planning process, however, significantly more cooperation is fostered toward achieving the specific objectives and general goals of a management plan. Consequently, resource management activities are more successful and more enduring.

Just as traditional linear planning has an identifiable sequence, modern hierarchical management also has a distinct planning sequence. The two models are marked by two vital differences. First, in the hierarchical model, consideration of biophysical conditions is on an equal par with goal-setting. Second, steps in this process are not linear; feedback loops in the decision pathway lead to a constant reevaluation of the goals as well as impacts and alternative strategies. The elements of hierarchical planning are best described in stages, where each stage includes a set of parallel activities and results of each stage are fed back into previous ones to maintain the balance between ecosystem integrity and human use (Fig. 1.1b).

- In stage one, for example, priorities are assessed. During this stage, managers and the participating community engage in preliminary discussions, surveys, and assessments to determine which uses or values have the highest priorities. The multiparticipant, interactive nature of this stage immediately sets a different tone for the ensuing planning, when compared with the more deterministic linear planning model.

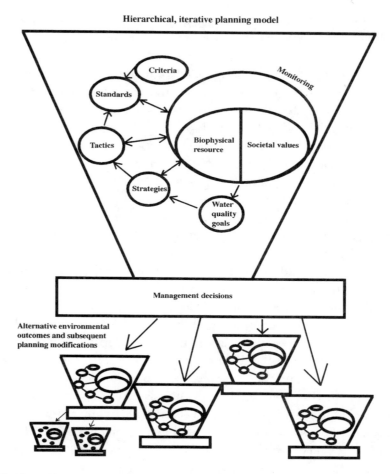

Fig. 1.1b Hierarchical, iterative planning model. In this diagram, societal values are composed of influences from history, economics, and culture as discussed in chapters 3, 25, and 26. The important considerations of the biophysical resource are discussed in chapters 7–13.

- Stage two is the point at which the preliminary discussions are crystallized into mutually agreeable, explicit statements of goals. Simultaneously, stage two is devoted to a biophysical assessment of the resources: are the stated goals suitable for the biophysical conditions, or should the conditions dictate a different set of goals? As the goals and the biophysical realities are compared, a need to reassess the stated goals often emerges. A feedback loop is created in which the stakeholders return to an assessment of priorities, with the new information in hand. The participants can then arrive at new goals and conduct new checks on the biophysical realities of the resource.

- Stage three is a time for integration and focusing. At this stage, policies and laws are written that express the goals of the participants, and preliminary strategies are set. This stage also includes the assessment of environmental impacts of the proposed strategies. Based on the language of the laws and policies, as well as on the results of environmental impact statements and reviews, managers and stakeholders may have to return to stage one to re-assess priorities and arrive at goals that have fewer negative impacts (or at lower costs), or they may return to stage two to refine their marriage of goals with the biophysical realities of the resource.
- Once goals, policies, and strategies meet the needs and expectations of stakeholders, the planning process proceeds to an implementation stage. As with the previous stages, implementation may bring to light a need for fine-tuning either the strategies, the policies and laws, or the goals themselves. At its best, this planning model would allow for sufficient flexibility such that feedback would be continuous among the stages, without unduly de-laying or interfering with management implementation.
- Finally, monitoring is a critical part of the hierarchical model, but it is dis-tinguished from the monitoring in traditional planning models by being integrated throughout the stages. Rather than referring to only a biophysi-cal assessment of whether the implementation has met its goals, monitor-ing in this framework refers to the continuous checks between and among participants. Do the goals reflect the needs and desires of the community? Can the resource support the stated goals? Will any environmental impacts outweigh the benefits of the proposed management actions? These ques-tions, which serve to link the stages in the process, are all part of the mon-itoring process. They complement the more familiar role of assessing changes in water quality parameters as a means to determine the success of different management strategies, and the newer role of monitoring the public's satisfaction with the management plan, and with its results.

In many ways, this model represents a marriage of two nearly parallel ap-proaches to resource policy (i.e., a policy approach and a scientific approach). Because it incorporates feedback loops and multilayered planning, it is well de-signed to detect ecosystem and social impacts while allowing for balance among management concerns. Thus, this model is attractive because it is sensitive to planning needs and accurately reflects social and ecosystem structures.

Traditional linear planning has a distinct advantage in facilitating efficient decision making. Because all decisions follow from the goal statement, man-agement planning can be as simple or as complex as a manager chooses to make it. In contrast, hierarchical planning loses efficiency as it requires increased

Fig. 1.2a Well-managed water resources, such as this Massachusetts coastline, present opportunities for diverse multiple uses while ensuring the integrity and sustainability of the water resource. (Photo by L. Vanderklein)

levels of evaluation and continuous reassessment. What it loses in efficiency, however, it gains in enabling sustainable management and improved water quality. It constitutes integrated thinking and may work precisely because it forces a multiscale view. As the relevant contexts expand, so does the potential for effective management (Fig. 1.2).

Because water quality management is on the cusp of a paradigm shift from traditional linear management toward hierarchical management, the literature is replete with examples of traditional planning and nearly silent on examples of hierarchical planning efforts. The remainder of this chapter is devoted to three case studies. The first looks at the consequences of linear,

Fig. 1.2b When goals focus on only one purpose, or a resource is used without consideration for the impacts of that use, water quality can become severely degraded. In this photo from a village in the Kathmandu Valley, Nepal, a local small lake is used for washing wool, washing clothes and dishes, and as a source of drinking water. It also receives run-off from construction activities and a dirt road. The lake supports no aquatic life, is shrinking from siltation, and is a medium for the transfer of diseases. (Photo by D. Vanderklein)

unidimensional planning to the Aral Sea region of Central Asia. The second reviews efforts by managers of the Laurentian Great Lakes to respond creatively to continued degradation of the lakes. The third stresses that water quality debates are ongoing issues for all societies, and that the structure of management planning itself will have a significant influence on the resolution of future water quality decisions.

Unidimensional Management and Tragedy: The Aral Sea

Among the most infamous examples of water planning gone awry is the truly tragic draining of the Aral Sea, a consequence of a single-minded focus on the goal of increased cotton production in Central Asia. It represents unidimensional, sectoral planning at its most extreme, and constitutes a severe cautionary tale for the sort of feedback-free and biophysically removed planning that has typified traditional water quality and water use planning.

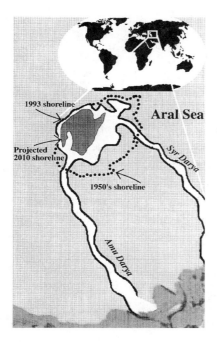

Fig. 1.3 Located in Central Asia, the Aral Sea is fed by two large rivers and serves a key role in maintaining the climate and productivity of area resources. As these rivers were canalized to promote irrigated agriculture, the volume of the Aral Sea has shrunk by two-thirds and its salinity has tripled.

The Aral Sea is located in the heart of Central Asia (Fig. 1.3). Historically, the Aral Sea was a vital feature of the landscape, and a critical oasis in a desert landscape. Its 64,500 km² area provided annual catches of as much as 50,000 tons of sturgeon, carp, and other important fish, wildlife pelts, and productive year-round pastures for grazing (Percoda 1991). The rivers that fed the Aral, the Amu, and Syr were used for nonintensive irrigation of grains. Ecologically, the sea was critical in regulating the regional climate: it interrupted the flow of cold north winds, and evaporation from its surface was precipitated thousands of kilometers away on the glaciers of the Pamir and Tyan-Shan mountains (Percoda 1991). As the waters melted, they provided fertile river valleys and ended up in the Aral Sea again.

In the years following World War II, the Aral Sea region (Khazahkistan) underwent rapid development. Economically depressed and dependent Russia placed increasing priorities on achieving independence in certain key commodities such as cotton. The Aral Sea region appeared to fit the requirements for intensive cotton production. Consequently, the area saw an unprecedented development of irrigation, a process involving the construction of numerous dams, reservoirs, and canals that increased the area of irrigated agriculture from 2.9 million ha in 1950 to nearly 7.5 million ha in the early 1990s (Percoda 1991). The sources of the irrigation were the tributaries of the Aral Sea: the Amu, the Syr, and the Vakhsh rivers. Consequently, water flowing into the Aral was reduced from its stable value of 50 km³ per year, to only 1–5 km³ through-out the 1980s. As predicted—and disregarded by planners of the agricultural development plan—the Aral Sea shrunk rapidly from its original height of 53 m above sea level to 40 m. In the process it lost, by various estimates, 60% to 70% of its volume and more than half of its surface area; as a consequence, salinity tripled (Percoda 1991; Tursunov 1989).

At the time the Aral Sea project was planned, Russian scientists accurately predicted that the effects of the irrigation withdrawals would be severe. Cultural biases of the time and political pressure to encourage the scheme led project sci-entists to see these consequences through a set of value filters that trivialized the implications compared with the benefits of increased cotton production. Per-coda (1991) quotes the former president of the Turkmen Academy of Sciences as saying, "I belong to those scientists who consider that drying up of the Aral is far more advantageous than preserving it. . . . Cultivation of this crop [1.5 million tons of cotton] alone will pay for the existing Aral Sea with its fisheries, shipping and other industries. Secondly . . . the disappearance of the sea will not affect the region's landscape." [p. 111]

Severe Impacts and No Feedback

Even if original planning processes ignored or vastly underestimated the implications of ecological changes that would result from the draining of the Aral Sea, feedback loops within a monitoring program might have prevented the severity of the effects by catching important indicators early on and sug-gesting modifications to the irrigation scheme. Typical of linear and sectorial-ized planning, however, few feedback loops were in place, and feedback that was received came from a different sector (e.g., ecological information) and was considered irrelevant by irrigation planners and managers (Tursunov 1989).

The consequences of the irrigation schemes are twofold. One set of conse-quences arises from the shrinking of the Aral Sea. Increased salinity, for ex-ample, had effectively ended the fisheries industry by the early 1980s. Other losses in economic vitality have since occurred as the Aral coast recedes: some former ports are now located as much as 80 km away from the current shore. Public health threats have also resulted from the shrinking sea. As the sea re-cedes, it leaves behind salt-encrusted banks that are quickly dried and carried in large regional dust storms between 150 and 300 km wide (Kotlyakov 1991). High incidences of brucellosis, tuberculosis, bronchial asthma, and other re-spiratory diseases are directly attributable to the highly saline dust that resi-dents breathe (Tursunov 1989). The transportation of the salts is not a purely local phenomenon. Although as much as 520 kg of salt fall on every hectare of land in the Aral Sea region, Aral Sea salt has also been carried to the Pacific and Arctic oceans and has been identified in the Ganges and Brahamaputra rivers of India (Percoda 1991).

Climate is also affected by the shrinking sea and salty dust storms. Because the sea no longer has enough surface area to interrupt the cold north winds and contributes less overall moisture to distant mountain snows, it no longer provides ameliorating temperature effects. Percoda (1991) reports that the absolute maximum average daily temperature is increasing as the sea shrinks. In addition, the salty dust storms that land on the mountain glaciers has promoted unprecedented glacial melting—12 times previous stable rates. The melting glaciers were overlooked for decades because the rivers still ran full. The source of the glaciers' water is not being replaced by new snow; rather, it is a mined resource. The further the glaciers melt, the more uneven the local weather will be.

The receding sea has also caused a substantial drop in groundwater level and a dramatic increase in groundwater mineralization. Groundwater in some areas has dropped as much as 8 m, and mineralization has tripled to 100 g per liter or more (Percoda 1991). The drop in groundwater has led to the drying up of tributary rivers fed by groundwater and the disappearance and dramatic reduction of riparian forest, wetlands, and the previously productive pasture land and meadowland. All of the dried areas have been replaced by salt beds.

The second set of consequences affecting the Aral Sea region involve the irrigation canals and the efforts to maintain cotton production in a rapidly changing landscape. For example, one of the most important design flaws in the Aral region irrigation scheme was the creation of very long, unlined canals. Because of the desert conditions and sandy soils, water seeps from the canals and evaporates rapidly. In some areas, only 20% of the water actually reaches the intended vegetation. The remaining water creates extensive salty flood plains along the canals, causing buildings and communication lines, orchards and vineyards to collapse as the ground under them sinks. In one district alone, more than 2000 ha of productive farmland has become salt marsh (Percoda 1991).

Cotton production has also suffered. Because of the seepage and salinization, cotton and rice production in the region requires two to four times as much water as in other areas. To maintain a low enough salt concentration in the soil, cropland must be flushed at least four times before planting. While the process removes 40 to 60 tons of salt per hectare, it also leaches salts and minerals that are required for productivity. To compensate for the loss, farmers in the region have turned to intensive use of fertilizers and pesticides at levels far above recommended concentrations. In some areas, pesticide use exceeds the national average by 20 times and some crops exceed health standards by two to four times the allowable levels of nitrate and pesticide (Percoda 1991).

The dramatic loading of pesticides and fertilizers has been correlated to the area's severe health crisis. The republic, for example, has the highest infant mortality rate in the country, and epidemics of yellow jaundice and intestinal diseases followed the increases in fertilizer application. The flooding of sewers and septic systems has also contributed to extremely high rates of bacterial diseases: typhoid in the area is more than eight times the national average.

Remediation and Future Planning?

The situation in the Aral Sea is clearly extreme and requires attention. Most observers agree that the situation can be improved if significant steps are taken to restore the flow of water to the Aral Sea. Some recommendations include reducing the volume of cotton and rice production, introducing widespread water-saving technologies, reconstructing irrigation systems (including lining canals), and ending the introduction of new irrigated areas (Tursunov 1989). Sufficient "waste water" from the irrigation system exists so that the Aral Sea could receive a flow sufficient to allow for at least partial recovery. Nonetheless, Tursunov (1989) reports that the planned reconstruction of the irrigation system calls for grading the fields and cutting thousands of kilometers of long canals. Thus, the solution preferred by officials is more of the same.

The crisis of the Aral Sea has been studied and debated by scores of scientists in the last decade. More than 100 Aral Sea conferences have been convened, and both scientific and popular journals continue to recount the story and possible solutions. To date the economic forces in the region, acting in concert with a weak tradition of integrated planning, have continued to prevent significant changes in the management of the area's water resources. At the current rate of desiccation, the Aral Sea will disintegrate into a series of shallow, unconnected lakes by 2010, and the region will become unsuitable climatically for cotton or rice production. The cost to ecological integrity and human health is already extreme. Action requires a commitment to an integrated management process that is currently beyond the reach of responsible parties.

Managing the Great Lakes: Innovation One Step Behind the Problem

In contrast to the Aral Sea, development along the Great Lakes shores was not a coordinated effort. Rather, as planning and development occurred at local levels along the coasts of the five Great Lakes, numerous "small" impacts accu-

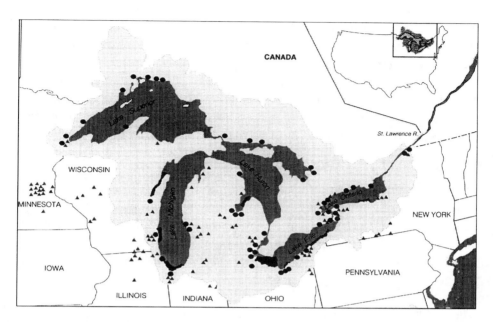

Fig. 1.4 The Great Lakes Basin forms a large section of the U.S.-Canada border and is home to more than 40 million people. The lakes and their drainage basin (designated with light shading) have been heavily impacted through the cumulative effect of numerous "small" impacts. In this figure, triangles denote toxic waste sites, circles denote areas of significant pollution, and dark shading at the edge of lakes (e.g., the southwest edge of Lake Michigan) denotes highly eutrophic areas. Toxic waste sites outside of the drainage basin are pictured here because subsurface pollution can affect water quality over a large area and does not necessarily correspond to surface drainage patterns. (Modified from Miller 1992; reprinted with permission courtesy of Wadsworth Publishers, Belmont, CA.)

mulated to create a series of water quality crises. What has distinguished the efforts of Great Lakes planners has been their interest in adopting new strategies in response to each new Great Lakes crisis. Perhaps more than any other system, the history of Great Lakes management reveals the evolution from sectorial, linear planning toward hierarchical, multidimensional planning.

The Great Lakes, which are home to more than 40 million people, have reached their present state as a result of a long line of impacts and management trials (Fig. 1.4). Like the salt oceans, these "sweet oceans" have always given the impression that they were constant and unlimited. Their sheer size—20% of the world's fresh water and 95% of the U.S.'s surface fresh water—long enabled people to disregard the effects of land use practices on the lakes themselves. In many places and communities, the myth of unpenetrable water quality still holds strong. This illusion creates an apathy toward deteriorating conditions, primarily because many communities cannot conceptualize that their small outputs could lead to significant basin-wide impacts. Other communities have accepted that the condition of the lakes has declined due to cumulative local impacts and are responding strongly to the challenges of improving conditions and preventing further degradation.

The five Great Lakes vary considerably in depth, total volume, and residence time. Even within a single lake, however, the morphometry creates numerous bays, inlets, and shoals that act largely as independent smaller lakes. Lake Erie, for example, has a shallow shoal at its center that alters circulation patterns and creates, in effect, a eutrophic lake within the larger lake. Similarly, bays such as Saginaw and Green Bay, although large, have little cross-flow with Lake Michigan and, therefore, constitute effectively isolated bays. The result is that local water quality can be severely degraded in these bays and inlets. From a management point of view, however, the flow restrictions at least create a concerned community and a natural point of focus for management activities. If the problem can be subdivided into natural segments, the problems and solutions may appear more manageable to the populace.

The Accumulation of Impacts

The history of Great Lakes water quality management has been as remarkable for the development of innovative management as for the blind trust in the fact that development would be relatively benign. Together these facts have combined to yield a series of successful campaigns against particular pollutants while the overall condition of the lakes has continued to degrade from multiple impacts. For example, nearly 140 exotic animals and plants have become successfully established in the Great Lakes ecosystem since the early 1800s, leading to dramatic alteration of food webs and community dynamics (see chapter 24) (Mills *et al.* 1993). Roughly 1200 miles of sanitary and shipping canals now criss-cross the Great Lakes, altering water flow and sediment input and facilitating the introduction of exotics (Ashworth 1987). More than 450 toxic substances have been identified in the Great Lakes, ranging from DDT carried by wind and rain from developing countries, to significant loads of PCBs, mercury, and other compounds from industrial outputs (Meybeck *et al.* 1989).

The development of the Great Lakes basin began in the early 1800s with the discovery of vast white pine forests. Ideal for everything from ships masts to railroad ties and building frames, the white pine forests of the Great Lakes basin eventually formed the building blocks of the U.S. industrial expansion. The extensive cutting, which left behind bare stumps from present-day western New York to Wisconsin, spurred significant changes in the Great Lakes ecosystems.

Higher temperatures in tributary streams, for example, have been blamed for the disappearance of the Atlantic salmon (Sly 1991), while heavy sedimentation of streams and shore areas introduced high loads of phosphorus into the area waters. Settlements followed the cleared lands, and by the 1900s large communities, including Detroit, Cleveland, and Chicago, had taken hold. The pace of development of these towns and cities left little time for city planning. Milwaukee, for example, went from a population of two families in 1833 to a population of nearly 50,000 in only 20 years. Similarly, Chicago's population erupted from 12 families between 1800 and 1830 to a population of 5000 by 1837, 200,000 by 1850, and 2 million by 1900 (Ashworth 1987). Spiraling community development along the three "Lower Lake's" coasts led to large accumulations of raw sewage along the coast—the same areas that were used for water consumption.

Sector by Sector, Reactive Management

The earliest organized efforts to combat water quality problems in the Great Lakes began with outbreaks of cholera and typhoid in many communities in both Canada and the United States. With death rates from these diseases climbing higher than 100 in 100,000 in some communities, the outbreaks dramatized the unsafe disposal practices of these early communities. The crisis led to the earliest water quality agreement between the United States (and its eight affected states) and Canada (and its two affected provinces). Central to the Boundary Waters Treaty of 1909 was the creation of an international coordinating body, the International Joint Commission (IJC). The IJC's mandate is rooted in the premise that the lakes are a common resource and that neither the United States nor Canada has the right to degrade their condition (Meybeck et al. 1989). Either government may refer water-related problems to the IJC, which then investigates the problem and recommends remedial actions. The cholera and typhoid outbreaks constituted the first referrals and, after study confirmed the unsafe sanitation conditions, the IJC made recommendations pertaining to the filtering and chlorination of municipal waters and the moving of intake valves. Successful eradication of the diseases followed in the 1930s.

Public health problems represented the majority of IJC referrals into the 1960s. During that decade, however, eutrophication of the Lower Lakes became an area of significant concern. Throughout the 1950s, scientific studies began revealing significant changes in the chemistry and aquatic community of Lake Erie: dissolved oxygen levels had dropped to 0 over a 2600 square mile area, and the abundant mayfly larvae had been replaced by a benthic community of oligochaetes, known as "sludge worms" (Ashworth 1987). Erie, the oldest and shallowest of all the Great Lakes (average depth 58 feet versus 200 feet or greater for the other lakes), was naturally borderline eutrophic. Billions of gallons daily of only minimally treated municipal and industrial sewage, sedimentation, and flow alterations caused the lake to become severely eutrophic, as well as heavily polluted with bacteria.

Although Erie was the bellwether, large bays in other Lower Lakes such as Green Bay, Saginaw Bay, Lake Michigan's shallow "south basin," and others were all beginning to experience severe eutrophication problems. An IJC referral subsequently studied cultural eutrophication in the lower lakes, and documented widespread and severe problems. This effort represented the beginning of experiments with "taking the problem to the people." The media became involved in outreach efforts and the U.S. Public Health Service conducted a series of hearings in Cleveland and Buffalo in 1965. Heated and contentious, the hearings publicized, but also polarized, the issues among municipal and industrial polluters. Municipal officials denied that their municipalities and industries were responsible for the degradation and blamed upstream communities. Denials notwithstanding (or perhaps because of them), public pressure to effect change mounted.

The debates set the stage for dramatic changes in water quality, both nationally and in the Great Lakes. In 1972, several acts were passed that expressed a commitment to water quality improvements. The U.S. Clean Water Act of 1972 designated federal water quality guidelines and standards and included a provision for penalties against polluters and a federal commitment to assist communities in upgrading and building sewage treatment plants. Canada passed a similar Water Act in 1972. In addition, the International Great Lakes Water Quality Agreement (GLWQA) was signed between the United States and Canada, establishing a framework for cooperative research and abatement programs. At a state level, several phosphate-limiting agreements were established between the states bordering Lake Michigan and Lake Erie, while states, provinces, and local communities all began instituting laws aimed at sewage effluent standards and the control of phosphates. Some industries even established self-regulation with respect to certain discharges; others took cities to court over new standards.

The phosphate reduction activities began to have noticeable effects by 1975, when the phosphorus loading levels were halved for Lake Erie and Ontario. By 1983, levels had dropped to approximately one-seventh of their 1972 rates. The reappearance of blue and algal free water is treated with caution by some experts. A coincidental increase in water level during the recovery period may have contributed significantly to the improved conditions. In addition, while total sediment load has decreased, other eutrophication indicators have not decreased accordingly. If water levels drop, the lakes could again become eutrophic (Ashworth 1987).

Great Lakes managers are contending with introduced exotic species, such as the alewife, that have altered the Great Lakes food chain; they must also deal with issues of pollution from agricultural land and other "nonpoint" sources, and with issues such as the increasing load of toxic compounds in the lakes. PCBs, for example, were used by Great Lake industries for more than 20 years before serious health and ecosystem-wide effects were noted. Managers must face down citizen apathy and ignorance as well. Many of the problems currently confronting the Great Lakes are difficult to detect: PCBs are odorless and colorless; changes in community structure may affect fishermen but are otherwise not obvious (nor are the implications clear to most shoreline residents); and toxic sediments are buried. Problems continue to build in the Great Lakes not because of institutional blindness, as in the Aral Sea, but because of piecemeal disregard for the implications of individual actions. Thus, until recently, management has been a process of playing catch-up with each new water quality assault that emerges.

Approaching Multidimensional Problems from Several Angles

Each "new" assault has, to date, been met with a new agreement, a new commission, and a new set of guidelines. This management structure has been termed by Meybeck et al. (1989) a "hierarchical morass of committees and responsibilities cascading downward from . . . the IJC" [p. 234]. However unwieldy, the IJC and the various Great Lakes Commissions and agencies have still distinguished themselves by continual innovation in water quality management. The evolution of approaches used in Great Lakes management has paralleled the level of water quality understanding on a scientific and social basis. Earliest efforts with typhoid, for example, were strictly a treatment of symptoms. Despite the effectiveness of the immediate restitution, the long-range implications were ignored. Nonetheless, the creation of an effective international commission and the political support it received were experimental and revolutionary for its day.

The crisis in the 1960s forced a broader understanding and broader controls of the water quality problem. Sweeping controls were a response to an emerging understanding of ecosystem connectivity. The blanket remediation efforts however were usually focused on one pollutant or affected population. Continued degradation, despite significant improvements, brought the IJC to a new focus on water quality management in the late 1980s. Instead of concentrating on single points or the lakes as a whole, efforts were directed to multiple problems within heavily impacted areas. Each of 42 areas was directed to design a "Regional Action Plan" (RAP) for remediation. Ecosystem-wide in focus, and recognizing the importance of incorporating social elements, the RAPs are as innovative today as the previous efforts were in earlier times. Whether they will be more successful as a whole has yet to be determined. Several RAPs are evaluated in more detail in chapter 28.

Future Planning Facing the Upper Mississippi

The previous two examples outlined consequences of unidimensional planning and attempts to respond to those consequences. These cases give the impression that many of the most important water quality decisions have already been made—that water quality is dominated by responses to previous decisions rather than a current planning for future concerns. In fact, both are important issues, and frequently the combination of the reaction to past practices and proposals for further modifications stirs the greatest debate. The final example in this chapter outlines an ongoing debate: a water quality decision that will come before the U.S. Congress and public in 1999. Whatever its final outcome, the debate surrounding the issues reflects the growing intolerance for environmental impacts, the pressures for multidimensional planning, and the need for new management and planning models to balance the conflicts of values and scientific uncertainty.

The debate pits commerce against ecology along the Mississippi River (Rebuffoni 1994). The Mississippi is an enormous and vital riverway. Its flood plains support some of the most productive agricultural land in the United States; its channels support a barge industry that reportedly saves the nation $1 billion in transportation costs each year (over other transportation such as rail, truck, and plane). The river is also a critical ecological resource, providing fish and wildlife habitat, hydrologic stability, and ecological integrity of forests, wetlands, and aquatic life along the nearly 6000 km course of the Mississippi and Missouri rivers. The controversy has emerged because the barge industry contends that new locks and dams must be constructed along the upper Mississippi to maintain the competitiveness of river transport. Environmental activities and area residents are concerned that the new locks would accentuate the already existing problem with sedimentation and pollution, tilting the balance of ecological integrity along the Mississippi to severe degradation (Fig. 1.5).

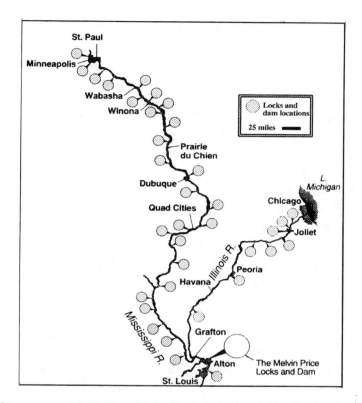

Fig. 1.5 Strong economic interests and a devotion to technology led to extensive damming along the upper Mississippi throughout the 1930s. Currently, the barge industry claims that the system of locks and dams is antiquated and inefficient; it is hoping that extensive renovations, such as have been undertaken at the Melvin Price dam near Alton, will ensure the continued profitability of the industry. Environmentalists claim that the existing system of locks and dams is already seriously damaging the ecology of the upper Mississippi and are resisting efforts to build larger dams at other existing dam sites. (Modified from Rebuffoni 1994. Reprinted with permission by the Minneapolis Star Tribune.)

The Economic Arguments

The barge industry, which is represented by the Midwest Area River Coalition 2000, claims that the existing network of locks and dams on the Mississippi and Illinois rivers is outdated and will not enable river transport to stay competitive in the next decades and century. Barges, for example, are becoming larger—typically nearly 1200 feet in length. Current smaller locks and dams require that the tows be disassembled, passed through the locks in stages, and reassembled for the journey. The process often takes upward of 2 hours, compared with 30 minutes in a lock that is long enough to accommodate the expanded barges. Preliminary suggestions by the industry and the Army Corps of Engineers propose that expanded facilities be constructed at three locks and dams on the Mississippi and two on the Illinois. Additional improvements are required on 11 other locks and dams below Dubuque, Iowa. Any increase in costs, the industry contends, would reduce the income of farmers whose products are shipped by waterways, diminishing the competitive edge of U.S. corn farmers in world markets (Rebuffoni 1994). The estimated $1 billion annual savings over other transportation means is a cornerstone of improvement arguments.

The Environmental Rebuttal

Resistance to the lock and dam improvement proposal radiates along many lines of argument. One view contends the existing waterway is already impacted by pollution, siltation, and the dredging that maintains the deep channels. Advocates of this position point to the condition of nearby waterways such as the severely silted Illinois and hypothesize that the Mississippi will suffer similar problems if damming and barge traffic expand. In addition, opponents of the expansion plan point out that the economic arguments of the barge industry are plagued by significant loopholes.

The economic loophole that troubles many revolves around the tradition of taxpayer support for the entire barge industry. Prior to 1985, the barge industry payed no money toward the construction or maintenance of inland waterways. Since 1985, members have had to pay a tax on fuel; the tax goes to a trust fund that is used to help finance new construction and major rehabilitation work. The fund will soon cover half of the costs for new and major reconstruction, but the rest is still borne by taxpayers (Rebuffoni 1994). In addition, dredging costs, which keep the barge channels open, are payed for solely by taxpayer money. In 1993, the dredging costs amounted to $430 million, $82 million of which was expended on improvements to the upper Mississippi. Opponents argue that if the industry itself had to bear more of the costs of upkeep and construction, it might find alternative, less environmentally harsh solutions to maintaining the competitiveness of river transport. The arguments in favor and in opposition to the expansion plan clearly reflect discrepancies in values between people directly dependent on the barge industry, such as farmers and the industry professionals, and those who live in the Mississippi basin but who are not directly associated with barge traffic.

The ecological arguments against expansion contend that the lock and dam infrastructure already in place along the Mississippi have created a serious ecologically degraded condition, separating some backwaters from the main channel and resulting in the sedimentation of the rest. As sediment fills the backwaters, biological productivity is decreased, as are diversity and wildlife habitat. In addition, the unstable backwaters are unable to provide flood protection. In essence, environmentalists contend that the upper Mississippi, although still ecologically viable, is at a critical juncture and that the assimilative and buffering capacity of the river does not exist due to the historical changes. Any new stress would strip the river of its ecological resilience.

The Resolution

The stakes are high enough in this debate that the Army Corps of Engineers has embarked on a multiyear, $44 million study of the feasibility of the lock and dam reconstruction, $13.5 million of which is earmarked for environmental studies. Whatever the results of the study, the conditions under which the decisions will occur are different than those in place when the first locks and dams were constructed in 1930. In that technological era, which was characterized by industrial and land expansion, environmental concerns and multiple uses of the river were minute factors in any equation. The decision making at the time, though not without resistance, was a unidimensional one in which the benefits of river transport weighed most heavily. In 1999, the competing interests are larger, more vocal, and more scientific. In the intervening years of study, environmental crises, economic crises, political winds, and advocacy organizations will all have significant influences over the values and interpretation of data that will be factored into the final decision on the dam.

Summary

- The practice of water quality management allows society to define and protect the uses it values from any water resource. This choice is always a cost–benefit decision. Uses protected also represent opportunities foregone and costs incurred.
- Traditionally, water quality management has been a unidimensional science, serving a single designated purpose. The emerging view is that any water body must meet simultaneously several goals (i.e., serve multiple uses).
- Effective management requires explicit recognition of *why* specific attributes are to be protected. A planning model guiding development of that mission statement makes effective management more feasible.

For Further Reading

Ashworth W. *The late, Great Lakes: an environmental history.* New York: Alfred A. Knopf, 1986.

Gerlach LP. Crises are for using: the 1988 drought in Minnesota. *The Environmental Professional* 1993;15:274–287.

Hartig JH, Vallentyne JR. Use of an ecosystem approach to restore degraded areas of the Great Lakes. *Ambio* 1989;18:423–428.

Meybeck M, Chapman D, Helmer R, eds. *Global freshwater quality: a first assessment.* Oxford/Cambridge: Basil Blackwell Ltd., 1989.

Norton BG, Ulanowicz RE. Scale and biodiversity policy: a hierarchical approach. *Ambio* 1992;21:244–249.

Percoda N. Requiem for the Aral Sea. *Ambio* 1991;20:109–114.

2

History of Water Quality Management: The Problem and Its Science

Overview

Pollution has always accompanied human civilizations. Throughout history, human civilizations have contended with water quality problems related to the disposal of human waste as well as problems stemming from the development of irrigated agriculture and mineral extraction. Over time, the nature of the dominant problem has changed as societies developed new techniques for water quality management and as new industrial and urban challenges to receiving streams emerged. In contrast, laws designed to protect the common resources have remained generally ineffective and problems have tended to concentrate as populations increased.

Water quality management is an intriguing mirror on society. The field evolves continuously, reflecting changes in social values, scientific understanding, and infrastructure development. While scientific developments control the level of possible understanding related to water quality questions, cultural dimensions dictate the shape of specific water quality controls. For example, England's water quality management has progressed steadily toward a focus on river basins and a reliance on privatization to control costs and effect controls. In the United States, water quality management has been defined by state–national tensions, and the approach to management has fluctuated as the dominant concerns of different historical eras found expression in water quality policy.

Introduction: Water Pollution Through Time

Pollution is as old as civilization and is one of its surest consequences. Water pollution, in particular, is a natural result of most human activities and land-use practices. The awareness of pollution is also an ancient companion to human civilizations, although the capability to effect solutions is a comparatively recent phenomenon. Archeological evidence details that the earliest civilizations relied on at least three strategies to combat the problem of accumulated waste. The first strategy involved centralized rubbish piles (tells or middens) that contained everything from pottery shards and broken tools to human waste and plant and animal remains. The second approach was maintaining a nomadic lifestyle, a system in which population numbers were low and waste and disposal problems diffused over a large area (Pointing 1991). Third, waterways have received and carried away the wastes of any population living near them.

All of these approaches—middens, nomadism, and using waterways to carry away waste—work well when populations are small and scattered. As populations grow, waste problems also grow, sometimes disproportionately to the population increases. Consequently, as centers of population grew throughout history, a variety of approaches were implemented to keep pollution and its inconveniences at a manageable level (the public health consequences of polluted waters were not recognized until much later). Rome's creation of aqueducts, for example, has been attributed to pollution of the Tiber (in 300 B.C.) and the search for cleaner sources of water (Pointing 1991). China developed, and maintained for centuries, a collection system where human waste was collected in buckets and used as manure on fields (Haslam 1990). Pointing (1991) notes a variety of other approaches used in early societies, including street sweepers, scheduled street washing, large refuse piles, and, in Egypt, construction of cities on hills that facilitated the draining of sewage into the Nile. Few of these approaches proved effective, and the sewage problems in nearly every city and town were often intolerable.

The most common responses to increasing pollution were fatalistic acceptance and attempts to procure fresh water from other sources. Some early societies operated open or partially covered sewers, but the sewer ended either in a mire or in the local river or lake. Most societies clustered as close to the river edge as possible, facilitating waste disposal. Once local supplies were polluted, elaborate delivery schemes were developed to bring in water from distant sources. As indicated earlier, Rome built aqueducts, while London laid pipes to bring water from Tyburn spring instead of relying on the polluted Thames (Pointing 1991). So many cities followed suit that circumventing the local problem was the primary approach to pollution management from the 12th century to the early 20th century. This technique is still used in many places.

In addition to the detrimental health effects of polluted drinking water that literally plagued societies, systems for supplying water for drinking and agriculture had other water quality consequences. In Mesopotamia, at least three separate civilizations rose, flourished under irrigation schemes, and collapsed as over-irrigation and poor canal designs led to the salinization of the soil, decreased crop productivity, and eventual agricultural collapse (Pointing 1991). Similar scenarios have been advanced to explain the fall of the Aztecs, Mayans, and ancient cultures of the Indus valley. While many factors contributed to the collapse of the Roman civilization, one important factor may have been the toxic effects of lead poisoning from their sophisticated lead-pipe plumbing network.

Crises and Management

By 1800–1850, water quality conditions in many parts of the world were terrible (Fig. 2.1). Streets were choked with sewage and garbage; cholera and typhoid epidemics were commonplace. Lamb (1985; citing Typhoid Fever, Budd 1874) reports the following for the condition of the Thames River in 1858–1859:

> For the first time in the history of man the sewage of three millions of people had been brought to seethe and ferment under a burning sun, in one vast open cloaca lying in their midst.
>
> The result we all know. Stench so foul, we may well believe, had never before ascended to pollute this lower air. Never before, at least, had a stink risen to the height of an historic event. Even ancient fable failed to furnish figures adequate to convey a conception of its thrice Aegean foulness . . .
> . . . At home and abroad the state of the chief river was felt to be a national reproach. 'India is in revolt and the Thames stinks,' were the two greatest facts coupled together by a distinguished foreign writer, to mark the climax of a national humiliation. [p. 4]

In the United States, the problems were equally severe. Conditions in Boston in 1885 (Fair *et al.* 1966, quoted in Lamb 1985) were described as follows:

> Large surrounding territories were at once, and frequently, surrounded and enveloped in an atmosphere of stench so strong as to arouse the sleeping, terrify the weak, and nauseate and exasperate everybody. [p. 4]

THE "SILENT HIGHWAY" - MAN.
"Your MONEY or your LIFE!"

Fig. 2.1 Far from a recent concern, severely degraded waterways have been a constant accompaniment to urbanization. This 1858 cartoon captures the severity of degradation in the Thames in the early industrial age. ("The Silent Highwayman": Punch Publications Ltd.)

The severity of situations, brought on by rising populations and inadequate facilities, coincided with increased awareness of the modes of disease transmission. The result was a crisis that brought the first effective improvements in water pollution.

Early Responses to Pollution Crises

By the late 1800s, it was generally agreed that "pollution" caused biological problems, including problems with significance to human health. The British were leaders in attempts to quantify and regulate water quality, establishing a series of laws and commissions to control and study water pollution. For example, disposal of human waste into London sewers, which were designed specifically for storm discharge, was prohibited until 1815, when population pressures forced officials to allow disposal. The Gas Works Clauses Act of 1847, on the other hand, banned industry from discharging industrial wastes to sewers (Haslam 1990). The Salmon Fisheries Acts, passed in 1861 and 1865, were also intended to halt pollution. These acts declared illegal any practice that resulted in a fish kill (i.e., anything was legal until it caused an impact, then it could be viewed as illegal in retrospect). Several other English pollution control acts and royal commissions followed. Two major obstacles hampered effectiveness of all of these acts and commissions: lack of a uniform definition of pollution and a confusing array of responsible (often overlapping) authorities.

The Need to Define Pollution

In an attempt to resolve the problem of a uniform definition for pollution, the 1898 Royal Commission on Sewage Disposal developed a measure called "Biochemical Oxygen Demand" (BOD_5) (Haslam 1990). Recognizing that oxygen in streams and rivers is critical for fish production, the commission developed BOD_5 as a measure of the oxygen-consuming ability of a water sample. The definition of BOD_5 was derived from typical British conditions: mean annual air temperature in England is 20°C; mean flow time of English rivers, from highlands to the sea, is 5 days; and the principal wastes impacting streams were organic, oxygen-demanding wastes like domestic sewage. The BOD_5 metric was defined as milligrams of oxygen consumed by one liter of sample water, incubated at 20°C for 5 days. The observed range of the new metric was from less than 1 (clean water) to 10 (a moderately impacted stream) to about 300 (raw sewage).

The commission also suggested that fish production in English rivers would be "un-impacted" if oxygen was maintained at a concentration of at least 4 mg/L. Discharge limits on an effluent could then be determined based on dilution and instream mixing. The commission used, as an example, a situation in which a river provided eightfold dilution to the effluent, and calculated the stream standards based on this dilution factor (Haslam 1990). Consequently, if a stream has an ambient oxygen concentration of 8 mg/L, a background BOD_5 of 2 mg/L, and an effluent will be diluted eightfold, then BOD_5 of an effluent must be maintained at or below 20 mg/L to maintain instream oxygen above 4 mg/L. The commission used the same logic to develop criteria for suspended solids, concluding that a criterion of <30 mg/L for solids would protect the aquatic resource (Haslam 1990).

The setting of these early standards illustrates a practice that has plagued water quality management since that point. In demonstrating the calculation of the standards, the commission used the example cited above, with eightfold dilution and stated ambient and load levels of oxygen. For the next 60 years, the fact that those conditions were merely examples was overlooked because government agencies were so relieved to have a "uniform" pollution definition. As a result, 20 mg/L and 30 mg/L became the standards for river pollution in Britain (Haslam 1990), despite significant differences in stream oxygen requirements and dilution factors. The same standards were then applied uniformly by hundreds of agencies and for essentially all municipal and many industrial wastewater treatment facilities worldwide. Only since the mid-1970s have agencies begun to develop and implement site-specific, instream, water-quality-based standards.

Institutional Roadblocks and Unclear Chain of Command

England's second major obstacle to resolving impacts was the number of boards and commissions with varying responsibilities. Throughout the 1800s, England was not only developing scientific concepts of water pollution, but also defining the administrative and legal consensus that would guide its water pollution decision making. For example, although local statutes had instituted bans on using rivers for waste disposal of different substances as early as 1388, not until 1859 was the common law tradition summarized in a legal decision involving competing interests on a river (Newson 1992). British cultural attitudes were further institutionalized 20 years later in the Rivers Pollution Prevention Act of 1876. Under this act, it became a national offense to dispose of wastes into rivers, but the municipal authorities maintained their traditional role as implementing authorities (Newson 1992). Consequently, more attention was paid to the trade-offs pollution prevention would mean to local development than to the effects on river quality. In addition, other boards and commissions—for example, those in charge of fisheries—had jurisdiction over the same stretches of river as municipal water boards. The duplication of efforts between successive administrative units and sectors led to uncoordinated and ineffectual management.

To a large degree, the management infrastructure improved in the 1930s when England began to adopt a more integrated river–basin approach to its water resources management. The effort was launched with the establishment of Fisheries Boards in 1923 and Catchment Boards in 1930; these boards were charged with overseeing the management of efficient land drainage (England is prone to waterlogged soils and flooding due to a high rainfall and clay soils) (Newson 1992). While the drainage efforts were later judged misguided and focus was eventually shifted to wetland retention, the administration of resources according to catchment, rather than municipal boundaries, was a critical development. It was followed in 1948 by the River Boards Act, which set up 32 boards in charge of licensing polluters and water resource abstractions. While the river board concept was expanded in 1951 to include Scotland and was adopted by France and Germany in the 1950s, effective pollution control suffered from licensing exemptions, secretive procedures, and the continued strength of development-focused interests.

Administrative roadblocks were not, however, unique to England and its European neighbors. Different roadblocks have confronted the more geopolitical focus adopted for water resource management in the United States, Canada, and other countries. In the United States and Canada, for example, states and provinces have relative sovereignty over their water resources, an arrangement that complicates management of transboundary water resources. Many areas in the world are now developing regional approaches (see chapter 16) that are based on several biophysical characteristics, rather than corresponding with hydrographic basins or political boundaries.

Growth in Scientific Understanding of the Basic Resource

The development of biological oxygen demand measures did not occur in isolation from other scientific advances of the time. Two parallel water-quality-related fields of science were developing quickly in the late 1800s: aquatic ecology and limnology versus sanitary engineering and impact assessment. One of the first seminal papers in limnology, for example, was published by Stephen Forbes in 1887. In that paper he described a lake as a community of organisms that developed through co-evolution and natural selection. He suggested that the community represents, and responds to, watershed characteristics. For example, Forbes postulated that lakes could be classified as either "watershed lakes" (i.e., those dominated by surrounding conditions) or "fluvatile lakes" (i.e., those dominated by upstream conditions). He also suggested that fish production and community composition should be predictable in lakes protected from sewage. The focus of his work was the belief that each lake was a microcosm, relatively isolated from its surroundings.

Between 1902 and 1909, two German scientists, Kolkwitz and Marsson, developed the saprobien approach to understanding water pollution, a technique they called "ecology of plant and animal saprobia." They suggested that several well-defined zones existed in a river polluted by sewage, with highly polluted zones a short reach below the effluent followed by successive zones of recovery (see chapters 6 and 17 for more detail and Fig. 17.3). According to Kolkwitz and Marsson, specific species of plants and animals were characteristically found in each zone and river quality could be classified by examining those biotic communities. This model was later developed into an indicator organism approach, which is widely used in water quality and many other fields of environmental assessment.

Through about 1935, scientists began to recognize and quantify impacts of various kinds of pollution. For example, scientists began studying impacts of irrigation, erosion, and sewage on fish and phytoplankton species (Thompson 1912; Dana 1916; Purdy 1916), as well as the impacts of land use patterns on hydrology (Bates and Henry 1928). At the same time, efforts were being made to understand conditions in natural, unimpacted lakes. For example, Birge and Juday (1911) studied ambient conditions in Wisconsin lakes and formed the basis for much of modern limnology. By the late 1930s, the "water quality" literature was sufficiently rich that Ellis (1937) was able to write a review of the effects of land use impacts (mostly erosion) on fish.

From the mid-1800s to the late 1930s, most limnologists and aquatic scientists concentrated on understanding dynamics of pristine conditions (e.g., Birge and Juday 1911) and species-specific changes to specific kinds of impacts (e.g., Kolkwitz and Marsson 1908; Purdy 1916). That scientific growth was complemented by the work of sanitary engineers who tried to generalize water quality impacts in designing water treatment systems. For example, in 1925, Streeter and Phelps modeled the decrease in dissolved oxygen (oxygen sag; see chapter 17) that occurred below sewage effluents, developing a hydrochemical modeling analog to the Kolkwitz and Marsson saprobien zonation. The Streeter–Phelps oxygen curves became the standard that the engineering industry used in designing sewage treatment plants.

A broad second phase of limnology and aquatic science began in the early 1900s with efforts to generalize from individual situations to classes of impact. While observations of impacts and pristine conditions continued, scientists began looking for concepts and theories that would apply across riverine systems or across lakes. In 1943, for example, Brinley wrote about responses of algae and fish to sewage and attempted to describe the food web of a polluted stream quantitatively. His work previewed some of the ideas that became evident in Lindeman's (1942) work on Cedar Bog Lake, Minnesota, in which he developed "food-web" ideas as a contrast to the widely accepted linear "food chains."

The growth during this period (1935–1960) represented a move toward quantitative descriptions of the community and the ways that the community reacted to categories of impact. While many initial efforts were hampered by limitations in the field of statistics itself (e.g., Patrick 1950), statistical tools were soon developed that met the evolving needs of the science. The 1950s constituted an era of significant growth for water quality science, marked by Tarzwell and Gaufin's (1953) first quantitative expressions of community level impacts and Doudoroff and Warren's (1957) creation of community-level biological indices (see also chapter 6).

The field of water quality management could develop fully only as these scientific advances preceded. Water quality decisions reflected not just the scientific understanding of the day, however, but also the social values and preoccupations of each era. The following section uses the history of water quality management in the United States to illustrate how scientific advances and social forces combine to create different water quality approaches in different periods of recent history.

Water Quality Management as a Reflection of Society

U.S. water quality history, like all management programs, has been, and remains today, a mirror of societal goals and concerns merged, to different degrees, with scientific understanding in the field. For example, in the early part of this century, the United States and much of the developed world became enamored with technology and made their natural resource decisions based largely on what was technically possible. While technology has never fallen from grace, the reliance on large-scale technological feats has been softened by parallel developments such as more global perspective and an emerging concern for the effects of technological solutions. The result is an increasingly integrated approach and larger-scale view of water quality management. In turn, the disciplinary boundaries of fields such as water quality are softening. No longer a purely technical field, water quality management is now social as well as scientific and technical. The progression of these changes is best seen in terms of major time blocks characterized by a dominant view of resource management. The following summary of water quality trends is focused on the United States, but may be broadly generalized to represent the trends in many westernized countries.

1930s, MAXIMIZE EFFICIENCY. In the 1930s, U.S. and western European societies were engrossed in maximizing productivity and efficiency. This era, which followed the U.S. Great Depression, was marked by rapid growth. Ecosystems that exported energy were viewed as wasteful. Their energy was harnessed and "made efficient" through dams, power plants, and production schemes.

1940s AND 1950s, TECHNOFIX. In the post-World War II years, technological solutions were offered for many needed or desired changes in U.S. and western European society. Many new technologies were available as spin-off products from the war effort. Although ecosystems were seen as resilient and self-purifying, it was recognized that the self-purification capacity was limited. When anthropogenic influences exceeded resilience, technological solutions were called into play. Waste treatment plants and stream channelization characterized the dominant paradigm (i.e., an engineering approach to managing water quality).

1960s, MAXIMIZE PRODUCTIVITY. As a more global perspective emerged, people became increasingly aware of population growth. Science introduced the Green Revolution. Productivity for human consumption dominated management philosophies. Ecological concepts of trophodynamics, efficiency of energy transfer between trophic levels, optimal exploitation of animal populations, and other global production concepts drove management goals. Ecosystems were viewed as production units in which primary and secondary production could be harvested for human uses.

1970s AND EARLY 1980s, MANAGEMENT AT THE POPULATION LEVEL. In the 1970s, U.S. and western European societies began to understand the effects of a variety of influences on individual populations. Scientists recognized and made public a variety of anthropogenic effects on plants and animals. In the early 1980s, globalization emerged as people recognized the interconnectedness of biomes and hemispheres. "Pollution control" replaced "increasing production" as a central goal of western society. Applied ecologists conducted thousands of studies designed to assess environmental (i.e., population- and community-level) impacts and thereby establish criteria and catalogue pollution effects. Notably, few scientists and even fewer managers looked at these studies from a synthetic, whole-ecosystem view. The focus of concern was trophic levels or discrete components (e.g., target populations) rather than whole-ecosystems.

LATE 1980s AND 1990s, MANAGEMENT AT THE ECOSYSTEM LEVEL. Many ecologists working in terrestrial and aquatic systems have identified emergent, ecosystem-level properties since the late 1960s. Not until the early 1980s did the whole-ecosystem approach gain sufficient momentum (i.e., a theoretical basis and support among scientists) to lend it credibility, however. Acceptance of new scientific ideas is a function of both timing and serendipity, both of which came together for E. P. Odum's 1985 paper on general ecosystem properties. His work represents the seminal cohesive work upon which much of the future management is, and will be, based (see chapter 15).

Agreement has not been reached among aquatic ecologists and managers about the utility or feasibility of managing at the ecosystem level. Some ecosystem ecologists suggest that ecosystem theories have general, but not practical, applicability. Others feel significant practical utility exists upon which managers can base decisions. Still others contend that current theories and data sets do not adequately account for the complexities that ecosystems exhibit in their response to, and recovery from, stress. As with any new paradigm, the scientific community will continue to debate the theories until a comfortable consensus emerges.

Water Quality Progress: Improvements in Conditions

Although countries have taken different approaches to water quality management overall, the issues that have received significant regulatory attention have improved. To date, a major focus of water quality programs has been on reducing river pollution from domestic sewage. As a result, the organic load (i.e., biological oxygen demand) of rivers has significantly decreased in industrialized countries since the 1970s, while it has increased in the rapidly industrializing countries of Asia and Central America. Among European countries, Switzerland, Denmark, the Netherlands, Sweden, and Western Germany have achieved a level at which nearly 100% of municipal sewage receives at least secondary treatment (i.e., 85% reduction of incoming BOD_5 and suspended solids).

Some progress has also occurred in reducing heavy metal concentrations in the waters of industrialized countries. For example, lead concentrations near the mouths of rivers in Canada, Japan, Belgium, the Federal Republic of Germany, and the United Kingdom generally decreased between 1970 and 1985 (World Resources 1992). In the same period, concentrations of metals in the Rhine River declined slightly after decades of increases. Industrialized countries have also made progress in reducing water quality impacts from forestry, in eliminating certain highly toxic pesticides such as DDT, and in achieving general awareness of water quality problems in the population. These improvements are encouraging, but tenuous. New agricultural and industrial chemicals continue to emerge more rapidly than regulatory agencies can study and regulate them. The number of affected streams and lakes is increasing as land use intensifies worldwide. An important problem in developed countries is the increasingly apparent effect of cumulative impacts, for which current management strategies are inadequate.

Fig. 2.2 Like many urban areas in developing countries, Djakarta, Indonesia, is experiencing rapid urbanization. Infrastructure development is not, however, keeping pace with the expanding population, and scenes such as this one are increasingly common. (Photo by D. Vanderklein)

In most developing countries, water quality is generally deteriorating, especially around urban areas. Population growth, urban growth, and industrial expansion are generally not managed by appropriate standards, enforcement, or infrastructures such as sewer systems and water treatment plants (Fig. 2.2). Meybeck *et al.* (1989) noted that only 10 out of 60 rapidly developing countries have established effective water quality laws, regulations, or enforcement infrastructures. In addition, the pace of development and pollution creates nearly impossible situations: the pollution that grew over a 100-year period in developed countries now occurs in barely a generation in developing countries (Meybeck *et al.* 1989). Consequently, few major cities in the developing world have adequate sewage treatment facilities, and municipal potable water supplies are often neither treated nor disinfected. For example, in 1990 only 41% of urban residents in Latin America and the Caribbean were connected to wastewater collection systems. Of this water, over 90% was discharged, untreated, to area waters: nearly one-fourth of the urban population have no acceptable sanitation options (Bartone 1990).

Similar problems have arisen in developing countries worldwide. Hjorth and Thi Dan (1993) cite figures showing that throughout Africa, only 52% of the urban population receives safe water and 40% has access to adequate sanitation; in India only 30% to 40% of the urban poor have access to safe water and fully 75% of all urban residents have inadequate sanitation options. Hanoi epitomizes another typical problem: nearly two-thirds of the city's sewers were built before 1954 and have not been maintained or upgraded since that date. Consequently, even the 20% of the population served by sewers receives inadequate waste disposal (Hjorth and Thi Dan 1993). Bangkok, one of the world's important centers of commerce, began considering options for a sewage system only in 1990 (World Resources 1992).

The challenges are daunting. Yet, this book highlights important and encouraging developments in social awareness and scientific understanding that indicate significant, global improvements are possible in the field of water quality. While our current understanding will undoubtedly not be the pinnacle for this field, it is a dramatic and critical stage at which sufficient technical, scientific, and social developments have combined to yield significant predictive ability; real and lasting improvements in water quality are now truly viable. Whether the societal and political will is strong enough to implement changes remains to be seen.

The Next Stage: Proactive, Integrated Management?

Our understanding of water quality as a science has progressed greatly since the earliest civilizations grappled with the effects of (mis)management. Despite the scientific and technical progress, management in developed and developing countries alike remains reactive, controlling effects after they have been identified. While many reasons lie behind this pattern of reactive management, one important factor is that policy makers and scientists have long seen the world at different scales, with scientists focusing on smaller-scale patterns and policy makers operating in a world of larger-scale patterns. (Throughout this text, the term small-scale is used to denote short-term and narrow spatial scales. Conversely, the term large-scale denotes events and patterns that occur across a broad spectrum of spatial and temporal scales [see chapter 7].) The emerging ecosystem-level view of ecology outlined earlier represents the first time that the scales of reference for scientists and managers are coalescing. This merger should enable water quality management in the coming century to take a more proactive stance in issues of planning and remediation (Fig. 2.3) (Slocombe 1993; Perry 1994).

This larger-scale, ecosystem-level perspective is critical. If managers are sufficiently proactive, focusing at the larger scale enables them to act in time to prevent or mitigate declines before ecosystem damage becomes too severe. It enables scientists and managers to cooperate across local, regional, and national

Fig. 2.3 Ecosystem-level planning may even be exercised in large-scale, technologically advanced projects such as this surge barrier in the south of the Netherlands. Periodic flooding of densely populated areas necessitated a flood barrier, but sensitivity to ecological impacts of a large-scale dike led to this system of flexible flood gates. Under normal conditions, the flood gates are open and water flows freely between the estuary and the open sea. When a significant storm surge is predicted, the flood gates are closed. Consequently, water quality of the estuary is maintained, but flood protection is also ensured. Although a planning and technological success, the cost of similar projects is prohibitive for most coastal nations. (Photo by D. Vanderklein)

boundaries. It also enables us to progress toward appropriate scientific and social solutions to ensure sustainable ecosystems for the 21st century.

The remainder of this book is devoted to communicating the new and emerging themes in water quality science. These themes include changes in the way standards are derived and set (chapters 5 and 6), increased awareness that effects of our management decisions have specific temporal and spatial dimensions (chapter 7), advances in the understanding of the basic ecology of the resources (chapters 9–13), new concepts regarding ecosystem-level response to stress (chapter 14), and approaching management from appropriate spatial and temporal scales (chapters 15 and 16). The social contexts within which water quality is managed are interwoven throughout each chapter, but are stressed in particular in chapters 3 and 4 and in the final section discussing policy applications (chapters 25–28).

Summary

- Water quality was actively managed by many ancient societies. Management as a science was virtually nonexistent between A.D. 500 and A.D. 1850, by which time environmental conditions were truly intolerable.
- Growth of management was impeded by lack of clear definition of items like "pollution" and by numerous layers of overlapping and contradictory policies and laws.
- More recently, the science upon which water quality is based has grown significantly, allowing more defensible decision making.
- Water quality management in the future will be characterized by more proactive decision making, addressing questions about how ecosystems function within landscapes rather than how populations function at sites.

For Further Reading

Barton CR. Water quality and urbanization in Latin America. *Water International* 1990;15:3–14.

Haslam SM. *River pollution: an ecological perspective.* London: Belhaven Press, 1990.

Hjorth P, Thi Dan N. Environmentally sound urban water management in developing countries: a case study of Hanoi. *World Resources Development* 1993;9:453–464.

Lamb JC. *Water quality and its control.* New York: Wiley, 1985.

Newson M. *Water, land and development: river basin systems and their sustainable management.* London: Rutledge Press, 1992.

Perry J. Water quality in the 21st century. *Limnetica* 1994;10:5–13.

3

Attitudes, Goals, and Management Strategies

Overview

Water quality management comprises a balance between meeting utilitarian uses of water and responding to each society's environmental and ecological values. Because the balance is different in each society, the approaches taken to water quality management can be described as lying along an attitudinal continuum ranging from utilitarian to naturalistic. Most societies are clustered at the utilitarian end of the continuum where the environment is valued almost exclusively for the goods and services it yields. Midway on the continuum are societies that value goods and services but also value ecosystem integrity and biophysical characteristics in their own right. This attitude is increasingly exhibited by several more industrialized countries. At the naturalistic end of the spectrum are societies that value all living creatures, and consequently their habitats, equally. No single country adheres to this view, although such a philosophy is practiced by certain societies, such as the Jains and some Buddhists. Sensitivity to these philosophical differences is a significant factor in developing water quality goals that are realistic and achievable for a given society. Thus, management goals might be largely anthropocentric (e.g., sustained economic growth) or resource-oriented (e.g., restore biotic integrity) but must always reflect the ways a society assigns social needs and values to a resource.

Societal Attitudes

The previous chapters outlined historical developments in water quality management and water quality science—that is, the development of a science-based field that has traditionally responded to concerns in individual administrative sectors (e.g., transportation, fisheries, potable water). Behind the science and the administration of water quality lie societal attitudes toward the priority values of water. The explicit recognition of the importance of values and goal statements in water quality management may represent the most unique aspect of the evolving management paradigm. As will be described throughout this book, the evolving paradigm has many component parts, including new scientific understandings and new theories about integrating management efforts. Perhaps its most dramatic departure from previous paradigms is recognizing that water quality management cannot be separated from the psychological and sociological dimensions of natural resource use. This and the following chapter discuss the role of people, their attitudes, and their water-use patterns in setting water quality priorities for management. The concepts discussed in this chapter will be revisited in chapter 26 in a more analytical framework. Here they are presented as an important conceptual base for ensuing discussions of policy formation and scientific inquiry.

Societal attitudes toward the environment can be seen as points on a continuum from utilitarian to naturalistic. Environmental sociologists have advanced many different terms and descriptions for the characteristics displayed at these points. Regier and Bronson (1992), for example, describe four world views, ranging from the "exploitist," which is purely utilitarian in its evaluation of natural phenomenon, to the "inherentist," who sees inherent value in natural phenomena, with the value independent of human interests. Between these two extremes Regier and Bronson (1992) describe the exploitist-biased "utilists," who value environment and environmental impacts only insofar as they affect sustainable uses of a valued resource, and "integrists," who adhere to the belief that the natural cycles and processes of environmental phenomena must be maintained for human and environmental quality to be realized.

Similarly, Meffe and Carroll (1994) describe four philosophical orientations that vary from the strictly anthropocentric view in which values are instrumental (utilitarian), through intermediate ethics of "stewardship" and "biocentrism" to the "ecocentric" in which the intrinsic environmental values dominate the relationship of humans to their surroundings. From a different perspective, Petulla (1985) describes environmental ethics from three primary viewpoints: the economic tradition, which is the most utilitarian and focused on describing costs and benefits of environmental goods and services; the biocentric tradition as espoused by Muir and Thoreau, which combines fear and respect for the power of nature; and the ecologic tradition, which is a scientifically derived, conceptual model of the ordering of nature. Another author, Colby (1990), describes an evolutionary relationship between philosophical approaches to the environment and their expression in management models. His thesis is discussed in more detail in chapter 26.

Although these and other authors use different terms to describe the psychosocial relationship of humans to their environment, the differences between the terms are often largely semantic. They demonstrate that the task of defining the relationships of societies to their environment is still a relatively embryonic one, riddled with new, old, and emerging terminologies. The fact that more than 100 books on environmental ethics were published in English alone between 1990 and 1995 demonstrates the growth and importance of this emerging recognition. It may also demonstrate that as societies change from one point on the continuum to another, citizens necessarily become introspective about the relationships between themselves and their environment.

In the following sections, each of these main points on the attitudinal continuum is described in more detail. The descriptions are then applied toward typical approaches to water quality management, revealing that the approaches adopted by different countries to the administration of water quality management reflect a great deal about the underlying attitudes of societal members toward water resources.

Attitude Clusters

The terms used above to describe environmental world views connote that societies have discrete and easily categorizable attitudes. Although convenient, this representation is far too simplistic to be useful for management planning. As reviewed by Callicott (1994) and Engel and Engel (1990), however, subtle yet significant differences exist in environmental attitudes within each society and even among different branches of the same religion (see chapter 25). To a manager planning specific actions at a local level, these subtle variations are of great significance, providing important clues to what motivates citizens to respond to and care for their local environment. Because of these variations in attitude, the following discussion presents environmental attitudes not as discrete and separate points, but as a continuum ranging from the strictly anthropocentric to strictly ecocentric. Along this continuum, societies tend to cluster at certain points, but both within each cluster and among different clusters can be found common elements and important subtle attitudinal distinctions that set the flavor of decision making related to natural resources.

The most anthropocentric, or utilitarian, cluster views natural resources as existing solely for use by humans. From the Bible to modern fiction, many western societies are schooled in this view of the environment. To a utilitarian, the goal of water quality management is to maintain quality appropriate for a mix

Fig. 3.1 The majority of the world's people live close to water, and follow a primarily utilitarian approach to their local resources. (Photo by J. Perry)

of human uses, such as clean drinking water, a fishery, or irrigation. By far, the majority of the world's societies take this approach to water quality use and management (Fig. 3.1).

Midway on the continuum are societies that place a high value in the utility of natural resources, but also place explicit value on the integrity of ecosystem functions such as nutrient cycling and the hydrologic cycle. In these societies, water quality management attempts to balance human uses against disruption of these processes, seeking an optimal balance of the two. Human uses are no longer the ultimate goal in this case, but neither is there specific valuation of nonhuman species. This approach has been adopted in the last 20 years by many industrialized nations such as the United States, Canada, and several nations of western Europe.

At the far end of the spectrum are societies that value aquatic (and other) ecosystems in their own right. For them, goals may be expressed as "integrity of the biotic community," including maintenance of community structure and function, preservation of selected species and habitats, and maintenance of diversity. While no nation reflects this attitude in its purest form, many individuals within "green" parties and conservation organizations continue to lobby for more naturalistic attitudes and legislation. As a lifestyle, this view is probably best represented by such Eastern religions as the Jains and some Buddhists who value all life equally (see chapter 25). In the United States and Norway, the "Deep Ecology" movement is the most vocal proponent of a naturalistic approach. Unique even among naturalistic ideas, Deep Ecologists feel that humans are "outside" of the natural order of things and place nonhuman values of ecosystems over and above human values.

The distinction among world views along this continuum revolves around how broadly or narrowly a society defines its scope of management. The more naturalistic the focus, the broader the mandate for nonspecific water quality protection and the more a society will manage by intent of the law instead of strictly following the letter of the law (Fig. 3.2).

Implementation: From Attitudes to Goals

Ecosystem integrity is a goal that may be used by any society at any time to indicate that the society places value on ecosystem health. The word "integrity" has different meanings to societies along varying points of the utilitarian–naturalistic continuum. To the most utilitarian and anthropocentric, the phrase "ecosystem integrity" may imply that pollution impacts are acceptable until an ecosystem can no longer absorb changes. In other words, human society can use an *assimilative capacity*. To the most naturalistic, ecosystem integrity connotes a hands-off policy of no allowable change. Throughout the planning process, these attitudinal differences are reflected in regulatory language and management prescriptions. The differences are most distinguishable in the approaches used to legislate policy.

Nearly every country assigns water quality management to a unique government department. Worldwide, responsible agencies wield widely varying amounts of regulatory and discretionary power. Despite these differences, only a limited number of legislative approaches have evolved to meet the diverse needs of different countries. In general, countries with a utilitarian view of the environment legislate for very specific impacts: the more specific the limits, the greater latitude to affect quality up to the given limit. A midrange approach uses legislation that regulates effects on all aquatic life. This broader approach still

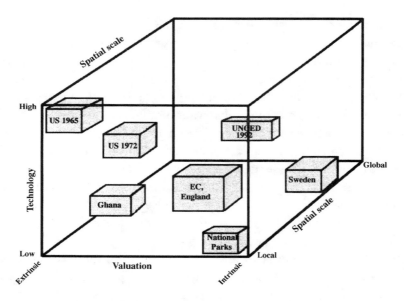

Fig. 3.2 Diagrammatic representation of the spectrum of attitudes that affect environmental management and regulation. Policies and goals align along three interrelated axes: the valuation of environmental resources (extrinsic versus intrinsic); the reliance on technology in relation to meeting resource goals; and the focus of programs, from local to global scales. The box defined by these three axes provides a useful reference point for visualizing the attitudes and approaches taken by different countries, different historical periods, and different programs and institutions. For example, Ghana concentrates heavily on extrinsic values of its resources, while Sweden approaches water resources from a highly intrinsic point of view. Even within the same country, societal changes result in legislation that aligns at a different point in the attitude "box." In 1965, for example, the United States relied heavily on technology to solve local problems of extrinsic value. The 1972 Clean Water Act reduced the reliance on technology and expanded the scale of concern spatially while recognizing new intrinsic values. Other notable comparisons are between programs such as the U.N.'s global focus, which recognizes important extrinsic values, versus the U.S. National Parks system with its local focus and intrinsic valuation of park resources.

allows for some contamination and pollution, but the protection to the environment is broadened by the lack of specificity in the legislation. At the naturalistic end of the spectrum lies legislation that disallows any effect on natural resources. This approach is the least specific legislatively, but the strictest in intent. Some examples of common approaches are described in more detail below.

Anthropocentric Goals

ENVIRONMENTAL QUALITY STANDARDS (BEST USE OF AVAILABLE ASSIMILATIVE CAPACITY). One of the most common legislative tools is establishment of quality standards (i.e., attributes of the environment that we wish to retain). Quality standards are, in effect, a regulatory tool that lists specific qualities associated with specific desired uses. To establish such standards, uses or values of the water resource are determined and then specific quality attributes are assigned to those uses. For example, a water body for use as a public water supply will have an associated list of attributes such as low turbidity or levels of heavy metals below some certain threshold. Consequently, quality standards are specific to a given use and represent the maximum allowable level of pollution.

Quality standards are based on assimilative capacity of a water body (i.e., the ability of the resource to receive wastes or impacts and maintain its structure and function within the bounds of our standards). An ecosystem can assimilate or diffuse some impacts without exhibiting unacceptable characteristics. Logically, larger bodies of water can assimilate more than smaller streams; fast-moving streams can assimilate more than slow-moving ones. Because water treatment facilities are extremely expensive, an enormous emphasis has been placed on the natural assimilative capacity of waters throughout history (i.e., "dilution is the solution to pollution"). This concept is still the primary "treatment" philosophy in many developing countries.

LIMIT VALUE APPROACHES. Another common legislative tool for anthropocentric goals is regulating the polluting technology instead of regulating the effluent level per se. In this case, the goal is to set limits based on operational characteristics of the discharging facility rather than on the level of contaminants in the receiving water. This approach results in more uniform standards, a feature that appeals to countries such as the United States with strong traditions of equality. Because limit value approaches ignore the attributes of the receiving water, however, they can be ineffective in achieving effluent quality control because they may over- or underregulate any given wastewater treatment facility.

In limit value approaches, discharge quality is determined on the basis of available treatment technology (and sometimes cost). Treatment philosophies that are based on characteristics of the technology include Best Practical Technology (BPT) and Best Available Technology (BAT). In the early 1970s in the United States, industries were required to have BPT waste treatment. In other words, each facility had to operate in such a way that its effluent quality was equivalent to, or better than, that of the average facility of their industrial class, using treatment technology that represented a reasonable compromise between effluent quality and cost. Thus, the Environmental Protection Agency was required to classify all industries into categories, determine average effluent quality for each class, assess costs and effectiveness of treatment technologies, and then specify quality characteristics associated with the "Best Practical" alternative. Later in the decade, industries were required to implement the more rigorous "Best Available Technology" for their industrial class. While the EPA used the same system for developing BATs and BPTs, the critical difference between the two standards is that BATs stress adoption of the best commonly available technology and consequently place less emphasis on cost factors.

Limit value approaches are used throughout the world because they are easier to implement than limits on receiving waters. In many developing nations and countries with strong central governments, however, they are seen as a constraint to industry and are not as widely adopted as are quality standards.

SUSTAINED ECONOMIC GROWTH. Another common approach for achieving anthropocentric goals is an economically based alternative to the resource-oriented goals espoused by many industrialized countries. This approach focuses on human *use* of the environment. In many developing countries, use (or exploitation) of natural resources is required to support a burgeoning human population (Falkenmark and Suprapto 1992). Although "exploitation" is not a long-term sustainable approach, a focus on resource consumption for human use is a viable, short-term management goal. It is adopted by many—if not most—new, rapidly growing societies.

Mixed-Value Approaches

"NONTOXIC TO AQUATIC LIFE," A CHOICE FOR COMPLEX EFFLUENTS. Midway along the valuation continuum, regulation seeks to ensure quality adequate for specific human uses while simultaneously ensuring that aquatic ecosystems are not compromised by these uses. One approach for accomplishing this goal is to establish water-quality-based criteria. The need for these standards arises because most municipal and industrial effluents are linked with complex waste streams representing many different processes within the facility. In this alternative to the BAT and BPT approach, water-quality-based criteria are developed such that the entire effluent, when mixed with appropriate quantities of receiving water, must be shown to be nontoxic to aquatic life. In this case, the management goal is to ensure that the quality of discharged material will allow certain designated human uses while sustaining specified forms of aquatic life.

An advantage of this alternative is that effluent limits are site-specific and, therefore, more accurately reflect the tolerance or assimilative capacity of the receiving water. Limits are determined on the basis of the actual mixture discharged by a given facility, given receiving water characteristics of the location.

Disadvantages of this approach stem from its specificity. Because standards are developed for a specific group of test species, it is difficult to incorporate other organisms into those standards. In addition, both effluent quality and in-stream, or "receiving water," conditions vary through time; it is difficult to factor temporal variation into regulations.

Biotic Integrity Goals

MANAGEMENT TO RETAIN UNIMPACTED CONDITIONS. At the naturalistic extreme, the goal of maintaining ecosystem integrity translates into managing for unimpacted conditions. At its strictest interpretation, this approach is unrealistic: no unimpacted ecosystems remain in the world. Even the most remote ecosystems have been affected by human society through a variety of means such as long-range transport of air pollutants. For better or for worse, we live on a managed planet whose human populations are growing at such a rapid rate that unimpacted or aboriginal conditions no longer exist, nor will they ever again. More relaxed interpretations, however, seek to approximate unimpacted conditions and to ensure the long-term survival of these pristine landscapes. This approach is seen in many wilderness areas.

ANTIDEGRADATION. "Antidegradation" also belongs on the naturalistic end of the attitude spectrum. According to this approach, management would ensure that water quality is not degraded from its present condition. Antidegradation, or preservation of present biotic integrity, is a realistic management goal for isolated areas such as parks and reserves. Attainment of such a goal requires very careful management and is often accomplished—albeit at considerable expense. While it is difficult to quantify biotic integrity, use of the term implies that structure (kinds and numbers) and function (rate processes) of the entire biotic community become no worse than their present condition. Maintenance of those processes and properties requires that all inputs and outputs to the system be maintained at an essentially constant level. For example, if a goal for a park is to maintain biotic integrity, human visitation must be kept to a level equivalent to that when baseline conditions were established. If human visitation is reduced, plants and animals that rely on those disturbances will be disadvantaged in the same way that other plants and animals would be threatened with increased human use.

Future Trends in Regulation

As stated in previous chapters, water quality regulations evolve continuously to reflect social and scientific developments. Continuing water quality degradation in many countries has dramatized significant oversights in legislation implemented to date. Interestingly, the importance of responding to these over-sights may be interpreted equally strongly from a utilitarian viewpoint, in which continually degraded quality erodes valued goods and services, or from a naturalistic viewpoint, in which degraded quality interrupts inherently valu-able natural cycles. One of the most important oversights involves the syner-gistic or antagonistic interactions between the effluent and the receiving water.

Effluent-receiving water interactions take place over fairly large distances, usually outside the realm of defined mixing zones. Consequently, a range of ef-fects may occur that are not adequately managed by available control strategies, including sublethal effects, short-term episodes, or long-term trends in effluent quality. Thus, synergistic impacts that degrade quality but that fall outside of most management structures include both synergisms between actual pollut-ants, and synergisms between any pollutant and natural variations in water chemistry and aquatic ecology. Effluent limits tend to be imposed broadly—for example, across all facilities of a given class of industry. As a result, these limits may not address industry-specific conditions nor have they traditionally con-sidered geographic differences among receiving systems.

Current trends in water quality management revolve around recognizing spatial patterns in water quality generated by a combination of natural condi-tions and historical influences. Warm-water, muddy streams will never support jumping rainbow trout. Nevertheless, many of our previous uniform-goal strate-gies would manage or regulate the water resource as if that end-result were ex-pected. An emerging alternative to this approach suggests regulations based on these spatial and historical conditions. Thus, the trend in the future is to design "best attainable quality" goals that are more realistic, less idealistic, and more regionally appropriate.

Critics suggest that best attainable quality, by having flexible standards, al-lows greater degradation of existing resources and does not promote restora-tion or improvement. Advocates counter that regionally specific goals and criteria are appropriate precisely because they are flexible: they offer the best hope to match biophysical goals and strategies with biophysical and social re-alities, thereby ensuring attainable standards. They also take advantage of local and regional institutions and give a sense of "ownership" of projects to the af-fected parties, a factor that has been widely acknowledged as a key to success in environmental management. Because managing for "best attainable quality" is still an emerging concept, water quality literature uses many different terms to approximate the new approach. The most common of these, and the one used most consistently throughout this text, is "regional management" or "ecore-gional management." The evolution of this approach is explored in chapters 14 and 15, and its application in chapters 16 and 27. Regional management is being implemented in most states of the United States as well as being attempted in parts of Europe (see chapter 16).

Regional Differences in Attitudes

As stated earlier, goals differ not only among countries, but also within countries, regions, and states. Using this knowledge is new, however, to most natural resource management. For example, in the early 1990s, several researchers began evaluating user perceptions of water quality to integrate these attitudes into improved management plans (Smith and Davies-Colley 1992; Thornton and McMillan 1989; Smith *et al.* 1991). Significant among these is an emerging body of literature describing important regional differences in perception and valuation of local resources (Eleftheriadis *et al.* 1990). Different expectations for resource quality may develop in areas separated by only a few hundred miles. For example, user perception studies in Minnesota have found that water quality rated as "acceptable" by the public in the southern region is unacceptable in the Northern Lakes and Forests region. The differences appear to be linked to typical natural variations in water clarity between the regions. Clearly, baseline regional differences have very real implications for user perception of value and the setting of water quality standards and management action in the state.

Differences in biophysical baseline values are only one part of a complex of social, cultural, and administrative factors that control regional and local attitudes toward water quality and, consequently, water management challenges. Confidence in the management agency is, by itself, an important determinant of people's perception of water quality problems. For example, Syme and Williams (1993) conducted a study of perceptions of drinking water quality in Australia. They found that people's judgment of drinking water quality was closely related to the perceived credibility of their water supply institution, as well as to the degree of control they felt they had over water quality and environmental problems.

Syme and William's study echoes concerns expressed a year earlier by Fawell and Miller (1992). These authors suggest that as environmental awareness has increased in the United Kingdom (and, it can be assumed, in other countries), people there have become less tolerant of environmental risks yet remain unable to evaluate the nature of those risks. Thus, public trust in water quality institutions has decreased, as has public confidence in the scientific determination of risks and acceptable standards. While perceptions of trust are not usually factored into analyses of environmental decision making, they clearly play an important role and can have a dramatic influence on the effectiveness of management strategies.

A Comparison of Goals and Management Strategies

Chapter 2 contained a brief comparison of the history of water quality management in England and the United States. Those examples are extended here to provide some insight into the ways in which variances in values and attitudes are expressed in water quality legislation, illustrating how different societies—even those with broadly similar goals—may take different approaches to legislating changing environmental values (Table 3.1).

For example, the United States is a large and young country that prides itself on its extremes. Right or wrong, the United States does things whole-heartedly. Consequently, U.S. residents expect immediate solutions to problems and the nature of these solutions reflect the paradigms of the day. England, on the other hand, is a more intimate country, both in size and its political structure. Maintaining decorum and traditions are important to English society. Consequently, change is expected to be more gradual and less "upsetting."

In addition, English barristers traditionally maintain close associations with industries and professional associations. In consequence, laws are enacted more quietly and with greater regard to economic impacts on specific polluters. Although progress is made in England's water quality management, river authori-

Table 3.1 Water quality management goals are highly diverse and usually reflect the mission and operating philosophy of the agency involved. Any given goal has associated limits that constrain management of the resource.

Exemplary Organization	Type of Goal	Limits
U.S. National Park Service	Antidegradation	Local scope
U.S. 1972 Water Quality Law (PL 92-500)	Utility for swimming, fishing	Allows degradation, difficult to assess
U.S. 1985 Water Quality Law	Biotic integrity	Difficult to measure
European Community	Assimilative capacity	Allows degradation, may result in ecosystem damage
Nigeria	Economic growth	Degrades resource, limits future options
Deep Ecology, some Norwegian watersheds	No detectable impact, protect intrinsic value	Precludes other uses, constrains economic growth
Ecosystem Management, some U.S. Forest Service locations	Sustainable ecosystem structure and function	Public perception is varied, thus support varies widely

ties generally refrain from prosecution when interests of a local authority or industry appear threatened.

Trends in England's Water Quality Laws

England has made important changes in water quality administration (see chapter 2) but has rarely deviated from the practice of regulating through water quality objectives. Thus, details of the administrative structure change in response to economic and social conditions but the overall goals have not changed. England's Water Act of 1945 established a framework of River Boards to study conditions in the nation's rivers. The River Board format persists today, but the boards provide water regulation rather than water quality management.

The present institutional implementation of the River Board framework has its roots in the 1963 Water Resources Act (Parker and Sewell 1988). This act sought to bring about greater coordination of the water industry, which had previously been managed under hundreds of small management units. The 1963 act consolidated agencies, introduced an economic basis for water management, and developed a national water policy.

Although these changes were initiated in 1963, they were not completed until the 1973 Water Act consolidated approximately 1500 management units into 10 Water Authorities virtually overnight. The Water Authorities represented a major departure from previous approaches: they were regional-scale, river-basin-based, multifunctional agencies operating on a source-to-mouth principle. This act also removed water management from local government: the new River Authorities were semi-autonomous, statutory private companies.

Parker and Sewell (1988) suggest that the strategic rationale for development of the River Authorities was increasing industrial and urban demands for water. They also note that the economic recession of the late 1970s halted these increases. Subsequent years have brought about a relaxation of monitoring and regulatory guidelines, although no changes in actual standards. Since the early 1980s, England has experienced a net decline in river water quality and a growing farm waste problem. Reductions in public investment have led to the demise of many sewage systems and declining public health conditions.

In the Water Act of 1983 and amendments in 1987, another new structure was imposed on water management institutions (Parker and Sewell 1988). The latest series of administrative changes spurred the privatization of water supply, sewerage, and sewage services through creation of 10 Water Services Public Limited Companies (PLCs). Other water resource management functions were transferred to a new National Rivers Authority (NRA). Some of the biggest challenges facing the new NRA and PLCs come from the European Commission (EC) and demands to comply with EC water quality directives.

Trends in U.S. Water Quality Law

Legislation is a much more public process in the United States than in England. Vocal interest groups and lobbyists use a variety of media to attempt to influence law making. In water quality control, this process has resulted in standards emulated around the world, but the goals of these standards have vacillated significantly from utilitarian to nearly naturalistic before finally settling on a midrange approach. In contrast to England, the United States has changed its goals but kept its administrative framework relatively constant.

Water quality laws in the United States actually predate the formation of the nation. European settlers in New England protected their resources with a series of colonial laws known as "the ancient charters." Based on common law rights, the charters decreed that lakes, bays, and rivers were available for common use. The charters forbade "corrupting" springs and other sources of water, prohibited poisoning water, and contained provisions against overfishing and the building of dams or weirs without permission. Although far-thinking, these early laws were gradually neglected or repealed as rivers evolved into a source of power for industrial use and more immediate needs gained prominence.

Official, national water quality control legislation in the United States began with the Rivers and Harbors Act of 1899. This major piece of legislation was prescient in that it chose to regulate practices that had negative impacts on water quality, rather than regulating the quality of the water itself. It declared illegal the discharge of any material into the navigable waters of the United States or into any tributary from which wastes might reach the navigable waters. Clearly concerned more with obstructions to traffic than with water quality, the prohibitions exempted urban run-off from streets and discharge from municipal sewage.

The 1899 act focused on prohibiting specific practices, based on their assumed impacts on water quality. In contrast, most U.S. water quality laws enacted between 1900 and 1972 focused on the expected results (i.e., they tried to regulate ambient water quality). The vast majority of that 70-year body of legislation placed limits on end-points and allowed dischargers the freedom to choose the best method of compliance. The targeted pollutants changed as impacts, populations, and industrial growth increased, but the approach to regulation remained constant.

With the passage of PL 92-500, the Water Pollution Control Act Amendments of 1972, U.S. water quality policy changed dramatically. This new law reflected the philosophy that any activity that affects water quality should be curtailed or controlled. At this point U.S. policy makers implicitly recognized that management on the basis of ambient conditions had failed for several reasons, including emerging awareness of bioaccumulation, nonpoint pollution, and the inability of ambient quality standards to address regional differences in quality.

For these reasons, the 1972 law turned to technology regulation as an answer to achieving clean water standards. The act's intent was to restore and maintain the integrity of the nation's waters through such goals as "fishable/swimmable waters" and "zero discharge of pollutants." While reflecting ambient standards, the act was written to be implemented almost entirely on the basis of attaining zero discharge. It also contained a long—if unachievable—list of strict expectations. Most significantly, all discharges were to be eliminated by 1985; in the interim, specific treatment practices were mandated for industries and municipal waste discharges. Although laudable, the goals were too ambitious to be achieved; despite $15 billion in investment, few significant improvements were made in the first decade after the act. Congress has since repealed the goal of zero discharge and rephrased the goal of biotic integrity.

Every 5 to 6 years since 1972, the Clean Water Act has come up for reauthorization (and amendments) before Congress. The reauthorization process gives the legislature an opportunity to modify approaches that are not achieving desired goals, alter strategies to reflect changes in administration, and incorporate new research directions, new goals, and new national and political priorities. For example, the Clean Water Act of 1977 represented a modulating swing in the water quality regulation pendulum: incentives were introduced to assist industry and cities in complying with fishable/swimmable goals. It also contained provisions for additional federal grants to municipalities for construction of treatment facilities and staff training. Industry, in its turn, was granted an extension of critical effluent deadlines. Lastly, in a continuing effort to define the role of states and the federal government, states were granted primary authority over water quality and water use.

Since 1972, reauthorizations and associated legislation have continued to strengthen the role of states in water quality management. They continue to search for ways to implement effective management for nonpoint source pollution. Designated industries occasionally receive variances, and specific standards change as science and society require them to. In addition, the use of incentive programs such as permitting discharge trading permits has increased for many areas of pollution control in the Unites States. Overall, however, the basic approach of the technology-based standards established under the 1972 law continue to operate, although the specific mandates of the act change in response to changing social priorities.

Summary

- Each society has cultural values that determine its attitudes and the ways it values natural (and other) resources.
- Those attitudes are expressed as explicit goals for water quality management. Such goals range from highly anthropocentric views (e.g., *The resource is there for humans to use as we wish*) to highly protectionist (e.g., *No impact is defensible*).
- In the future, water quality management goals will be developed in ways that more explicitly recognize temporal and spatial differences among water bodies and that recognize the impacts of multiple uses on water bodies. Goals must acknowledge that no water body can serve all ends equally.
- The evolution of decision making that is becoming prevalent worldwide will result in goals that are expressed on a much more local scale. That trend will create conflicts as science increasingly recognizes that cumulative effects and large geographic scale changes dominate biophysical water quality patterns.

For Further Reading

Callicott JB. *Earth's insights: a survey of ecological ethics from the Mediterranean Basin to the Australian Outback.* Berkeley: University of California Press, 1994.

Falkenmark M, Suprapto RA. Population-landscape interactions in development: a water perspective to environmental sustainability. *Ambio* 1992;21:31–34.

Petulla JM. Environmental values: the problem of method in environmental history. In: Bailes KE, ed. *Environmental history: critical issues in comparative perspective.* New York: University Press of America, 1985;36–63.

Regier HA, Bronson EA. New perspectives on sustainable development and barriers to relevant information. *Environmental Monitoring and Assessment* 1992;20:111–120.

Syme GJ, Williams KD. The psychology of drinking water quality—an exploratory study. *Water Resources Research* 1993;29:4003–4010.

Thonton JA, McMillan PH. Reconciling public opinion and water quality criteria in South Africa. *Water South Africa* 1989;15:221–226.

4

Global Water Resources and How They Are Used: The Expression of Goals and Objectives

Overview

The crux of all water quality discussions is that we use and depend upon water for nearly every aspect of our lives. Clean water is vital for domestic uses ranging from transportation and industry to food production and ecological well-being. Unfortunately, these uses are not always compatible nor are they valued equally within or among societies. Uses reflect societal and cultural values. The central aim of water quality management is the protection of the most highly valued uses. This chapter reviews the distribution of water on the planet, traces how heritage and location (e.g., arid or mesic) often dictate water-use patterns, and describes how those patterns are translated into protected uses by management agencies. The chapter also reviews water-use patterns of municipalities, agriculture sectors, industry, and other major water users around the world.

Introduction: A Watery Planet

How much water is really available and why is water use a concern? For most people in the "developed" world, making water available is as simple as turning on a tap. Easy and safe water access is, in many ways, inexorably linked with our concept of development. Easily available tap water creates an illusion among many world citizens that the resource is infinite. It is not, nor is water equally available or potable around the globe. Globally as well as regionally, water use and management are influenced by inherent patterns of water distribution and availability. Use patterns integrate value decisions and ecological possibilities for different water bodies and consequently form a fundamental tenet of water quality management discussions. Many of the concepts discussed here are revisited in chapter 25, when the discussion is framed from a perspective of how use and values determine water quality decision making.

Global Patterns

While the earth is known as the *water planet,* the vast majority of that water is unavailable for human use. More than 97% of the earth's water, for example, is saline ocean water. Another largely unavailable reservoir of water is the 2% of the earth's water frozen in polar icecaps and glaciers. Of the remaining 1% of the earth's water, more than half (0.6% of the total supply) is contained in groundwater (Shiklomanov 1993). There is 125 times as much groundwater as there is water in rivers and streams. Miller (1994) notes that if all the earth's water were equivalent to 100 liters, fresh water would comprise only 3 liters of the total and available fresh water would amount to only $\frac{1}{2}$ teaspoon. These figures have been displayed graphically in many different ways in efforts to dramatize the fact that an extremely small percentage of the world's water is available for human use (Fig. 4.1).

If water were a nonrenewable resource such as oil, gas, or coal, the entire resource would be exhausted in less than a single day. Total water available for public water supply in streams and rivers of the world represents about 12 mi³, or about 12 times the volume of San Francisco Bay. The population of the United States uses that quantity of fresh water each month. The sum of all of the water in the world's lakes and reservoirs represents the quantity of water used by the United States in 25 years. We can sustain those levels of consumption only because water is continually recycled and reused. As a result, rivers can supply us with two-thirds of the water used in the world even though they represent only 0.03% of available water supply. Only a fraction of the water withdrawn for use is actually lost to the system from which it was drawn. The rest is returned and withdrawn again by

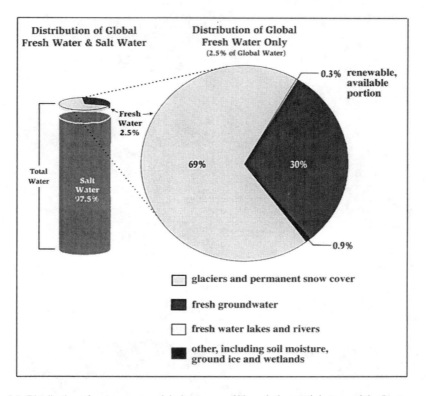

Distribution of Global Fresh Water & Salt Water

Distribution of Global Fresh Water Only
(2.5% of Global Water)

0.3% renewable, available portion

Fresh → Water 2.5%

Total Water

Salt Water 97.5%

69%   30%

0.9%

☐ glaciers and permanent snow cover

■ fresh groundwater

☐ fresh water lakes and rivers

■ other, including soil moisture, ground ice and wetlands

Fig. 4.1 Distribution of water among global reserves. Although the earth is termed the "water planet," barely 0.3% of that water is available for human use (Engelman and LeRoy 1993; reprinted with permission by Population Action International).

downstream users, returned and withdrawn again, and so forth along the length of the river.

We are all downstream from someone else. The central goal of water quality management is to maximize uses of a water resource while not degrading quality to such a degree that other significant benefits are precluded.

Regional Distribution

The "½ teaspoon" of available fresh water discussed above is distributed over the entire planet, but not uniformly. Some regions are water-rich, while others are water-poor (Fig. 4.2a). Similarly, human population densities are not as closely correlated with water availability as they perhaps "should" be. Especially in industrialized countries such as the United States, it has become possible to "make deserts green" and to establish large population centers

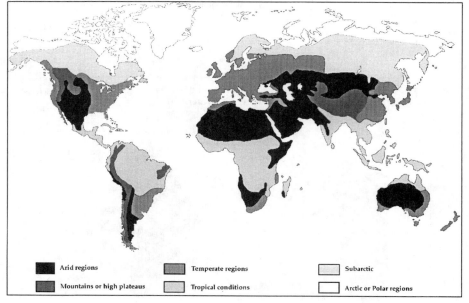

a

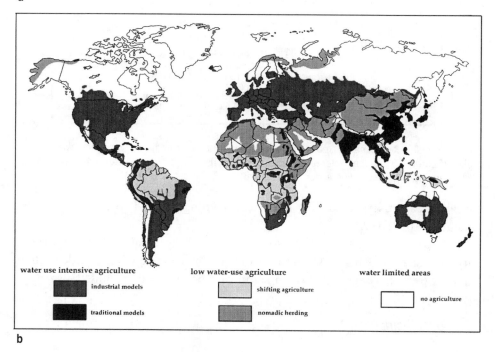

b

Fig. 4.2 (a) The distribution of water around the globe is uneven and does not correspond to politically defined national or regional borders (modified from Miller 1992; reprinted with permission of Wadsworth Publishers). (b) While in many areas agricultural productivity is limited by climate, large areas of the world have been transformed into highly productive zones through the use of water-intensive irrigated agriculture (modified from Miller 1992; reprinted with permission of Wadsworth Publishers).

(e.g., Las Vegas, Los Angeles, Phoenix) in areas where locally available water is sufficient for only a small fraction of the current population (Fig. 4.2b). Large networks of aqueducts now cross the United States and other countries to bring water to these population centers. Controversies erupt frequently over issues such as the effect water withdrawal has on a given water resource or if rich and politically favored communities should have the power to harvest distant water sources.

Figure 4.2a also shows that water supplies do not follow national, state, or territorial boundaries. In fact, major rivers often serve as the dividing line between adjoining nations or states, confounding water management by necessitating international and intranational cooperation. In some areas of the world, water availability constitutes a central issue in society (Fig. 4.3). In other areas, availability is taken for granted but quality is a central concern (Fig. 4.4). In still other areas dependent on navigable water, keeping waterways clear of debris may be the major concern. In all cases, water quality management becomes a balancing act between quantity and the quality required for primary uses.

Fig. 4.3 In many arid areas of the world, such as the Sahel in Africa, water quality concerns are secondary to concerns of water availability. (Photo by J. Perry)

Fig. 4.4 Thailand, a country rich in water resources, relies heavily on the assimilative capacity of its rivers and canals. Here, water availability is taken for granted, but awareness is only beginning to grow that water quality may suffer even when quantity is plentiful. (Photo by D. Vanderklein)

Using the Water We Have

Identifying the uses for which a water resource will be managed is the first, and most important, step in managing water quality. Through the process of assigning different uses to water bodies and thereby setting priorities for water resource allocation, all remaining steps in water quality management become relatively well identified. Designating those uses is highly controversial in most parts of the world, however, because uses are often competitive or mutually exclusive. Thus, the process of use allocation involves setting priorities and valuing one goal more highly than another. Establishing uses and resolving the conflicts that result constitute very challenging elements of the management cycle.

Significance of Water to Human Societies

An adequate supply of "clean" water is one of the most basic human needs— and one that is not met for more than half of the world's population. According to various estimates, one-half to two-thirds of the world's population does not have access to adequate quantities of safe drinking water or sanitation. The United Nations declared the 1980s to be the "Water and Sanitation Decade" and directed the World Health Organization to carry out necessary actions to ensure that the world's people were provided with water and sewage facilities. The task was much larger than anticipated. At the end of the decade, large areas of the world and most the world's population still received inadequate water quantity and quality. The implications of this supply gap are enormous. Higher child mortality, for example, is closely linked to inadequate supplies of clean drinking water (see chapter 17, Fig. 17.2), as is greater incidence of many debilitating diseases ranging from malaria to typhoid. Food supply and adequate nutrition also depend on adequate water, and associated costs in human lives, human potential, and economic costs to arid countries are incalculable.

Some of the popularity of Frank Herbert's *Dune* novels has been attributed to the conceptual departure of Dune (a water-scarce planet) from much of the earth. Human societies have traditionally evolved around water bodies (e.g., along river corridors, and ocean and lake shores); the drier the environment, the greater the requirement for technological intervention, from deep wells to aqueducts. Comparing Figs. 4.2a and 4.2b, it becomes clear that climate exerts a significant influence on use patterns. Superimposing a socioeconomic map on these patterns would dramatize even more clearly the ways in which water availability can constrain the rate of development and spatial distribution of many societies in the world today.

Use Conflicts

Water is unique among our natural resources in the degree to which it is reused. As discussed in the introduction to this chapter, the limited amount of water that exists is continually being reused. The same water may be used in one location for recreational uses and later used to generate electricity or float a barge down a river. The same water might be used yet again as it is withdrawn for a public water supply or an irrigation system. Only a few uses actually *consume* water (i.e., remove it from the system so that it is no longer available to downstream users). Figure 4.5 illustrates a typical water-use cycle and demonstrates our dependency on multiple uses of water.

Because water is truly a common resource, conflicts in its use result when changes imposed by one user preclude later use by another, or when a given application itself precludes a simultaneous user. Conflicts arise over water quantities, water facilities, and changes in water quality. Disputes over quantity are most relevant to the few water users that consume most of the water they use, making the water unavailable to others. Irrigated agriculture, for example,

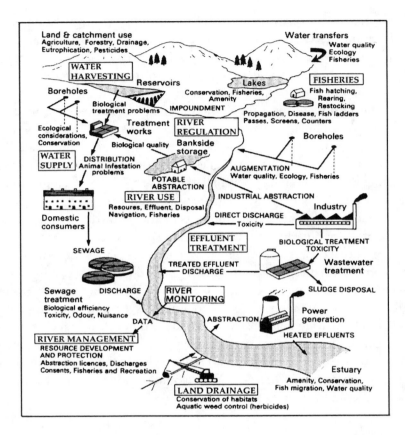

Fig. 4.5 A typical water use cycle, demonstrating the multiple ways in which a single body of water may be used in a watershed (after Newson 1992; reprinted with permission of Routledge Press).

consumes 55% of the water it uses (see chapter 19). The consumptive nature of irrigation, therefore, limits many simultaneous users of the same resource. Municipal facilities such as cities consume 21% of water they withdraw. In contrast, industry, which withdraws very large quantities of water, consumes only about 3% of that water (see chapter 20). Unless unacceptable changes in quality occur, many industrial users could benefit from the same water resource.

Industrial needs, however, often require the physical alteration of a stream or river, and these alterations can be a source of conflict for other water users. For example, building dams can generate controversy because the nature of the facility precludes canoeing or kayaking along the river, and may influence fishery stocks as well. In addition, because most water uses change the quality of water returned to the hydrosphere, downstream uses are restricted to those compatible with the water quality discharged by upstream users. This upstream–downstream relationship is the most familiar avenue for water quality disputes.

Designation of which water uses will be protected or maintained for any given resource is the most important decision made in any water management program. That decision should be undertaken very carefully, because it dictates all other aspects of management. The following section discusses ways in which those decisions are made.

The Concept of Protected or Beneficial Uses

Because many water uses conflict with one another, management agencies must establish priorities among uses (and, therefore, among users). This process establishes the possible uses of a water body and then makes conscious decisions about resolution of potential conflicts. A *beneficial use* is one that has value to some subset of society. Typical examples include domestic water supply, irrigation, waste disposal, fishing, and navigation. *Protected use* is a more restrictive term; it results from a political or managerial decision that assigns a higher priority to one (or more) beneficial uses than to competing or conflicting uses. Designation of a protected use suggests that other uses—whether of adjacent land, water, or atmosphere—will be controlled in order to protect the designated values. For example, if fishing is a *protected use* on a given stream, then logging and road building in the watershed must be conducted in such a way that stream sediment loads, turbidity, and temperature changes do not adversely affect the fish population.

A *beneficial use* is a generic attribute suggesting that certain characteristics must be present if the water is to be used for some purpose. In contrast a *protected use* is a statement of specific intention to constrain management practices.

Water-Use Patterns Around the World

Water use falls into several major classes, each of which is associated with certain quantity and quality requirements. These classes include water for human drinking and cooking, waste disposal (including waste heat and other waste products), crop production (including aquaculture, livestock, herbaceous crops, and trees), industrial use (e.g., mining and manufacturing), recreational use, navigational uses, and ecological values such as production and survival of natural lake, riverine, or wetland communities. The quantity of water required for activities within each of these classes is influenced by several things. On a major scale, geographic variables such as climate, temperature, precipitation, and water availability significantly influence water use. Out of necessity, arid or xeric areas generally have lower water use per capita or per land area than do more mesic areas. This observation is not true, however, when an arid area has been irrigated for large-scale agriculture. Agricultural uses of water are enormous, and some of the most productive lands are located in normally arid areas such as California's Central Valley. In the moist areas of North Carolina (in southeast United States), for example, irrigated agriculture requires approximately 595 m³ of water per hectare per year. In contrast, in the arid areas of California and eastern Washington (in western United States), irrigated agriculture requires more than 13,000 m³ of water per hectare per year. Arid areas may also have high water use rates if economic vitality of the state or country (e.g., oil-producing areas) enables heavy subsidy of water use rates and supplies.

Culture and religion can also play a significant role in water use patterns (see chapter 25). In Islamic areas of the world, religion dictates that people wash five times each day, or once before each prayer session. That requirement places pressure on water resource managers to supply adequate quantities and quality of water to the population, which in turn dictates growth patterns and pressures on water supply facilities. In general, however, increasing availability of water has the universal effect of increasing water use, especially for domestic purposes.

Traditionally, water is most used for domestic applications such as drinking, cooking and bathing, industry, and agriculture. In most developed areas of the world, recreational and ecological uses of the water place increasing competitive pressure on these traditional uses.

Domestic, Municipal, or Public Water Supply

One of the first areas to be addressed in economic development is water supply infrastructure (Fig. 4.6). Municipal water serves many purposes. The most obvious may be potable water, but only 40% of "domestic" water in the United States is used in homes and only about 20% of that amount is used for drinking and cooking. The rest of the water supplied to homes is used for waste disposal, bathing, and landscaping. The remaining 60% of domestic water goes to institutions. These percentages vary widely between urban and rural areas, and among the different countries and regions of the world. Urban areas exhibit higher institutional use, while rural users require proportionately more water for consumption. Similarly, citizens of developed countries allocate more water to institutional use and less to homes (Table 4.1).

Many different variables affect the rate of water use. During the year, warmer and drier times will have higher water use rates. In relatively arid areas such as the southwestern United States, domestic water use during the summer averages 2.3 m³/person/day, with much of that amount being used for cooling and landscaping. In the winter in the same areas, water use drops to 0.4 m³/

Fig. 4.6 Even a communal supply of clean, piped water can be an important asset to developing communities. The provision of clean water and the provision of adequate waste disposal usually occur separately as infrastructures develop, however. Consequently, while many Nepali villages such as this one have enough piped water, the open, slow-running sewer is not adequate to remove the waste; as a result, water-borne diseases remain a major problem throughout the region. (Photo by D. Vanderklein)

Table 4.1 Water is used for domestic needs, industry, and agriculture, among other purposes. The percentage of total water withdrawal used for any specific use varies among regions of the world, and is a product of socioeconomic conditions, tradition, culture, and water availability. Europe and North America are highly industrialized, while Asia and Africa use water primarily for agriculture. (Data from World Resources Institute 1993.)

Region	Domestic	Industrial	Agricultural
World average	8%	23%	69%
South America	18%	22%	60%
Europe	11%	55%	34%
North and Central America	9%	42%	49%
Asia	7%	8%	85%
Africa	6%	4%	90%

person/day (Tchobanoglous and Schroeder 1987). During the day, water use in a suburban home fluctuates from near zero in the early morning hours to nearly 100 liters/hour for short periods in late morning and evening. The vast majority of water in the home is used for waste carriage. Toilet flushing and bathing together account for nearly 75% of all domestic water use.

Many other variables also influence municipal water use around the world. Average total domestic water consumption in the United States in 1980 was 693 L/person/day (Lamb 1985). That per-capita rate increased 67% between 1950 and 1980 and does not consider the 300% increase in water consumption for hydroelectric power that occurred in the same period. In Baghdad in 1990, water consumption was only 500 L/person/day, a rate 28% less than the U.S. rate of 10 years earlier. Immediately after the Gulf War, the quantity of water available in Baghdad dropped to 15 L/person/day (National Public Radio, March 5, 1990).

The downstream impacts of municipal water supply and waste treatment are generally twofold. Withdrawals for municipal water use represent abstractions—quantities of water taken from the stream channel and diverted to another use. Water supplies are nearly always drawn from upstream, high-quality surface water sources or from groundwater. Although this use on the downstream community principally impacts water volume, the percentage of available water removed is usually very small. In contrast, downstream impacts of municipal wastewater are numerous. Generally, they include increases in organic matter and reduced stream oxygen concentrations, increases in nutrient loadings, and increases in some toxic substances like chlorine and chloramines. Most of these impacts are relatively well controlled by most municipalities around the world.

Industrial Water Supply (e.g., Manufacturing and Mining)

Most industrial water is used for cooling, although many industries consume large quantities of water in their manufacturing practices. Within limits, cooling represents a relatively benign water resource use. A typical one-megawatt steam generating plant, for example, produces one-third of its energy as electricity and two-thirds as heat (33% efficiency) and discharges water to a receiving stream. If this plant were limited to discharging water that caused no more than a 0.3 °C change in the temperature of the receiving stream, the plant would require 1.6 m³/sec (138,240 m³/day) of cooling water, a quantity equivalent to the requirements of a city of 628,000 people (Tchobanoglous and Schroeder 1987). All but 3% of that water would be returned to the stream from which it was taken. The power plant would *consume* only approximately 4000 m³ of water per day. In general, industrial production requires enormous amounts of water. Countries that prioritize industrial production, therefore, face numerous trade-offs in areas where industrial requirements compete with other supply needs (Table 4.2).

In most developed countries, industrial water uses are closely monitored and regulated by government and are usually operated to have minimal impact. However, in cases where industrial effluents are not well managed, water quality impacts are of three general classes: heat, algicides, and metals. As noted, many industries use water for cooling, discharging very large quantities back to the surface water. The heat loads of those waters can be significant, raising in stream temperatures by several degrees centigrade in isolated cases. Industries must expose water used for cooling to the atmosphere so that evaporation can remove the waste heat. That process usually involves exposing shallow, wet surfaces (e.g., cooling towers) to sunlight, resulting in large accumulations of algae on the cooling surface. Industries use a variety of algicides to control that problem, but discharge of those algicides to surface waters causes significant instream impacts in isolated cases.

Industries consume much smaller quantities of water in their manufacturing processes than they use is cooling. These *process waters* are the ones most closely regulated and most likely to cause a downstream impact. The contaminants commonly found in industrial wastewater include more than a dozen heavy metals and other toxic compounds (see chapters 17 and 20).

Table 4.2 Agriculture and industry are major users of water. Various industrial processes and agricultural crops require different amounts of water, ranging from small quantities such as 10 gallons required to process one gallon of gasoline, to more than 2000 gallons required to grow one pound of cotton. (Data from U.S. Geological Survey).

Product or Crop	Gallons of Water Consumed
One automobile	100,000
One pound of cotton grown	2000
One pound of aluminum refined	1000
One pound of grain-fed beef grown	800
One pound of rice grown	560
One pound of synthetic rubber	300
One pound of corn grown	170
One loaf of bread produced	150
One kilowatt of electricity produced	80
One pound of steel produced	25
One gallon of gasoline produced	10

Agricultural Water Uses

Agriculture is by far the largest water user in the world today. Vast areas of the world are already irrigated, and irrigation development continues to increase in an attempt to meet the world's growing food demands. Much of the water applied to agricultural crops is consumed through either evaporation or plant growth. Because irrigated agriculture occupies such a large land area, the quantity of water consumed is dramatic: irrigated agriculture in China, for example, consumes a quantity of water each year equivalent to two and a half times the mean daily flow of the River Nile. Water consumption rates range between 0.5 and 1.0 m per season (i.e., 5–10,000 m^3/ha/yr) depending on the crop, temperature, and length of the growing season (Table 4.3).

Although irrigated agriculture is, proportionally, an enormous drain on water resources, water also plays a role in the production of nonirrigated crops. For example, water is required for aquaculture in ponds, rice paddies in riverine environments, wetlands, and coastal areas, and for nonirrigated cropland and forestry. Agricultural waters are primarily taken from surface waters, although groundwater pumping is extensive in western, economically developed countries. Excess waters are released back into many streams and rivers; wastewater injection wells (i.e., those that deposit wastes directly into the subsoil and possibly to the aquifer) are also quite common in isolated areas.

Table 4.3 **The amount of water withdrawn for various uses is a product of water availability as well as cultural factors such as the degree of industrialization. More industrialized countries consume more water per capita than do less industrialized countries. (Data from Gleick 1993).**

Country	Available Water (km³/yr)	Fraction of Available Water Withdrawn per Year	Per Capita Withdrawal (m³/person/yr)	Domestic %	Agriculture %
Central African Republic	141	0	27	21	74
Libya	0.7	404	623	15	75
Canada	2901	1	1752	11	8
United States	2478	19	2162	12	42
Afghanistan	50	52	1436	1	99
Bhutan	95	0	15	36	54
Iraq	100	43	4575	3	92
Albania	21	1	94	6	76
Bulgaria	205	7	1600	7	55
France	185	22	728	16	15

Beyond the sheer volume of use, agricultural uses of water are critical because of the often significant changes in downstream water quality due to fertilizers, herbicides, erosion, and stream diversions (which may affect water volume). Thus, the source of agricultural water is as important a consideration as is the return water. Agriculture has the largest water quality impact of any water use (see chapter 18).

Recreational Uses

The importance of specific recreational demands on water resources generally coincides with increases in a country's socioeconomic status. Even where not stated directly as a use, many people in developing countries employ bodies of water for swimming and as gathering places. Thus, while the quality of these waters may not be considered for recreational values per se, the incidence of disease transmission in these common waters is still high and of primary concern for public health.

In developed countries, however, recreational values of water including aesthetic values are increasingly important and increasingly specific. Two generalizations can be applied to recreational water demands. First, recreation often uses water "in place." Examples include fishing, boating, aesthetics, and bird watching. Second, recreational uses have specific quality requirements such as clarity or fish production and generally do not cause significant downstream impacts. Of course, exceptions to that generality exist, such as in areas of high recreational use in particularly fragile ecosystems. An interesting phenomenon in many developed countries is that many recreational users are members of fishing groups, hunting associations, or advocacy groups. Each of these clubs exerts significant influence on water quality planners because it represents an organized use group that can exert both political and social pressures in local areas.

Ecological Water Requirements

Concern for ecosystem values as a water use is a relatively new phenomenon in water quality management. Only since the 1970s has our awareness of nonhuman uses begun to develop. Many opinions exist as to the priority that the variety of "ecological" water uses should be given in decision making.

An example of an ecological use is protection of biotic diversity, an issue whose significance is rapidly gaining importance for human society. Its primary anthropocentric value lies in the fact that we cannot predict the future utility of a resource given today's knowledge. Science could potentially discover highly valuable uses for some aquatic plant or animal. Conservation of the current aquatic environment's diversity increases the probability that some organism of future value will be protected and promotes the overall sustainability of that aquatic resource.

Another example of an ecological use given high priority is fish passage. Great amounts of public funds are spent each year to ensure (1) that adequate water volume remains in the stream to sustain "normal populations" of aquatic organisms and (2) that migratory species such as salmon can pass. Sometimes a productive use of the water resource is lost (e.g., certain kilowatts of power generation) to allow other productive uses, such as maintenance of a mayfly population. In other cases, funds are spent to mitigate an ecological loss such as when migrating salmon are transported around dams or funds are invested to restore a damaged ecosystem.

Perhaps because ecological valuation is a relatively new area, it is rich in potential conflicts. For example, in the San Joaquin Valley in California, wetland issues capture the complexity of water resource management planning. Agricultural interests in the late 1800s created wetlands along the lower delta of the San Joaquin estuary. These low-lying areas require extensive pumping to prevent saturation, which would change the plant and animal community. Today, these wetlands are valuable, highly productive nature conservation areas that harbor many species of birds and wildlife. They are clearly not "natural" areas, however. Proponents of conservation of the wetlands suggest that water should be released from upstream dams to hold back saline intrusion. Antagonists claim that use of "valuable" upstream water from reservoirs cannot be justified to protect an area that is neither natural nor "useful."

Establishing *Beneficial* and *Protected Uses*

The process of establishing protected uses for a water body is straightforward but contentious. As a first step, water bodies are classified based on their biophysical attributes (e.g., chemistry, biology, physical properties). Integral to the classification process, the water quality manager refers to criteria that describe the quality attributes necessary for each use (see chapters 5 and 6). The results of this classification establish the uses that could possibly be sustained on each water body. At that point, societal input from the general public, politicians, and other stakeholders establishes priorities among the possible uses. After priorities are set, the highest priority uses are designated as protected; the *protected* list will also include uses that do not conflict with ones already designated. Policies and standards are implemented to protect those designated uses, management practices are implemented to comply with those policies, and monitoring programs are implemented to assess compliance.

Conflict Resolution in Establishing Protected Uses

Because so many water uses conflict with one another, societies need ways to resolve conflicts among users. The traditional means for resolving water use conflicts (though perhaps not always equitably) is through legal action. For example, a society may pass a law dictating that one particular water use be given a higher priority than another. In the United States, water for domestic consumption has the highest priority. In watersheds where the water is to be protected for domestic use, other uses become secondary and are subject to the quantity and quality restrictions of domestic use.

In other situations, however, the case is less clear. For example, the city of Los Angeles desperately needs to increase its available water supply. In the early 1930s, an engineering group designed a system that would draw water from the Mackenzie River in Canada and transport it several thousand kilometers through pipes, canals, and river channels to Los Angeles. The implications in terms of water loss in the Mackenzie drainage, interbasin transfer of organisms, water quality, and quantity changes in the intervening basins are herculean. Those obstacles have been disempowering to date. Similar designs to bring water from the western slopes of the Rocky Mountains to Denver and from the Great Lakes to Chicago are less grandiose and have met with varying degrees of success. These kinds of conflicts will surely increase throughout the world in the next 25 years.

Another emerging strategy is conflict mediation. In this approach, two or more sides meet to discuss and find mutually agreeable solutions without resorting to legal challenges. In 1995, for example, the city of New York and residents of the upstream areas that supplied New York's water resolved a long-standing and bitter controversy through conflict mediation. A significant opportunity exists for research and practice in conflict resolution in water use allocation; that opportunity will only grow for the foreseeable future.

Summary

- Although water is widely available on the earth's surface, the quality of a given water body constrains its utility for any given purpose. To explicitly

recognize and manage those quality differences, we define *beneficial uses* (i.e., uses to be protected) for each water body we manage.
- Each beneficial use (e.g., drinking water, irrigation, salmonid spawning) has associated water quality attributes.
- Establishing designated uses is a critical first step in formulating water quality management policy, steps toward implementation, and means for resolution of conflicts.

For Further Reading

Leopold LB, Langbein WB. *A primer on water.* Washington, D.C.: US Geol. Survey, 1963.

Shiklomanov IA. World fresh water resources. In: Gleick PH, ed. *Water in crisis: a guide to the world's fresh water resources.* Oxford University Press, 1993;13–24.

Solley WB, Chase EB, Mann WB. *Estimated use of water in the United States in 1980.* Washington, D.C.: US Geol. Survey, Circular 1001, 1983.

Wolman MG. Changing national water quality policies. *Journal of the Water Pollution Control Federation* 1988;60:1774–1781.

5

Developing Standards from the Tradition of Toxicology

Overview

Water quality decision makers must balance society's needs for environmental quality with the needs for economic productivity. Policies and regulations are the institutional mechanisms for enforcing that balance, but criteria ensure that scientific principles are infused into the management process. Criteria represent the scientific community's best judgment about the impacts of specific conditions on specific aspects of a water resource (usually biological effects). In other words, they are use- and stress-specific tolerance limits that can be established for any compound in (e.g., a chemical), or characteristic of (e.g., flow rate or temperature), a body of water. Chemical criteria are based on rigorous, repeated toxicological experiments that demonstrate clear and predictable responses of species to precise changes in the water body. With the help of concentration/response curves, management agencies are better positioned to judge harmful levels of given pollutants in different water bodies, and to factor these limits into management decisions. The values scientists obtain from toxicology experiments, however, are not translated directly into policy. Instead, these values must go through several filters before they are transformed into enforceable standards.

While chemical criteria are objective and quantified, their use is often not representative of biological responses in actual streams and lakes. Consequently, some countries are moving toward balancing toxicologically derived criteria with criteria developed from biological assessments. Other countries do not have such a mechanism for incorporating toxicology values into unique standards. Consequently, many countries rely on "borrowed standards"—that is, guidelines established by such agencies as the U.S. Environmental Protection

Agency (EPA), World Health Organization (WHO), or the European Economic Commission (EEC). Still other countries do not set standards, or set them in such a way that they are unenforceable.

Introduction: Science and Decision Making

Natural resource management decisions can be made with or without a scientific basis, but throughout history, the most sustainable decisions have had underpinnings of scientific veracity. This pattern holds true whether the scientific appraisal is incorporated into policies based on quality standards and assimilative capacity or "no degradation" policies. Clearly, early societies understood general relationships between water quality and disease and enacted legislation based on that understanding. Steadily deteriorating water quality serves as a testament to the fact that legislation without a scientific base is usually inadequate. Not until the 1890s, when the first concrete scientific evidence emerged that sewage-polluted water was the carrier for infectious diseases, did water quality regulation begin to have a scientific point of reference. It took another 60 years, and accrual of a significant body of scientific evidence, before public awareness led to demands for water quality improvements.

Scientific evidence is not the unequivocal answer to sustainable natural resource policy. Uncertainty forms an integral part of science and provides the means through which scientific progress is made. Advances in science and the accumulated "facts" we now hold arose through debate and different interpretations of scientific evidence. In part this trend is due to shifting values and shifting scientific paradigms, but it is equally attributable to the uncertainty inherent in many scientific studies. The choice of study subjects, the temporal scale of a study, the assumptions incorporated into models, difficulties with extrapolating across species (e.g., from mouse cancer studies to human cancer risks), and large variations in even the best data sets can all lead to contradictory conclusions on environmental impacts of any given stress (Lemons 1995). After accumulated studies, however, some scientific "truths" have emerged beyond the realm of likely doubts. For example, smoking is known to increase the risk of lung cancer, just as ingesting pathogens in the water is known to lead to an array of gastrointestinal diseases. On the other end of the scientific spectrum, no consensus has been reached on issues such as the possible human health effects of electromagnetic fields (Strauss and Bernard 1991). In between these extremes lies a wide continuum of scientific "facts," scientific "best guesses," and scientific "gut feelings"—all of which influence policy.

Although debate rages about the actual role of science in environmental policy making, science does have an important and central role to play in most developed countries. Through experimentation on toxicity of different chemicals, identification of sensitivity of different organisms to certain pollutants, and quantification of "sublethal" effects of specific chemicals, science influences policy by providing the basis for setting of standards and other regulations. In a now-classic example, scientific studies of bird shell thickness and rates of successful hatches were influential in demonstrating that DDT, a commonly used pesticide in the 1960s, had a serious negative impact on nontarget species (specifically, raptors such as the eagle and osprey and fishing birds such as the pelican). The irrefutable evidence led to the ban on DDT in the United States. While few cases are as clear-cut as this one, the DDT example does demonstrate that scientific studies on toxicity influence policy-making circles.

Biomonitoring and Toxicity Testing: The Means to Scientific Knowledge

Assessing potentially harmful compounds or conditions is an enormous undertaking. Current estimates suggest that more than 100,000 synthetic compounds are produced and used worldwide (Nash 1993). Only 1% or 2% of these compounds have received any toxicity testing: 26 chemical groups are known to be human carcinogens, approximately 600 are known rodent carcinogens, and more than 2000 are known mutagens (Lave and Ennever 1990). Synthetic compounds are only a portion of the suite of potentially toxic substances, however: naturally occurring metals, when concentrated, can have lethal or sublethal effects, as can oxygen and nutrient levels, sediment concentrations, temperature, and water flow rates (see chapter 17). For example, the problem of concentrated metals can be acute. Mason (1991) reports that the human-induced rate of environmental contamination for several important heavy metals is many times greater than the natural geologic rate: 2.3 times greater for mercury, but between 12 and 13 times greater for copper and lead, and 110 times greater for tin. Because scientific studies and resultant water quality standards can address only a small proportion of the potentially harmful substances, choosing those substances to be targeted becomes an important task.

Several techniques have been developed for assessing the impact of different compounds and conditions on water quality and aquatic communities. Generally, these approaches can be classified into two categories: observational bioassessments that assist with the detection of general conditions, and experimental bioassays that quantify specific effects from specific conditions (Ghetti and Ravera 1994). Although many different forms of bioassessments exist, each enables scientists and managers to detect changes in the structure and function of an aquatic ecosystem and thus to monitor "ecosystem health." Bioassessments do not provide information on which potential stress factors are causing water quality deterioration or improvements, but do provide critical clues as to the location of problems or unusual conditions that warrant further detailed exploration. The detailed exploration is carried out through experimentation on "suspect" compounds or conditions to determine their concentrations, toxicity, and mode of action. Biomonitoring and bioassessments will be covered more extensively in chapter 6. The remainder of this chapter will concentrate on bioassays and toxicity testing, and the process by which these are translated into enforceable standards.

Specification of Parameters and Test Conditions

Toxicity testing and the development of criteria usually begin with observation of a problem or potential problem (e.g., loss of key species, behavioral changes, high pollutant concentrations, health effects). Controlled experiments follow to determine the scope and nature of the problem. The managing agency can then balance criteria for different uses. Consequently, toxicity testing comes into play at several times in the policy-making process; compiled information from the existing pool of scientific research may provide the initial basis for establishing standards. Later, toxicity testing becomes part of the feedback loop between monitoring and policy revisions; when monitoring indicates unusual or "red flag" conditions, toxicity tests can define the problem element and assist policy makers in revising standards to fit new information (Fig. 5.1).

   An example of the role of toxicity testing comes from the Kesterson wildlife refuge in California's San Joaquin Valley. In the early 1980s, scientists and managers at the Kesterson Reservoir conducting routine monitoring activities began to notice an unusually high number of deformed waterfowl chicks and high mortality of migratory waterfowl. The observation sparked a series of studies that eventually identified high concentrations of selenium in the reservoir, determined that this compound in the concentrations present was toxic to waterfowl, and traced its source to agricultural drainwater that leached selenium and other elements from soils in the normally arid area (Branch *et al.* 1993). Subsequent toxicity studies were also conducted on fish and humans. Interestingly, the first toxicity tests, in 1982, were generally discounted because procedures for detecting selenium at the levels reported were not yet accepted practice. Consequently, not until 1985, when more sensitive assay techniques were devel-

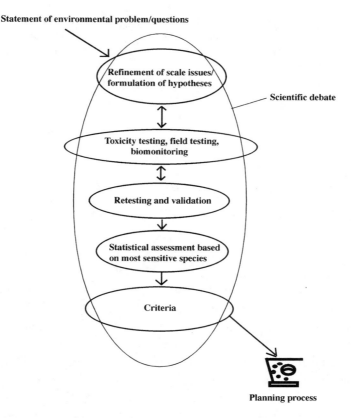

Fig. 5.1 Scientific input is a critical and methodical part of the policy process. It is the one element of policy determination that is intended to be objective and quantified. Even the scientific process, however, is subject to the interpretations of different scientists, the pull of different primary belief systems, and limitations of technology. Consequently, even this most objective arm is open to debate and must be modified as new information is received. The best possible objective criteria are fed into the policy loop (as outlined in Fig. 1.1b) where they are weighed against economic trade-offs, social values, and existing regulatory considerations.

oped, was the presence of selenium in the water, fish, and bird tissues fully accepted and action taken to modify the flow of agricultural drainage waters that concentrated the selenium.

Phrasing Criteria to Meet Different Goals

In the Kesterson example, studies showed that the effects of selenium on water fowl were detectable at a different level than the concentration that caused effects on fish and shellfish. Consequently, no single criterion, or "level of

Table 5.1 Substance-based criteria lists enable policy makers to compare all of the known substances that affect any given water use, such as potable water. This table also highlights that standards, even for such a basic use as public drinking water, can vary widely. (Data from Gleick 1993)

Material (mg/L)	World Health Organization	European Community	United States	Canada
Arsenic	0.05	0.05	0.05	0.025
Fluoride	1.5	1.5	4.0	1.5
Nitrate	10	50	10	10
Chloride	250	25	250	250
Sulfate	400	250	250	500
Trihalomethanes		0.001	0.1	0.35

quantified effect," exists for selenium. Rather, there is a suite of effect-specific criteria for all compounds, with each criterion being relative to the nature of the receiving water and the desired water use or affected organism. Differences among these use-specific criteria are critical to the eventual choice of management standards for different policy preferences. If, for example, a given criterion is chosen as a management standard because an estuary will be managed for a duck habitat, that standard will preclude other uses that are more sensitive to selenium. If the risks to agricultural production from restoration efforts are calculated to outweigh the benefits to wildlife and health, then the scientific results may be considered less important in the final standard-setting process.

The same data may be recombined and rephrased in many ways. Substance-based criteria, for example, identify the differential sensitivity of different uses/organisms to effects by a given substance (Table 5.1). It is more common, however, to express water quality criteria from the perspective of given uses such as drinking water or aquaculture. This approach entails constructing tables comparing all known quality criteria for a given use (Table 5.2). Specific criteria for each use will be the same on both substance-based and use-based tables, but use-based tables are often more appropriate for monitoring and regulation. The tables are most powerful when used together. For example, lawmakers may wish to establish a series of standards to protect trout fishing. A use-based table would enable lawmakers to take into account the scope of compounds known to affect this fishery. Such a table would not yield sufficient information about how these compounds affect other uses, so cross-referencing both tables is necessary to ensure application of the most appropriate standards.

Table 5.2 Quality criteria considered safe for various forms of aquatic life may be expressed as the maximum safe level for a species or the maximum level of a substance deemed to protect the aquatic ecosystem. (Modified from Leeden *et al.* 1990)

Maximum Safe Level of Copper Sulfate for Various Fish Species (mg/L cu)		Threshold Concentration (i.e., Levels Considered Safe) for Aquatic Life	
Trout	0.14	Total dissolved solids (mg/L)	≤2000
Carp	0.30	Conductivity (μmhos/cm)	≤3000
Suckers	0.30	Temperature, warm water fish (°C)	≤34
Catfish	0.40	Temperature, salmonids (°C)	≤23
Pickerel	0.40	Dissolved oxygen (mg/L)	≥5
Goldfish	0.50	Free ammonia (mg/L)	≤0.5
Perch	0.75	Arsenic (mg/L)	≤1.0
Sunfish	1.20	Cadmium (mg/L)	≤0.01
Black bass	2.10	Free chlorine (mg/L)	≤0.02
		Mercury (mg/L)	≥0.01
		Silver (mg/L)	≥0.01
		Zinc (mg/L)	≥0.10

Development of Toxicity Relationships

Developing toxicity or impact criteria is a well-established, but sometimes torturous, process. It rests on a few concepts. *Toxicity,* for example, implies that a compound or element has a negative influence on an organism—that is, an influence producing an effect outside that organism's normal range. The most common effect used to evaluate toxicity is *mortality.* Mortality is a useful endpoint in studies because measurement is unequivocal and easy, and has clear significance to society. Many other *sublethal* effects exist that, although more difficult to measure, have important ramifications for water use. Human health criteria for lethal levels of lead, for example, are not useful. Not only are those levels rarely encountered, but lead affects health—in particular, brain and nerve development in children—long before it approaches lethal levels. Thus, our human health criterion for lead is one below which no detectable health effects exist.

Acute Versus Chronic Toxicity

Toxic compounds or stresses may cause mortality through high-pulse, short-term events or they may effect nonlethal changes in organism function through long-term, continuous exposure over the life cycle of an organism. Acute toxicity caused by high-pulse, short-term events is one of the easier values to test because experiments are themselves short-term, the concentrations of the compound are very high, and the end-point (mortality) is known. Testing for chronic toxicity is similar in approach but the experiments must last longer (ideally several generations), must test for a variety of effects, and use far lower concentrations of toxicants. Typically, chronic toxicity values are at least an order of magnitude smaller than those for acute toxicity, and are often far smaller. For example, Laws (1993) compares the lethal concentrations of four elements for goldfish to the sublethal concentrations at which learning is impaired for those same four elements. The lethal concentration is 48 times greater for selenium, 273 times greater for mercury, 320 times greater for arsenic, and 1570 times greater for lead.

Both chronic and acute values are equally important for understanding aquatic health and water quality effects, and both are factored into standards. In the United States, for example, standards typically will be phrased with two values. The first pertains to levels that would cause acute toxicity in a small number of species: it is typically listed as a value (concentration) that should not be exceeded for more than 1 hour every 3 years. The second value corresponds to the lower chronic effects level, and is typically phrased as a value that is not to be exceeded for more than 4 continuous days every 3 years.

The Range of Sublethal (Chronic) Effects

As indicated in the previous paragraphs, many compounds affect aquatic life long before they cause mortality. Sublethal effects of different compounds are numerous and often species-specific. Mason (1991) summarizes studies of sublethal concentrations of bleached kraft mill (paper processing) effluent on three species. In pinfish (*Lagodon rhomboides*) growth was not affected, but the percentage of body fat was altered; in sockeye salmon (*Onchorhynchus nerka*) growth was reduced; in coho salmon (*O. kisutch*) growth was increased. Sublethal concentrations of mill effluents in this situation may not kill populations of fish, but rather exert selective pressures on individuals and species, causing an altered composition of aquatic communities.

More celebrated cases involve sublethal concentrations of pollutants that cause behavioral modifications and physiological changes. For example, sublethal concentrations of copper cause dominant sunfish to increase their dominant behavior; subacute concentrations of DDT caused lack of coordination, alteration in feeding, and other behaviors in *Therapon jarbua*; detergents damaged catfish taste buds, impairing their ability to find food; swimming and orientation may be reduced by a variety of oxygen-demanding effluents (Mason

1991). Impaired reproduction is another significant sublethal effect noted for many substances.

Toxicity Measures

The effect of a toxicant (i.e., any material that produces toxicity), depends upon its concentration and the length of time it interacts with an organism (as well as on biotic and abiotic interactions, which are discussed in more detail below). Examples of toxicants include many compounds in industrial wastes, chloramines in municipal wastes, and oil and gas in urban run-off. After toxicants enter the aquatic environment, they interact with the water, its constituents, and with one another, and affect biological processes. In the environment, the toxicant acts upon biological receptors and may be transformed by that interaction. Chemical and physical characteristics of the environment also change the toxic material. Therefore, issues such as distance from point of release, mode of exposure (e.g., through the skin, food intake, or gill exposure), and length of time the toxicant has been in the environment all influence toxicity.

Toxicity is determined experimentally based on dose–response or exposure curves (Fig. 5.2) (see highlight). Animals or plants are exposed to known con-

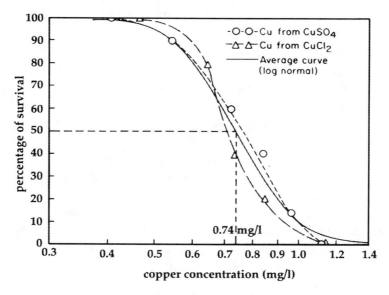

Fig. 5.2 Dose–response curve demonstrating the acute toxicity of copper to the common bluefill (*Lepomis macrochirus* Rafineque) after 96 hours of exposure. The dose–response curve is typically sigmoid-shaped, and allows simple calculation of the concentration corresponding to 50% survival (0.74 mg/L). (Simplified from Trama FB 1954; reprinted with permission from Notulae Naturae of the Academy of Natural Sciences of Philadelphia.)

centrations of the compound for specific amounts of time, and their responses are measured. As indicated, the initial response in a screening test is usually mortality. Phrasing the results of those toxicity tests uses certain conventions, such as the level at which 50% of the population died, or median tolerance level, rather than the concentration at which all died. Using many individuals and successively lower concentrations of a toxicant, researchers can determine the median tolerance limit within a specified timeframe. (If time is allowed to vary, a median value—in this case termed *incipient lethal level*—may be calculated by log transformations of the data; see Laws 1993 for details.)

In early toxicity tests with mammals, the term LD_{50} (lethal dose) was used to indicate that 50% of the test population died within a specified timeframe when exposed to a particular dose. Subject body weight influences determination of a lethal dose, however. For aquatic organisms, the actual dose in the body is rarely determined; instead, organisms are exposed to a concentration of the compound in water. Therefore, aquatic toxicities are usually expressed as LC_{50}, or the lethal concentration at which 50% of the test population died.

Mortality of individual species, although dramatic, may be less sensitive to ecosystem impacts than a change in community dynamics. As described above, sublethal effects are often the most important in water quality as they lead to subtle, but critical, changes in community structure, organism dominance, and resistance of both organisms and the whole ecosystem to further stress. These concerns led to development of the term EC_{50}, or the concentration at which 50% of the population exhibits a specific effect. EC_{50} measures may be developed for a wide range of a population's life history characteristics, such as reproductive failure, impairment of learning, impairment of taste, swimming orientation, or dominance behavior.

An EC_{50} is also useful in that it enables comparisons between similar experiments. Because researchers frequently address different questions about test populations, managers must be cautious about comparing these results. For example, one investigator may focus on ways that heavy metals affect reproduction; another investigator may be interested in behavioral changes that resulted from exposure to those same heavy metals. While these investigations may have important management implications, they measure different effects and the respective EC_{50} measurements cannot be compared.

Biotic and Chemical Interactions; Toxicity Complications

Toxicity studies aim to identify relatively simple and direct cause–effect relationships between a toxicant and some end-point such as mortality or a behavioral or physiological change. That simple relationship is often confounded by a diversity of interacting factors. For convenience, these factors may be divided into three classes: environmental factors such as temperature and water hardness; biotic factors such as the life stage of an organism or its acclimation to given conditions; and chemical interactions among two or more compounds.

ENVIRONMENTAL FACTORS. One of the most difficult aspects of toxicity determination is testing for behavioral differences of the toxicant under varied environmental conditions. Environmental conditions can affect not only the behavior of the toxicant, but also its diffusion from, and concentration in, an area. Three of the most common environmental influences are temperature, water hardness, and pH.

The moderating effects of temperature are often biological in origin, with increasing temperature usually increasing the sensitivity of an organism to toxic effects. Most organisms have a range of temperature tolerances for different activities. As temperature rises, so does the physiological activity of an organism. As physiological activity increases, so does respiration and oxygen demand. As a consequence, the rate of exposure to any toxicant present in the water will increase: the toxicant is pumped across the gills at a higher rate, or taken up into plant tissues at a greater rate (Laws 1993). In addition, the physiological resistance of an organism may be lowered by temperature stress, making it less able to combat a second stress.

Acidity is another environmental factor that frequently influences the toxicity of a compound. Acidic water has more hydrogen ions available, and the chemical reactions among these available ions and other compounds frequently increase the toxicity of different compounds. In some cases, toxicity is increased as pH increases (i.e., as water becomes more basic). Ammonia (NH_3), for example, is a highly toxic compound, unlike ammonium ions (NH_4^+). Increases in pH cause ammonium ions to lose their extra hydrogen, thereby converting them to the toxic form of ammonia. Although water at pH 7 might be harmless, the same water at pH 9 would be highly toxic to most aquatic organisms (Laws 1993). In other cases, decreased pH may mobilize compounds that were otherwise harmless. Aluminum toxicity, for example, is a common problem in acidified lakes (see chapter 22). The effect of pH on heavy metal toxicity has been traced to differences in metal speciation and binding strength (Mason 1991).

Water hardness also affects toxicity, particularly toxicity of heavy metals. Water hardness results from dissolved inorganic salts in the water, especially salts such as calcium carbonate. Because it offers extra surfaces to which the metals can bind, hard water frequently reduces the toxicity of heavy metals (Fig. 5.3).

BIOTIC FACTORS. The most obvious biotic factor influencing toxicity is the variety of genus- and species-specific tolerances. Some species are extremely

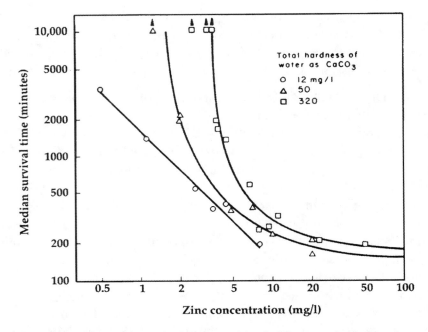

Fig. 5.3 Effect of three levels of water hardness on toxicity of rainbow trout. In this example, increasing water hardness decreased the toxicity of the water. At a zinc concentration of 4 mg/L, median survival time was approximately 400 minutes for the lowest water hardness, approximately 600 minutes for an intermediate hardness, and over 10,000 minutes in extremely hard water. (After Lloyd 1960)

sensitive to certain toxicants or other water quality conditions, while others can tolerate extreme conditions of the same compound. The range of tolerances can be small for universal toxicants, but is more commonly quite broad. For example, the range among tested species for tolerance to the compound dieldrin is a factor of approximately 70 (Laws 1993) (Fig. 5.4). These ranges are critical in determining final criteria, as the most sensitive 5% of species usually determines the safe criteria. The ranges are also a key factor in bioassessments (discussed in chapter 6), as they allow researchers to judge water quality by the presence or absence of known pollution-tolerant or -intolerant species.

Even within one impacted species, different life stages may complicate the task of analyzing toxic impacts. For example, most organisms are generally more susceptible to toxicants during early life stages. Consequently, a criterion that would be sublethal for most of an organism's life may be highly toxic to eggs or hatchlings. Toxicity rules for the U.S. EPA require that chronic toxicity testing be conducted throughout the life cycle of an organism, or, if the life cycle is more than a year, at least throughout the several months of acute sensi-

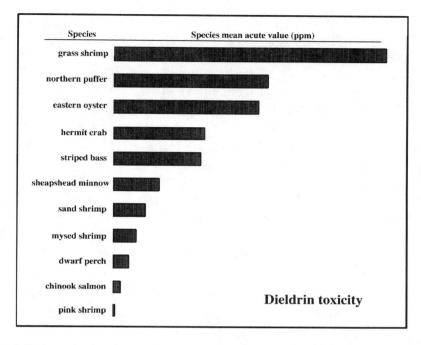

Fig. 5.4 Each species varies in its tolerance to a given toxin. Here, different species reflect a wide range of tolerances to dieldrin. Maximum allowable values (standards) are set using the most sensitive 5% of the species. (Data from EPA 1980)

tivity (Laws 1993). While these rules establish practical testing conditions, they may still be too short-term to detect chronic stress upon generations of aquatic organisms.

Acclimation is another important aspect of toxicity. For many toxicants, exposure to low levels causes the organism to develop a tolerance to that level of the compound or condition. According to Laws (1993), acclimation in fish has been demonstrated with ammonia, cyanide, pH, phenol, synthetic detergents, zinc, and temperature. While the U.S. EPA does not consider acclimatized animals in its determination of criteria, acclimation frequently occurs in testing situations because of the nature of the tests, and is difficult to factor out when determining acute and chronic toxicities.

CHEMICAL FACTORS. The last complication in toxicity studies is that different toxicants react with one another. The reactions may, or may not, alter the toxicity of each compound separately. In general, compounds found in the same body of water may interact to create any of three different conditions. Antagonistic effects occur when two (or more) compounds counteract one another's individual toxicity so that the overall toxic impact of the compounds together is lower than that of any individual compound. No interaction occurs when two compounds continue to act as independent agents: toxicity may be greater because more toxic compounds are present, but the compounds themselves do not alter one another's toxic properties. Additive or multiplicative effects occur when the compounds interact chemically to make their total effects greater than the sum of their individual effects. Additive (multiplicative) effects may be slightly, strictly, or supra additive, creating a variety of possible effects ranging from minor elevation to dramatic increases in toxicity (Fig. 5.5).

Accounting for these synergistic or antagonistic effects is extremely complex because the effects are often species-specific, relative to the life stage of the affected species, and variable according to the ratio and concentrations of the toxicants. Although synergisms can radically alter the toxicity of a compound, the U.S. EPA currently ignores these possible effects in establishing criteria, largely because of inadequate data and the inability to extrapolate from one study or

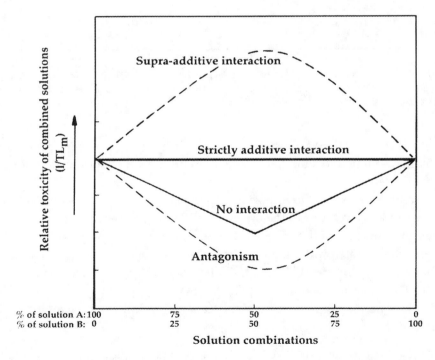

Fig. 5.5 Few toxicity tests can adequately determine synergistic effects between compounds. This figure illustrates that interactions among compounds may create significantly different effects than the influence of any single toxicant. Although some interactions may produce no net effect on toxicity, others may decrease toxicity and still others may increase toxicity. Although this figure portrays four possible reactions, these lines are meant to represent possible points along a gradient of interactions. (Modified from Warren 1971)

species to another. Some examples of chemical interactions can be summarized from Mason (1991): "Calcium is antagonistic to lead, zinc and aluminum. Copper is more than additive with chlorine, zinc, cadmium and mercury, whereas it decreases the toxicity of cyanide." Laws (1993) notes that some pesticides are supra additive, while the toxicity of cadmium is decreased by other metal ions, such as copper and zinc.

Evolving Tests for Evolving Needs

As described above, the actual response of a particular species to a particular toxicant is often highly predictable and relatively easy to measure in simplified laboratory conditions. Under field conditions, however, interactions among compounds, the environment, and the biological systems of the affected organisms can invalidate the simple assumptions of laboratory experiments. This section explores a few of the evolving techniques used to create more realistic tests. Developing appropriate criteria from these approaches has proved difficult exactly because field conditions are more realistic: they are less controllable and hence it is more difficult to isolate cause-and-effect relationships. In many ways, these field tests approach the same questions as laboratory tests, but from a different angle. Laboratory tests ask and answer the linear questions of cause and effect and generate measurements that should protect ecosystem health. Field tests use a more circuitous route, using correlations and regression analyses to address links between environmental conditions and ecosystem "health" (see chapter 15). These measures can suggest avenues of concern and possible needs for specific enforceable criteria.

Static and Flow-through Bioassays

A bioassay is a test in which one or more organisms are confined and subjected to test conditions. These types of tests are routinely performed in the laboratory but may also be completed in the field. In a static bioassay, the water remains with the organisms throughout the test; in contrast, a flow-through bioassay uses moving water.

Some better-known static bioassays include work done by the U.S. EPA in the 1970s using a green alga *Selanastrum* sp. The EPA developed a standard test protocol that involved a specific test culture of the alga, known light and nutrient conditions, and standard techniques for measuring responses. The *Selanastrum,* or algal, bioassay was used to establish effluent limits and environmental criteria for many different compounds.

Flow-through bioassays are used for organisms that require moving water such as some fish species. Field bioassays (e.g., fish in a cage in a river above and below an effluent) have been used as tools for environmental monitoring but rarely as tools for developing criteria. More recently, these field bioassays have been used to develop criteria for mixed or complex effluents.

Microcosms, Mesocosms, and Whole Ecosystems

In the search for increased realism combined with reasonably controlled situations, many studies have used small, semi-isolated field chambers. A typical *microcosm* might be a 4–5 liter floating container housing one or a few smaller organisms. Because the experimental water is the same as the lake or stream, chemical conditions approximate actual conditions. In addition, depending on the design of the containers, ambient light and temperature may reflect actual conditions. Test conditions such as pH can be altered, and the system is small enough to allow for many replications.

A *mesocosm* is a mid-size vessel (e.g., 1–5 m in diameter) much like a plastic bag suspended in the water. Sometimes it is open to contact with the bottom; at other times, it is sealed from the bottom. In either case, mesocosms are several orders of magnitude larger than microcosms. They allow for within-vessel chemical, physical, and biological interactions but do not allow larger-scale interactions such as those occurring in a natural body of water. A distinct advantage is that they are large enough to provide some natural reactions but remain small enough that they can be replicated within a single lake.

The search for realism is never complete until a *whole ecosystem* is studied. A few analyses have manipulated entire ecosystems. For example, watersheds throughout the world have been experimentally logged or treated in other ways, and streams draining the watersheds studied to measure ecosystem response. Grasslands and forests have been subjected to radiation and other pollutants. Some of the best-known whole-ecosystem manipulations in the aquatic field have been conducted at the Experimental Lakes Area in Canada. At that site researchers manipulated pH, watershed forest cover, and nutrients over a 20-year period. The information provided through these studies is invaluable. It is not, however, universally applicable. Because environmental conditions vary significantly from lake to lake, stream to stream, and region to region, specific results of whole-ecosystem manipulation cannot be extrapolated from one system to another. In addition, the experiments are extremely expensive, require a long-term commitment, and are not replicable.

Each of the experimental systems has advantages and disadvantages. Laboratory bioassays are relatively inexpensive and may be done locally. Although they can produce quick assessments of effects of a given substance, they cannot, however, indicate how substances act together or how qualities of the receiving water change the relationships. At the other end of the spectrum, whole-ecosystem manipulations produce extensive data sets but are usually too time- and energy-intensive to provide adequate guidance for managers. Also, because mesocosms and whole-ecosystem manipulations look at larger pictures, it is more difficult to pinpoint cause-and-effect relationships. Their more realistic attributes also make them more difficult to translate into effective standards.

Nonetheless, as described in more detail in chapter 15, these tests have provided significant insight into the ways ecosystems function. While they may not provide enforceable criteria, they can provide important guidance for ecosystem assessment and monitoring and, from there, important feedback about the need for new or revised criteria. Each testing system is inadequate by itself, but when used in combination, the picture provided by these systems becomes increasingly clear and defensible.

From Toxicity to Standards

After toxicity relationships are developed, they can be incorporated into water quality laws and standards. Standards are the legal expression of criteria—that is, the enforceable numbers for allowable concentrations of pollutants. They are a unique arm of the law, however, because they must remain flexible enough to allow for incorporation of new information. Scientific understanding about effects of various compounds, their distribution, and effects in the environment are continually expanding. Also, new industrial and manufacturing processes and new technologies frequently lead to development of new compounds that are, or might be, released into the environment. Criteria must, therefore, be continually modified to reflect new studies and new conditions.

Criteria: Accumulated Scientific Knowledge

Criteria encompass the accumulated scientific knowledge about a given pollutant—the summation of existing studies relating life history and survival characteristics to different concentrations of a pollutant. Inherently specific, they are the keys to enforceable legislation and the objective bases for pollution control regulations. Notably, they are not equivalent to the toxicity studies discussed above. Standards emerge as the results of toxicity studies are put into a policy filtering process that includes risk assessment, cost–benefit analyses, and political trading. In translating toxicity tests into criteria and standards, it is necessary to make broad statements from often limited and specific information. Thus, a mechanism must be constructed to extrapolate the effects from one study species to the next, and from one pollutant to the next as well as to account for different procedures, units, and different tolerances of test species. In addition, standards and final criteria must reflect the economic and logistic reality of achieving certain water quality values.

For example, research has shown that lead levels higher than 10 μg/L lead to an increased probability of adverse impacts on human health: the logical criterion for lead would be at or below 10 μg/L. Achieving a 10 μg/L standard for drinking water may be technically and economically infeasible over a large area, however. Instead, a typical scenario is that policy makers face a choice: either one-third of the people in an area can receive water with lead levels below 10 μg/L, or the entire area can receive water with lead levels no higher than 25 μg/L. In this case, the final standard for lead in drinking water may be above the recommended criterion, but at the level that will afford the best protection to the largest number of people. It assumes that achieving health and societal benefits of increased water supply to the whole area outweighs costs of impaired water quality (i.e., the difference in health due to the additional 15 μg/L lead). In practice, standards for drinking water are usually set well below the final criterion value, accounting for a safety factor (see below) and other possible sources of exposure to the compound.

The Language of Risk

Establishing "acceptable risk" is one of the touchiest issues in standard setting and raises the question of "how safe is safe enough" (Lave and Ennever 1990). Risk assessment is most clearly developed in human health policy. For example, the U.S. Congress draws a distinction between carcinogens in food additives (none is permitted) and carcinogens that are natural substances in foods (the amount must be below an ambiguously defined "unreasonable risk"). In foods, therefore, different substances are regulated on a different basis.

In addition, the term "acceptable risk" begs the question, "acceptable to whom?" With human carcinogens, for example, what constitutes a regulatory risk: increased cancer incidence of 1 in 1,000,000, or 100 in 1,000,000? The issue is of significant importance. It not only defines how many cases require regulatory attention, but also affects the economic use and production of regulated chemicals. Differences of perspective also control the approach taken to how much risk is acceptable. Lave and Ennever (1990) trace, for example, the evolution of the U.S. EPA standards for benzene, a solvent used in many applications but closely linked to leukemia. When the Occupational Safety and Health Administration (OSHA) proposed lowering the benzene standard from 10 ppm to 1 ppm, the agency was sued simultaneously by a petroleum workers union, which felt that the 1 ppm standard was insufficient to protect workers' health, and the American Petroleum Institute, which claimed that OSHA did not have compelling evidence to enforce even a 10 ppm standard. Each group weighted economic and health risks differently.

In water quality, public health concerns are an important dimension of standard setting. The concentration of mercury, PCBs, and other compounds in fish, for example, or lead in drinking water have human health implications. The more difficult challenge, perhaps, is defining acceptable risk for aquatic health and determining whether aquatic health means the risk to a fish species, or risks to the functioning of the aquatic community (see chapter 15). Risk assessment in water quality can be even more complicated because it evaluates not just the nature of the hazard and the toxicity of a given compound, but also the availability of the compound as it is transported down river, adsorbed into sediments, or modified by geologic conditions. These issues are factored into the standard-setting process to produce an estimate of the probability of risk and the magnitude of the resulting impacts (Kaplan and McTernan 1993).

Because risk is a matter of perception, it is difficult to anticipate how communities will respond to perceived or potential threats. In the selenium example at Kesterson wildlife refuge, for example, researchers and policy makers were surprised by the lack of concern evinced by local communities for potential human health threats posed by the selenium. Health officials believed a significant threat of bioaccumulation arose through consuming fish and waterfowl from the reservoir. Many were also concerned about potential dangers from consumption of crops irrigated with the selenium-rich waters (Branch *et al.* 1993). Rather than focus their concern on the potential health threats, however, local residents and some local water districts became more concerned about the risk posed to their livelihood by the efforts taken to reduce selenium poisoning. Reducing selenium in the ecosystem could only be achieved by eliminating drainage of agricultural water into the reservoir. In turn, that action would require terminating irrigation to area farms. One study reported that nearly 50,000 acres of land in the San Joaquin Valley could be taken out of production because of the change in drainage (Branch *et al.* 1993). The forecasts of economic consequences were apparently more important to area residents than potential threats to their health: the risks of job loss were not acceptable, while the risks to health were more acceptable, given the trade-offs.

Data Aggregation and Safety Factors

Before criteria can be established, the available data must be aggregated and assembled into a meaningful form. In this phase, species-specific differences and effects of pH or water hardness are synthesized into a meaningful "bottom line" toxic value. For the U.S. EPA, tested species are typically grouped by genus, and the values for different genera are averaged. These values are then ranked to estimate the concentration of the toxicant that would be acutely toxic to no more than 5% of the species. In essence, the most sensitive species defines the criterion.

The process is similar for calculating final chronic values. Because fewer chronic effect tests are carried out, the final chronic value is often calculated as a ratio of the acute values to the available chronic values. Criteria are established differently for plant susceptibility and for "residue" values, or the probability of bioaccumulation in the food chain. Calculation of residue values includes many assumptions, including lipid content of the organism and the "bioconcentration factor" (the concentration of a toxicant in one or more tissues divided by the average concentration of the toxicant in the water) (see Laws 1993 for a full discussion). The secondary criterion referred to earlier is the final acute value, divided by a safety factor of two, and the lower of either the final chronic value, the final plant value, or the final residue value.

When environmental standards are established, they are usually set at a more stringent value than strict application of the criteria would suggest. This legal buffer, or *safety factor,* is based on the logic that scientific results are variable and results of exposures in the field may not closely parallel those obtained in a laboratory. They are also based on the fact that some effects (e.g., lethality below 10%) are difficult to measure. Therefore, when a regulatory agency reviews relevant toxicity test data and LC_{50} measures, it adds a safety margin. In developing a standard from a criterion, an agency might include a tenfold safety factor and make the "standard" 1 µg/L based on a criterion of 10 µg/L. Tenfold factors are common in public health criteria for which adequate studies have been conducted; safety factors may be as large as 1000-fold if there are inadequate long-term studies of acute effects (Laws 1993).

Criteria in Action: Water Quality Requirements for Different Uses

Because criteria and standards ultimately reflect values and uses, they are phrased to accommodate different purposes and different uses. Among the most common are public health criteria, fisheries or aquatic health criteria, and agriculture criteria. As the Kesterson wildlife refuge example illustrates, these criteria may periodically come into conflict. Water quality managers and decision makers must balance these conflicts, setting standards that support the highest-priority use without discounting the possibility of a changing set of priorities and values.

Public health is the principal concern in many water quality management situations, and public water supply is the principal public health issue in many of those cases. Qualities of the public water supply and standards for controlling quality drinking water vary considerably around the world. These variations reflect different priorities placed on public health, environmental health, and economic productivity. With some compounds such as arsenic, widespread agreement exists about what constitutes a "safe level" for public consumption. In other cases, such as the organic chemical trihalomethane, a 3500-fold difference exists between the most conservative and the most liberal standards (see Table 5.2 on page 85). The discrepancies between these values are difficult to explain. In some countries, standards are set without scientific input; frequently, the numbers for these standards will be quite low. In the former Soviet Union, for example, environmental standards were so strict that they were unenforceable and well below measurable values. As a result, standards effectively did not exist and none was enforced. Other countries may set high standards, even in the face of contrary scientific evidence, out of concern for short-term economic productivity.

Large differences in standards can also result from the calculation of safety factors, or assumptions pertaining to the quantity of a given substance to which populations are exposed. For example, calculation of public health standards in the United States begins from the concentration of a substance at which there is no observed effect, or the lowest observed effect. Once this value is established, exposure to the substance is calculated according to its concentration in significant food supplies and the rate of absorption from that food, the exposure from drinking water and the amount of water consumed per day, and other sources of possible exposure such as air. Tables that calculate the exposure, uptake from the exposure, and absorption rate from the source can then yield absorption subtotals for the different sources of exposure. In sum, these absorption subtotals should be at or below the level of lowest or no observed effect. At each stage of these calculations, assumptions are made about the exposure level.

Cadmium standards, for example, involve factoring in estimates of the amount of shellfish and fish consumed by the average person, and the bioconcentration factor for those different food sources. Because fish and shellfish are relatively small portions of the average U.S. resident's diet, cadmium ingested from these food sources is calculated to be relatively small in the United States. Consequently, more "leeway" exists for drinking water standards for cadmium and the present standard is, according to Laws (1993), debatably high. In another country or a specific community within the United States, these calculations might appear very different. If people consumed significantly more fish or rice (as, for example, in Japan), the allowable drinking water standards for cadmium would have to be significantly lower to yield the same total absorption value.

Fisheries and agricultural uses have historically constituted the other two most common categories of criteria. Fish-related water quality criteria can be presented in either of two ways. The first relates to specific toxicities of a compound for specific species. The second relates to values for maintaining a "healthy aquatic environment." For example, determining how much copper to allow in an effluent could be based on criteria for the principal fish species that live in the water body to which the discharge occurs. Alternatively, the regulatory authority may examine and adopt a list of criteria expected to protect all aquatic life of a region (Table 5.3). Agricultural criteria, however, involve a different set of variables, those that affect growth and production of crop plants. Consequently, agricultural criteria are more concerned with variables such as sodium concentration, the proportion of calcium and magnesium cations that mitigate the effects of sodium, and the presence of trace elements that limit plant growth. The sodium absorption ratio (SAR; the ratio of sodium to calcium plus magnesium) is an agriculturally specific water quality criterion developed to express sodium content relative to those mitigating factors, and to

Table 5.3 Agricultural uses of water are primarily oriented toward protection of crop production. A "threshold of concern" would suggest that a farmer might want to increase the use of dilution water applied to a crop. In contrast, when levels exceed the criterion, that water should no longer be used as an irrigation source. (Modified from Leeden et al. 1990)

Variable	Threshold of Concern	Criterion
Total dissolved solids (mg/L)	500	1500
Conductivity (μmhos/cm)	750	2250
Sulfate (mg/L)	200	1000
Chloride (mg/L)	100	350
Sodium absorption ratio (SAR)	6.0	15
Arsenic (mg/L)	1.0	2.5
Boron (mg/L)	0.5	2.0
Copper (mg/L)	0.1	1.0

quantify the degree to which sodium in an agricultural water supply threatens agricultural production.

One of today's most demanding issues in water quality management is development of ecosystem health criteria. Those criteria are intended to protect ecosystems when other criteria values are not exceeded. Ecosystem-level criteria are exceedingly difficult to define because ecosystems are extremely complex and vary greatly through both space and time. In sufficient quantity, even the most innocuous anthropogenic material will have deleterious effects on one or more species in an ecosystem. Similarly, below some lower threshold even the most toxic material (e.g., dioxin) has no detectable effects. For the foreseeable future, there will be no useful criteria for ecosystem health per se, because the definitions are elusive and the scientific community continues to debate what constitutes ecosystem health (see chapter 15 for more discussion and references).

The Expression of Standards

Throughout the world, standards are expressed in one of three forms:

- Regulations imposed on discharged material itself (i.e., *end-of-pipe standards*);
- Regulations imposed on discharging industries or facilities (i.e., *technology-based standards*); and
- Regulations based on receiving water quality itself (i.e., *assimilative capacity standards*).

Just as the standards themselves reflect social values and conditions, so does the form that those standards take. Different legal frameworks lend themselves to different enforcement systems. Countries vary widely in their dependence on the industries and social systems most impacted by these different forms.

End-of-Pipe Standards

End-of-pipe standards impose quality limits on material discharged into the environment by specifying characteristics that must be met at the discharge point. Design engineers, legal specialists, and administrators often favor these kinds of standards because they are uniformly applied across different subsets of industry and across wide geographical limits. For the same reasons, they are often favored by professional organizations involved with treatment design and enforcement. This professional and administrative support, coupled with uniformity among dischargers, makes end-of-pipe standards relatively easy to adopt. Consequently, they have been broadly applied throughout the world.

Unfortunately, end-of-pipe standards have serious limitations:

1. They do not account for the fact that significant ecological effects may exist at pollutant concentrations below the level of chemical detection.
2. They do not account for possible synergistic effects among a variety of pollutants in an effluent.
3. They are not sensitive to diverse ecosystems and do not account for the fact that specific toxicity depends on the environment into which a pollutant mix is discharged.
4. They do not take into consideration that ecological conditions are a product of thousands of simultaneous effects in a given location.

Thus, end-of-pipe standards have administrative advantages but do not always result in desired degrees of protection for the aquatic environment.

Technology-based Standards

Technology-based standards provide an alternative format for development of standards. As implied by their name, these phased-in standards are set for different classes of industry according to the nature of their waste treatment technology. Specific treatment technologies are defined for each industry or class of industry and phased to become more controlling over time.

The first phase of such a program is usually imposition of Best Conventional Technology (BCT). BCT is the treatment system that results in the best performance (e.g., highest-quality effluent) among those treatment systems that are widely adopted and used in the industry. Thus, the model for performance is what is already in place.

The second phase, commonly imposed after several (e.g., 5) years of BCT standards, is Best Practical Technology (BPT). In this case, treatment technology has been widely tested and available and does not impose undue economic hardship on industry. BPT treatment requirements are stricter than BCT requirements but emphasize economic security for the industry rather than improvement of environmental quality.

Finally, after some additional time (e.g., 5 more years) a third and final phase is imposed. In Best Available Technology (BAT), the treatment system should remove the greatest quantity of pollutant, independent of cost considerations.

One of the technology-based standards' greatest strengths is that they control emission of pollutants that were unknown when standards were developed. Its proponents also claim that technology-based standards continually force development and adoption of new technologies for new treatments. Opponents argue that they actually encourage suppression of new treatment technologies because new technologies would be expensive and the user community would be unwilling to spend more to cover these costs. Still others claim that these standards lead to inefficient or ineffective use of new technologies rather than effective use of well-known technologies.

Further controversy about such standards centers around effectiveness of the controls themselves. Many claim that technology-based standards over-control industries because they are phrased in terms of "maximum possible control" and make the assumption that discharge of any material is bad for the environment. In contrast, others claim that these standards are a "friend" to industry and that they do not force adequate control. These differences of opinion reflect the fact that all management decisions are a balancing act. Complete agreement on the nature or form of any standard is a rarity.

Technology-based standards, like end-of-pipe standards, offer clear administrative advantages but may not always lead to the desired degree of environmental protection. Because they fail to consider any characteristic of the receiving environment, they may fail to protect sensitive environments. They also do not consider synergistic effects of chemicals.

Many developing countries employ technology-based standards but never go so far as implementing BAT approaches. These countries contend that BAT is most appropriate where the environment is highly valued and/or the economy is strong (i.e., in more developed nations). Where economies are struggling and industrial pollution is not widespread, as in some developing nations, strong feeling exists that BAT standards are unnecessarily burdensome. In these nations, the best balance between cost and protection is argued to be achieved through BPT. This situation may change as the pace of industrialization and intensification of industrial pollution continues in these countries.

Assimilative Capacity Standards

Assimilative capacity standards developed from failures of both end-of-pipe and technology-based standards to protect environmental quality adequately. Based on the specific qualities of each "receiving system" (e.g., stream or lake), implementation of this form of standards requires the management agency to inventory its water resources and classify each resource by intended use and desired quality (see chapter 6). These classifications guide managers in seeing

---

Technical Highlight: Determining Dose–Response Curves

Dose–response curves are determined from test populations. Responses of these populations to successive concentrations of a substance are measured and expressed as a plot of the relationship between concentration (on the x-axis) and response (on the y-axis). These relationships are not linear; they will be sigmoid or S-shaped. The characteristic sigmoid shape results because some lower threshold exists below which no identified response occurs, and some upper threshold beyond which all of the organisms are dead. Below the first threshold and above the second, the curve is flat because a change in a dose does not result in any response in the population. The convention of using a 50% effect rate for LD_{50}, LC_{50}, and EC_{50} arises from the statistics used. An LD_{50} (or LC_{50} or EC_{50}) measurement is used to imply population-wide effects, but it really represents effects on individuals in the test population. To apply experimental results to the whole population, we need to estimate the probability that other individuals of the same species would react in a similar way. This estimate is achieved through expressing "confidence limits" around the dose–response curve. A confidence limit is a statistical measurement of the degree of variability around any point on a graph. Confidence limits around sigmoid curves are most narrow (i.e., estimates are more precise) near the median. Therefore, we can estimate the LD_{50} (probable death of 50% of the individuals) much more precisely than we can estimate the LD_{10}.

---

appropriate, allowable impact levels. Dischargers (point and nonpoint) are then identified. Those that are causing the system to exceed standards are identified and enforcement or management action taken to bring about compliance. Finally, a monitoring program is implemented to ensure that: (1) management agents continue to meet compliance goals; (2) compliance goals adequately support designated uses; and (3) uses designated for a water body continue to reflect society's values.

Summary

- Water quality standards are most defensible when they are based on scientific data (although they are not always developed that way).
- Standards are legal statements proscribing certain conditions for the water resource and allowing penalties if conditions are violated. They are legally binding but do not necessarily have any basis in scientific reality.

- Criteria describe scientific relationships between water quality conditions and given uses. They are based on scientific findings but have no legal standing.
- Criteria are based on toxicity or dose–response relationships in which the response of individuals or communities is measured against standard test conditions. These relationships may be laboratory-based (with high replicability and known variance, but low realism) or field-based (with low replicability, high cost, and high realism).
- The most effective standards are based on frequently reviewed criteria, which in turn incorporate both laboratory and field test conditions.

For Further Reading

Branch KM, Orions CE, Horton RL. Risk perception and agricultural drainage management in the San Joaquin Valley. *The Environmental Professional* 1993;15:256–273.
Lave LB, Ennever FK. Toxic substances control in the 1990s: are we poisoning ourselves with low-level exposures? *Annual Review of Public Health* 1990;11:69–87.
Lemons J., ed. *Scientific uncertainty and its implications for environmental decision making.* Blackwell Science (in press).
Patrick R. What are the requirements for a successful biomonitor? In: Loeb SL, Spacie A, eds. *Biological monitoring of aquatic systems.* Boca Raton: Lewis Publishers, 1994;23–39.
Strauss SH, Bernard SM. Power and the people. *The Environmental Forum* 1991;8:11–15.

6

Classification and Environmental Quality Assessment: The Search for Ecologically Accurate Aquatic Metrics

Overview

Criteria derived from toxicology studies provide quantitative measures of chemical impacts on biological resources. Toxicological studies cannot accurately test for the diverse and interacting influences in a "real" stream, lake, or estuary, however, nor can they account for natural regional variations in stream or lake biota. Consequently, while enforceable standards have until recently been based primarily on chemical criteria, water quality managers have turned instead to a system of bioassessment. Because biological systems are both the resource of greatest concern and the integrator of chemical, physical, and biological interactions, using the biota themselves to indicate water quality impacts is an elegant concept. It is not, however, a concept that lends itself to a simple, universal model. Hundreds of biotic metrics are in use in different countries, and even the most commonly used models have their detractors.

While toxicology begins with the premise that a given compound will have a given effect, bioassessments begin with the premise that sites can be compared to a "nonimpacted" or "minimally impacted" reference site. Using a combination of biological metrics, varying classes of impact can be assessed and roughly measured.

Classification of water resources consists of the pictorial representation of chemical or biological assessments—that is, the summarization of data in a graphic format. Classification may be noted in a variety of contexts: for individual factors such as dissolved oxygen or phosphorus concentration, for overall "pollution" impact (e.g., from several different biotic metrics), or even for flow characteristics that may affect water quality. The variety of detail used in classification schemes, including the reflection of spatial and temporal scales, is defined by the goals of the agency or scientist using the system as well as by the role of the classification system in a monitoring program. For example, classification may be used to design bioassessments and monitoring programs or to illustrate the results of monitoring efforts. Either way, the utility of this approach lies in its graphic depiction of changing conditions along different spatial and temporal scales of concern in water quality management.

Introduction: The Search for Ecological Significance

As outlined in chapter 5, toxicology has been the preferred and traditional means of establishing water quality criteria and their associated standards. In several areas, however, toxicological studies cannot provide accurate portrayals of ecological conditions in a real body of water. For example, they cannot test for synergistic effects among different chemicals, replicate the mixing of water in large bodies, or test the impact of changes in one species on the larger community. Natural conditions of water bodies (e.g., water hardness or pH) also affect toxicology, and accounting for these factors is extremely difficult. As toxicology studies have come under increased scrutiny, and optional means for evaluating ecosystems have come into the forefront, it has become increasingly apparent that toxicology-based criteria alone are a weak basis for ecologically sustainable standards.

Chemical criteria and toxicological studies clearly serve a vital role. Nevertheless, consensus is emerging that the biggest threat to water quality and aquatic communities does not come from any single regulatable toxicant. Rather, it comes from an array of multiple stresses such as flow restriction and alterations of channel structure, erosion, introduction of exotics, the multiplicity of toxicants, and the proliferation of nutrient inputs. Consequently, water quality managers and policy makers need tools that can reflect the response of biological communities to these multiple threats, indicate areas of greatest concern, and point to areas for remediation. They need, in essence, to use the biological communities themselves to judge their own health, as chemical measures alone appear to be inadequate to protect the range of aquatic communities. Karr (1995), for example, notes that the proportion of Ohio's waters assessed as degraded by chemical criteria was 50% lower than the proportion assessed when biological assessments were used instead. Karr (1995) also notes other cases in which chemical criteria alone provided too narrow a base for decision making. For example, in one area, chemical criteria indicated the need for expensive denitrification waste treatment plants that later failed to create improvements in the biological assemblage. Rather, chlorine in the effluent of those plants became a major factor degrading the aquatic community. Lyon and Farrow (1995) echo these concerns and imply that the reliance on chemical criteria has created a focus on point-source effluent, while more significant improvements can be achieved with a focus on nonpoint sources of pollution and the monitoring systems needed to understand them.

Bioassessments Through Time

The use of biological cues to indicate environmental conditions is an old concept. Links between human health and water sources have been noted, and utilized for planning purposes, at least since 100 A.D., when Roman officials monitored human health to assess the quality of water in the aqueducts (Ghetti and Ravera 1994). The ability to measure and monitor the organisms in the water itself first emerged with the invention of the microscope. Consequently, not until the mid-1800s were specific links made between bacteria in water and human health, and not until the turn of the century did impacts of water quality on the aquatic organisms themselves receive scientific scrutiny.

The earliest water quality bioassessments related to the effects of sewage on stream life. Kolkwitz and Marrson (1908) traced the process of decomposition in a stream below sources of sewage, identifying distinct "saprobic" (e.g., pollution) zones. For example, immediately below the sewage, the water was "polysabrobian," or devoid of aquatic life. This zone was followed by two progressively cleaner, more colonized zones termed mesosaprobian and oligosaprobien, respectively. In their extensive studies, Kolkwitz and Marrson eventually catalogued the tolerances of more than 800 plant and animal species within these saprobic zones (Davis 1995). Their studies not only introduced the concept of water quality zones and the ability to classify rivers according to their level of pollution, but also initiated an understanding of indicator organisms for those different zones.

Refining the understanding of saprobic zones and their indicator organisms occupied many scientists in the early 1900s. The system was used extensively in the United States and Europe, forming the basis for studies of the effects of sewage from large cities such as Washington, D.C., and Chicago on downstream aquatic life. Studies such as those conducted by Richardson (1922) and Davis

(1995) along the Illinois River used saprobic systems to document hundreds of miles of polluted riverways south of Chicago. Richardson's studies also improved the understanding of indicator organisms and the complex interactions occurring as decomposition preceded. Consequently, certain taxa were recognized as indicators, reducing the reliance on cataloguing numerous specific species, and between four and seven saprobic zones were recognized rather than the three originally identified by Kolkwitz and Marrson.

The final stage in the saprobic system came with efforts to create numeric indices of water pollution from the observations of indicator organism abundance. For example, Davis (1995) reports one of the first numeric indices, by Wright and Tidd in 1933. Using the pollution-tolerant oligocheates (aquatic earthworms), they classified the degree of pollution according to the density of oligocheates per square meter: 1000 per square meter was negligible pollution, 1000–5000 per square meter indicated mild pollution, and more than 5000 per square meter represented severe pollution. Other studies determined that the relative abundance of several taxa was more important than the density of any given taxon (Davis 1995).

Beyond the Saprobic System: Biotic Indices and Multiple Metrics

In the 1950s, serious criticism was launched toward the saprobic system. While the system clearly yielded important information, refinements in aquatic ecology began to point out areas that were not addressed by the saprobic system. For example, the system reflected only response to sewage; by the 1950s, however, many different kinds of pollution were known to affect water quality and aquatic life. A second criticism reflected the quantitative bias of the age—that the indicator organisms were not studied well enough to serve as reliable quantitative measures. The third criticism, also reflective of changing values, proposed that the species used in the saprobic system were not economically important and, consequently, were ill-chosen. Lastly, and perhaps more fundamentally from an ecological point of view, criticisms were raised because the saprobic systems used a few indicator organisms rather than the composition of aquatic communities as a whole. As understanding of aquatic communities progressed, many observers felt that pollution monitoring and classification must include a broader array of species and taxa filling different functional roles.

Despite its critics, the saprobic system fostered an array of pollution studies and attempts to classify polluted water. It also had led to the development of numeric indices. As studies in pollution and aquatic ecology advanced, several distinct types of indices came into use. Reflecting the community-based science of the time, for example, Patrick (1950) designed a graphic and numeric index that utilized seven taxonomic groups, some reflective of "clean water conditions" (e.g., fish, diatoms and green algae, insects, and crustacea) and others reflective of polluted waters (e.g., blue–green algae, oligocheates, protozoa, and certain snails and worms). As suggested by earlier studies, waters were classified based on the relative abundances, rather than absolute abundances, of individuals in these groups (Davis 1995).

Throughout the 1950s and 1960s, many different important indices were proposed and tested. These measurements included indices based on indicator organisms, relative abundance of important taxa, overall community diversity, biomass, and finally, diversity and abundance of certain sensitive taxa. A few of these indices are described briefly below. For more information, see Davis (1995), Karr (1991), and Ghetti and Ravera (1994).

SAPROBIC INDEX. Despite the criticisms leveled at the saprobic system, it remains a powerful tool in monitoring and is still in use. Modern application of the system, however, utilizes a weighted numeric index to calculate pollution stress. The development of the saprobic index introduced several important changes to the original saprobic system, such as the incorporation of relative abundances of indicator organisms. The saprobic index assigns a numeric weight to each saprobic zone. Thus, organisms found in the oligosaprobic zone are assigned a number "1," polysaprobic organisms are assigned a "4," and organisms in the intermediate zones are assigned numbers accordingly. These indicator values are then multiplied by an abundance value that indicates whether a given indicator species occurred rarely, frequently, or abundantly at the monitoring site. The weighted averages yield a number between 1 and 4, with 1 to 1.5 indicating only slight impurity and 3.5 to 4 indicating very heavy pollution.

One more important refinement of the saprobic index involved the incorporation of an indicator weighting system. Prior to this point, one important criticism of the saprobic system was that the assignment of saprobic value in different zones seemed arbitrary when the same organisms occurred across all, or several, zones. In answer to this criticism, Zelinka and Marvan (1961, Davis 1995) categorized the different indicators according to their abundance in different saprobic zones. Those highly specific to certain zones received the highest indicator values (up to 10). The index was improved significantly by incorporating an indicator weight, a modification that Davis (1995) credits with ensuring the continued use of the saprobic index throughout Europe and other parts of the world.

PERCENT EPT. An alternative to the saprobity index that follows the same conceptual development is "percent EPT" or percent of the collection that consists of organisms of the orders Ephemeroptera, Plecoptera, and Trichoptera. These three orders of insects are widely distributed in cold, high-quality waters and are extremely sensitive to organic pollution. They are readily collected and identified and their biology is well known. Therefore, in a "high-quality" stream, 60% to 80% of the macroinvertebrate fauna will be from these three orders. All indices suffer problems, however. In the case of percent EPT, problems emerged because these three orders of insects are not diverse in lakes or estuaries, and are much more widely distributed in the northern hemisphere than in the southern hemisphere. Consequently, these organisms are highly useful for only a small subset of the world's ecosystems. In addition, they represent responses to oxygen-demanding waters more precisely than other types of impacts and, therefore, may not reflect a broad range of impacts and stressors.

PERCENT TUBIFICIDS OR NUMBER OF OLIGOCHEATES. In the tradition of Wright and Tidd and the "original numeric index," some indices utilize the presence of pollution-tolerant organisms rather than the dearth of pollution-intolerant organisms. Because pollution-intolerant taxa are usually replaced by pollution-tolerant taxa, indices can reflect the proportion of a group of tolerant species that commonly substitute for sensitive species. This system is similar to percent EPT, in that it uses taxonomic groupings to evaluate water quality changes. It differs in that the measured group consists of known opportunists, rather than known sensitive species. Also similar to percent EPT, these indices are often difficult to interpret because the tolerant taxa—oligocheates, tubificids, and chironomids—are highly diverse and nonspecific: they live in a variety of environments, from very clean to very polluted. In addition, because these taxa occupy different habitats, sampling differences can greatly bias the index.

DIVERSITY INDICES. Patrick's (1950) use of seven taxa and Richardson's (1928) observation that relative abundances of taxa were more important than absolute numbers initiated a school of indices that is used extensively today. One of the most widely used measures has been the diversity index, which is a measure of the evenness of the distribution of aquatic organisms. Because impacted systems are less diverse than unimpacted systems (i.e., systems with lower water quality will have fewer species and a larger representation by a smaller number of species), a measure of diversity should yield significant information about the presence of multiple kinds of pollutants. The number of species in a collection is termed its *richness* and the distribution of individuals among the species is termed its *evenness*. In 1950, Shannon and Weiner, in independent studies, expressed diversity indices as a combination of both richness and evenness in one quantitative measure. Wilhm and Dorris (1966) were the first to use such indices in an aquatic study, describing the severe decrease of diversity below a pollution source in an Oklahoma stream and gradual recovery toward high diversity.

Like all indices, diversity indices have been criticized on a number of fronts. For example, important community changes can occur from a shift in species dominance or actual replacements of one species with another. Because diversity indices treat the specific species as irrelevant, they may miss important ecological clues. In addition, diversity indices may appear skewed depending on the season of collection. Because they measure pure abundance of species, such indices are not inherently sensitive to the large variations in these numbers throughout the year. Some researchers found that these omissions and masks yielded counterintuitive findings—that is, an undisturbed wilderness stream had the same diversity index as moderately polluted urban rivers. One resultant modification of the diversity index incorporated the biomass of the individuals caught as a weighting factor for their health and importance in the food web. Despite the fact that these indices were calculated for thousands of studies performed in the 1970s as part of Environmental Impact Assessments, diversity and stability are not necessarily correlated. Nonetheless, the diversity index remains a tool of choice for many water quality managers.

BIOTIC INDICES: INVERTEBRATE TOLERANCE SCORES. Another major school of biotic indices utilizes a similar approach to the diversity index, but relies on a restricted group of organisms—those insect groups found in the benthos. Benthic insects (macroinvertebrates) have emerged as some of the most sensitive indicators of pollution. When the principles of the diversity index are integrated with the principles of indicator organisms, the resulting tool becomes more specific and ecologically accurate than its predecessors. First developed by Chutter (1972) for sewage impacts in South African waters, this approach assigns tolerance values to insect species and then calculates an average tolerance among all animals in a collection. Thus, tolerance values become a weighting system.

Chutter's model was first adapted for use in North America (primarily in the Upper Midwest) by Hilsenhoff (1977). Both these authors use tolerance values that primarily reflect tolerance to organic pollution (e.g., sewage). Chutter called the measure the Biotic Index. Hilsenhoff's model, which became known as Hilsenhoff's Biotic Index, is calculated as the sum of the number of individuals of each taxon multiplied by the taxon tolerance value. When that sum is divided by the total number of individuals in the sample, it yields a number between 1 and 5 (later modified to between 1 and 10) that reflects macroinvertebrate diversity ranging from 0 (clean) to 5 (highly polluted). Both indices require identification to a genus or (preferably) species level.

Hilsenhoff also published the Family Biotic Index (FBI) for use in areas where fauna are not well known. The Hilsenhoff technique has been applied to a geographic analysis of water quality in a series of "high-quality" trout streams and was highly correlated with land use practices.

INTEGRATED, OR MULTIPLE, METRIC INDICES. Among the many problems associated with individual indices, two consistent problems emerged. First, no index could work well in all situations. Second, species-specific identifications required trained professionals; volunteers and other nonprofessionals would be unable to assist in monitoring efforts. As the search continued for widely applicable, simple yet accurate metrics, a new school began to emerge that utilized several metrics rolled into one.

One of the earliest of these indices was advanced by Karr (1981). Termed the Index of Biotic Integrity (IBI), it concentrated on the abundance, trophic composition, species composition, and condition of fish, evaluating the range of ecological levels from the individual through the population, community, and ecosystem scales (Karr 1991). Each of these characteristics was measured and weighted differently, depending on its ecological relevance, then summed to produce an IBI that rates river segments from excellent to extremely poor (i.e., devoid of fish) (see IBI highlight).

An important innovation of the IBI is that the value assigned to each of the 12 metrics is framed in comparison to a minimally impacted regional reference site. As mentioned earlier, many of the biotic indices faced criticism because they were relevant for only small geographic areas. Managers and policy makers, however, continued to search for a more unifying approach that could reflect natural variations rather than be invalidated because of them. The idea that different aquatic (and terrestrial) ecological regions exist has been widely adopted since it was first illustrated in the late 1970s (see chapter 16). Thus, natural conditions in Rocky Mountain streams will be considerably different from those of Midwest tributaries of the Mississippi River. In fact, more than 170 aquatic ecoregions have been identified for the United States (chapter 16). Because the IBI is calculated differently in each region, it remains sensitive to natural variations in water quality and aquatic communities.

In addition, by incorporating numerous indices, the IBI effectively addressed the problem of limited utility. Among the 12 metrics were those identifying the presence of important indicator organisms such as pollution-tolerant sunfish, the relative abundance of long-lived versus short-lived species, the proportion of fish occupying different trophic positions, and the number of fish in ill health. Consequently, it incorporated measures of diversity, dominance, abundance, and indicator organisms.

Other multiple metrics have been adopted for other taxa, such as the benthic invertebrates used in Hilsenhoff's Biotic Index. Karr (1991) mentions that the Invertebrate Community Index has been developed by the Ohio EPA. The U.S. EPA has developed a similar, but even further simplified, "Rapid Bioassessment Protocol" (RBP) consisting of eight metrics to be gathered from each of two different stream habitats.

A variety of multiple-metric approaches are also used to assess ecosystem-scale environmental quality. For example, one approach (termed AMOEBE) used in the Netherlands measures a broad range of biotic and abiotic variables, evaluates their condition, and creates a schematic map of actual versus desired environmental quality. The schematic map consists of a circle representing acceptable bounds for each variable. Each variable's actual state is represented graphically, like spokes of a wheel. Spokes that are significantly impacted extend out from the circle, indicating nonacceptable ecosystem quality. By evaluating a cross section of species and key biophysical features, researchers can estimate the impact of policy alternatives. Although not a quantitative, deterministic model, AMOEBE is useful as a coarse-scale descriptor and is a powerful illustrative tool (Nip and Udo de Haes 1995).

The IBI, ICI, and RBP measures have proved to be powerful models, and have now been adapted for use in states ranging from California to Maine. Nonetheless, they too have been criticized on several fronts. From an ecological point of view, for example, some have questioned the apparent arbitrariness in the assignation of values and metrics. In addition, these metrics have been criticized for not considering economic values and economically important species in the formation of the indices. From a social and economic point of view, all biological metrics are caught in a continuous debate between choosing indicator species and community structures that reflect purely ecological values versus choosing indicators that are more closely integrated with the economic choices society must make. The level of debate between economic values and "ecological ones" rises and falls with economic times, and can be expected as a constant companion to water quality decisions. Because water quality decisions are made in the economic and social arena, in the final analysis any index must go through its social/economic relevance filter. Managers must recognize that economic viability will be a continuous criticism of any monitoring efforts, from toxicology to biotic indices.

Adoption of Biological Indices

Changes in environmental management often turn into a contest between policy formation and "on the ground" practices. Sometimes policy drives field-level science, and sometimes the reverse occurs. In the case of bioassessments, tools are being adopted faster than the policies that require (or accept) them. Consequently, in the United States, the EPA has recognized the value of biological criteria but so far has only encouraged the states to enact biocriteria; EPA has given no national level direction.

Biocriteria, whether descriptive or numeric, set guidelines for evaluating bioassessment results (Southerland and Stribling 1995). In the United States, Ohio was the first state to develop specific methods and procedures for biological criteria. Ohio's 1987 efforts were followed by numerous workshops, protocols, and policy recommendations at the national level encouraging states to adopt biological criteria and some form of RBP. According to Southerland and Stribling (1995), current EPA policy calls for states to incorporate biological criteria into their water quality standards.

Forty-seven states, the District of Columbia, and Puerto Rico employ biological monitoring in their water quality programs. Tremendous variation exists in how well developed each program is, as well as in how well biological criteria are incorporated into state protocols. Most advanced in establishing biocriteria are Maine, Ohio, and Florida (Southerland and Stribling 1995). Each of these states have actual numeric biological criteria incorporated into their standards.

Florida, for example, uses an invertebrate diversity index to mandate that diversity must not fall below 75% of reference measures. Florida's biocriteria are strictly constructed, however, and are consequently difficult to apply to a broad range of problems.

In contrast, Maine and Ohio have searched for flexible and more comprehensive incorporation of biocriteria into their existing water quality programs. Maine has designated four "life use" classes for the state's waters and set "measurable ecological attributes" for each class. Southerland and Stribling (1995) offer the example that the standard for Maine's highest water quality class (AA) states that "aquatic life shall be as naturally occurs" such that there will be taxonomic equality with minimally impacted reference sites, numerical equality with reference sites, and the presence of pollution-intolerant indicator taxa.

Ohio's biocriteria regulations are woven throughout the state's water quality regulations. For example, for each of Ohio's ecoregions, there are specific criteria for IBI, ICI, and a modified index of well-being (MIwb), all of which must be met before full attainment is achieved. Ohio's program has also evaluated state resources extensively, studying pollution sources ranging from confined spills to nonpoint impacts, municipal effluent, and habitat modifications. Several studies have indicated the need for more stringent standards than had been applied under toxicology criteria, a process that was upheld in state court.

While other states are not as advanced in their use of biocriteria for regulation (27 have biocriteria in the development phase), biocriteria are still being used in water quality management in most states. Arkansas and Texas, for example, use biocriteria to assign "aquatic life use" classifications in different regions of the states. Other states, such as Connecticut, New York, Nebraska, and Vermont, are using biocriteria as part of a comprehensive evaluation of impairment and compliance with existing permits and regulations.

The variety of approaches exhibited by different U.S. states is also evident among different countries in a region. Among the countries of western Europe, for example, there are notable preferred classification systems as well as a broad range of differences among adopted systems. As Table 6.1 indicates, most countries use both biological and chemical/physical indices and they divide their water bodies into a limited number of distinct classes. (When actually assigning waters to these classes, waters are sometimes classified for uses higher than they currently meet; the classification scheme thus serves to force improvements in water quality.) What is not evident from the table is that managers also must demonstrate a clear correspondence between classes and goals (e.g., attainment of designated uses) and that these schemes usually contain some capability for displaying supplementary information about specific lakes and streams.

Chemical/Physical Indices

Biological assessments represent an important scheme used to classify water resources and water quality priorities. In addition, chemical and physical indices have also proved important in the classification of water quality. In the geologic category, for example, watersheds may be rated by "sensitivity indices," which predict a watershed's sensitivity to inputs through either surface run-off or groundwater flow. These indices generally incorporate soil characteristics, slope, vegetation, and geologic parent material.

Creating a Viable Index

All indices—biological as well as chemical/physical ones—have been criticized on the basis of ecological arbitrariness, and addressing these concerns is one of the more serious problems for water quality managers. As described here, the appearance of arbitrariness arrives through several channels: the necessity of simplification, the incorporation of "scientific best judgment," and a certain degree of uncertainty of how to aggregate data from an assessment. Creating any index requires choosing indicator organisms or chemical/physical structures, weighting these parameters, and aggregating them in a meaningful way; there is room in all three steps to create an index with limited ecological relevance. To

Table 6.1 Water quality may be classified by use or by a suite of chemical variables and/or biological variables. Many countries have developed their own water quality metrics for use in classifying water bodies.

Country	Chemical Variables	Biological Variables	Number of Classes
Belgium	Oxygen balance (i.e., % DO saturation, BOD_5, ammonia)	Belgian biotic index[1]	5
Denmark	Organoleptic (visual, odor),oxygen balance (DO, BOD_5, ammonia, COD), "worst condition" among 31 variables[2]	Saprobic index, pollution tolerance of taxa after Hynes 1960	5–6
France	"Worst condition" among 31 variables[3]	French Biotic Index[3]	5
Germany	DO, BOD_5, ammonia; four lake trophic classes	Saprobien system, after Kolkwitz and Marrson 1908	7
Greece	None	None	N/A
Ireland	None	Modified Trent Biotic Index[4]	
Italy	None	None	N/A
Luxembourg	Oxygen balance	Belgian Biotic Index[1]	5
Netherlands	Oxygen balance	Belgian Biotic Index[1]	5
England and Wales	Oxygen balance, toxicity to fish, potable water criteria[6]	Trent Biotic Index[5]	5
Scotland	DO, BOD_5, ammonia, pH, nitrogen, phosphate, temperature conductivity, fecal coliform bacteria[7]	Trent Biotic Index	4

[1]LaFontaine *et al.* 1985
[2]Danish National Agency for Environmental Protection, 1970
[3]Duport and Margat 1983
[4]An Foras Forbatha 1984
[5]Woodwiss 1964, 1978, 1980
[6]U.K. National Water Council 1981
[7]Scottish Development Department 1976, 1981

be useful, the index must be sensitive to changes in water quality. The index value should change only when water quality changes are relevant to designated uses, however. For example, if an index value changes in response to changes in dissolved oxygen, and if the lower limit for "acceptable" oxygen conditions is 4.0, the index should cease to change if oxygen reaches 4.0 or lower. The index must be structured so that various constituents can be differentially weighted, because not all variables will have equal importance for use of the water resource. A quantitative relationship should exist between changes in the index and changes in "quality" for a given designated use. Preferably the index will approach some conceptually identified level (e.g., zero) at the point at which designated uses are no longer supported.

Mathematical expressions of water quality have evolved from subjective classification systems to highly quantitative and specific applications. Early indices, based on a limited number of variables such as dissolved oxygen and biological oxygen demand (BOD_5), were calculated using simple arithmetical averaging. Later indices were calculated using more sophisticated techniques such as multiplicative aggregations, geometric approaches, and observation rankings. Because many water quality variables are non-normal in their statistical distribution, some authors have presented nonparametric versions of indices while others have proposed complex multivariate models. Although these complex systems are relatively powerful, they begin to deviate from some of the original reasons for using an index (such as simplicity and ease of calculation, ability to explain clearly water quality levels to nonspecialists). Their use thus presents a dilemma to water quality managers.

Non-numeric Indices

Numeric indices are only one class in an array of classification schemes. Another important tool in this array is the use of integrative analysis rather than indices for single designated uses. Bhargava (1983), for example, developed a water quality index to classify reaches of the River Ganga in India. Use of the Ganga is particularly intensive, including industry, bathing, agriculture, irrigation, fish culture, recreation, and public water supply. To develop his indices, Bhargava incorporated nine water quality variables (i.e., temperature, turbidity, oxygen, BODs, chloride, conductivity, hardness, coliform bacteria, and nitrate nitrogen). He combined those relevant to a given designated use. For example, temperature is used only for fish and wildlife, and conductivity is not used for bathing or industry. Rating curves were then developed for each of the variables, and overall index values were generated with a geometric mean aggregation.

With modifications, this model has been employed elsewhere to classify other water reaches with multiple uses. For example, it has been used to assess water quality of residential canals in the center of Baghdad (Al-Ani et al. 1987). The canals are used for irrigation, livestock, public drinking water, fisheries, and aquatic life. Consequently, the protection of irrigation, livestock, and aquatic life uses was of paramount importance. Following Bhargava's model, different variable sets are included for each use, which generates unique Water Quality Indices (WQIs) for each use. Because water quality in Baghdad varies significantly by season, separate indices were calculated for winter and summer. Thus, for each location on the canal, and for each of two seasons, the manager has an index value for each of three designated uses. Using this system, Al-Ani *et al.* (1987) determined that significant land use changes occurred between their stations 2 and 4 along the canal, and that those conditions dramatically affect the suitability of canal water for livestock. For example, chloride was 73 mg/L at station 2 and 138 mg/L at station 4, while nitrate was 0.04 mg/L at station 2 and 2.25 mg/L at station 3 in the summer (Al-Ani *et al.* 1987). Safe concentrations returned beyond station 5. Using this information, a manager could suggest that the canal water was unacceptable for livestock watering at stations 3 to 5 in the summer, and that the problem probably resulted from sources of chloride and nitrate. That assessment could then be used to guide management actions.

Future Trends for Chemical/Physical Indices

One of the latest trends in the field of water quality indices goes beyond classification of water according to designated use to assess attainment of that designated use. This approach includes implicit incorporation of the water body's assimilative capacity and, therefore, marks a major improvement in utility of these indices for management purposes. This family of indices produces an acceptable index value until conditions degrade to such a degree that the designated use is not supported. In the United Kingdom and Australia, these new indices have been applied to river and stream management. In the United States, they have been applied in the Great Lakes in an index that addresses compliance with the goal of nondegradation.

The Australian model uses a unique approach. All of the water quality indices presented to this point incorporate some form of aggregation that allows the final index value to mask the worst measured condition. This problem is known as "eclipsing" because the index tends to hide the potentially most valuable piece of information—the identity and magnitude of the one variable that precludes safe use of the water. The larger the number of variables included in index calculation, the more effective this masking becomes because weights assigned to each determinant become smaller with an increasing total number of determinants.

The Australian system uses an alternative aggregation method called "minimum operator." The philosophy of a minimum operator is that the variable causing the poorest water quality will be the one that most effectively controls use of the water resource. Therefore, the matrix selected for determining a water quality index should be the one that displays the poorest or controlling water quality level rather than the one (or more) that shows average water quality condition. In such a minimum-operator metric, new determinants can readily be added anytime (and others eliminated), weights are not required, no upper or lower limit exists to the number of variables used in classification, and locations can be compared using different numbers of variables at each station.

Classifying and Measuring Water Resources

Classification may logically be discussed at the end of a discussion of criteria and standards, or at the beginning of it. In part this flexibility exists because classification is not a static tool; rather, it is a continuous and cyclical process based on fluctuating conditions of water quality. Consequently, bioassessments are based on previous classifications of water resources and regional patterns and impacts. In turn, the bioassessments yield new classifications of water resource classes. Classification can facilitate the setting of standards, but once established, compliance with those standards must be monitored. If necessary, standards can be revised or enforcement initiated if new classifications reveal nonattainment of specified levels.

Water quality classifications take many forms, as illustrated in Figs. 6.1–6.3. They may reflect the spatial distribution of single variables such as dissolved

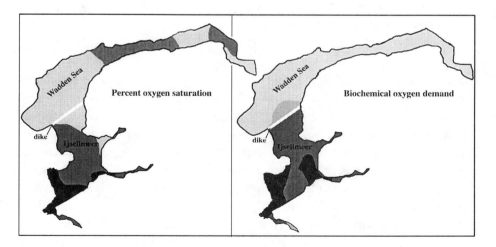

Fig. 6.1 Classification according to discrete variables provides detailed information on spatial variability in parameters of water quality concern. In this example, two parameters—percent oxygen saturation and biological oxygen demand—are compared for the Ijsellmeer and the Wadden Sea off the coast of the Netherlands. Parameters are most concentrated in the darkly shaded areas, grading to only minimal concentrations in the lightest areas. (Maps excerpted and redrawn from Newman 1988)

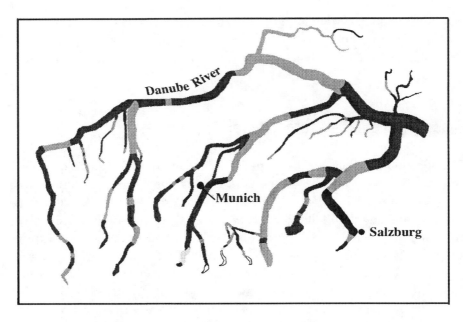

Fig. 6.2 Classification systems may also represent overall conditions, rather than specific parameters as in the previous figures. In this example based on a portion of the Danube River through Germany, water quality measures have been combined and weighted to yield overall condition statements for different segments of the river. Black is used here to designate the most polluted state, or class IV sections; dark gray represents class III, heavily polluted; light gray represents class II, moderately polluted; and the lightest color corresponds to class I, unpolluted segments. (Based on Newman 1988)

oxygen or phosphorus within water bodies (e.g., Fig. 6.1). Visualizing problems in this way is an essential step in focusing local efforts as well as in assessing national-level priorities and tracing patterns that may assist management efforts. For example, if phosphorus loadings appear to correspond with streams in agricultural areas, managers can narrow the causes to agricultural practices. If phosphorus levels appear to follow patterns of large municipal areas, managers can turn their attention to phosphorus concentrations in municipal effluents.

Another type of classification system, illustrated in Fig. 6.2, aggregates all pollution data to identify general water quality classes. Although some specificity is lost in these classification schemes (it is no longer possible to determine primary sources of pollution and their spatial variation throughout an area), the broad picture often serves as an important management tool. Lumped classifications, for example, provide an easier tracing of temporal trends in water quality degradation or improvement, and may isolate smaller-scale spatial trends (areas of special concern that may require special attention).

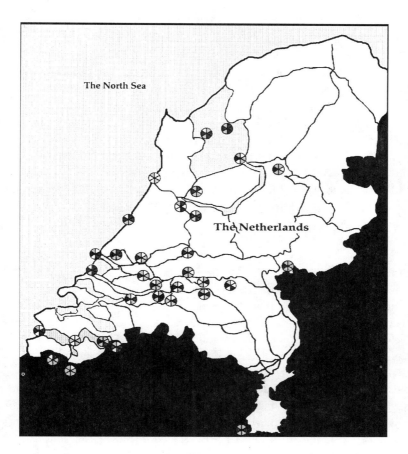

Fig. 6.3 In an effort to retain information about variance in individual variables, pie charts have been used to show spatial variation in the concentration of various compounds. Note that more detail is available, but spatial patterns are less clear than when using classification according to discrete variables. (Simplified and redrawn from Newman 1988)

A third aggregation system attempts to integrate both of those strategies mentioned above. Utilizing pie charts, bar graphs, or other graphic data, these classification schemes attempt to illustrate how both relative sources of pollution vary within a planning area, and how aggregate pollution stress varies (Fig. 6.3). Although these classification maps require the most careful scrutiny, they provide a great deal of invaluable information in one simplified format.

Lastly, classification schemes can isolate important spatial patterns in water quality which may facilitate management planning. For example, water quality in an estuary is significantly affected by flow patterns in that estuary. Flow

Focus: Calculation of a Macroinvertebrate Index of Biotic Integrity

Hilsenhoff's (1982) Biotic Index serves as an example of a macroinvertebrate-based index. It is based on the logic that species that are intolerant of certain forms of contamination will not be found at contaminated sites. Given large amounts of field data, the tolerance scores for a taxon can be generated based on the frequency with which it is found in polluted waters. Using insects, amphipods, and isopods, Hilsenhoff assigned each species an index value ranging from 0 to 10, with 0 being most intolerant and 10 being most tolerant. Index values for several hundred Midwestern U.S. species have been published. When an individual or group cannot be identified to a species, a value for the genus (or even family) is used. In fact, a separate index called the Family Biotic Index (FBI) relies solely on family level identifications.

Samples are collected from rocky and sandy substrates using a kick net or D-frame net. The scientist picks the first 100 organisms from the net, striving for variety in the organisms selected. Because invertebrate taxonomy is extremely difficult, certain complex groups (e.g., Coleoptera) and organisms with small body size (i.e., <3 mm) are typically excluded from this count. Selected organisms are identified and tabulated. The index is calculated by multiplying the number of individuals of a species times the tolerance value for that species, then dividing by the total number of individuals in the collection (usually 100). Thus, the index is really an average tolerance value. Hilsenhoff suggests that the index is highly useful for organic impacts like sewage or for impacts of dams. It is somewhat useful for heat, heavy metal, and toxic impacts if specific tolerance values are developed. It must be seasonally calibrated by sampling all sites at similar and representative temperature and horologic conditions (e.g., winter, late spring, and late summer).

restrictions create distinct water quality zones, and treating the basin as a whole might yield less success than treating the basins separately.

Managers will be asked to analyze their data and classify water bodies in many ways. Typically these distinctions will be made according to existing water quality, designated uses, and management goals. Some classification schemes such as ecoregions, state, federal, or European Community water classes are geographically broad. They are intended to compare and contrast large areas and compliance with associated overarching management goals. Others have been developed and used for detecting differences among a small number of stations on a water body, primarily for use in impact assessment.

Focus: Calculation of a Fish-Based Index of Biotic Integrity

Fish communities integrate properties of the entire watershed. Fish are comparatively long-lived, widespread, and relatively easy to identify. They live in a wide variety of habitats and occupy a range of trophic levels. The fish community is often economically important and its significance is readily grasped by the public. For these reasons, fish are often seen as valuable indicators of the health of an aquatic ecosystem. The Index of Biotic Integrity (IBI; Karr 1981) is a measure of the "health" of the entire fish community.

In calculation of the IBI, each of 12 attributes is evaluated for each collection from the fish community. The attributes are scored on a scale of 1 (worst) to 5 (best) according to specified criteria for each score.

After collection, the index is calculated by summing the score for each of the 12 metrics. The IBI ranges from 12 to 60, with the following classes assigned to IBI scores:

Excellent	57–60
Good	48–52
Fair	39–44
Poor	28–35
Very poor	<23

A sixth "no fish" category is possible. If the IBI for a site falls between classes, averages of several samples can be calculated or an informed biologist can be asked to provide interpretation (Fausch *et al.* 1984). This metric varies with stream size, so reference values must be developed from "ideal" or unimpacted streams in the same area as the sample stream.

Summary

- Classification of water bodies allows the manager to implement strategies for groups of similar resources. Classes are usually developed according to management goals, designated uses, and/or existing water quality.
- Chemical/physical indices are mathematical expressions of several different variables combined into a single score. Indices must be based on variables relevant to the intended use of the index (e.g., variables that would reflect failure to support a designated use).

- A common problem with indices is eclipsing, a by-product of the averaging process that inadvertently masks problem variables. One solution to eclipsing is use of a minimum operator, a technique that results in a low index value if one or more variables lies outside of a given range.
- Biological indices allow the manager to describe the biotic community with one value. Diversity indices, indices of biotic integrity, and simple richness indices (i.e., numbers of taxa) are examples of such indices.
- A serious limitation with the use of indices or other classification schemes is the false sense of security they provide. These schemes mask the variation inherent to natural populations and ecosystems, making choices seem more clear-cut than they are.

For Further Reading

Cairns J Jr, Smith JP. The statistical validity of biomonitoring data. In: Loeb SL, Spacie A, eds. *Biological monitoring of aquatic systems.* Boca Raton: Lewis Publishers, 1994;49–68.

Courtemach DL. Merging the science of biological monitoring with water resource management policy: criteria development. In: Davis WS, Simon TP, eds. *Biological assessment and criteria: tools for water resource planning and decision making.* Boca Raton: Lewis Publishers, 1995;315–326.

Hilsenhoff WL. *Using a biotic index to evaluate water quality in streams.* Wisconsin Department of Natural Resources, Madison, Technical Bulletin 132, 1982:22 pages.

Hilsenhoff WL. Rapid field assessment of organic pollution with a family-level biotic index. *J N Am Benthological Soc* 1988;7:65–68.

Power M, Power G, Dixon DG. Detection and decision-making in environmental effects monitoring. *Environ Mgt* 1995;19:629–641.

Yoder CO. Policy issues and management applications for biological criteria. In: Davis WS, Simon TP, eds. *Biological assessment and criteria: tools for water resource planning and decision making.* Boca Raton: Lewis Publishers, 1995;327–344.

7

The Role of Scale Issues in Water Quality Management

Overview

Natural resources operate over a wide range of space and time scales. Spatial scales range from individual organisms through communities, habitats, watersheds, and regions, and finally to the globe; temporal scales range from nanoseconds to epochs. That scales exist is obvious, yet scientists and managers are only now learning how scale-dependent processes relate to one another and how a given scale defines scientific and management questions. An observer's scale of interest, for example, will determine the variables and questions used to understand a system.

Scale issues are at the root of an important shift in ecology and natural resources management. Historically, the natural world was seen as linear and controlled by physical attributes of the environment; managers held smaller-scale views tied to specific physical and geological attributes of a water resource. The emerging view begins with questions of scale, and then uses these questions to define relevant parameters to test or manage. Scale issues are presented in this chapter as a framework for understanding the dynamic processes at work in the water cycle, rivers, lakes, groundwater, coastal regions, and wetlands. This scale-sensitive perspective forms an important basis for further discussions of land-use impacts on water resources, management decision making, and monitoring.

Introduction: A Growing Awareness of Scale Issues

Natural phenomena occur over a wide range of spatial and temporal scales, from global patterns to microhabitats, and from geologic time (millions of years) to diel (24-hour) fluctuations. At each scale, a different set of constraints and controls affect behavior and structure of natural resources. A common pitfall of science is to sample variables at one scale and assume that results explain variation at other scales. Similarly, a common error of management is to plan actions at one scale and overlook the effects these activities have on higher and lower scales.

Ecology has lagged behind earth and atmospheric sciences in recognizing and accepting the scale dependencies of natural processes (Wiens 1989). Bounded by tradition and long focused on experimentally manipulatable (i.e., small-scale) biophysical variables, ecology seemed immune from the scale issues that complicated other sciences. Wiens (1989) suggests that ecologists have been slow to recognize the importance of scale dependency because ecological processes generally are comprehensible on human time frames, unlike physics, astronomy, or geology. Much of ecology is intuitively familiar, which may have slowed recognition that familiar, tangible scales represent only part of a larger set of scale-dependent processes. Furthermore, in attempts to reduce variability and improve cross-study comparisons, ecologists have developed standard procedures that ignore important scale issues (e.g., using quadrants or study plots of a standard size). Breaking out of a monoscale view requires new techniques and new perspectives; that transformation is precisely what has occurred in ecology in the past decade.

Currently, the importance of scale issues in ecology is not only recognized, but is also considered a primary issue. Initially, the acceptance of scale issues developed from advances in testing and modeling. Technology enabled ecologists to measure properties on much smaller and larger scales. Significant changes in theories of nutrient cycling have emerged, for example, from improvements in ability to measure microbiological processes. Similarly, computer-aided analysis of satellite images (using geographical information systems [GIS]) has revealed important (large-scale) global patterns in natural resource distribution. Currently, important new understanding in regional patterns of natural resources comes less from new technology than from differences in questions being posed.

Social pressures have also contributed to the shift in thinking about ecological scales. Societal demands for multiple resource management and politically expedient decisions have encouraged scientists and managers to think critically about the effects of scale. In turn, scientists and managers are developing knowledge on how information gathered at one level can be extrapolated to another level. Nonetheless, ignoring scale effects is still common practice. Typically, community ecology studies are conducted on plots no larger than 1 m in diameter, despite significant differences in types of organisms studied and in ecological questions posed (Wiens 1989; Levin 1992), a fact stemming from the mindset that "real science" means manipulative experiments. May (1988) admonishes that "we should resist the temptation to find some problem that can be studied on a convenient scale, but rather should strive to identify the important problems and then ask what is the proper scale on which to study them" [pp. 356–357].

The Scales in Question

Temporal and spatial scales determine which questions can be answered in any given situation, but the manager's scale of interest determines which questions are asked. The understanding of the manager (or scientist, policy maker, or concerned citizen) about the interaction of these scale-dependent processes will largely determine the utility of information and the sustainability of management decisions. The lesson taught by the expanding understanding of scale-dependent processes is that no single correct scale exists with which to pose a question or formulate an answer. Scales are, indeed, all relative. Nonetheless, without attention to scale issues, information can be misleading and decisions based on that information are, therefore, likely to fail. Experiences of ecotoxicology and the need to scale up from microcosm experiments to macrocosm and whole-system experiments (as outlined in chapter 5) provide examples of studies that are accurate, but misleading, because of a discrepancy between the scale of study and the scales of interaction in the "real" world. Below, arranged from large scale to small scale, are examples of spatial and temporal scales and water quality questions answerable at each.

- The globe: A global-scale focus has been a powerful force in clarifying the impacts of local activities on global processes (e.g., global increase of carbon dioxide, changes in the ozone layer, and other atmospheric properties) and, consequently, on other communities around the globe. Few management questions are addressed on a global scale, but the awareness of the connection enables managers to better frame regional action plans.
- Biome, on the order of 2×10^{12} m^2 (e.g., 3 to 5 in the United States): Questions such as ozone layer thinning, acid deposition, global change, sea level changes, and desertification.
- Regions, on the order of 2×10^{10} m^2 (e.g., 2 to 3 in Minnesota or in England): Questions such as regional nonpoint source water quality control standards, transboundary pollution in air or groundwater, and impacts of one country's action on the air quality of another country.

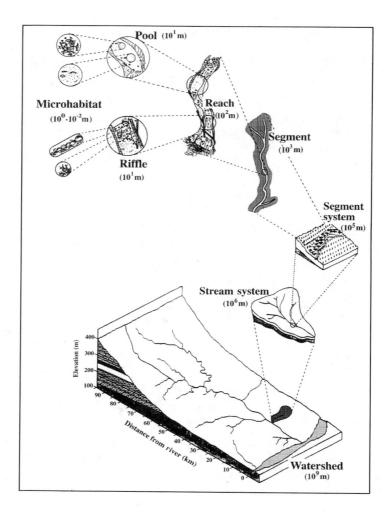

Fig. 7.1 Aquatic ecosystems function at a variety of spatial scales ranging from regions (not pictured) and watersheds to specific reaches and microhabitats within a stream lake or wetland. At each scale, different biological, geological, and hydrological forces predominate. For example, in a segment system, the topography of the surrounding landscape is especially important; in contrast, at the segment scale, the linear forces of waterflow through the riparian zone becomes more important. (Modified from Frissell *et al.* 1986; Troelstrup and Perry 1990).

- Subregions, on the order of 10^9 m^2 (e.g., Minneapolis–St. Paul or Delhi metropolitan area): Questions such as effectiveness of a class of land-use management practices.
- Small watersheds, on the order of 4×10^8 m^2: Questions such as effects of a particular land use action (e.g., a specific clear-cut or agricultural practice). Watersheds and six successively finer spatial scales are illustrated in Fig. 7.1.

- A stream reach and its riparian zone, on the order of 10^4 m^2 (e.g., a 100-m section of stream and its immediate area): Questions such as impacts of a specific point source.
- Habitats, on the order of 10^2 m^2 (e.g., a few meters of stream channel or a small bay or wetland): Questions such as health of a particular plant or animal species population.
- Microhabitat growth and function of an individual organism.

Temporal Scales (Hierarchical, from Largest to Smallest)

In water quality management, temporal scales reflect acuteness and persistence of a given water quality impact. For example, effects of acute toxins are assessed on a scale of days. Functional characteristics of an ecosystem (e.g., changes in decomposition or primary productivity) or effects of seasonal stresses such as agricultural run-off are addressed at the scale of seasons within a year. Information collected among a series of years will answer questions about the effects of a stress on migratory fish populations. Cumulative impacts of chronic stressors such as climate change, nutrients, or exotic species will be detectable at a scale of decades. The more acute a stress, the more quickly its impact will be felt and the finer the necessary scale of observation. The more persistent a stress, the longer the necessary scale of consideration.

As discussed by Wiens (1989), the ability to predict ecological phenomena depends on the relationships between spatial and temporal patterns. The relationship between spatial and temporal scales is particularly important in natural resources management because policies tend to apply over large areas but are frequently based on only short-term studies. Consequently, it is important to recognize that as spatial scale increases, the relevant time scale must also increase because at larger scales processes operate at slower rates and have larger time lags and indirect effects are more apparent. As Wiens suggests, "any predictions of the dynamics of spatially broad-scale systems that do not expand the temporal scale are pseudopredictions. The predictions may seem to be quite robust because they are made on a fine time scale relative to the actual dynamics of the system, but the mechanistic linkages will not be seen because the temporal extent of the study is too short" [p. 389].

Observer Scale

Our understanding of scale effects does not occur independently of the observer. As mentioned earlier, managers, scientists, and policy makers all have a scale of interest. An industry manager may be concerned only with the section of river affected by an effluent plume; a city water quality manager is concerned with the cumulative effect of many effluent plumes, a concern that encompasses a larger area and longer timeframe. Similarly, state-level managers may be concerned with how the run-off from cities affects regional groundwater and rural supplies. At each level, the manager or scientist chooses a scale of interest at which to describe the water quality patterns and processes.

Because the study scale will reflect only selected patterns, choosing this scale carefully—and being aware of its limitations and competencies—is critical. For example, in one critique of the field of limnetic primary production, Carpenter and Kitchell (1987) note that "even when lakes lie in similar watersheds, experience the same weather and are sampled by standardized methods, their productivity varies about tenfold" [p. 417]. They also comment that the many physical and biological processes on different time scales that could contribute to this variability have not been addressed by limnologists: "Generally, we sample regularly (weekly, biweekly) and calculate annual production, thereby limiting our analyses to one or two time scales. The unexplained variability that we observe may be caused by a host of dynamic processes occurring at time scales that have been neglected" [p. 317].

Following a similar line of logic, Crowl and Schnell (1990) tested whether spatial scale and sampling resolution affected the conclusions of multiple regressions and principal component analyses in their study of snail size and abundance in streams. Their study revealed that snail size and abundance could be attributed, variously, to crayfish abundance and algal biomass (at large scales among streams), flow rate and substrate particle size (at smaller scales within streams), and/or microhabitat such as pools and riffles (at the smallest scale within streams). Even more than the understanding this study promotes of snail distribution, it is important in the way it identifies the effects of the observer's scale of interest and choice of sampling regime on the conclusions drawn from ecological studies.

Recognizing that researchers have addressed the same question from different scales of reference can resolve many cases of scientific controversy. For example, control of species distributions in coral reefs (a fine-scale question) is strongly influenced by stochastic events; species composition of a local fish community is often unpredictable. Reef systems on multiple atolls (a coarser-scale view), however, reveal predictable patterns of community composition. Divorced of their scale dependencies, these facts appear to contradict one another. When seen as small- and large-scale patterns, together they reflect a more complete picture of reef ecosystem dynamics (Weins 1989). They also demonstrate a critical issue in scaling: results at one scale generally cannot be translated to another scale. In this example, it would be a mistake to assume that, because random events in a single reef make fish distribution unpredictable, fish distributions in a region are also unpredictable. The larger scale buffers effects of random events. Thus, coarse-scale distributions become more predictable. The scale of analysis must be chosen carefully because it determines the level of understanding.

The professional or personal orientation of observers also influences the way a system is understood. For example, a hydrologist and a water quality manager evaluating the same river will be concerned with different variables and likely would divide the river at different spatial scales. A "stream reach" for the hydrologist may be the distance that determines flood potential. Riparian land use patterns and their impacts on a given stretch of river might determine a "reach" for the water quality manager. Thus, although concerns often intersect (land use both determines flood potential and is, in turn, determined by flood potential), different observers use different criteria for seeing and explaining the natural world.

Translating Information Across Scales

Understanding that scale effects exist is one important step to better management. Learning how to think across scales offers the key to putting that understanding to use. Scientists continually search for the most basic relationships among phenomena so that they may extrapolate findings across a broad range of situations. However, mechanisms frequently operate at different scales than those upon which the patterns are observed. For example, distinct patterns may be observed in invertebrate distribution in a stream. That patterning is not explained until the level of observation is scaled up to include the correlation of invertebrates with different substrate types and the correlation of different substrates with land use in those river sections.

Recent advances in hierarchy theory offer one way to think about, and navigate among, scales. Hierarchy theory states that coarse-scale processes constrain patterns at finer scales, demonstrating that coarse-scale processes are not merely aggregations of fine-scale ones—instead, they are more than the sum of their parts (Fox 1992; Muller 1992). In the example above, it would not have been instructive to collect more invertebrate samples without increasing the scale of samples to include substrate and riparian land use patterns. In this case, the larger scale of land use constrains the smaller scales of invertebrate distribution.

In other cases, fine-scale processes aggregate together to form a coarse-scale pattern. In the reef atoll example given earlier, the large-scale predictability of fish distribution results from an aggregation (averaging out) of small-scale, stochastically controlled patterns. Overall, the numbers and habitats of given species can be safely predicted, although it may not be possible to find a given species at any given point. Indeed, Wiens (1989) has assembled a model to illustrate the relationship between the spatial–temporal scale of an analysis and the predictive power of that analysis (Fig. 7.2). Hierarchy theory is a way of organizing thoughts about these scale effects. It helps to explain those cases where fine-scale processes aggregate to form coarse-scale patterns and situations in which coarse-scale patterns effectively constrain development of fine-scale patterns. Several "rules of thumb" in hierarchy theory are particularly useful from a management perspective:

- It is necessary to use multiple scales to understand any environmental process (that is, conclusions from one scale do not explain processes at other scales).
- Predicting upward is difficult (i.e., to make predictions about a higher (coarser) level based on information gathered at a lower (finer) level; downward prediction will lack specificity but should successfully identify relevant variables.)
- Higher levels effectively constrain patterns and processes at lower levels.

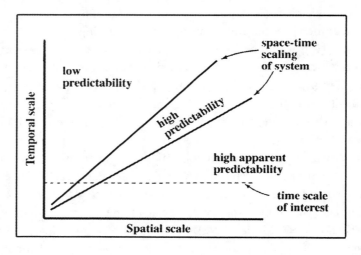

Fig. 7.2 A relationship exists between predictive power and the spatial–temporal scale at which an analysis is conducted. As we include greater (i.e., coarser) spatial and temporal scales in our questions, the precision of our impact predictions declines. In most water quality studies, relatively short-term predictions have received most attention. Managers have been willing to accept the imprecision associated with coarse spatial scale to gain predictability over short timeframes. (Redrawn from Wiens 1989).

Hierarchy theory offers improved guidelines for understanding the effects of scale on ecological processes and can help guide scientists and managers in the inevitable process of extrapolating scientific information across scales. Hierarchy theory also helps to explain some of the intricate connections between scale effects. For example, primary productivity of algae is related not only to phosphorus and nitrogen concentration but also to nitrogen : phosphorus ratios. This subtle, but significant, difference represents an important refinement of previous knowledge, and was facilitated by a sensitivity to hierarchies of scales across gradients of chemical concentrations in a lake.

Effective water quality management requires attention to multiple scales. The art of simplifying (to finer scales) and aggregating (to coarser scales), as well as of recognizing parallel scales, is key to achieving adequate understanding of ecosystems and their management needs. Hierarchy theory shows that aggregation is the more difficult task—and the one most frequently facing managers. Managers are often confronted with many fine-scale studies on which to base a coarse-scale management plan. Levin (1992) suggests that knowing how to sift through the noise to find the pattern is one of the great challenges of ecology. Similarly, it is one of the key skills with which sustainable management decisions are made.

Scale Effects in Water Quality

Spatial dynamics of the landscape strongly influence water quality; slope, soil chemistry and erosivity, geography, latitude, temperature, and land use patterns have all been correlated to water quality sensitivities. Similarly, water quality depends on temporal influences such as the duration of a given land use. As a result, water quality issues are multidimensional, influenced by atmospheric (e.g., rainfall), surface, and subterranean (e.g., aquifers and springs) conditions. As illustrated in Fig. 7.3, each dimension, or scale, is governed by different properties and affected differently by stress. Consequently, effective management can continue only if these scale and process differences are recognized and incorporated into management decision making.

Streams provide a useful illustration of water quality variability, as well as the way in which the observer's perspective influences the way a system is viewed. To most of us, streams, lakes, and other natural features seem fixed in time and space. We may recognize and come to expect seasonal variability in water level, but we always expect the river to follow the same course, and the rapids and pools to be in the same place year after year. We may expect that individual trees along the bank will die and others will grow, but we also expect

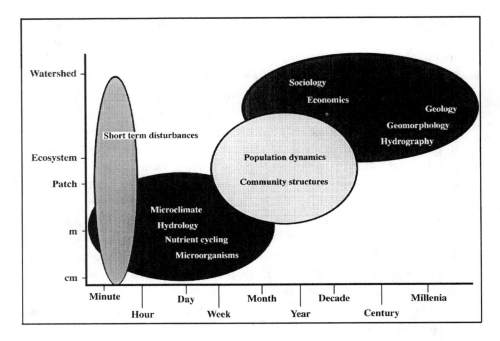

Fig. 7.3 Society's decisions about management of a water resource are influenced by the spatial and temporal scale at which that resource is valued. Some resources will respond to stress at different scales than others, making prediction and assessment a difficult science. (Modified from Muller 1992).

that the forest type will remain constant within our lifetime. Overall, our perspective constrains our observations. In short, we see and expect spatial and temporal variability consistent with the scale of the distances we travel and the timeframe of our observations. Few of us would recognize the same stretch of river if we saw it 200 years ago, however, or if we saw it from an airplane or a fish's perspective. These other temporal and spatial scales are less intuitive but no less valid.

As hierarchy theory demonstrates, patterns we see on the most familiar and intuitive scales are, in fact, constrained by large-scale processes. Long-term climate patterns constrain yearly ones; yearly patterns constrain daily ones. Long-term geomorphic dynamics constrain land use patterns; land use patterns constrain water quality. Streams, rivers, and lakes undergo tremendous natural changes, maturing over time and along the course of the stream valley. Managers must take into account the nature of these changes and the scales of time and space over which they occur.

Management Implications

Our understanding of how scalar differences in our perspective control our interpretation of water quality is still in its infancy. Relatively few studies have been conducted to identify controlling variables and the different scales at which we interpret them. Nevertheless, such knowledge holds tremendous potential for management and modeling. Standards for some variables may be set based on broad regional patterns of geology, while others may require localized standards and management regimes. Conceptual diagrams such as Fig. 7.4 are increasingly useful for compiling our available knowledge and presenting it in a way that can assist scientists, managers, and monitoring specialists in asking and answering water quality questions at the appropriate scale. These and other water quality monitoring and management variables are discussed below.

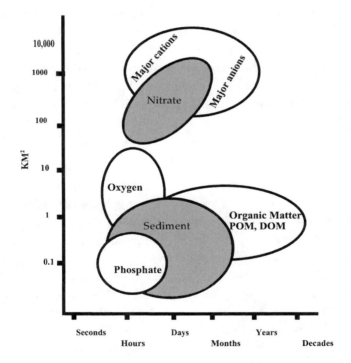

Fig. 7.4 The variables most commonly measured in water quality analyses vary on different temporal and spatial scales. For example, nitrates that are controlled by subsurface geology and climate vary at the watershed and among-years scale. In contrast, phosphorus is controlled at a more local- and short-term scale (e.g., among weeks and among stream reaches).

INVERTEBRATE COMMUNITY STRUCTURE. Invertebrate community *structure* is a valuable water quality monitoring tool that is highly correlated with regional changes in surface geology and watershed morphology. Consequently, community structure appears to be influenced primarily at regional and watershed levels—that is, at spatial scales of hundreds of square kilometers. Invertebrate community *function,* on the other hand, is responsive to smaller-scale differences in the landscape. Specifically, community function, as judged by percentage of specific feeding groups, is more highly correlated with local-level, reach/riparian processes on a scale of hundreds of square meters.

SUBSTRATE CHARACTERISTICS. Stream substrate controls diversity and distribution of invertebrates and fish and is strongly influenced by land use. Consequently, this intermediate level scale serves both as an indicator of land use and as an indicator of invertebrate population. For example, in studies in Southern Minnesota, Troelstrup and Perry (1990) showed that macrophyte abundance is low on rocky substrates and that rocky substrates are more abundant at agricultural sites with high current velocities. Macrophytes tend to be more abundant at forested sites with lower current velocity. In other words, macrophytes vary at a scale of a few kilometers in concert with changes in land use practice. At a finer scale, such as hundreds of meters, wood and leaf substrates are highly correlated with the amount of reach and riparian forest above, and next to, a site. Temporal scales are also evident in these substrate patterns. Leaf substrates are strongly seasonal, varying according to the month of the year while sediment-dominated substrates vary at a more coarse scale of decades, reflecting temporal scales in riparian land use patterns.

WATER CHEMISTRY. The water chemistry exhibited at any one point in a stream or lake is the result of a surprising number of scale-dependent processes. For example, specific conductance varies at a regional scale (e.g., hundreds of square kilometers) with changes in subsurface geology and regional land use. Nitrogen and pH reflect finer-scale land use patterns varying at a watershed-level scale of tens of square kilometers, an order of magnitude smaller than conductance. Alkalinity, turbidity, and temperature are highly correlated with reach- or riparian-level management, reflecting even finer-scale patterns of hundreds to thousands of square meters. Some of the scale-dependent variability in lake water chemistry is illustrated in a 1993 study of lakes in Scotland. In contrast to the homogenous view of lake chemistry, George (1993) elaborates on large-scale patterns in which land use or the shape and alignment of basins leads to irregular mixing of water chemistry. For example, Loch Lomond exhibits a clear north–south gradient in chlorophyll-*a* despite the lack of clear boundaries between lake basins. Meso-scale patterns also exist in lakes—for example, with respect to plankton patches; micro-scale patterns are evident in the transport of phosphorus in lakes. Overall, George determined that the large- and small-scale variations may result from a cascade of effects in which uneven physical mixing affects water chemistry, which then constrains biology. In meso-scale effects, however, George identifies that the chemical gradients were produced by biological patterns that had been independently generated by physical mixing.

Phosphorus and Chlorophyll-*a*: An Example of Scale Effects

As outlined above, the scale of study, components of variance, and even choice of study unit and analytical technique are all important in water quality and comparative ecology. In practice, managers and scientists often lose their frame of reference when extrapolating patterns across scales. For instance, a strong interlake relationship between lake chlorophyll-*a* and phosphorus concentrations has been extensively documented. These interlake studies however cannot be used to explain intralake variability in chlorophyll-*a* concentrations. Lakes respond to reduced phosphorus loading at different rates, so it is inaccurate to transfer patterns from one scale (interlake) to another (intralake). At an interlake scale, phosphorus concentration correlates to many other factors (e.g., specific conductance, mean depth, catchment geology) that are important to algal growth and add explanatory power to the observed correlation between total phosphorus and chlorophyll-*a* concentrations. Within a lake, however, this covariation is partially masked. Thus, these phenomena are under scale-dependent controls.

A Scale Example: Macrophyte Biomass

The pattern of macrophyte biomass provides another example of how incorporating scales has facilitated an improved understanding of the natural world. Until recently, substantial disagreement existed about factors controlling biomass of submerged macrophytes in lakes. Light and day length appeared to be important variables, as did lake chemistry and the size and structure of lakes. Rather than a jumble of unpredictable influences, these factors can be seen as scale-dependent processes that operate at different hierarchical levels. At the largest scale, a great deal of variability in submerged macrophyte biomass worldwide can be attributed to differences in latitudinal position of the lakes. Models based on latitudinal differences alone have little power to explain regional differences in macrophyte biomass, however. Instead, within the constraining latitudinal scale control on day length and temperature, and regional patterns of geology are apparent. Once inside a regional context—e.g., at a lake-by-lake scale—macrophyte biomass is related to local differences in littoral (shore) morphometry. It should be stressed that these scales are not alone in operating on macrophyte distribution: they are simply the factors that have been identified by a researcher's scale of interest. Land use patterns and resultant differences in lake chemistry and temperature are known to influence macrophyte growth, and introduced predator species may play a role as well. These processes may or may not fit into the scale hierarchy presented above.

Studies of stream biota have found similar hierarchical patterns constraining and regulating growth of macrophytes. For example, Corkum (1992) identifies biomes as the largest level of organization. She also states that "macroinvertebrate community composition may be a function of both longitudinal gradients and lateral (vegetation away from the river channel) dimensions" [p. 112], thus identifying within-biome scales of concern. The complex of spatial scales influencing stream macrophyte growth is also acted upon by temporal dimensions. For example, in a related study, Richards and Minshall (1992) evaluated the effects of wildfires on stream macrophyte and found that distinct differences persisted for as long as 5 years following a disturbance. Recognition of the influence held by these interrelated spatial and temporal patterns makes the evaluation of water quality impacts even more complex. As disturbances, water quality impacts have their own spatial and temporal dimensions, but operate within preexisting spatial and temporal dynamics.

Highlight: Groundwater Quality, Scales, and Management

Managers commonly turn to models to help them incorporate the many factors influencing management decisions. In groundwater modeling, scales are determined by characteristics of the groundwater system, by availability of data, and by management goals. Thus, groundwater modeling is a multidimensional process involving natural and human-induced influences (e.g., climatic effects, pumping, deep-well injection, agricultural irrigation, and drainage).

Anthropogenic influences generally affect local- and intermediate-scale groundwater quality; large, regional-scale phenomena are generally of natural origin. Some anthropogenic quality changes also vary at regional scale, such as: (1) groundwater withdrawal for irrigation; (2) nonpoint pollution caused by fertilizer, herbicide, and pesticide use in agriculture; (3) acidification of groundwater from acid precipitation; and (4) quality changes resulting from urbanization.

Modelers may incorporate a variety of hierarchies in their models. For example, using administrative elements such as townships, counties, states, and river basins may prove the best approach for directing optimal management choices. Such an approach does not follow natural boundaries and elements, so chemical and biological details will often be lost in an administrative hierarchy.

Another modeling option incorporates scale-dependent biophysical variables. For example, industrial and municipal water supply wellfields may influence an area of less than 1 km^2 to more than 10 km^2. Private, single-household wells have a small area of influence (often less than 100 m^2) but the combined drawdown of many such wells can cause serious aquifer depletion, especially if operated in tandem with agricultural use. Point source contamination such as spills, leaching, and tank failure often occurs at local scale (100–1000 m^2). Unmanaged, however, the area affected by a spill can become quite extensive (1000–10,000 m^2).

Information must also be organized according to temporal scales. Incidental local situations (e.g., construction site dewatering, chemical spills) usually exist on scales of weeks to months. Seasonal effects are related to agricultural uses and use of aquifers as thermal energy sources or storage. Mid- to long-term scales (1 to 20 years) apply to many wellfield operations, dewatering of mining sites, and local pollution problems. Longer-term effects (20 to 100 years) hold special interest in regional water resource development, hazardous waste displacement, and regional nonpoint pollution. Historical periods (100 to millions of years) have less immediate use to most managers. Nevertheless, groundwater aging may be used to help determine recharge potential of a given resource.

In practice, modelers must integrate as many of these scales as practical to achieve a model that has geophysical, biological, and administrative validity. For example, a model of the South Platter River in Colorado was used both to evaluate effects of policies and legislation on water level and quality and to allocate irrigation water according to availability and water rights. Achieving the second goal (i.e., water allocation) requires daily simulations for surface water and a weekly simulation for groundwater located further away from the river. The scale for the policy model and the model used in evaluating new legislation (e.g., formulation of new water distribution rules) is more coarse because long-term effects are the primary focus. A weekly simulation is used for surface water and a monthly one for groundwater.

Summary

- The scales at which we pose our research and management questions constrain our answers. Water quality has long been constrained by the view that relevant issues were site-specific and operated at the "less than 1 year" time scale. We are now recognizing that many scales are involved and that these scales must be explicitly recognized in any water quality management strategy.

- Hierarchy theory provides a framework that allows us to examine relationships among processes operating at different scales. It suggests that we must carefully consider the scale at which a parameter is controlled and the associated variables controlled by a given parameter.
- This chapter presents several examples of water quality variables that operate at specific scales and that are controlled by, and/or control, variables at other scales. It also discusses the management implications of those relationships.

For Further Reading

Fox J. The problem of scale in community resource management. *Eviron Mgt* 1992; 16:289–297.

George DG. Physical and chemical scales of pattern in freshwater lakes and reservoirs. *Science of the Total Environment* 1993;135:1–15.

Minshall GW. Stream ecosystem theory: a global perspective. *J N Am Benth Soc* 1988; 7:263–288.

Wiens JA. Spatial scaling in ecology. *Functional Ecology* 1989;3:385–397.

8

Water and the Hydrologic Cycle

Overview

Water's unique and vital role in nature stems from its wide range of unusual physical properties that define ways in which water can be used, affected, and treated. In water quality management, some of the most important of these properties include the following:

1. The wide range of temperatures at which water remains liquid.
2. Unusually high boiling and freezing points.
3. Attainment of maximum density at 4°C, and a decrease of density as water cools between 4°C and 0°C.
4. An ability to hold a relatively constant temperature, thus buffering temperature changes (i.e., its high specific heat).
5. Low viscosity that decreases with increasing pressure.

Other critical properties of water derive from its distinctive chemical structure and its unique, high density. The hydrogen bonds themselves make water molecules "dipolar" (slightly positive at one end and slightly negative at the other), which enables water to dissolve more substances than any other liquid on earth.

These properties define possible human uses of water and explain why water can be found as a gas in the atmosphere, as a liquid both above and below ground, and as a solid wherever the temperature falls below 0°C. Water is cycled among these water "reservoirs" in a process called the hydrologic cycle: water is evaporated from surface waters or transpired from plant surfaces, released into the atmosphere, carried by wind, and returned to the earth as precipitation.

Globally, the amount of water returned to the atmosphere equals the amount that falls to earth as precipitation. Nevertheless, significant variations in temporal and spatial distribution of evaporation and precipitation result in occurrence of xeric (dry) and mesic (wet) areas as well as floods, droughts, and other climatic extremes.

Introduction: Water as a Unique Substance

Water's unique physical properties help to define its range of human uses. For its molecular size and weight, water has anomalous thermal properties, densities, viscosity, and electrical conductivity, and is an unusually effective solvent (Table 8.1). Most water properties are taken for granted, but each is remarkable for defying a different set of chemical rules. Without these anomalous properties, water could not be supplied through pipes, would not exist in solid, liquid, and gaseous forms, and would not be continually cycled from the atmosphere to the earth and back via precipitation and evaporation (e.g., no hydrologic cycle would occur). Some of these anomalous and critical properties of water are described below, followed by a review of the hydrologic cycle and its relationship to water quality management.

Table 8.1 Water is a unique substance with many specific properties that have enabled life to evolve to its present form on earth.

Property	Value
Freezing point	0°C
Density of ice at 0°C	0.92 g/cm^3
Density of liquid water at 0°C	1.00 g/cm^3
Heat of fusion	80 cal/g
Boiling point	100°C
Heat of vaporization	540 cal/g
Critical temperature	347°C
Critical pressure	217 atm
Electric conductivity (25°C)	4×10^{-6} siemens/m
Dielectric constant	78.5
Surface tension (25°C)	71.97
Sound velocity	1496.3 m/sec
Specific heat	1 cal/g/°C
Viscosity (20°C)	1 centipole (= 1 millipascal/sec)
Molecular weight	18

Thermal Properties

Typically, a liquid's boiling and freezing points are relatively close together. Substances with higher molecular weights generally have higher freezing and boiling points. Water is unusual in this regard because it has an extremely high boiling point and a relatively low freezing point considering its low molecular weight (Fig. 8.1). If it followed the pattern of other substances, water would boil at −91°C and freeze at −100°C (Lyklema and van Hylckama 1979). The high boiling point reflects the fact that water molecules are bound with strong attractive forces, higher than in any other liquid. Thus, dissociating water molecules is difficult, and water occurs as a liquid over a wide range of temperatures. In turn, water can support an extraordinary diversity of life, and most of the earth's surface remains covered in water in its liquid form.

One of water's most important physical properties becomes evident as water cools toward its freezing point. Like other liquids, water becomes more dense as it cools. At the critical temperature of 4°C, however, water molecules

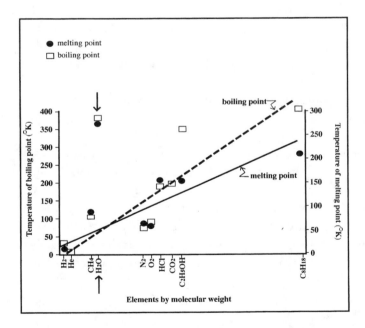

Fig. 8.1 Among the many unique chemical and physical properties of water are notably higher melting and boiling points than those of other compounds with a comparable molecular weight. For example, water has a molecular weight of 18 but its boiling and freezing points are most similar to those of octane (C_8H_{18}), which has a molecular weight of 114. These unique properties define water's role on earth and the ways that life has evolved to rely on those properties. (Data from Stowe 1987)

change shape and water expands (or becomes less dense) by approximately 11% as it cools to the freezing point. The fact that water is most dense at 4°C, rather than at 0°C, means that ice floats and insulates water below it. Thus, lakes and ponds do not freeze from the bottom up, a fact that holds tremendous significance for life in aquatic systems. The expansion of ice relative to liquid water also accounts for the physical force associated with weathering rocks and bursting pipes.

Another interesting physical property of water is that the temperature at which melting of ice takes place is relatively high, but energy needed to convert ice to liquid once melting has begun is relatively low. Distinct from other liquids, much of the orderly structure in ice is not destroyed upon melting.

Water is also unique because it requires so much energy to heat up a water molecule, but it sustains heat well once that goal is reached. This high specific heat, or heat capacity, is a by-product of water's highly structured molecules and contrasts with elements such as mercury, which has a very low specific heat and so has value as a temperature gauge. High specific heat means that large volumes of water such as oceans or large lakes have relatively constant temperatures throughout the year. Even smaller aquatic habitats are buffered by water's high heat capacity, experiencing ranges of 25°C or less despite air temperature ranges more than twice as broad. In addition to its role as a climatic buffer, high specific heat is also critical in heat-regulating mechanisms of higher organisms, especially warm-blooded animals.

Based on water's strong molecular bonds, water would be expected to have a high viscosity (i.e., it would be "thick" and flow slowly). In fact, it has a low viscosity; the strong molecular coherence does not confer a lack of mobility. Consequently, blood moves easily in bodies and aquatic organisms may move quickly through water at any depth. Another important and distinctive feature is that water's viscosity decreases as pressure increases. Nearly all fluids move more slowly under pressure, but water flows faster (Lyklema and van Hylckama 1979). Consequently, water supply systems may use pressure to distribute large quantities of water over long distances. Without this decreasing viscosity feature, water supply would be a virtual impossibility. This characteristic also enables continuous flow of nutrients (and pollutants) in waters through both surface currents and deep ocean or benthic (i.e., deep) lake currents.

Chemical Properties

The critical physical properties of water are enabled by its deceptively simple atomic structure. The water molecule consists of three atoms—two hydrogens and one oxygen. Hydrogens are attached to oxygen at an angle, much like

Mickey Mouse ears. The two hydrogens are covalently bonded (i.e., each hydrogen shares an electron with the oxygen atom), which gives the molecule overall electric neutrality. The oxygen nucleus has a greater affinity for electrons than the hydrogen nucleus, however, so the oxygen side of the molecule has a slight negative charge and the hydrogen side has a slight positive charge (i.e., molecules are "dipolar"). This polarity causes hydrogen of one molecule to be attracted to oxygen of another, bonding through a "hydrogen bond." The angles at which oxygen and hydrogen are held in the water molecule (105°) and the hydrogen bonding scheme produce a tetrahedral, crystal structure for ice. Moss (1988) notes that other substances typically have 12 neighbors in their crystals, making the solid more dense. The fact that ice has only four neighbors explains its low density, its ability to float on liquid water, and its insulating properties.

WATER AS A SOLVENT. A central theme of water quality management science is that practices can alter water chemistry and, therefore, change the potential uses of water resource as well as altering environmental conditions for the aquatic and shore-line communities. This altered chemistry stems from the dipolarity of water molecules and water's resultant ability to attract and dissolve charged molecules. Water is known as the "universal solvent"—more compounds will dissolve in water than in any other known substance. Water can only dissolve ionic (charged) crystals such as salts, however; it cannot dissolve chemically neutral substances such as fats or oils. (One reason that oil spills are an unusual source of pollution is that water molecules cannot dissolve the ionically neutral oil molecules.) For neutral substances to dissolve, it is necessary to use detergents, which essentially polarize the neutral compounds so that water can dissociate them. Many organic compounds contain slightly charged (polar) groups within their chains, so they demonstrate some solubility in water. Most atmospheric gases are nonpolar, so their solubilities in water are low. Their solubilities vary with temperature, pressure, and their concentrations in the atmosphere.

The degree to which compounds will dissolve in water depends on their charge and the size of the ion. The attraction is greater for highly charged molecules, but decreases as ionic size increases (Moss 1988). Highly charged, moderately sized, compounds such as sodium, potassium, and calcium bind easily with water molecules and are brought into solution quickly. Other compounds (e.g., aluminum and silica) have high charges, but small ionic radii. In these compounds, the charge attracts the oxygen atom so closely that a hydrogen bond is weakened and a hydrogen ion ejected into solution. That ion then forms a metal hydroxide with a charge insufficient to attract a water molecule.

Consequently, hydroxides are precipitated. Iron and manganese may behave in either way. Still other compounds such as nitrate, carbonate, and sulfate carry very strong charges, attracting the oxygen atom. With these compounds, both hydrogens are ejected into solution and will form hydrated hydrogen ions that remain in solution (Moss 1988).

Inorganic composition of natural water reflects the differential solubility of elements, as well as availability in the soil of elements for potential solution. It also reflects interaction among ions themselves. For example, calcium and carbonate are each highly soluble without the presence of one another, but readily precipitate as $CaCO_3$ if mixed. Typically, natural waters contain Na^+, K^+, Mg^{++}, Ca^{++}, HCO_3^-, SO_4^-, and Cl^- (Moss 1988) (Table 8.2).

Table 8.2 A variety of constituents are found in most water bodies. Some of these materials are dissolved and are used by plants and animals in growth. Others are suspended in the water column. Human impacts often cause these constituents to fall outside their natural ranges, impairing future uses of the water body.

Constituent	Source	Typical Range	Significance
Silica (SiO_2)	Dissolved from parent material	5–30 mg/L	Required for diatom growth; may cause industrial scale if over 300 mg/L
Sulfate (SO_4)	Parent material industry acid deposition	5–150 mg/L	Associated with H^+; may cause acidification; causes diarrhea if >250 mg/L
Oxygen (O_2)	Atmosphere, photosynthesis	5–15 mg/L; 0–20 in impacted conditions	Required ≥4 mg/L for respiration of most animals
Hardness ($CaCO_3$)	Parent material; refers to Ca and Mg or cation component	25–300 mg/L	Higher hardness mitigates impacts of heavy metals; at high levels precipitates out to form carbonate scale
Alkalinity ($CaCO_3$)	Atmosphere, parent material; refers to CO_3 or anion component	5–150 mg/L	Higher alkalinity buffers acidification changes; associated with increased primary productivity
Chloride (Cl)	Parent material Cl_2 from waste treatment	0.1–15 mg/L	Cl ion dangerous to water supply if >150 mg/L; Cl_2 molecule highly toxic if >0.5 mg/L
Suspended material	Erosion, stream bed	5–25 mg/L in streams; <5 mg/L in lakes	Obstructs light, settles and smothers benthic plants and animals
Dissolved solids	Parent material, erosion	150–300 mg/L	May deposit to form scale if in excessive amounts

Hydrologic Cycle

The hydrologic cycle is the central concept in hydrology. It describes how water evaporates directly from surface water or transpires from plant surfaces, is transported to the atmosphere, and returns to earth as precipitation. At its most general, the hydrologic cycle outlines the interrelatedness of all water sources, as well as continuous movements of water through vapor, liquid, and solid phases (Fig. 8.2). Thus, on a global scale, water is evaporated from oceans, rivers, and lakes, returned to the atmosphere, and then "rained" down on earth at some distant point. At smaller temporal and spatial scales, however, the movement of water in the hydrologic cycle is more complex, with water moving at different rates through the different water "reservoirs" of atmosphere, ground, and surface waters. How long water remains in each reservoir (i.e., its residence time) and the spatial extent of a given reservoir both have important implications for regional and local water cycles and water quality.

The components of the hydrologic cycle are: precipitation, evaporation, transpiration, infiltration, percolation, and run-off. Each of these elements, and their importance in water quality management, is described below.

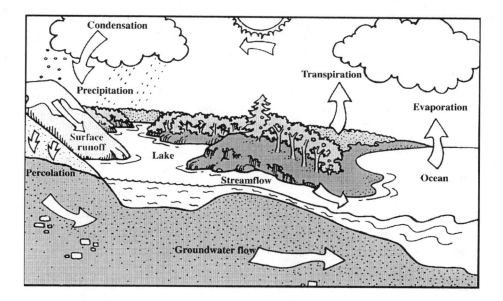

Fig. 8.2 The hydrologic cycle includes condensation from the atmosphere, run-off to surface and groundwaters, transit (with varying storage times) to the oceans, and evaporation back up to the atmosphere. (After Environment Canada 1990).

Precipitation

The earth's surface receives atmospheric water in the form of rain, dew, snow, frost, and hail. Precipitation plays a dual role in water quality management. In some cases, as with acid precipitation, it plays a direct role: chemical reactions in the atmosphere turn sulfur emissions into acids and precipitation causes the acids to "rain down" on land and water surfaces. More commonly, precipitation plays an indirect but critical role in determining distribution of arid and moist areas, and in determining the intensity of floods and droughts. Consequently, despite the global average value of 1000 mm of precipitation per year over land surfaces (Shiklomanov 1993), water quality managers are typically more concerned with regional and local variations from that average.

Evaporation

Through evaporation, a liquid is changed into a gas. Globally, the most important form of evaporation takes place from seas and oceans. From a water quality perspective, however, evaporation is critical because it concentrates pollutants in a body of water. For example, a common water quality problem worldwide is salinization of soils and surface aquifers from irrigated agriculture. This process occurs as excessive volumes of water are evaporated from the soil surface and the natural salts are left behind. Through time, these salts become sufficiently concentrated to lower agricultural productivity. While approximately eight times as much water is evaporated annually from oceans as from land, the ocean's volume ensures dilution of pollutants (Shiklomanov 1993). In contrast, in regions with high evaporative demand, evaporation can effectively concentrate salts, toxins, and other compounds, rendering water unhealthy and lowering agricultural productivity.

Transpiration

Transpiration is the process by which water vapor passes through a living plant and enters the atmosphere. It is an important part of the hydrologic cycle but is not a significant consideration in water quality management.

Infiltration

After water lands on soil surfaces, it infiltrates or soaks into surface soil layers. Plant roots take up some infiltrating water and eventually release it into the atmosphere through transpiration. In water quality management, infiltration represents an important means by which some pollutants are carried into groundwaters and others are "filtered out" and bonded to soil surfaces. Pollutants that reach the groundwater system typically persist for long periods and are distributed slowly over large distances.

Percolation

Percolation is the downward flow of water through soil and permeable rock formations to the water table. It occurs much more quickly than infiltration and thus is much more important for transmitting surface pollutants to groundwater.

Run-off

Run-off is the portion of precipitation moved by gravity in surface channels or depressions. It is a residual quantity, representing excess of precipitation over evapotranspiration. Run-off is an important consideration in water quality management because it carries both dissolved and suspended particles from land. Everything from soil particles to heavy metals, pesticides, and herbicides are carried in run-off. Approximately 47,000 km³ of water per year (35% of all precipitation) is returned to oceans through run-off (Shiklomanov 1993).

Reservoirs and Residence Time

Atmospheric water, surface water, and groundwater are all part of the hydrologic cycle, but each of these different reservoirs has a unique potential for water quality degradation and improvement. A large part of that potential relates to how quickly or slowly water is replenished in the reservoir (i.e., residence time). The world's oceans represent the reservoirs with the largest residence time and hence, the smallest potential for degradation in the short run but the greatest risk for long-term degradation. Water may remain for hundreds of years in the ocean and travel for hundreds to thousands of miles before being recycled to atmospheric water and distributed elsewhere. The oceans' sheer volume makes them resistant to pollution impacts (they dilute inputs to a nondetectable level). Long residence time of oceanic waters means, however, that any pollutants and their effects persist for long periods.

In contrast, water in a lake has a shorter residence time and smaller spatial distribution. It is more susceptible to pollution because the diluting volume of water is lower. At the same time, the lake will recover relatively quickly from any effects (once the source of the effect is removed) as the lake water is renewed. Similarly, atmospheric water is cycled back to the surface relatively quickly, so if a pollution source is removed, atmospheric water quality will recover quickly. Because it travels large distances, atmospheric water diffuses some pollutants and redistributes others.

Atmospheric Water

On average, atmospheric water has a residence time of approximately 10 days. While that residence time is among the shortest of all hydrologic reservoirs, atmospheric water is also carried the farthest distance during that short time: air masses can be hundreds of kilometers wide and can carry water at speeds exceeding 100 km/sec. Thus, water-borne pollutants may not persist long, but they may be carried to areas far from the source.

Atmospheric water, as the water available for precipitation, is also vitally important in water quality through its control over arid and mesic areas, floods, and droughts. Lastly, global atmospheric water has vital implications for regulating the earth's heat balance. In particular, short-wave radiation from the sun passes through clouds and heats the earth's surface (Fig. 8.3). Heat leaves the earth's surface as long-wave radiation, however, and in that form clouds absorb heat and reradiate most of it back to earth. Global warming predictions have recently concentrated attention on the critical role of clouds in the global climate cycle (see chapter 23).

Hydrospheric (Surface) Water

The hydrosphere encompasses all surface waters, from glaciers and oceans to lakes and streams. While nearly 80% of the earth's surface is covered with water, in chapter 3 it was noted that only a minute percentage of that water is available for human use. There is general agreement that available fresh water represents less than 0.3% of global fresh water reserves. Total water "available" to humans for uses like hydropower, fishing, and drinking is only about 250,000 km^3 (Shiklomanov 1993).

Residence time of hydrospheric water varies tremendously. For Antarctic ice, it may be as high as 10,000 years; in contrast, residence time in a stream or wetland may be as short as hours or days (Speidel and Agnew 1982). Among available sources of water, average residence times (renewal rates) vary from approximately 11 months for fresh water lakes to 7 to 17 days for rivers and

Fig. 8.3 Clouds play a critical role in the hydrologic cycle, as well as in controlling the temperature of the earth's surface. (Modified from color photo by Ahrens 1994; reprinted with permission)

streams. A longer residence time connotes greater potential for a pollutant to persist and concentrate in any one area. Nonetheless, significant pollution can occur in reservoirs with short residence times if the source of pollution is continuous. For example, a small stream may have a residence time of only 2 days. If a textile mill uses the full stream water volume for dye washes daily, the stream will not be clear of that pollution until 2 days after the mill stops operation. Even then, pollutants trapped in the sediment probably mean that pollution will persist. Impacts of a contaminant or pollution source must be judged in the context of the residence time of receiving waters.

Lithospheric (Ground) Water

The lithosphere consists of water in soil and rocks under the soil. It is the largest reservoir of fresh water available for human consumption, although exact measurements of its volume are elusive. Estimates of groundwater volume (water in the saturated zone) range from 7 million km^3 to 330 million km^3 (Speidel and Agnew 1982); 8 million km^3 to 10 million km^3 appears to be a common estimate of fresh groundwater reserves (Shiklomanov 1993). Long considered an inherently safe water source, groundwater is now recognized to be highly vulnerable. Groundwater moves extremely slowly, often on scales of tens of meters per year. Consequently, groundwater is renewed very slowly and is subject to build-up of contaminants. Residence times of groundwater discharges to the sea are approximately 4000 years; the potential for pollutants to accumulate in this water is extremely high.

Spatial Distribution of Water

Precipitation is distributed unevenly over the globe, with large-scale patterns of rainfall determined by global circulation patterns such as trade winds and jet streams. Trade winds, for example, move from east to west on either side of the equator. In between them lie doldrums in which air is rising and wind velocities are low. Between 30° and 70° N or S, winds generally blow from west to east (westerlies), moving atmospheric water toward the poles. These large-scale patterns may be obscured, however, by wind and rain of local origin.

Topography and local wind patterns frequently determine small-scale patterns of precipitation. For example, in the western United States, air is forced upward as it crosses coastal mountain ranges and the Rockies. As the air rises, it cools and can hold less moisture. A band of increased precipitation results on the western edge of both mountain ranges. On the eastern side, "rain shadows" with areas of deficient moisture become evident. Local differences in evapotranspiration are also critical in determining water availability. Seasonal precipitation distribution and air temperature patterns, in part, determine evapotranspiration. Thus, if precipitation occurs largely during winter, as in the western U.S. mountains, much of the precipitation becomes run-off or infiltrates into the soil. Summer precipitation, however, occurs when evapotranspiration is at a maximum. Much of this precipitation returns to the atmosphere without significantly affecting soil reserves.

Aridity Indices

A region's aridity is a function of not only how much rain falls, but also the timing of rains. Even an area that receives significant rainfall can have important seasonal water deficits. For example, 80% of Asia's water run-off occurs over a period lasting only 6 months. In Australia, 70% of run-off occurs in just 3 months (Shiklomanov 1993). The remaining months can be marked by severe water shortages. Water budgets and water indices (e.g., aridity indices) are two tools used to gain a more accurate view of water supply in each region or country (Fig. 8.4).

As described by Shiklomanov (1993), indices may be based on total precipitation (e.g., total km^3 per season or per year), run-off per unit area (e.g., km^3 per year per km^2), or per unit population (e.g., m^3 per year per capita). Each in-

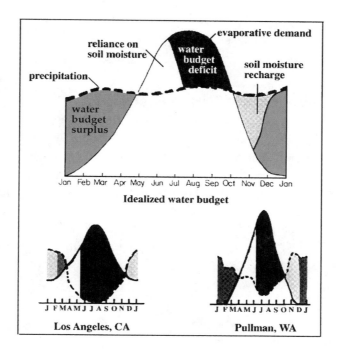

Fig. 8.4 Water budgets provide a graphic depiction of the distribution of moisture throughout the year in any given area. An idealized water budget includes periods of deficit, surplus, soil recharge, and soil depletion. The idealized budget is rarely achieved, however. Examples from Los Angeles and Pullman, Washington, illustrate the range of potential deviation from the "ideal" water budget. In Los Angeles, the period of water deficit is long under the influence of high summer evaporative demand and low precipitation. The period of water surplus is barely 1 month long. In Pullman, summer evaporative demand is high, but precipitation is more uniformly distributed throughout the year, with adequate recharge in the fall and a 3–4 month period of water surplus.

dex yields a very different picture of water availability. Based on absolute volume, Brazil may be the country with the largest water reserves in the world. Its total run-off constitutes more than 20% of all global renewable water resources—nearly twice as much as the former Soviet Union and nearly four times more than China or Canada. Total run-off is, however, a less useful measure in water quality management than measures such as regional or per capita availability, which reflect demand. Using these indices, Norway may be the most water-rich nation, with the world's highest run-off per unit area (1,250,000 m³ per year per km²) and highest water availability (98,000 m³ per year per capita).

Population density plays an enormous part in regional and per capita water availability. China, for example, has the third largest percentage of global run-off, yet run-off per unit area is significantly lower per capita; the country's water availability is, in fact, only 2.5% of that of the most water-rich countries. A rapid population increase in much of the developing world is projected to put enormous pressures on available water, making total run-off less important than run-off per capita (see chapter 25).

General Implications for Water Quality

This chapter has given an overview of the properties of water and their relevance in water quality management. This chapter and the preceding one on scale issues also provide an introduction to the following chapters, which focus on the world's hydrospheric and lithospheric reservoirs: rivers and streams, lakes, groundwater, coastal zones, and wetlands. While these subsequent chapters explain the ecology and water quality concerns of each different resource, the themes outlined here are evident throughout all spheres of water quality management. Specifically:

1. The potential for water quality degradation and subsequent recovery is related to the residence time of water in a given reservoir, as well as to the volume of water in any given body of water.
2. Physical, chemical, and biological processes, including water cycling, occur at different spatial and temporal scales, so the scale of interest must be matched with the scale of mechanistic control.
3. Managing for an average condition (e.g., of rainfall or flow rate) means managing for the condition that never exists; a tenfold variation from year to year is common in many parts of the world. Management must take a broad enough view to include the range of possible conditions.
4. Just as all water and its inputs are in constant flux, so are the values that define water use, quality concerns, and policy directives.

Summary

- Water is a unique substance. Its thermal properties, physical properties such as cohesion and specific heat, and chemical properties such as dipolarity result in a compound that allows, and controls, life on earth.
- Water passes through a hydrologic cycle, which includes the following phases:
    Precipitation, which varies widely around the globe and among years, but is relatively predictable at the scale of regions and decades.
    Evapotranspiration, which is driven by solar energy and available water, again relatively predictable only at the scale of regions and decades.

Infiltration and run-off are processes driven by gravity and controlled by vegetation, soils and geology. They are predictable on a finer scale, such as watersheds and subdecades.

Storage includes surface water (a transient and variable resource) and groundwater (a resource that, in most cases, changes very slowly and over large spatial areas).

• We do not manage the hydrologic cycle, but our management is controlled by it. Indices allow us to express spatial and temporal patterns that can be used to develop and generalize management strategies.

9

Rivers and Streams: One-Way Flow Systems

Overview

Although rivers constitute only a minute proportion of global fresh water reserves, they are unusual from a water quality perspective because so much of their water is actually used. Rivers have long attracted civilizations because of their productive banks, the power-generating abilities of moving water, and their ability to carry away wastes. Their one-way flow, linear form, fluctuating discharge, and unstable channel and bed morphology distinguish rivers from other aquatic environments. An important result of the one-way flow is that rivers are very open ecosystems, receiving more resources from "upstream" than they recycle internally. Another significant element is the fact that organisms receive an energy subsidy from the current. If these organisms can hold still against the current, their food will be delivered and their wastes removed. Consequently, the biotic community gains only a temporary hold on nutrients. Rivers are also characterized by a high proportion of inputs from the watershed; they have relatively high inputs of allochthonous (terrestrial) material and relatively little autochthonous (in-stream) production.

The narrow and linear form of rivers results in very close links with the surrounding terrestrial ecosystem, closer than those formed for a lake or ocean. Thus, stream reaches reflect closely the type and degree of land use in their watershed. River water quality may be degraded through any of a variety of impacts, including erosion, control of natural erosion, inputs of chemicals or energy, and changes in surface flow or groundwater quality. These impacts are then transferred downstream. Because river waters are renewed relatively frequently, however, the potential for recovery at each point is great.

Introduction

Water quality in rivers and streams is critically important. Rivers have provided the basis for cities, towns, and entire civilizations for millennia. At the same time, changes to those same rivers, waterway pollution, and collapse of river functions have led to the demise of many river-dependent communities, species extinction, and both permanent and temporary changes in biological assemblages. Understanding and managing rivers is complicated by the uniqueness of each stream reach. Differences in geology between different rivers and within the same river lead to different water quality impacts; differences in the flow speed, scouring, stream temperature, nutrients, chemistry, sediment load, and size all influence the degree to which human activity affects a river or stream, and constrain the ways in which those effects can be reduced.

River ecologists have developed a variety of classification and descriptive schemes with which to define the various river forms. These approaches reflect their developers' focus—for example, geologists classify streams by their relative age in geologic time, biologists by their productivity, and engineers by their physical structure (e.g., as it relates to the need for flood control measures). Just as no ideal river exists, no ideal classification or descriptive approach has been determined. The water quality manager is interested in how humans modify water chemistry, flow, and biology and in features such as resilience, assimilative capacity, stress response, and corrective measures. Consequently, both the water quality manager and scientist must understand rivers from all perspectives: flow characteristics, chemistry, biological communities and food webs, and temporal and spatial changes in rivers. Despite the relevance of these parameters to water quality, this chapter cannot provide an exhaustive review of river ecology, chemistry, and flow characteristics. Instead, it gives an overview of features most salient to water quality for natural resource managers. More extensive references are included at the end of the chapter for those wishing to pursue this topic further.

Streams and Rivers on Earth

The amount of fresh water in streams and rivers has been estimated at only 0.006% of global fresh water reserves (Shiklomanov 1993). Even within this tiny fraction, enormous global variability in the distribution of streams and rivers creates a patchwork of irregular river conditions and management considerations. For example, the Amazon alone drains one-sixth of the globe's run-off. The Amazon's run-off is more than 10 times that of the Mississippi and almost five times as much as the Congo. The Nile is the longest river in the world, outstripping the Amazon by several hundred miles, but has a flow more than 200 times smaller than that of the Amazon (Gleick 1993).

From a water quality perspective, viewing water distribution in terms of water availability may prove more useful than measures of total run-off. Availability and use confer development-related stresses on these resources. Availability may be evaluated in many ways. One common computation looks at total run-off by area. This measurement may be calculated on a coarse scale (e.g., over continental areas) or on a finer scale (e.g., within continents or within regions of a country). For example, continental "specific run-off" shows that Europe, Asia, and North America all discharge approximately 10 liters/second/square kilometers (L/s/km^2). Africa's specific discharge is only half this amount, however, while South America's is twice as high. Australia, a desert-dominated continent, has only one-tenth of the run-off of Europe, while Oceania's run-off is five times greater (Shiklomanov 1993).

When computed on a finer scale, it becomes evident that these coarse-scale differences may have little relevance for water quality management. For example, although North America as a whole has a specific discharge of 10.7 L/s/km^2, Canada, the United States, and Mexico all experience very different degrees of water use and need. Figured on an area basis, the United States and Canada have similar mean annual river run-offs (248,000 m^3/km^2 for Canada and 207,000 m^3/km^2 for the U.S.). On a per capita basis, however, Canada has more than 10 times the water availability of the United States (Shiklomanov 1993). At increasingly smaller foci, the differences between regions within a country and specific sites within a region yield a patchwork of river patterns with much greater relevance for local management.

As in all hierarchical systems, the regional and continental variability in river run-off constrains water resources management and defines an overarching set of water quality concerns in different regions. Below this level, a myriad of other biophysical, social, and demographic factors control the potential uses of different river resources and the sensitivity of a given river to water quality degradation and recovery. To name but a few examples: topography defines conditions of flow rate and substrate that regulate the biological community of streams and utility for different purposes; substrate geology may regulate river water chemistry; the distribution of populations along a river reach defines the intensity of use. At each scale, a river's use must be consistent with its ecologic potential. The remainder of this chapter is devoted to a discussion of the conditions and controls of river ecology that are of concern to water quality managers. Before discussing river ecology per se, the following example from India is offered as a reminder that river use and water quality impacts occur in a human context, and that both water quality impacts and their solutions vary according to patterns of use and regulation.

The Yamuna River, which flows through Delhi, receives untreated and only partially treated wastes from more than 7 million residents and their factories, changing the river from a class B (clean) river to a class E (highly polluted) river in a matter of miles (Gopal and Sah 1993). Managing the water quality problems of the Yamuna is a problem of sewage and water supply infrastructure, education, and poor regulative capacity: India's Water Prevention and Control of Pollution Act of 1974, which regulates effluent standards, does not cover the municipal bodies responsible for discharge of domestic sewage (Gopal and Sah 1993). Mountainous stretches of the same river, however, are not subject to such intense use, nor does the degradation there indicate an infrastructure problem. In the mountains, the Yamuna suffers from soil erosion and siltation resulting from deforestation and intensive livestock grazing. Managing water quality in the upper basin, therefore, requires intensive education of individuals, changes in cropping and grazing patterns, and promotion of alternative fuels and improved forestry practices. Judging water quality at each stream reach requires an understanding of the baseline river ecology; designing management plans, on the other hand, requires building on this knowledge and looking at the social structures that influence land use.

The Ideal River: A Useful Framework That Does Not Exist

There is no idealized river. Nevertheless, imagining an idealized river serves as a useful point of departure for discussing river flow and ecology. As described by Moss (1991), the ideal stream

> . . . lies in uniform geological terrain; it is most erosive in its head waters where the rocks and boulders left from a previous spate create an uneven or turbulent flow over a rough stream bottom. As the downward erosion of the stream bed continues, however, the slope of the bed is progressively reduced, and the stream's ability to erode its bed declines. The overall slope of the bed flattens until it reaches, over a sufficient distance, the horizontal when the stream (then usually called a river, but there is no absolute distinction) meets the sea. [p. 61]

Along the way, the catchment area of the stream increases, as does the water supply to the stream/river. In addition, the water movement becomes less turbulent and fine sediment begins to be deposited among rocks. At the lower reaches, the bed becomes dominated by fine sediment, with bare stones existing only in the center of the channel where flow is swiftest and most turbulent. In streams with a highly permeable substrate, a zone called the hyporheos, pictured in Fig. 9.1, interacts with the water column to supply energy, nutrients, and propagules. Streams underlain by impermeable geology such as basalt or sandstone have a minimal or nonexistent hyporheic zone. In the lowest reaches, the channel may widen over a flood plain as the amount of normal flow increases. Such a channel moves from time to time as the river erodes soil from outside its bends and deposits it on the insides (meanders).

This smoothly grading, continuously widening, progressively less turbulent model of a river is, in practice, continuously interrupted. Geological variation alters the erodibility of the river bed, land forms such as basins or uplifted areas alter the model, and human management such as channelization and dams alters the idealized progression. In addition, the distance from a headwater stream to a flood plain river may take effect over hundreds or thousands of

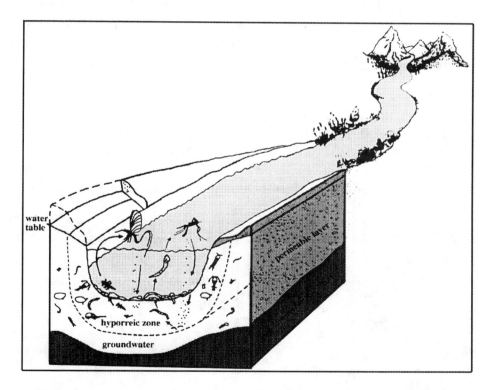

Fig. 9.1 The hyporheos is a zone of water and substrate underlying a river bed. Significant fluxes of matter and energy often occur between the hyporheos and the benthos. (Redrawn from Williams 1994)

miles, or in an area of high or low rainfall. As illustrated in Fig. 9.2, each geographic and structural difference creates a unique river. In turn, each different river has a uniquely structured series of biological communities and environmental conditions. Bearing in mind this cautionary precept, the discussion that follows concentrates on commonalities between river and stream systems, rather than their differences.

Ecology of Upland Streams

The ecology of streams and rivers is subject to the influence of flow, turbulence, light, and temperature regimes. For example, the fast-moving water in upper stream reaches limits the ability of streams to support phytoplankton growth. Instead, these systems are typically colonized by small, attached plants such as liverworts and mosses, which in turn house a community of small invertebrates such as mites, nematodes, and certain larvae. In the slower-running portions of streams and rivers with fine sediment build-up, conditions might support increased macrophyte growth. The ability of rivers to support different plant communities underscores one of the most important considerations in river ecology: the input of carbon into the system.

Carbon Inputs

Because rivers support little autochthonous (in-stream) production, terrestrial vegetation (allochthonous production) constitutes a critical part of river ecol-

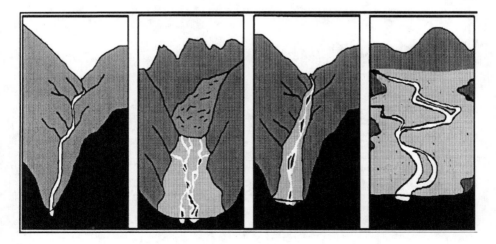

Fig. 9.2 Exemplary river transects show a variety of geomorphic and structural variations that influence the hydrology and ecology of streams and rivers. (Redrawn from Naiman *et al.* 1992)

ogy. In a scenario typical of many rivers, the combination of shading by over-hanging trees and fast water flow prevents algae and moss from growing on rocks. Consequently, internal primary production (P) rates remain low. On the other hand, respiration (R) can be quite high because sufficient carbon is supplied to the system through litter fall that is broken down in-stream. In these streams the ratio of P to R is less than one, indicating that respiration exceeds in situ primary production.

In larger channels, two related events can change the dynamics of stream ecology. First, less vegetation overhangs the stream, allowing for more light and increased growth of mosses and algae. Second, the decreased overhang results in decreased input of terrestrial vegetation. In this model, P:R ratios may be greater than one, as in situ primary production exceeds respiration.

Another important source of carbon into stream systems involves dissolved organic matter (DOM), which enters the streams and rivers through leaching and groundwater. Winterbourn and Townsend (1991) estimate that more than 70% of the annual energy flux in some streams comes from DOM. DOM is an important resource for bacteria and fungi, and consequently it is a vital, albeit more indirect, source of energy for animal consumers higher in the food web.

Early stream nutrient theory saw streams and rivers primarily as nonreactive nutrient conduits (Minshall 1988). Evidence compiled over the past two decades illustrates that organic matter and nutrients are instead continually "spiraled" in the stream. Carbon and nutrients flow to downstream reaches, but along the way they are stopped, metabolized, and then rereleased into the downstream flow (Fig. 9.3). This awareness grew out of an increased appreciation for the role of retention devices within streams. Fallen trees and logs trap leaves and smaller branches, act as sediment traps, and alter flow patterns. These in-stream dams provide a stable substrate that invertebrates can colonize or that can be used as "refuge" by mobile organisms. Thus, these structures serve as a base for the processing and release of nutrients and organic matter. Consequently, nutrient movement is controlled not just by downstream flow, but also by the nature of retention devices and the biotic communities that inhabit them. The term "nutrient cycling" has been applied to this process of in-stream nutrient cycling followed by transport downstream, followed by uptake, cycling, and transfer.

The Breakdown of Carbon Inputs

Invertebrates make up an enormous part of the biotic community of rivers. As a result, significant research has been devoted to their role in the trophic hierarchy, feeding habits, and survival methods in swift-running streams. One of the most useful concepts to emerge from these studies is that of the functional feeding guild, which was first proposed by Cummins in 1973 and discussed in some

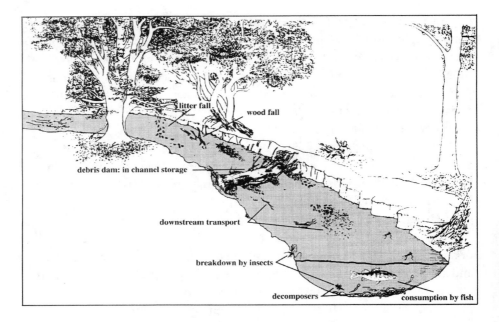

Fig. 9.3 Organic carbon is one of the principal vehicles through which energy is processed in stream systems. Organic carbon enters streams from the riparian zone through a variety of routes such as litter fall and surface run-off and is often held in place by retention devices (e.g., trees or rocks). Once in the stream, the carbon is broken down by insects (see Fig. 9.4) and incorporated into the aquatic food web. The rate of processing is controlled by the ecology of the plants and animals in the stream, which is in turn a function of climate, topography, and geological variables. (Modified from Gregory 1992)

detail in Winterbourn and Townsend (1991) and Moss (1991). Primarily descriptive, these guilds group organisms according to their feeding habits. For example, shredders literally "shred apart" the leaves that fall into upland streams and begin the process of breaking down the terrestrial inputs into a form usable by the aquatic community. The role of functional feeding guilds in the breakdown and processing of this material is uniquely important in stream ecology. This process is described below and presented diagrammatically in Fig. 9.4.

BACTERIA. When a leaf falls into a stream, several processes are already at work. Bacteria on the leaf surface have already begun the process of degrading the leaf. Within the stream, additional colonizing bacteria attach themselves to the leaf and begin the process of "conditioning" the leaf for further consumption.

FUNGI. Following bacterial conditioning, fungi represent the next step in colonizing leaf litter. These free-floating hyphomycetes latch onto the leaf surface and continue the process of leaf conditioning. Importantly, these fungi take in

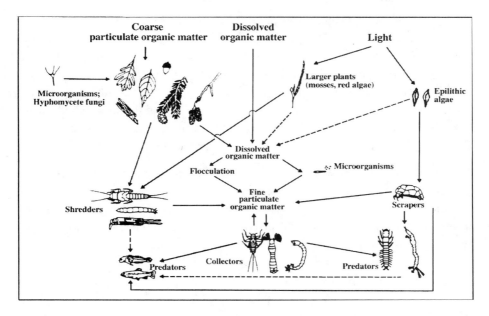

Fig. 9.4 The processing of organic carbon begins with chemical leaching and physical breakdown of litter fall and wood fall. That step is followed by bacterial and fungal colonization, a process that facilitates consumption and breakdown by insects that shred or "scrape" the material into fine particles. Fine particles are readily consumed by other feeding groups of insects such as collectors. Predators complete the processing of carbon by eating the fine particles as well as the insects themselves. (After Moss 1988)

nitrogen from the water column and increase the nitrogen : carbon ratio of the colonized litter. The improved quality litter then becomes an important food source for invertebrate animals such as "shredders." Leaves and large leaf particles are termed coarse-particled organic matter (CPOM).

SHREDDERS. The shredders are often the first invertebrate colonists of leaf litter. Shredders, which may include larvae, nymphs, and Crustacea, bite out the softer parts of the leafs, typically leaving the vascular skeleton behind. Moss (1991) comments that shredders accelerate by a fifth or more the process of degrading and decomposing leaf litter. Their important role is influenced, however, by the type of organic matter available to them. Shredders show a distinct preference for certain species of leaf (e.g., elm over maple, maple over alder or oak), as well as for fungi-colonized leaves over those not yet conditioned.

COLLECTORS, SCRAPERS, AND CARNIVORES. After the fungi and shredders have contributed to the breaking up of leaf litter, the fine material remaining [e.g., abraded material, shredder feces, and other fine particulate organic material (FPOM)] is available for filter, deposit, and scraper feeders. As their names imply, each of these groups employs a different feeding technique. For example, scrapers such as snails and limpets rasp at rock surfaces to extract debris. In contrast, mayfly nymphs and some caddis-fly larvae have stiff bristled mouthparts with which they scour rocks. Filter feeders gather organic matter from flowing matter by trapping them on fine hairs located on either mouthparts or legs, while caddis-fly larvae construct nets between stones and the stream bed (Moss 1991). Although the distinctions between these functional feeding guilds are useful, feeding strategies overlap considerably between members of these guilds.

CARNIVORES. Leaches, insect larvae, and water mites represent the invertebrate carnivores of stream systems. They, in turn, are potential prey for carnivorous fish and other top predators such as birds. Algal grazers, detritus feeders, insect and plant feeders, and zooplankton feeders are found in the reaches dominated by these communities.

Attempts to Find a Unifying Theory

Differences between ecological relationships in different rivers and streams have confounded river ecologists, and led to the search for a unifying theory of river production and food web organization. The river continuum concept is one theory that has attracted significant support. The concept, which was developed through observations of North American streams, is adequate to describe the progression of processing of allochthonous (terrestrial) organic matter in streams. As described by Moss (1991), the river continuum concept sees:

> . . . an orderly sequence in which CPOM is first processed by shredders,
> which provide food for collectors downstream; the life histories of different
> animals in the two groups are seen to be synchronized to take advantage of
> the annual pulses of incoming organic material—the collectors timed slightly
> later than the shredders. In the upstream sections, the ratio of in situ photo-
> synthesis (P) to total community respiration (R) is found to be less than one,
> indicating a dependence on the allochthonous matter to feed the community.
> Further downstream, when the stream widens and is not so darkened by
> overhanging forest, so that the stone surfaces may be colonized by algae and
> mosses, the P : R ratio may be greater than one. At the later, plains stage, when
> the river becomes large and is more likely to collect inorganic silt, which de-
> creases light penetration, the P:R ratio may fall again. [p. 73]

In ecology, the search for a single theory that accounts for processes in all of the diverse ecosystems of the world is often disappointing. The river continuum concept, however appealing and valid in certain areas, is inadequate to explain all global river systems. It has been disproved in New Zealand streams, in particular, where steep slopes and heavy and irregular rainfall scour the stream bottoms. The concept is equally unlikely to apply in many parts of Asia or Africa that experience highly seasonal and torrential rains.

Ecology of Lowland Rivers and Flood Plains

When a river reaches its lowland section several important physicochemical characteristics distinguish it from upland feeder streams. For example, water depth and siltation will increase while turbulence decreases (in particular, at the edge where waters move more slowly). In addition, slower-moving rivers and their channels retain water for longer periods so that plankton may develop in the water. The deeper and more stable the water in these lower reaches, the more the biological communities will resemble a lake.

Submerged plants are one feature that sets "lowland" rivers apart from their faster-moving, shallower counterparts. Plant beds house a complex community—trapping silt and leaf litter, providing a substrate for fish eggs and invertebrates, and hiding prey and predator alike. In addition, they contribute to an increase in organic matter for bacterial decomposition and so may provide a habitat that can contribute to denitrification of waters.

Changes in composition and density of submerged plants can indicate changes in water quality. For example, increased nutrient leaching into a river system (e.g., from agricultural fields) can encourage plant growth in previously marginal areas; increased sedimentation from upstream erosion can either smother plant beds or make rivers shallower and more susceptible to plant growth. Different metal and chemical tolerances can also promote the growth of some species over others, providing early clues to changes in water chemistry and flow.

One of the most important features of lowland streams is their flood plain and the dependence of the stream and flood plain on periodic floods. Much as the river continuum concept outlines a linear, longitudinal descriptor for riverine processes, the flood-pulse concept expands the scale of reference to include lateral and longitudinal dimensions (Johnson *et al.* 1995; Bayley 1995). Annual floods are typically seen with both temperate and tropical streams. In the flood process, aquatic animals migrate onto the newly available flood plain habitats, feeding on the diverse resources and using sites for spawning. Johnson *et al.* (1995) predict that organic matter from upstream areas will be less important to these stream communities than the material produced and consumed in the flood plain. Advocates of the flood-pulse concept suggest that the timing of spawning has evolved to take advantage of the annual pulse, as it provides for greater protection for fry. As the floods recede, nutrients and organic matter form the flood plain flow to the main channel and the flood plain itself receives a new layer of rich sediment.

Taken in combination, the river continuum and flood-pulse concepts begin to explain important processes acting on large rivers. Yet even together the two concepts tend to be more descriptive than mechanistic, and more focused on how physical conditions control biology than on which biological interactions control community structure (Johnson *et al.* 1995). Caution is also required when using these concepts because they describe rivers that have not been modified by dams, levees, and channels. These structures are ubiquitous, however, and have enormous influences on both channel structure and the movement of water through the streams. Flood pulses are often eliminated by dams, and water courses below dams frequently exhibit greater short-term variability than undammed streams (Minshall 1988; Sparks 1995). In addition, dams have increased the height of annual floods so that more catastrophic floods and fewer predictable, replenishing floods occur, again raising issues of aquatic and terrestrial equilibrium and disequilibrium.

Disturbance Regimes and the Organization of Communities

As illustrated by the discussion of annual floods, streams are dynamic resources experiencing regular disturbances. Whether bank erosion, flooding, or landslides, disturbances are an important part of river ecology—albeit a confounding factor for water resource managers. A central question of water quality management was posed by Stanford and Ward (1992): "How much disturbance can occur in a catchment before ecosystem resilience . . . is exceeded and ecosystem structure and function are permanently altered? . . . How much is attributable to natural interannual variation?" [p. 100].

Disturbances can also be large-scale phenomena such as the geologic cycles of glaciation and thawing. Consequently, some streams are younger than others. In particular, Arctic and high-altitude streams are young in geological terms, having only emerged from glaciers within the last 6000 to 10,000 years. Their youth may partially explain the low diversity found in these streams. Another explanation is that the climates in and around these streams remains harsh and fewer species can adapt to these conditions. For the water quality manager, the salient point may not be the true explanation for the low diversity, but simply the existence of latitudinal and altitudinal differences between streams, implying that Arctic and alpine streams may be more easily affected by decreased water quality, and have much slower recovery rates.

Disturbance regimes often underscore the importance of competition, competitive exclusion, and stochastic variations in the structure of communities. For example, in relatively constant conditions, community structure tends to be controlled by competition for niche space. In areas dominated by disturbances, communities may reflect less influence from competitive pressure and stronger influence from a successional stage. In rivers and streams where disturbance is frequent, organisms are washed away and room is made for recolonizing organisms. These communities are likely to have fewer species, with each species occupying larger niches. Recolonization after (spatially) minor, but frequent, disturbances can be extremely rapid, especially if tributaries are relatively unaffected. For example, in a desert stream in Arizona in which 98% of the algae and invertebrate standing crop was washed away in a summer flood, recolonization was rapid and successional patterns were evident over a period of weeks.

Two other types of disturbance effects on biological assemblages play less important roles in stream ecology and stream water quality management. First, "founder" species may gain the upper hand in a newly disturbed area. In these situations, typified by coral reef fishes, no succession occurs because each species is equally capable of colonization. Therefore the species established first becomes dominant. Second, some mobile organisms—notably, those of some lakes and oceans—vary their territory significantly in response to variable conditions (e.g., the vertical migration of plankton in response to large diurnal variations). To some degree this sort of migration may be found in the mouths of rivers as some species of fish migrate in response to salinity intrusion or to power plant thermal discharges. In general, however, this effect is a minor control of community composition in river systems.

Degradation as a Disturbance

Water quality degradation acts upon community composition in rivers in two ways. First, water quality "events" such as toxic spills, sewage overflows, or other one-time-only events may act as a disturbance within the river. Thus, these events may set back the successional stage of a community, may act selectively on one species and change niche relationships, or may induce mobile organisms to move to new areas. In all cases, recovery from these isolated events tends to be rapid as recolonization from upstream sources facilitates a return to previous conditions. The disturbance extent and location will influence recolonization, however. The higher up in a headwater such an event occurs, the longer the recovery period will be because the source of recolonizing organisms is reduced. Water quality disturbances may also occur at different spatial scales than typical disturbances in a given stream or river. For example, a pesticide spill may eliminate invertebrates from all rocks in a headwater stream, whereas a typical flood in the area would eliminate only invertebrates on the smaller, more abraded rocks. Perhaps more commonly, water quality disturbances such as oxygen sags below an effluent discharge will affect a small area and create a more patchy riverine community.

Water quality degradation may also affect community composition by accentuating the impacts of natural disturbances. For example, siltation from non-point-source run-off may diminish the number of suitable substrates for colonizing invertebrates, causing a slower recovery from disturbance, or one characterized by a different community assemblage. In a second example, pollution might alter niche dynamics by eliminating a pollutant-sensitive species from a river reach and causing a subsequent increase in dominance of other species. In all cases, the water quality manager must evaluate not only the impact of water quality degradation on a stream community, but also how the water quality disturbance relates to natural disturbance regimes.

Around the world, dams have been built to control the episodic nature of stream flow and to ensure better year-round water supply. In these cases, the biological assemblages of rivers will change toward a more stable regime whose community composition is dominated by competition among species and small niches. Many systems are disturbance-adapted, however. In such cases, the disturbances may affect one set of human values negatively (e.g., year-round water availability), while affecting others positively (e.g., a fishery or flood plain fertility). The conflicts between what constitutes a negative water quality impact will then be heavily influenced by the need to balance competing river uses. The effects of impoundment on downstream (and created lake) water quality has received extensive study and is covered in chapter 20.

Unique River Characteristics from a Water Quality Perspective

From a water quality perspective, important river characteristics include the following:

1. Unidirectional flow.
2. Short residence time of water in each reach.
3. River channel instability.
4. Fluctuating discharge.

5. The close connection between rivers and the surrounding terrestrial eco-system (and related determination of water chemistry).
6. Their dependence on energy, carbon, and nutrient subsidies (e.g., al-lochthonous material) from the surrounding watershed.

The importance of terrestrial input was discussed earlier. The first five charac-teristics, detailed below, distinguish water quality sensitivities of rivers from those of more confined water bodies. Within each river, however, differences of geologic composition, chemical buffering ability, rates of flow, and pollutant load and composition affect water quality.

Unidirectional Flow

It is axiomatic that water flows downhill. Just as the huge volume of the ocean has determined the way that people treat oceanic waters, the fact that water flows in one direction has framed the approach people have taken to rivers, waste management, and river pollution. The one-way flow of rivers has made them, historically, an inexpensive and maintenance-free sewer. Combatting the attitude that "what I can't see, isn't there" continues to plague river water qual-ity management efforts. The water quality importance of unidirectional flow has three components:

1. The psychological component outlined above, which posits that pollution "disappears" in streams.
2. Water movement, oxygenation, and action by microbial communities does confer some, but not an inexhaustible "self-purification" capacity.
3. Quality along a river varies with distance from the pollutant source, type and intensity of polluting activity, hydrogeological features such as the buffering capacity of the channel substrate, and the flow rate.

As a result of the third factor, impacts may "disappear" some distance downstream just as they may accumulate through synergistic and compounded impacts.

Residence Time

The residence time of water in any given location affects its susceptibility to degradation. In fast-moving waters, cleaner water quickly replaces affected water. In addition, a pollutant has less time to become incorporated into the biological community. Slower-moving waters are more susceptible to degrada-tion because sediment is dropped in these "deposition zones" and becomes the feeding substrate for the biological community. The sediment has a significantly longer residence time than the water in the channel, so any pollutant or nutrient carried by the sediment remains in the system at that location. In addition, slower-moving waters are more turbid (the water is "cloudy" with sediment). Thus, light penetration is decreased and invertebrates may be smothered with light silt.

Channel Instability

Channel instability can create water quality problems when land use aggravates instability or when attempts are made to control the banks. In the first case, erosion and periodic floods change the course, character, and flow of rivers. In the second case, attempts to control natural channel instability by straightening river reaches, lining banks with concrete, and creating check dams can each create its own unique water quality challenge. For example, channelization increases bank stability but also increases run-off volume and decreases natural buffering capacities. Despite the difficulties with channel instability, many riverine communities depend on periodic flooding and deposition of nutrient-rich silt. The challenge of management is not to stabilize the channels, but rather to check increases over some "natural" level (which varies from continent to continent, from region to region, and from river to river).

Fluctuating Discharge

Fluctuations in precipitation over time and over space create dramatic differences in water availability, pollution susceptibility, and management concerns. For example, long drought periods in the arid West of the United States have led to the creation of extensive irrigation networks and resultant complications from salinization. In addition, while availability of water in rivers changes seasonally, water use needs remain nearly constant throughout the year.

Uneven rainfall also affects the concentration of pollutants in rivers. In periods of low water flow, pollutants will be more concentrated than during periods of high flow. High rainfall in many cities, however, taxes the sewage treatment systems and results in the release of untreated sewage into area waters (see chapter 20). High rains can also lead to siltation of waters during episodic rainfall, just as lack of rain can increase solute concentration during drought periods. Because streams and rivers are also fed by groundwater, precipitation patterns and soil depths and permeability will also affect water infiltration and, therefore, stream water quality.

The River as Part of the Landscape

The river is a catchment. As such, it has its own unique ecology, but that ecology is inextricably linked to the surrounding landscape. Land use activities, from forestry to agriculture to urbanization, will all have an impact on nearby rivers and streams. Forestry activities, for example, may temporarily increase sediment loads in the river, decreasing invertebrate communities, and may create longer-term increases in stream temperature, causing changes in fish populations. In another example, an agricultural basin may be subject to increased sediment loads during fallow periods and nutrient increases during growing seasons as fertilizer is washed and percolated into the river. Virtually all land use activities are reflected, in one way or another, in changes in the water quality of the catchment water. Catchment morphology, area, and primary vegetation also are important factors in determining stream (and groundwater) water quality.

In addition, overland flows may affect water quality of a river or stream. Overland flow into a river channel represents only a small fraction of water that flows into a river but it carries sediment loads from cultivation and urbanization activities such as road grading. In contrast to overland flow, most water reaching a river arrives via percolation through soil. These sources carry leachates into the stream, such as nutrients, pesticides, or herbicides, as well as toxic chemicals from improperly lined dumps or landfills.

Spatial and Temporal Variations in a Water Quality Context

Minshall (1988) notes that both spatial and temporal scales in river ecology vary by 16 orders of magnitude. Rivers change and evolve on levels ranging from microhabitat and microprocessing scales of seconds to tectonic events encompassing hundreds of millions of years. Debris from large-scale disturbances may persist for as long as 400 years and cause continuous changes in river channels. Long-term community succession changes have been demonstrated on scales of years to tens of years, and short-term succession may occur in a few days to a few months. While water quality management frequently focuses on small- or midscale effects, significant long-term effects can, and probably do, result from all of the channel modifications, land use changes, and accumulated community impacts.

Despite the wide range of scale processes acting on flowing water, water quality managers are most often concerned with the midrange of effects and changes. Processes occurring on a time scale of less than 15 minutes, or at a distance of less than 500 m, reflect important processes in river ecology. Overall, however, they contribute less to the science of water quality management than the next tier of temporal and spatial scales. Similarly, the longer-term processes (greater than 100 years), although providing an important framework within which management can proceed, are usually too long a consideration for management activities. Temporal and spatial scales of consideration to water quality management are presented diagrammatically in Fig. 9.5 and may be summarized as:

- 15 minutes to several days, over distances of about 5 km: Important properties related to the speed and distance of flow (i.e., how fast a pollution spill can carry downstream). Issues at this scale include spills, point source discharges, and acute biological responses of relatively immobile organisms such as plants, bacteria, fungi, and many insects.
- 1 day to several months, over distances up to 100 km: Concerns related to diel oxygen dynamics, point sources such as logging activities and road building, biological responses in many insects, and responses in some fish.
- 1 month to 1 year, over distances up to 500 km. Important seasonal processes of rainfall, spate, and drought. Concerns related to medium-term and more continuous sources of pollution (e.g., agricultural practices or waste disposal/processing); changes in community function; recovery from short-term impacts and biological responses in fish migrations.
- 1 year to tens of years, source to sea or 1000 km: Cumulative effects of multiple pollutants and impacts (e.g., bioaccumulation, increased siltation from agricultural development upstream in combination with increased urbanization downstream, increases in flow, storm sewer flows of oils and urban chemicals); long-term changes in community structure and function; recovery from impacts.
- 100 years and more, on a biome scale: Long-term processes of 100 years or longer are vital in understanding natural variation in streams, recovery processes, and long-term impacts such as climate change. They provide an important frame of reference for water quality managers, but generally are too long-term to be subject to management practices.

Specific Water Quality Impacts

On a global scale, it is difficult to quantify the causes and sources of river water quality degradation. Even where statistics are available, they do not tell the full story of water quality deterioration. All pollutants are not equal. For example, although siltation affects almost half of the river miles in the United States, the

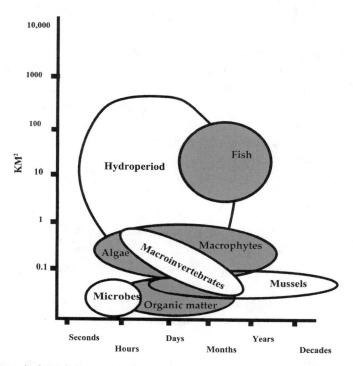

Fig. 9.5 Stream hydroperiods are highly variable and are controlled by geomorphic and climatic influences. Fish communities are relatively mobile and thus vary significantly in both space and time. Stream macroinvertebrate communities, in contrast, are relatively stable over the decades and within a scale of a few square kilometers.

ecological impacts of siltation are often more subtle, and more temporary than the effects from metal pollution or acidification. These latter impacts affect fewer river miles but have a more significant impact on the ecology and recoverability of the affected river. Because the impacts of these pollutant sources are discussed in chapter 17 and specific sources are covered in chapters 18–24, the following paragraphs serve as only a cursory mention of the kind and degree of impacts facing river water quality managers.

Organic Pollution

People have relied on river water "self-purification" for thousands of years. As the repository for sewage, rivers have a long history of successfully processing and breaking down the wastes associated with humans and their early industries. The fast-moving stream pictured in Fig. 9.6 is typical. It has a short renewal time, can break down contaminants physically, and is well aerated. The population expansion and concentration in the past 200 years, however, have severely strained rivers' ability to process wastes. The result of organic pollution is twofold: (1) a decrease in dissolved oxygen concentrations below the effluent plume and subsequent decreased ability of the water to support aquatic life (see chapter 17), and (2) the transmissions of human pathogens in water. As with most stresses, the overall effects of organic pollution are exhibited through the elimination of sensitive species, increases in some tolerant species, and related changes in community structure. Effluent pretreatment and proper dilution are two extremely effective approaches to reducing the impact of effluents on riverine communities.

Sedimentation

As indicated earlier, sedimentation is the most common source of water quality degradation. While point source pollution has been the focus of water quality controls in the past 20 years, only recently has the magnitude of nonpoint sources been recognized. Sedimentation in water has two effects: a purely physical process that blankets the substrate and organisms, clogging gills and feeding mechanisms, and the provision of a chemical carrier for pollutants. Sedimentation may also reduce in-stream primary productivity by preventing light infiltration. While severe deposition may alter the benthic community

Fig. 9.6 The assimilative capacity, or "self-purification," of rivers is facilitated by their short renewal time combined with their capacity for physical breakdown of pollutants and the continuous reaeration of waters. (Photo by D. Vanderklein)

structure through reduced diversity and increased dominance of a few species, moderate or light sediment loads will settle relatively quickly and the impact will be moderate and relatively local. Sedimentation management generally seeks to reduce sediment run-off by altering agricultural or forestry practices or by constructing settling ponds.

Thermal Pollution

Thermal pollution can result from either the reduction of shading over a river or stream by logging activities or the presence of a heated effluent from power plants. In either case, the impacts of thermal pollution tend to be localized and relative to the season, stream size, and latitude. For example, when river or stream waters are already warmed (i.e., by summer temperatures or tropical conditions), many species are operating near their thermal maximum. In these conditions, increases in water temperature may easily stress fish and other biota to lethal levels. For example, the disappearance of the Atlantic salmon from the Lake Erie basin has been attributed to massive, region-wide logging that raised the temperature of spawning streams to lethal levels for the salmon. In conditions where waters are cooler than the preferred range for some species, increases in water temperature may attract fish, which can create a problem if they are sucked into intake valves or suffer cold shock during plant shutdowns.

Less dramatic, but perhaps more insidious, are the more common effects of sublethal increases in temperature. Many effects fit into the "sublethal category." For example, a rise in temperature induces higher respiratory rates in organisms, inducing greater physiological stress. Furthermore, temperature increases reduce the solubility of oxygen in water, resulting in even less oxygen to meet the higher oxygen demand. Sublethal temperature increases may also induce early spawning in some species. If spawning occurs too early in the spring, the fry will have insufficient food and will be even more susceptible targets for predators. Largemouth bass, rudd, and pike-perch are among the species for whom early spawning has been noted as the result of sublethal temperature increases.

Techniques for River Water Quality Management

Efforts to improve river water quality generally fall into two categories: non-structural techniques that do not involve physical changes to the river, and structural changes that more directly change the flow and conditions of the river (Novotny and Olem 1994).

Under the heading of nonstructural changes, the historical favorite is benign neglect. Given sufficient time and a cessation of the causal agents, most rivers will recover from degradation without further human intervention. This technique is effective, but requires patience, a long-term frame of reference, and, most importantly, an actual cessation of the pollution. A second nonstructural approach is action through legislation and administrative procedures. This strategy includes regulating flow requirements, establishing water quality standards, identifying best management practices, undertaking public education efforts, and altering water rights laws to account for primacy of minimum ecological requirements. Other nonstructural means involve renegotiating water release schedules for dams to approximate seasonal cycles of water flow, the purchase of flood plain easements and other "conservation buy-out" efforts, and fencing of creeks and rivers to prevent heavy livestock or human use.

Structural management techniques may range from efforts to improve species-specific habitats to recreation of historical channel morphology. Some of these activities include changing in-stream woody debris characteristics, altering stream bank vegetation, and deflecting currents. The risk of any structural management effort is that improvements for one species or value invariably degrade quality for another species or value. For example, stream improvement projects aimed at trout and salmon production have been successful in many rivers, but at the expense of the beaver population. In another example, engineering efforts on the Blanco River of Colorado were designed to improve the capacity of a stream to carry high flows. A channel installed by the U.S. Army Corps of Engineers instead destabilized the banks, causing a loss of riparian vegetation and resulting in considerable sedimentation. This example is typical of the difficulties faced in river improvement efforts; because rivers are highly dynamic, it is difficult to anticipate fully the effects of structural changes.

Summary

- Rivers are dynamic systems whose properties are determined by their watersheds and riparian zones. Stream ecosystems are locally highly resilient, recovering relatively quickly from local-scale impacts. On the other hand, watershed-scale disturbances have long-term impacts on all aspects of riverine ecosystems.
- Stream channels represent ecosystems strongly dependent on the energy of flowing water. Management actions are often targeted specifically to maintain that property (e.g., retaining minimum stream flow, minimizing chan-

nelization, maintaining retention devices such as large wood, minimizing flow disruption from storage devices like dams, and maintaining intact riparian zones).
- Riverine systems vary predictably among upland, midreach, and lowland areas. Most upland rivers are heterotrophic, relying on energy, carbon, and nutrients from riparian zones. Lowland streams are more often autotrophic, relying on nutrients and energy from upstream sources and in-channel processes. These distinctions dictate variations in ecological properties and processes that demand different management strategies for different river reaches.
- Disturbance (i.e., an external force causing the system to function outside the bounds of its "normal" range) is an integral component to all current river ecosystems. Effects of disturbance are relatively predictable and represent the core of water quality management.

For Further Reading

Bayley PB. Understanding large river-floodplain ecosystems. *BioScience* 1995;45:153–158.

Hunsaker CT, Levine DA. Hierarchical approaches to the study of water quality in rivers. *BioScience* 1995;45:193–203.

Novotny V, Olem H. *Water quality: prevention, identification and management of diffuse pollution.* New York: Van Nostrand Reinhold, 1994.

Sparks RE. Need for ecosystem management of large rivers and their flood plains. *BioScience* 1995;45:168–182.

Stanford JA, Ward JV. Management of aquatic resources in large catchments: recognizing interactions between ecosystem connectivity and environmental disturbance. In: Naiman RJ, ed. *Watershed management: balancing sustainability and environmental change.* New York: Springer Verlag, 1992;91–124.

10

Groundwater and Water Quality: Water to Live on

Overview

Groundwater is the underground portion of the global hydrologic cycle. Although in many regions of the world it is the single most important reservoir of water, its fragility has only recently been recognized. Groundwater derives from precipitation that filters through soil and reaches the water table, the point at which all spaces between soil particles are filled with water. Areas within this saturated zone are called **aquifers**. Groundwater flows at many different levels and rates depending on the configuration of the water table and on the permeability of the soil and substrate it flows through.

Because groundwater receives water from surface sources, any substance that leaches through the soil to the water table, or that seeps or flows into rivers, lakes, wells, or other points of access becomes a source of contamination for this resource. Of primary importance for water quality management is that most groundwater reservoirs have long residence times and pollutants consequently become concentrated there. When that water is tapped for potable water supplies or irrigation, or when it recharges the water of a lake or stream, the concentrated pollutants represent an important concern for human and aquatic health. A variety of activities can result in groundwater pollution: highway salting, industrial waste, septic tank overflow or leakage, agricultural inputs, and urban waste. Few of these sources can be eliminated, but groundwater protection can be increased by improving construction of waste facilities, improving maintenance of underground pipes and wells, reducing unnecessary chemical inputs, and improving zoning. Protection can also be achieved through groundwater flushing, "venting," filtering, and decontamination.

Introduction: The Importance of Groundwater

Groundwater contamination is one of the most serious threats to public health faced in centuries. In part this concern has emerged because traditional land use management did not consider groundwater effects or recognize the integral links between groundwater quality and land use practices. While groundwater is now recognized as being closely tied to land use, that knowledge has come too late to prevent a growing groundwater contamination problem. Altering polluting practices, sewer lines, landfills, and settling ponds to stem the flow of pollutants into groundwater is an enormous undertaking: sudden and dramatic changes are likely to occur in contamination rates. The complicating factor in groundwater management and improvement is that groundwater has a residence time that varies between 2 weeks and 10,000 years; 300 years is a commonly cited average. Because of the extremely long residence times, even if all pollution were stopped today, it would take many decades for even a significant portion of the groundwater to become clean. Fig. 10.1 presents some of the spatial-temporal challenges in managing groundwater contamination.

Groundwater is clearly a vital resource, accounting for approximately 30% of global fresh water reserves, and constituting by far the largest reservoir of available fresh water (see chapter 8) (Shiklomanov 1993). While no reliable global estimates have been calculated for the percentage of fresh water supplied by groundwater, the rates for individual countries indicate a heavy, but variable use. Kiribati, in the Pacific, relies on groundwater for 100% of its water, for example, and many European countries, including Austria, Denmark, Italy, and Portugal, rely on groundwater for more than 90% of their potable water. In contrast, some countries as geographically diverse as Japan, Spain, Norway, and the United States depend on groundwater for 20% to 50% of their fresh water supply (Gleick 1993). The importance of groundwater can be demonstrated by the U.S. example: overall groundwater provides more than 50% of the nation's drinking water and supplies virtually all of the drinking water for rural communities (EPA 1992). On a finer scale, however, even those averages misrepresent local and regional management challenges. For example, according to the U.S. EPA (1992), nine states alone accounted for nearly three-fourths of the groundwater withdrawn in the United States, most of which was used for irrigation. In the eastern United States, the reliance on groundwater for drinking water is much higher than in the west.

In the United States, as in most developed countries, incidents of groundwater pollution are increasing. In 1991, the number of incidents of groundwater restrictions (i.e., where groundwater did not meet standards) stood at

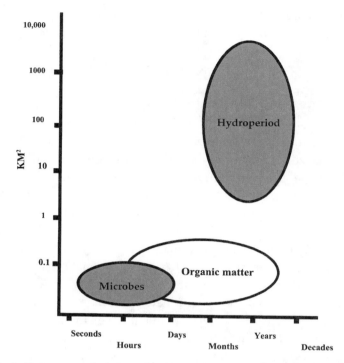

Fig. 10.1 Scale issues are particularly pertinent in groundwater quality management because of the enormous range of residence times, geographic extent of aquifers, and time lags between contamination and evidence of pollution. Groundwaters routinely fluctuate very slowly (on the order of tens of decades), and most aquifers cover relatively large spatial areas. Exceptions (not shown) include Karst aquifers and very shallow aquifers that might vary 10 times more quickly and at a spatial scale an order of magnitude smaller. Microbial communities and organic matter dynamics in groundwaters are also comparatively slow (compare with Figs. 9.6 and 12.4).

nearly 1300, a 37% increase from 1989 and nearly four times the number reported in 1987 (EPA 1992). In addition, more than 14,000 well closings were reported in 1991. While these numbers represent a small percentage of groundwater used, their increases are alarming and have made groundwater protection a priority concern.

Overall, the five major sources of groundwater pollution reported in the United States are underground storage tanks (39%), abandoned hazardous waste sites (17%), municipal landfills (16%), agricultural activities (15%), and septic tanks (14%). Taken together, municipal, industrial, and other landfills account for nearly 30% of the groundwater pollution in the United States. These figures may be typical of other countries with similar restrictions and building codes, but dramatic contamination in other countries from Eastern Europe to

Asia results from unlined, unregulated, and unmonitored disposal of toxic materials and industrial and municipal waste.

The Hydrology of Groundwater Flow

Groundwater, as the subsurface reservoir of the hydrologic cycle, includes both the "unsaturated zone," in which spaces between sediment particles contain air as well as water, and the "saturated zone," in which all spaces are filled with water. The water table is the top of the saturated zone, and infiltrated water slowly reaches the water table by gravity. Once all of the interstitial spaces are filled, the reservoir is called an *aquifer*. The term is a very general one, however, and aquifers can be deep, shallow, open at many places to surface water, or opened at only one narrow point (i.e., *confined*). In water quality management, the openness of an aquifer determines the likelihood that an aquifer will become contaminated and will, in turn, contaminate other water bodies.

Groundwater usually flows down a gradient linked to the slope of the water table but it also occurs in many layers in the earth, which do not always parallel surface flow. Some of the important flow characteristics are pictured in Fig. 10.2. Like surface water, groundwater flows toward, and eventually drains into, streams, rivers, lakes, and oceans. In fact most of the water in rivers and lakes originates from groundwater run-off; surface run-off alone contributes very little to most of these waters (Vanek 1987). Once groundwater reaches these surface reservoirs, the water cycle is completed as water is evaporated from the surface and returned to the atmosphere, only to return once again as precipitation. Surface outlets are an important point of pollution exchange between surface and groundwater reservoirs.

The integral links between groundwater and surface waters have been fully understood only recently. For water quality management, accounting for those links is vital. Because of the lag time between scientific knowledge and implementation of policy, the traditional view that surface and groundwater flows remain separate is still incorporated into most countries' legislation.

From a water quality perspective, two features of groundwater hydrology are especially important in determining risk and recovery potential. The first is *flow rate,* which determines the speed and spread of a contaminant plume as well as how much filtering will take place by subsurface particles. For example, aquifers with a fast flow rate will generally exhibit a long, narrow pollution plume, while slow flow rates will be reflected in plumes that spread laterally. The second important feature involves the *recharge* (and *discharge*) areas for each

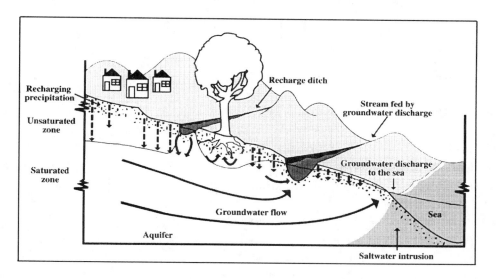

Fig. 10.2 Groundwater flows are controlled by infiltration rates and by substrate permeability. The susceptibility of a given aquifer to pollution is often determined by its openness to the surface. The unsaturated zone refers to the upper region of the soil in which interstitial spaces are partially filled with water and partially filled with air. The saturated zone refers to the region at which all interstitial spaces are filled with water. (After Environment Canada 1990)

aquifer. In these areas, water enters and leaves the aquifer. The location of recharge areas is vital for determining susceptibility to pollution and likely direction of contaminant plumes.

Flow Rate

In the subsoil, the flow rate is slow and is governed by permeability of the material and the hydraulic gradient, although soil permeability is often cited as the primary control. For example, given the same hydraulic gradient, water moving through low-permeability clay soils may move only 1 mm per day, while groundwater movement through medium-permeability sandy soil groundwater could amount to 50 cm per day. In gravel, this rate can increase to approximately 20 m per day (Barberis 1986). Movement is even faster in fractured or fissured substrate. In addition, fractured substrates provide only minimal filtering of surface run-off. The implications for water quality are directly related to these flow characteristics: contamination plumes may advance slowly or relatively quickly depending on the soil and subsurface rock, spreading laterally if the water movement is slow.

Differential flow rates, different types of aquifers, and varying flow directions complicate the water quality manager's job in identifying and controlling sources of groundwater pollution. Additionally, groundwater typically moves in parallel paths or layers, with little mixing between them. Consequently, some contaminant plumes may travel slowly in deep layers, while others will be largely confined to fast-moving surficial or shallow aquifers (Fig. 10.3).

Recharge and Discharge

Recharge is the water that seeps into an aquifer. It occurs intermittently, following periods of rain and snowmelt; it takes place where permeable soil or rock allows water to seep into the ground. Importantly, recharge may occur over only a very small area or may extend over many square miles. For example, valley aquifers may receive recharge from surrounding hillsides in addition to the water that falls in the valley itself. Logically, the larger the recharge area, the greater the chance that pollutants will seep or leach into the underlying aquifer. Nonetheless, large recharge areas also confer greater dilution potential.

Discharge points are areas where groundwater leaves the ground, such as streams, springs, or lakes. The relationship between recharge areas and discharge points is vital to a water quality manager because it usually indicates the direction and speed of flow in an aquifer. It allows the manager to estimate how a pollutant will enter the groundwater, where it will flow, and where it will

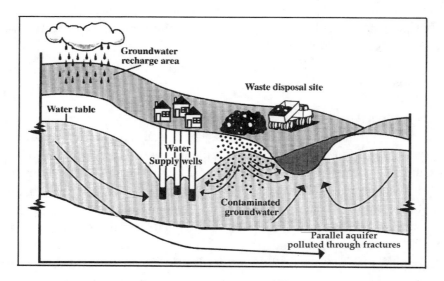

Fig. 10.3 Contaminants are routed to local groundwaters from surface sources as contaminants leach into groundwater or enter through recharge zones. Deeper and more regional groundwater quality is controlled by permeability (e.g., fracturing) of aquacludes. (Modified from Environment Canada 1990)

resurface. Characteristics of groundwater flow are not constant, however. The direction and speed of a contaminant plume can vary with the season or annually as subsurface flow responds to periods of drought or high rainfall. Consequently, a contaminant plume may discharge to a hillside spring in winter but into a lake in summer. Similarly, a lake may serve as a point of discharge if the water table is high, or a point of recharge if the water table is low (Vanek 1987). Whether a lake represents a discharge or recharge point will influence the concentration of contaminants in the groundwater as well as the behavior of pollution plumes.

Wells

Wells represent a unique part of the hydrologic cycle of groundwater because they are built by humans. In addition to being the access point for groundwater, wells influence subsurface water flow, principally when pumping from wells lowers the nearby water table. The resulting depression in the water table, called the *cone of depression,* influences contaminant flow in two ways. First, it may induce recharge from nearby streams or lakes, potentially drying up these areas and drawing potential pollutants into the well. Salt water intrusion, for example, results when the water table of coastal aquifers is lowered and seawater is drawn into the aquifer and well. Second, the cone creates a gradient that can draw contaminants directly into the well.

   Wells also influence the hydrologic flow, and in particular the flow of pollutants, because they create new recharge points and because the bore hole may create conduits across previously separated aquifers. These problems are particularly acute with unsealed wells in which inputs from the land surface (e.g., pollutants) can be carried into the groundwater.

Sensitivity to Pollution Derives from Geology

As described, groundwater flow, soil type, aquifer characteristics, recharge, and drainage area all play a role in determining the destination and pathway of a pollutant. Sensitivity ratings combine these physical characteristics to estimate susceptibility of groundwater to contamination. Typically, sensitivity estimates are based solely on geophysical features, although recent evidence indicates that incorporating land use practices into these ratings would improve their utility (Geier 1994).

   Examples of geologic substrates and their importance to water quality (listed from most to least sensitive) are discussed below.

Karst Formations

Karst formations are carbonate rocks in which carbonic acid in groundwater has dissolved the substrate. Karst areas are characterized by large fissures such as sinkholes, caves, allogenic streams, and springs. Each of these structures increases the interaction between the land surface and underlying aquifers, making these areas extremely sensitive to contamination. In agricultural areas, surface run-off often feeds directly into aquifers via these sinkholes, resulting in high nitrate and pesticide concentrations in groundwater.

Surficial Aquifers

Another aquifer type that is highly sensitive to contamination is the surficial aquifer. As the name implies, these aquifers lie close to the surface, although they can be overlain by a variety of surface geology. The variable geology results in a range of sensitivities but, in general, the shallowness of surficial aquifers results in high sensitivity to pollution. Deeper aquifers are usually better protected from contamination because of the filtering effect of the many layers of soil and rock. Surficial aquifers often experience little, if any significant, filtering.

Rock Subsurface

Traditionally, it was believed that thick layers of soil and rock effectively filtered pollutants. In areas with this geology (i.e., areas of volcanic origin), underground aquifers were considered safe from surface sources of pollutants. Recent evidence contradicts this idea. For example, in Hawaii, concentrations of several pesticides have been found more than 400 feet deep in Oahu's groundwater. Evidence suggests that these chemicals were carried downward by percolating water from overlying pineapple fields. The fact that fissureless rocks are rare implies that the presence of a rock substrate does not protect against groundwater contamination.

Clay Layers

Many water tables are overlain by a layer of clay or other less permeable formation. The clay layer acts as a barrier (called an *aquiclude*) between the surface and the water table. Infiltrating water may be "perched," traveling on top of the deeper, regional groundwater aquifer. In this case, the deeper aquifer is relatively resistant to water quality impacts.

Confined Aquifers

As mentioned above, confined aquifers are susceptible to pollution or contamination only through their recharge area, or through fault lines or other breaks in the impermeable upper layer.

Major Sources of Groundwater Pollution

Groundwater contamination comes from a variety of sources, as noted in the introduction to this chapter. Simply stated, any substance has the potential to end up in the groundwater, either via leaching through the soil or seeping through fissures, wells, lakes, or rivers. For example, population explosions of gypsy moths in the eastern United States even resulted in fecal contamination of many area wells. Other natural contamination derives from the chemical and physical processes that transfer impurities to water from the atmosphere, biosphere, or lithosphere (earth's crust). For example, the leaching of soluble salts from the soils and rocks that surround groundwater alters its mineral content. The range of natural variation in groundwater chemistry is enormous. In some places in the Canadian Shield, total dissolved solids are commonly around 25 mg/L, while in some deep saline waters in the interior plains of Canada, this concentration ranges as high as 300,000 mg/L (Environment Canada 1990). The range of possible dissolved solids in groundwater is also far broader than in surface waters, even though groundwater areas are less variable over time than surface waters.

In water quality management, these natural sources of contaminants are generally minor in comparison with contamination from anthropogenic sources. The origin of anthropogenic pollutants has a major influence on management options.

Diffuse and Point Sources

The origin of groundwater pollution is generally characterized as either being from diffuse (nonpoint) or point sources. As the name implies, diffuse pollution originates from diverse sources and is dispersed through a variety of pathways in the landscape. Farms are widely cited as examples of diffuse sources because agricultural pollutants (e.g., fertilizers and pesticides) are spread out over a large area and the rain water that leaches them may drain to different water bodies. Water from diffuse sources of pollution may percolate slowly over the course of surface run-off or may drain into a stream and enter the groundwater through infiltration.

In contrast, point sources of pollutants are concentrated and readily identifiable (e.g., run-off pipes, industry storage lagoons, spills). Because point sources are more easily controlled and monitored, they have been the focus of nearly all groundwater laws worldwide to date. Only recently has the control of nonpoint sources been incorporated into legislation.

Some major groundwater pollutants and their sources are as follows:

- Leaky storage tanks and vehicle spills result in a huge release of hydrocarbons (e.g., crude oil, gasoline, creosote) to the groundwater. In fact, hydrocarbons have polluted more of the U.S. groundwater drinking supply by volume than any other class of chemicals (Knox 1988). Because the use of these materials is so ubiquitous, it is difficult—if not impossible—to control a significant portion of these spills and leaks (see chapter 20).
- Agricultural activities have led to severe problems with nitrate and increasing problems with chemical residues of pesticides and herbicides. Livestock waste is another source of groundwater pollution (see chapter 19). The number of fertilizers, herbicides, and pesticides is growing exponentially, and in general the public is poorly informed about connections between use and water quality (see chapter 19). Increased awareness of recharge areas and limiting use in these areas, encouragement of integrated pest and nutrient management, and increased emphasis on proper disposal of pesticides and rinse water are all critical. Because use is so widespread, regulation and monitoring of these substances is less feasible than education, incentive, and outreach efforts.
- Industries are point sources of a variety of pollutants. Brines from oil and gas production, heavy metals, and other industrial pollutants are showing up in groundwaters in increasing quantities. Other industry sources are underground injection wells for industrial waste and leaks or spills of chemicals at manufacturing and storage sites, equipment cleaning sites, and industrial sludge disposal sites. In the United States and other industrialized countries, reasonable (if not fully implemented) controls have been placed on point sources. In rapidly industrializing nations, however, typically no controls restrict the release of the potentially polluting substances.
- Urban waste finds its way into groundwater through landfill leachate, leaky or poorly sited septic tanks, and dumpsites. Bacterial contamination of groundwater is a severe problem in some urban areas. Other urban wastes enter the groundwater through leaky sewer lines, graveyards, road salt stor-

age areas, salted highways, highway and railway accidents, and surface run-off. In addition, suburban wastes may enter aquifers through unlined wells.

- Increasing salinity is one of the most significant and widespread forms of groundwater pollution. It is caused primarily by the effects of irrigated agriculture, although salt intrusion into overpumped aquifers is common in coastal regions. Mining, oil field brines, and highway de-icing salts are also major contributors to groundwater salinization.

    Salinization results from several different processes. In inland areas, salinization generally occurs as infiltrating rain water dissolves minerals from soil and rock. Long residence times permit greater degrees of mineralization. Consequently, important features of aquifers used for potable supplies are relatively high permeability and relatively low residence times. Arid climates are susceptible to salinization because of the effects of low rainfall on leaching and recharge. Infiltration from rainfall may be negligible and leaching is not effective in diluting soil salt solutions enriched by evaporation. Any infiltration that does reach the groundwater table will be highly mineralized. Similarly, in poorly drained areas, evaporation can produce significant increases in salinity.

    Lastly, irrigated agriculture raises groundwater salinity by introducing problems of poor drainage and increasing evaporation (see chapter 19). Generally an inefficient process, irrigation causes evaporation from irrigated land and infiltration from canals. Waterlogging and other results of improper drainage lead to elevated water tables and elevated water salinity. Up to half of the world's irrigated land is affected by salinity to some extent, and productivity is seriously affected on at least 7% of the total land (Meybeck *et al.* 1989). Chapter 19 discusses some of the management options for reducing salinization caused by irrigated agriculture, such as lining canals, creating horizontal canals, flushing salts, and implementing drip irrigation.

- Among the more significant point sources for contamination of groundwater are municipal landfills and industrial waste disposal sites. Wastes such as hazardous material originate from improper storage, handling, use, and disposal as well as from spills and leaks. Wastes from landfills (or storage lagoons) may result from poor siting, improper liners or liner failure, and poor management and control of allowed materials.

    Other waste problems originate from septic tanks, primarily due to poor installation and maintenance. Even properly sited tanks may fail due to use of septic tank cleaning additives that erode pipes, disposal of other corrosive chemicals such as paint thinners in the septic system, or overloading the system with a garbage disposal unit. When any of these problems occur in or near sand and gravel aquifers, the potential for widespread contamination is great. Septic tanks and leaks and spills of petroleum products also serve as significant sources of groundwater pollution. Generally, management for these sources includes improvements in siting, maintaining and lining of landfills, and proper maintenance and siting of septic systems.

- Infrastructure problems are also increasing as underground pipelines are corroded, shaken loose by road traffic, displaced by roots, or simply poorly installed. As mentioned previously, wells may contaminate groundwater when leaky casings act as conduits for pollutants, and they may even act as conduits between aquifers. Such cases indicate design flaws that call for increased vigilance in design, monitoring, and maintenance of all underground lines.

Management of Groundwater Quality Impacts

Despite the different sources of groundwater pollution, their impacts on water quality are often similar. For example, salinization can result both from irrigation and from industrial or mining wastes. Similarly, oils, toxic compounds, bacteria, viruses, nitrates, and metals may all enter the groundwater from a variety of sources. For each of these contaminants, there are three main management alternatives: (1) reducing use of the pollutant (i.e., education); (2) improved confinement; and (3) treating, flushing, or venting the groundwater. Of these options, education is the most feasible for reducing the use of nonpoint sources; on the other hand, education efforts on such a wide scale and for such diverse audiences is a difficult and very long-term commitment. To reduce point sources, different countries take their own regulative approaches, which range from national level monitoring and fines to local level management committees that operate through consensus (see chapter 25).

Efforts at improved confinement include lining irrigation canals and monitoring and maintaining storage tanks and septic systems. Because the majority of groundwater pollution occurs through system failures, there is enormous potential in improving confinement. As with education, however, this approach is only partially a matter of infrastructure support and largely a matter of user education, making it a critical tool but one that is difficult to implement.

The third category, treatment, is in many ways the least preferred, as it amounts to cleaning up existing pollution rather than preventing it. In the long run, cleanup is far more expensive and technologically complex than prevention. Nonetheless, given the barriers to effective prevention, cleanup technology represents one of the most vital tools in groundwater management.

Many future cleanup efforts will probably utilize hydrocarbon-consuming bacteria (Knox 1988). These species degrade the hydrocarbons into carbon dioxide and methane gas. They are naturally occurring forms of bacteria, but are usually limited by oxygen levels. Scientists are experimenting with augmenting oxygen to facilitate in situ consumption of hydrocarbon pollution.

Groundwater flushing is another approach to cleansing groundwater sources. Utilized in cases of resistant pollutants, this process involves wells that pump clean water into aquifers and others that pump out polluted groundwater. The latter can then be treated and returned to the ground. The difficulty of identifying the boundaries of polluted aquifers complicates this process.

When contamination is confined to the unsaturated zone, *venting*—a process similar to water flushing—has been successful in cleaning the pollutants. Venting works by pumping air into the ground above the water table, thereby releasing trapped hydrocarbon pollutants.

Other treatment methods include the more familiar processes of filtration, disinfection (usually with chlorine), and water softening to remove various forms of pollutants. Using alternative sources of water is also a time-honored approach to reducing immediate risk. Few of these technological tools will prove completely effective without improved groundwater models. Improved models of groundwater flow are serving a critical role in predicting the speed and direction of contaminant plumes so that appropriate remedial action can be taken.

Groundwater Management Strategies

Policies and administrative approaches to reducing hazards are the last category of tools available to managers. They include zoning laws, improved monitoring and enforcement, and improved training procedures for those in jobs with high pollution risk. Overall, groundwater policies follow three basic protection strategies: nondegradation, limited degradation, and differential degradation. These three strategies differ about what is "acceptable risk" and who should bear the cost. Nondegradation places high value on a clean groundwater source and expresses a willingness to devote money and resources to maintaining that quality. Limited and differential degradation both try to maintain some minimum safe level of water quality while still allowing potentially polluting activities to continue.

A Brief Survey of Groundwater Management Institutions

United States

Throughout the United States, states determine for themselves the appropriate goals for groundwater, the appropriate role of government, who should pay, and who decides these issues. While state jurisdiction over groundwater is a long-standing tradition, increasing awareness of the transboundary nature of aquifers has introduced a viable federal role in groundwater management. Justifications exist for both a strong federal role and strong state roles. Arguments for a federal role include the need for coordination and uniformity across state and regional boundaries and the trans-state nature of many contamination problems. In addition, Duda's (1987) analysis of institutional barriers to groundwater management presents a case for strong federal leadership. Nonetheless, the diversity of state groundwater problems, historical dominance of states in land use issues, and the vacuum produced by lack of federal response to early detection of contaminated groundwater all point to a continuing strong state role in groundwater management.

Currently, the approaches taken by different states express wide diversity of views toward groundwater and reliance on groundwater. For example:

- Arizona is currently the only state to have a comprehensive groundwater policy.
- New York continues to rely on regulation of land use management to protect groundwater.
- Wisconsin has implemented a system of groundwater quality standards.
- Connecticut has experimented with strict "polluter pays" liability.
- Iowa has implemented a nondegradation policy.
- Colorado developed a protection plan with 14 agencies organized under the lead of the Colorado Department of Health.

Many states experience similar groundwater risks but differences in weather, economic activity, and local geology create unique concerns in different regions. For example, Table 10.1 illustrates that saline intrusion into groundwater is of primary concern in northeastern states because of the reliance on highway de-icing in the winter, but in south central and southwestern states, which have milder winters, highway salts do not pose a major concern. Similarly, mining is an important activity in the northwest with serious groundwater implications, but is rarely a concern in south–central portions of the United States. Interestingly, the logical policy direction is not always the one pursued. For example, Newson

Table 10.1 The principal sources of groundwater quality impairment vary among regions of the United States, primarily due to density of urban and industrial areas. Check marks indicate an important source; x indicates a source of secondary concern; ✦ indicates a source of tertiary concern; circles indicate unimportant sources for a given region. (Modified from Novotny and Olem 1994.)

Source	Northeast	Northwest	South Central	Southwest
Septic tanks, cesspools	✔	✔	✔	✔
Surface impoundments	✔	✔	✖	✦
Landfills	✔	✖	✖	✖
Spills	✔	✖	✖	✦
Highway de-icing	✔	✦	●	●
Surface discharges	✖	✔	●	✔
Land disposal of wastes	✦	●	✦	✖
Mining	✖	✔	✦	✖
Irrigation return flows	✦	✖	✖	✖
Agriculture				
Fertilizers	✦	✖	✦	✖
Feedlots, barnyards	✦	✖	✦	✦

(1992) notes that 27 states have statutes enabling state-level intervention to prevent "groundwater mining" (withdrawal over and above recharge), but that the states with the largest need for irrigation—and hence the largest problem with groundwater mining—continue to rely on a "good neighbor policy" to regulate use.

England and Wales

The response of the United Kingdom to land and groundwater pollution has been described as primarily reactive (Griffiths and Board 1994), with a great reliance on planning but little tradition of holding polluters responsible for their actions. Recent legislation provides the authority to prosecute offenders who pollute (knowingly), but the power of this legislation remains untested in the courts.

Two other important forces act on U.K. groundwater legislation. First, the United Kingdom is one of few industrialized countries that has not experienced a major groundwater (or land-based pollution) incident. These incidents, such as have occurred at Love Canal in the United States, Georgswerder in Germany, and Lekkerkerk in the Netherlands, have served as catalysts in the affected country (and its neighbors) to implement groundwater legislation and monitoring programs (Griffiths and Board 1994). England's deep, confined aquifers may be partially responsible for their "luck" in this regard. Nonetheless, most analysts feel it is only a matter of time before England, too, experiences a devastating accident.

As noted, however, most other industrialized nations have had galvanizing accidents. Consequently, the European Community (EC) drinking water directive sets strict standards for groundwater quality. The need to comply with EC directives constitutes another important influence on U.K. groundwater policy. For example, England's reliance on dilution and dispersal for landfill waste has been challenged by EC directives (Griffiths and Board 1994), as has its approach to, and standards for, nitrate in drinking water, and the subsequent need to manage nitrate inputs (Harris and Skinner 1992). (The relationship between the EC and member states is discussed in more detail in chapter 25.)

Much as states of the United States retain control over groundwater regulation, the tradition of relatively independent institutional authorities has been important in U.K. groundwater management (Harris and Skinner 1992). The Severn Trent Regional Water Authority was particularly influential in establishing groundwater policy guidelines through the publication of its aquifer protection policy in 1976. Although concentrating primarily on point-source pollution, the policy was groundbreaking in stating the need for both source and resource protection.

Although regulatory agencies have concentrated on the point sources, diffuse pollution presents the biggest overall threat to groundwater quality in England and Wales. Nitrate, solvents, pesticides, and chloride contamination have affected far more groundwater resources than have landfills, septic tanks, or other point sources of pollution.

Spain

Most of Spain differs from many of its EC neighbors in having a geology more similar to that of North Africa than Northern Europe. The country is largely arid, with deep surface deposits that commonly require wells more than 500 m deep. The geography is variable, however, and does include smaller Karst regions marked by increased pollution sensitivity. Because fresh water is scarce in many areas, user competition is severe, which may explain the historical concentration on water supply issues rather than water quality concerns. In fact, water quality problems are not widely acknowledged in Spain. According to Custodio (1991), the overwhelming concern for groundwater is drawdowns, some of which are on the magnitude of hundreds of meters. Custodio also reports a telling joke in Spain—that Spain's groundwater cannot be polluted because the percolation rate of pollutants is less than the drawdown rate of the water table.

Despite claims that Spain's groundwater is not polluted, Custodio reports convincing evidence to the contrary. Along the intensively cultivated coastal plains, the situation is particularly serious, with some wells measuring nitrate levels 50 times higher than the World Health Organization's recommended standard of 10 mg/L. Wells have been closed in some areas due to high levels of nitrates derived from agricultural production. In other areas, agricultural land uses cause sulfate and bicarbonate increases in well water due to the use of fertilizers. Data on pesticide movement through Spanish soils are regarded as confidential and are not available to the public. Other significant problems result from food processing industries, in particular production of olive oil, wine, and alcohol; several of these plants are located in the highly sensitive Karst regions of Spain. In addition, groundwater in Barcelona and Castillo de la Plana suffers from historical release of waste from closed metallurgical and leather factories. Barcelona's aquifer also suffers from disposal in gravel pits of chlorinated solvents from chromic acid production. Municipal trash has been discarded in gravel pits as well, due to delays in siting a new landfill. Further problems throughout the country arise from untreated municipal effluent, salinization, and mining spoils.

Despite these incidents, there has yet to be a comprehensive assessment of Spanish groundwater quality. The few known incidents suggest serious problems, but groundwater quality has been overlooked in Spain's water resource management. In fact, Custodio suggests that in many ways Spain's groundwater policy is simply not to study the situation.

A Case Study of the Tennessee Valley

The summary that follows (excerpted from Duda and Johnson 1987) illustrates many problems common to many groundwater pollution cases. First, groundwater pollution often "sneaks up" on people. Because of the long travel times of some groundwater, pollutants commonly surface after significant delays, and over wide areas. Consequently, identification and cleanup are difficult processes. A second problem is the difficulty encountered when responsible institutions are either poorly coordinated or no clear line of authority is established. A third issue is the need for proper zoning in areas with particularly high sensitivity to groundwater pollution.

Geology

More than half the Tennessee Valley is composed of Karst formations, the type of aquifer most highly vulnerable to pollution. Surficial aquifers underlie much of the rest of the region, with only a few aquifers being deeper and protected by overlying layers of clay.

Institutions

Throughout the 1980s, more than 60 federal and state agencies were involved in groundwater management in the region; in one state alone, 14 different state agencies were involved. Thus, fragmentation of responsibility among the agencies, duplication of effort, and unaddressed pollution sources were identified as problems. No programs existed for comprehensive monitoring, classification, or standards systems. Institutional barriers to effective management have been maintained through political resistance to change, traditional land use values and practices, inadequate funding, emphasis on economic development, and public apathy.

The Source of the Problem

Most pollution sites result from industrial waste disposal, leaking storage tanks, or oil and gas extraction operations (Fig. 10.4). Coal mining and oil and gas operations throughout Appalachia have led to serious degradation of portions of aquifers. Iron, dissolved solids, and acidity make the waters from these aquifers unsafe to drink. Oil and gas extraction has contaminated water supplies with brine, oil, gas, sulfur and surfactants. In addition to these causes, area farmers employ two practices that contribute to groundwater contamination. The first is drilling drainage wells to dispose of storm water and its pollutants. The second is filling sinkholes to plant crops. Not surprisingly, extremely high bacterial levels have been found in groundwater of some agricultural areas.

Wastewater disposal is the most widespread source of groundwater contamination in the Tennessee Valley. In particular, malfunctioning and poorly sited septic systems represent a major source of contamination. For example, Tennessee regulations allow blasting in rock for the placement of septic tank field lines, a practice that introduces human wastes directly into the groundwater.

The Dimensions of the Problem

According to Duda and Johnson (1987), 13 of 17 sites tested by the Tennessee Valley Authority in the late 1980s did not meet state drinking water standards because of bacterial concentrations. At 10 sites, anthropogenic organic chemicals were detected; public health criteria were exceeded at five of the sites. Statistics such as these are found throughout the region. For example:

- 75% of public water supply springs in a four-county area of western North Carolina were contaminated with fecal bacteria.

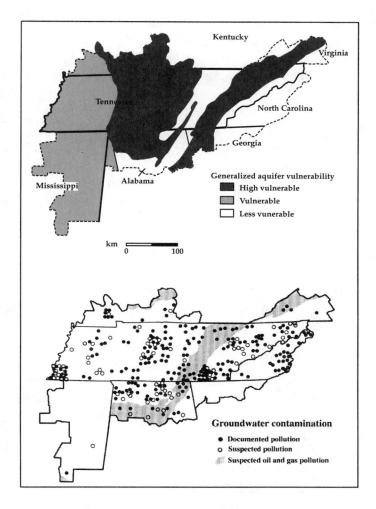

Fig. 10.4 Aquifers in the Tennessee Valley region differ in their susceptibility to contamination, depending on subsurface geology. Groundwater contamination patterns follow areas of geologic susceptibility. (Redrawn from Duda and Johnson 1987)

- In suburban Nashville and Knoxville, up to 80% of the wells in subdivisions are contaminated with bacteria.
- Toxic and explosive fumes have been detected in 51 locations in Bowling Green, Kentucky.

Policy Options

Duda and Johnson (1987) suggest several policy alternatives for the Tennessee Valley. As indicated earlier in this chapter, education is among the most powerful—

if difficult to implement—policy areas for groundwater protection. This theme is prominent in the following recommendations for the Tennessee Valley:

- Encourage local initiative and streamline institutional responsibility: Local communities are experiencing vastly different water quality problems and a program that enables local initiative may be the most effective approach to achieving potable water.
- Institute a policy of nondegradation of groundwater: Because the groundwater is dangerously polluted already, further degradation would be a serious hazard. A formal policy of nondegradation would provide the legal base for dramatic improvements in zoning laws, site cleanup, and other necessary actions.
- Work for groundwater to become a priority issue: Until groundwater is acknowledged as a critical issue, it will not receive adequate funds or legislative force to achieve the necessary management activities.
- Establish unified management of surface and subsurface flow: The Tennessee Valley is still being managed as though groundwater and surface water were separate features. That strategy is clearly unsustainable. By recognizing the connection between the two resources, management of groundwater quality may receive greater attention and influx of technical support.
- Restructure traditional land management practices: Traditional land management practices are one reason area water quality is suffering. A campaign to educate residents about the consequences of their traditional practices could significantly reduce further pollution.

Case Study: The Upper Rhine Aquifer

The following case study, excerpted from Meybeck *et al.* (1989), discusses the problems experienced in the heavily settled Rhine Valley. The upper Rhine aquifer is the most important fresh water resource in Western Europe. Almost 40 million people live in the Rhine basin. It is the site of a large part of European industry and almost 20% of the world's chemical industry. The aquifer provides more than 60% of the total needs for domestic, industrial, and agricultural uses in the region. The quality of its water is steadily deteriorating, however, as increasing pollution from industrial development and inputs from intensified agriculture find their way into the aquifer.

As in the Tennessee Valley, the diversity of responsible institutions complicates effective management of the upper Rhine aquifer. The fact that the aquifer and the affected region cross so many international borders aggravates the institutional complexity. Until recently, deterioration of the aquifer was also attributable to ignorance. For example, mining spoils were not regulated because their effects were not known. According to Meybeck *et al.* (1989), the region currently suffers most from momentum: development patterns in this densely populated farming and industrial area are resistant to change.

Geology

The upper Rhine aquifer falls along the Rhine rift and is composed of very permeable alluvial deposits that have accumulated along the fault trough. Flow rate is between 3 and 8 m per day. Overlain with only a few meters of shallow, permeable soils, the aquifer is vulnerable to chronic and accidental pollution. In fact, the only significant difference between the aquifer and surface water contamination (e.g., in the Rhine River itself) is the time scale involved. Contamination is delayed in the aquifer, sometimes by several years, as pollutants filter through alluvial deposits.

Institutions

The Rhine River and its underlying aquifer have not suffered from lack of attention, but certainly suffer from lack of coordinated action. The list of responsible institutions is long, and the exact responsibilities difficult to decipher. Because the Rhine aquifer affects Switzerland, Germany, France, and the Netherlands, significant efforts have been undertaken to institute international monitoring and regulation. In 1950, the International Commission for the Protection of the Rhine Against Pollution began studies to determine the origin and importance of different pollutants. The EC has a role in protection and management of surface and ground waters of the Rhine, as does the International Committee of the Water Authorities of the Rhine Basin. Two other international commissions, the International Commission on the Hydrology of the Rhine Basin and the Council of Europe, share responsibility for managing the resource. Beyond these commissions, each country has regulatory authority; France alone has three principal authorities for managing the waters of the Rhine River in the Alsace region. Detailed studies of the aquifer have been ongoing for more than 15 years.

Pollutants of Concern

To date, nitrates and chlorides have been the most important sources of pollution. Nitrate use has doubled in the last 10 years as grassland was converted to maize, lands were irrigated, and industrial and urban developments continued. Estimates are that, at the current rate of pollutant increase, water in the entire upper Rhine aquifer will be unfit to drink by the year 2000. Current nitrate concentrations are illustrated in Fig. 10.5.

Chlorides originate from potassium mine spoils. Both exploitation of the potassium and elimination of the waste are major sources of pollution to the aquifer. Leaching of the spoil dumps occurs on a massive scale, and sodium chloride concentrations in the water are near the saturation point. On average, 170,000 tons of salt filter into the aquifer every year. The dumping of salt in surface waters was halted in 1974 and pollution control wells have since been drilled downstream from the spoil dumps. These measures have led to a decrease in mineralization, but aquifer contamination is still significant and continues to work its way through the aquifer. Other sources of chlorides are the dumping of brine directly into the Rhine, the Lorraine salt mines in France, and industrial development in Germany.

Since the 1980s, pollution by organochlorines from insecticides has also become a concern. Currently, this pollution is localized but it has important economic consequences because it leads to deterioration in the quality of the water in urban areas.

Dimensions of the Problem

- In the Netherlands, chloride levels have increased from 60 mg/L^{-1} in 1930 to more than 170 mg/L^{-1} in 1975.
- Between 1980 and 1983, more than two tons of organochlorine products were removed from the aquifer from a single pollution control well (whose filter was placed 28 m deep) below an abandoned solvent dump.
- In 1986, a fire occurred in a chemical factory storing 1250 tons of toxic substances (i.e., insecticides, fungicides, herbicides, mercury-containing compounds). Many chemicals were washed into the Rhine with water used to extinguish the fire. Within weeks, organophosphorous compounds contaminated groundwater wells; within a month, mercury was evident. Contamination from this one source was evident throughout the late 1980s (i.e., for at least 5 years).

Management Options

- Pollution control wells continue to be an effective, if stop-gap method of removing pollutants from known point sources.

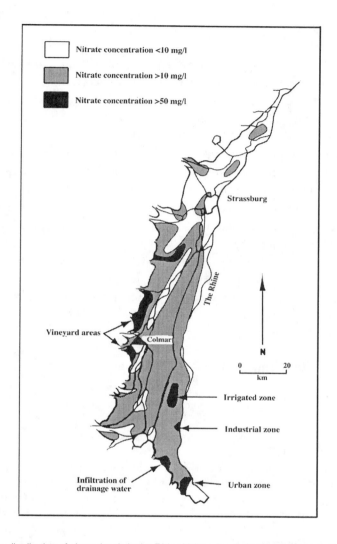

Fig. 10.5 The distribution of nitrate levels in the Rhine Valley closely follows land use patterns, with elevated nitrates evident in industrial, agricultural, and urban zones. (Simplified from Meybeck *et al.* 1989)

- Land use zoning may play a major role in slowing the rate of land conversion and nitrate pollution as well as in protecting sensitive recharge areas from pollution.
- Cleanup and confinement of all abandoned mine spoils.
- Improved international coordination and improved local authority over polluters.
- Stricter enforcement of brine dumping and other related practices.

Summary

- Groundwater was long regarded as comparatively infinite in supply and as a resource relatively immune to large-scale impacts (although local impacts were widely recognized). We now recognize that groundwater systems are often highly fragile, recover very slowly from impact, and often are of surprisingly limited quantity. This new understanding dictates a new paradigm for managing anthropogenic impacts to groundwater—a paradigm of very long timeframes and much more carefully considered decisions.
- Groundwater sensitivity is determined principally by geology, which in turn regulates geohydrology. Impacts can come from relatively direct sources (e.g., unsealed wells, underground storage tanks) or diffuse sources (e.g., land use practices).
- Cleaning a contaminated groundwater resource is a very difficult task, often requiring removal (i.e., pumping the water to the surface for treatment) and reinjection. Prevention is far more effective than remediation in groundwater management.

For Further Reading

Duda AM, Johnson RJ. Targeting to protect groundwater quality. *J Soil Water Cons* 1987; 42:325–330.

Environment Canada. *Groundwater—nature's hidden treasure.* Water series #5. 1990. Cat No En 37-81/5-1990E.

Geier TW, Perry JA, Queen L. Improving lake riparian source area management using surface and sub-surface runoff indices. *Environ Mgt* 1993;18:569–586.

Griffiths CM, Board NP. Approaches to the assessment and remediation of polluted land in Europe and America. In: Eden G and Heigh M, eds.: *Water and environmental management in Europe and North America: a comparison of methods and practices.* New York: Ellis Horwood, 1994;11–19.

Newson M. *Land water and development: river basin systems and their sustainable management.* London: Routledge Press, 1992.

Vanek V. The interactions between lake and groundwater and their ecological significance. *Stygologia* 1987;3:1–23.

11

Coastal Zone Water Quality Management

Overview

The coastal zone is a vital transition area between land and sea that is affected by both terrestrial and oceanic forces. In rivers, water flow is one-way; in lakes, the basins are largely stationary; in groundwater, movement is constrained by geology. In contrast, coastal zones are areas of dynamic water movement that integrate all of these modes of water movement and add the extra dimensions of tidal forces and salinity. Consequently, fresh water flows into coastal zones via rivers and groundwater tidal flows, waves, and the unique physical properties of saline water create a complex and fragile balance in coastal zone ecology.

Water quality management in the coastal zone has traditionally concentrated on inshore areas such as estuaries, bays, and lagoons; these areas are heavily populated and ecologically sensitive. Problems associated with anthropogenic influences such as waste disposal, development, industry, and agriculture include transmission of disease, deoxygenation, nutrient enrichment, toxicity, and interference with aesthetic values. Areas of special concern include estuaries and coral reefs, whose unique ecology requires special management techniques. Because coastal zone management is intersectoral, many countries have adopted unique approaches to coastal zone management. Only a few of these include specific provisions for, and attention to, the vital quality of coastal waters.

Introduction: Unique Challenges of Coastal Regions

Coastal regions are among the most productive ecosystems in the world, exemplified by the fact that coastal habitats provide feeding and reproduction ground for approximately 90% of the world's marine fish catch (World Resources 1992). The marine catch, in turn, constitutes more than 85% of the world's total fish catch. Coastal zones are also typically the most densely populated areas: two-thirds of the entire world's population lives within 60 km of an ocean, and in areas of Asia, Central America, and the Caribbean, population density along the coasts has reached as high as 2000 people per square kilometer (Lundin and Linden 1993). In the United States, the population density of coastal areas is five times the national average, and it is projected that 75% of the population will live within 75 km of a coast by 2010, up from 50% in the early 1990s (Hanson and Lindh 1993). While the attraction of the coastal regions stems from its diversity of services, the stress on coastal zone ecology is enormous and the continuation of those services is in doubt. Marine fish catches have already begun to decline, and certain coastal habitats such as mangrove swamps have declined by more than 50% globally.

Coastal zones include areas that vary dramatically in their geological formation, hydrology, ecology, and human use; they incorporate shore areas of the Pacific and Atlantic Oceans as well as the myriad of more confined seas such as the Mediterranean, Bering, South China, and North Seas. The variations in size, confinement, salinity, fresh water sources, and tides all connote very different sensitivities to water quality degradation and potential for improvements. Rocky shores, for example, are traditionally given less attention in water quality management because they receive fewer direct inputs, have greater flushing capability, and require a nonlocal management perspective. Instead, water quality managers have concentrated activity on estuaries, bays, lagoons, harbors, and other inner waters around which most human settlement and use in centered. Besides supporting highly concentrated human use, these systems are very sensitive and easily disrupted by changes in water flow, sedimentation, temperature, nutrients, or toxic chemicals.

This review is limited to those features of the coastal zone that inspire water quality concern (i.e., how water moves into coastal zones, human impacts, and effects of those impacts on coastal systems). The reader interested in more information on coastal ecology and management is referred to the large volume of literature on the subject (see the list of references at the end of this chapter).

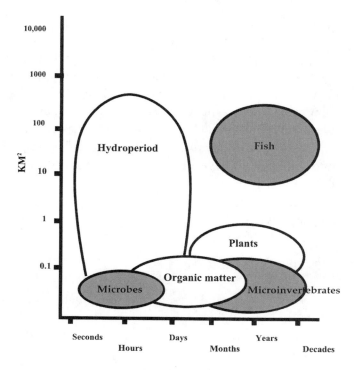

Fig. 11.1 Dynamics of coastal ecosystems are dominated by hydroperiods. Coastal waters are relatively uniform over large spatial scales, but vary with daily and seasonal tides. Most biological communities of interest to coastal zone water quality management are associated with near-shore conditions and thus vary more in space (i.e., among bays and regions) than through time. The exception is fishes, which are associated with open water areas and vary widely through both time and space. These scale issues complicate management, but recognition of them may also provide an important key to planning and implementing improved coastal management efforts.

A Continuum of Influences

The coastal zone is a land/oceanic continuum ranging from deep water, saline, wave-adapted coral reef systems to upper estuary, shallow, fresh water-dependent systems, and encompassing all areas in between. Within this continuum, an extraordinary diversity of community structures, representative species, and controlling environmental conditions occur over a very small area. Management concerns within this zone are equally diverse as illustrated by the spatial–temporal dynamics of representative components of the coastal zone shown in Fig. 11.1. Approaching coastal zone management requires subdividing the transition area into at least three sub zones: (1) estuaries and other "inshore" areas;

(2) coastal groundwater; and (3) seashore areas. A fourth subzone—deep off-shore areas—is an emerging concern, with dumping of plastics and accumulation of toxic compounds potentially threatening large areas (Laws 1993; World Resources 1992). To date, most of the scientific studies on deep offshore areas contain conflicting evidence and offer few management alternatives except for halting dumping at sea. Because the ecology of deep offshore zones is still largely unknown and the magnitude of pollution problems still speculative, this chapter will concentrate on the comparatively better studied and more intensely used inshore, groundwater, and seashore subzones. The boundaries of these subzones are not absolute or clearly defined, but the convention of dividing the coastal zone into these subregions assists in clarifying different water quality concerns derived from variations in ecological conditions and land use stresses.

Seashores, the Land–Ocean Boundary

The seashore area is a narrow band representing the actual landward extension of salt water. Both the value of this zone and its ecology are most closely related to ocean characteristics such as coastal currents, tidal flow, and wave action. Seashores, however, receive fresh water and land-based input from groundwater, rivers, and overland flow. The flow paths, rates, and quality of these waters influence the ecology of seashore communities and control the susceptibility of these communities to changes in water quality.

Water quality of seashore areas is of concern primarily from three perspectives: (1) recreation; (2) fisheries; and (3) ecological integrity. The first two concerns center around use patterns and resultant value of the resource for human use. Ocean beaches are one of the most popular vacation destinations worldwide, and commercial and recreational fishing (including shellfish) are multibillion-dollar industries. The third perspective, ecological integrity, is impossible to value objectively, but it is increasingly recognized as a vital concern.

Inshore Areas

In the inshore subzone, fresh and salt water mix and land-based forces of fresh water run-off become increasingly important. From a biological point of view, fresh water dilutes the salt and thereby enables the existence of an enormous variety of life forms. The moderated temperatures, shallow depths, and sheltered geography characteristic of inshore areas such as bays and estuaries represent important factors in their enormous productivity and diversity. These same features make inshore areas highly valued for human settlement because they are sheltered, fed by fresh water, and highly productive. Inshore areas are among the world's most densely inhabited ecosystems.

Inshore water quality issues include accumulation of toxicities, waste disposal, changes in salinity, and temperature alteration. Urbanization, shipping and boating, dredging and other development activities, creation of dry land, and industrialization are some uses that create water quality concerns in these sensitive ecosystems.

Coastal Groundwater

Coastal groundwater is a very different "zone" than inshore or seashore areas. Subject to far different spatial and temporal scales and environmental conditions, it is nonetheless a vital, albeit vertically defined, subzone on which human coastal communities depend. Salt water is not potable; a 2% salt water concentration is considered undrinkable by humans. Consequently, most coastal zone surface waters are nonpotable, and humans have a tremendous dependence on coastal zone groundwater for habitation needs. In many regions dependent on local supplies of fresh water, coastal groundwater is the only source of potable water. In other areas (e.g., the Netherlands), shallow groundwater aquifers provide water for both domestic and industrial use. Coastal groundwater quality management must address all of the issues of leaching and percolation of pollutants that concern inland groundwater managers. It must also address the significant problem of salt water intrusion.

The most common cause of salt water intrusion is overpumping of coastal aquifers, spurred by growing populations and increasing demand. Another cause of increased salt water intrusion involves the lowering or destruction of coastal dunes. Because dunes serve as recharge zones, they act to maintain high water tables. Their destruction lowers the water table and enables salt water to move inland. Development activities may also cause saline drinking water by draining coastal wetlands and dredging navigation channels. These activities may breach the subsurface brackish "interface zone" and leave local, but severe, increases in salinity.

Water Quality Risks: Effects of Flushing, Sediment Type, and Latitude

Two features—capability of flushing and sediment type—explain a tremendous amount of the susceptibility of each of these three coastal zones to decreased water quality. A third—latitude—reflects differences between productivity in temperate and tropical regions and subsequent differences in the valuation of coastal ecosystems.

Flushing

Generally, the more open a coastal system is to the flushing effects of waves, tides, and currents, the less susceptible it will be to decreased water quality as dilution and dispersal will buffer the impacts of pollutants. The opposite is also true: the more enclosed a body of water, the less it will benefit from tidal flushing and the more susceptible it will be to anthropogenic inputs and modifications. Consequently, even small pollution loads can degrade the environment of a shallow, enclosed estuary, while a larger, open estuary might resist degradation from substantially larger inputs because those contaminants would be flushed out to sea by tidal currents (World Resources 1992).

Coastal zone water quality management relies on the assimilative capacity of the ocean and the flushing effects of tidal action. Because oceans constitute more than 97% of the total volume of water on earth, their assimilative capacity taken as a whole is enormous. An emerging reality in water quality management, however, is that polluted coastal waters tend to stay nearer to shore than once believed. Consequently, the effective assimilative capacity of ocean waters is much smaller than its potential capacity.

In addition, the tendency of polluted waters to stay near shore contradicts the traditional assumption that tidal flushing always brings cleaner water into the coastal zone. Improved analysis in combination with increased accumulation of pollutants now suggests that some ocean currents may serve as much to redistribute pollutants as to dilute them. For example, Central American coral reefs exhibit heavy metal concentrations that are too large to be of local origin. Instead, currents from northern South America (off of Brazil, Venezuela, and Colombia) apparently carry their industrial pollution to Central American reefs and, potentially, beyond. On a more local scale, deep currents have been shown to be responsible for moving deposits of wastewater sludge back on shore, as occurred along Long Island, New York, in the late 1980s. Thus, a long-term view of oceanic currents increasingly challenges the idea that flushing is always positive.

Sediment Type

Sediment type is another critical determinant of coastal water quality. Coastal zones are generally dominated either by sands that easily percolate and transport waters or by fine sediments that tend to trap pollutants, especially toxics.

Sediment type is also an important feature in coastal erosion. According to Hanson and Lindh (1993), 20% of the world's coastal areas are sandy, and more than 70% have shown a net erosion over the past decades. Some of the erosion is dramatic, with coastal retreat of 25 m per year documented in Togo and retreats between 6 and 10 m per year being more commonplace in all continents. In the United States, only 12% of the coastline is considered stable. Some of this erosion may be the consequence of sea level changes, while the majority is more immediately attributable to the alteration of sediment flow patterns via construction of reservoirs, dams, and shoreline development. The long-term water quality implications of erosion are large, as these areas play an important role in groundwater recharge.

Latitude and Productivity

Not all coastal zones are alike, and differences in water flow rates, seasonal temperature fluctuations, tidal movement, and coastal geography all influence the relative importance and fragility of different coastal ecosystems. One of the largest scale differences is that distinguishing temperate zone coastal ecology from that of tropical zones. As described by Lundin and Linden (1993), the seasonal cooling of temperate waters creates an environment in which surface waters cool, eliminating temperature and density stratification in temperate oceans. Much as in a large lake (see chapter 12), this transformation enables surface and deep nutrient-rich waters to mix, providing nutrients to surface waters so that they can support blooms of microscopic plankton. In turn, the plankton blooms support enormously productive offshore fisheries and loosen the dependence of the temperate zone oceanic fish on coastal ecosystems.

In contrast, ecology of tropical oceans and coasts is controlled by persistent stratification of ocean waters. The stratification leads to little mixing of deep and surface waters and, consequently, tropical oceans have very low levels of primary production. In these areas, the dependence of human and aquatic populations on the shallow, productive near-shore areas is greater. According to Lunden and Linden (1995), more than 90% of the fish catch in the tropics is taken in shallow coastal waters, and the vast majority of commercially important fish depend on tropical coastal ecosystems such as mangrove forests or coral reefs. Some countries such as India and Thailand have lost more than 80% of their mangrove forests, however (World Resources 1992). Countries that have lost more than 70% are too numerous to list, but include Asian countries such as Bangladesh and African nations such as Somalia, Guinea-Bissau, and Ghana. The average loss of pre-agricultural mangrove area in all tropical countries substantially exceeds 50%. Habitat destruction of coastal regions is, according to Yap (1992), the dominant marine environmental problem in East Asia. Given the magnitude of this problem, the importance of halting the destruction of tropical mangrove forests becomes more than a paternalistic "north–south" environmental issue and can be seen as an issue vital to the health and viability of these ocean-dependent states and their inland neighbors.

Coastal Zone Uses and Their Impacts

Generally, human uses of coastal areas can be divided into three categories. The first, which comprises low-impact uses that depend directly on high water quality, includes uses such as recreation, fishing, some aquaculture, conservation, and potable drinking water. The second category includes uses that generally have a heavier negative impact on the water and can continue to be productive even with poorer water quality. Included in this category are mining, transportation (shipping), construction, industry, and urbanization uses. Urbanization pressures may be especially acute in developing areas in which waste is not treated before release to coastal waters. A third category includes those land uses such as forestry and agriculture that are located on or near coastal areas. They are major contributors to nonpoint pollution of coastal waters, even though their own dependence on the coastal waters is often more indirect.

The distinction between these categories is not always absolute. Heavy recreational use of fragile ecosystems can, for example, severely degrade resources. In general, however, water quality impacts result more from nutrient enrichment, organic matter input, siltation, and habitat destruction than from heavy recreational use. Commonly cited coastal area water quality concerns are: litter from ships, boats, or shore; sewage, including pathogens and heavy metals; petroleum; sedimentation; fertilizers and pesticides; and nutrient inputs (Windom 1992).

In developing nations, untreated domestic and industrial wastes are expected to constitute the major sources of coastal pollution for many years. Unlike the highly industrialized world, developing nations face acute shortages of research funds, equipment, and qualified environmental scientists to conduct regular pollution monitoring programs. In addition, most developing nations are experiencing high population growth rates (commonly 2% to 3% annually, as compared with 0.9% in the United States). Some coastal cities such as Dar es Salaam have a growth rate of nearly 5%, with population doubling every 7 years (Machiwa 1992). This rapid rate of urbanization without parallel infrastructure improvement leads, inevitably, to increases in pollution.

Ocean shipping and local dumping have long been viewed as the main culprits in coastal pollution, but more recent studies and estimates indicate that upstream run-off and atmospheric sources of pollution contribute the largest share of coastal pollution: fully 80% of marine pollution stems from land-based sources that reach the oceans via the atmosphere, direct discharges, and upstream run-off (Windom 1992). Some discrepancies in the source-input data indicate that, at smaller scales, this average may misrepresent the importance of distant sources. For example, in U.S. estuaries, 70% to 80% of the nutrient pollution derives from coastal sources, with phosphorus inputs being largely point sources and nitrogen inputs being more diffuse (World Resources 1992).

Management of these pollutants may be approached on a variety of spatial and temporal scales. At local scales, heavy metal pollution may be critically important, as seen in Japan's infamous experience with mercury poisoning of Minimata Bay in the 1950s and 1960s (Nash 1993). Groundwater quality is also likely to be a local-scale phenomenon that follows patterns of geology and coastal development. Windom (1992) suggests that litter and oil pollution are more midscale phenomena—unlikely to have wide regional or global significance but demonstrating a significant impact to specific species and industries. According to Windom and other members of the Joint Group of Experts on the Scientific Aspect of Marine Pollution (GESAMP), the three sources likely to be of large-scale importance in the future are nutrients, synthetic organic compounds, and sewage (Windom 1992). Their significance results from the escalating use of fertilizers and pesticides, from population increases and from the incomplete knowledge of the pathways of these compounds in marine environments.

The following sections focus on selected coastal land use practices that are significant from a water quality perspective. The discussion is meant to be illustrative, not exhaustive.

Agriculture

Agriculture is responsible for a significant amount of coastal zone water quality problems because of its association with herbicides, pesticides, and fertilizers. The magnitude of agricultural sources relates to the increased use of agricultural inputs (see chapter 19) and to the proportion of coastal land devoted to agriculture. In the United States, one-third to one-half of the land in coastal counties is used for agricultural purposes, a proportion that varies regionally; similar percentages are found throughout the world. Major coastal water quality concerns resulting from those agricultural areas include inputs of organic matter, fertilizers and insecticides, alteration of run-off patterns, loss of coastal wetlands, and sedimentation. A few of the most important issues are discussed below; management options for reducing agricultural impacts are discussed in chapter 19.

- Insecticides such as DDT and HCH are banned in most developed nations because of their long-term persistence and potential for bioaccumulation. These compounds are still in use for crop protection and public health (e.g., control of disease vectors) in many less-developed nations. In fact, use in those areas is often indiscriminate and widespread. High concentrations of these insecticides have been reported in the Bay of Bengal and along the coastlines in Latin America and Africa.
- Sedimentation from cropland yields four times more sediment to coastal public waters than any other erosion source.
- Feedlots are often placed along streams that discharge into coastal waters. Often they introduce animal wastes directly into surface waters.
- Irrigation stresses coastal systems because the diversion of river water can alter timing and the supply of fresh water to coastal water basins, disrupting salinity, nutrient, and circulation patterns (see the Aral Sea example in chapter 1).
- Fertilization is important to many farms because soil nutrient deficiency may limit yields. Nitrogen is a key nutrient for algae production in brackish water, however, and many coastal systems are nitrogen-sensitive. Excessive amounts of nitrates from agricultural run-off leads to eutrophication of coastal waters (see chapter 21).

Industry

The coastal zone provides ready sources of power, proximity to ports, and important raw materials. Thus, it is a desirable location for many industries; in fact, 40% of all manufacturing plants in the United States are located in the coastal counties. These coastal industries are dominated by paper, chemical, petroleum, and primary metals production.

Heavy metal pollution from coastal industry effluents has been responsible for dramatic, if localized, health problems. Two notable cases occurred in Japan: more than 100 deaths from Itai Itai disease (cadmium poisoning) and mercury poisoning of Minimata Bay that affected nearly 2000 residents through consumption of contaminated fish between 1960 and 1990. At Minimata Bay, pollution resulted from the intentional dumping of approximately 150 tons of mercury into the bay. Despite cessation of dumping, cleanups took decades and included dredging and vacuuming contaminated sediment. Cleanup efforts benefited from a large storm in 1982 that caused sufficient erosion and subsequent sedimentation to bury a significant portion of the remaining contaminated sediment. The storm's economic value, in terms of cost savings for cleanup activities, were estimated at $96 million (Kudo and Miyahara 1992).

While industrial waste in most developing countries is secondary to issues of sewage outfall, industrial pressures on coastal areas is an increasing concern in Asia (Yap 1992) and Africa (Machiwa 1992). Heavy metal pollution along China's coast may be among the highest in the world. According to Platte (1995), more than 1500 tons of heavy metals wash into the Yellow Sea annually, where they concentrate in the surface layers of the sediment. Dangerous levels of cadmium, copper, and mercury have all been reported in crustaceans and bivalves from the Yellow Sea. Dar es Salaam Harbor, on the other hand, suffers from the dumping of metallic wastes, dyes, and corrosive acids (Machiwa 1992). Lack of regulation and lack of awareness of the effects of wastes are problems on nearly every continent.

Urbanization

Urbanization and development along coastal zones affects coastal ecology in many ways. From a water quality perspective, sewage remains one of the most critical issues, but is only one of a complex of problems that coastal zone managers and communities must address (see also chapter 20 for a discussion of management options).

SEWAGE OUTFALL. Ocean dispersal of sewage is an ancient concept and is still practiced extensively. Although debate continues over its impact if systems are designed appropriately, all sewage outfalls place demands on the assimilative and distributive properties of the receiving water. As discussed earlier, offshore disposal is based on the concept that the ocean has a nearly unlimited assimilative capacity, but both deep ocean and tidal currents may limit the spatial diffusion of pollutants.

To disperse the large quantity of effluent produced by many communities in developed countries, sewage outfall pipes have become longer (pipes upward of 6 km are increasingly common) and dispersal areas have increased. These deep outfalls take advantage of different salinity gradients to reduce the chance of waste recirculating to the surface. Deep sea waters are typically more saline and, therefore, more dense than surface waters. Consequently, piping an effluent below a *salocline* will reduce its vertical circulation. Nonetheless, upwelling along the coasts does bring up deep ocean waters, which some observers suggest will act to return the pollution back to the surface. Whether the pollution is effectively buried or not, questions remain as to impacts of these effluents on ocean benthic communities and on the real assimilative capacity of the ocean. Along the New York, New Jersey, and Southern California coasts, the quantity of sewage has overloaded the assimilative capacity of the receiving water and benthic dissolved oxygen levels have fallen below the minimum that can support aquatic life. In addition, poorly understood ocean currents have washed wastes to shore in many areas.

In many developing countries, sewage is discharged with no treatment and inadequate funding for sewage disposal results in near-shore deposition, rather than deep offshore disposal. Ironically, in some areas such as the Azores, the pollution disposal problem is partially a result of a burgeoning tourist industry. Tourism, in turn, suffers when amenity and health risks increase (Depledge *et al.* 1992). Sewage-related degradation has been noted in all continents, including such diverse coastal zones as those along the South China Sea, the Philippines, Singapore, Malaysia, Indonesia, Sri Lanka, the Persian Gulf, Caribbean, Colombia, and Cuba (Lunden and Linden 1995).

LANDFILLS AND SOLID WASTE. Leachate from inadequately lined landfills or dumps along the coast may carry large amounts of heavy metals, bacteria, insecticides, cleaning products, and other pollutants to coastal waters. Although these materials have serious consequences for human and coastal ecosystem health and have been documented as pollutants of noncoastal groundwater, the flow of groundwater to coastal areas has received comparatively little study. Landfill leachate has, however, been identified as a major source of PCB pollution in coastal areas.

URBAN RUN-OFF. In an unpaved area, less than 10% of rainfall may become run-off, while the remainder is taken up by plants or filtered into the soil. In a highly urbanized area, as much as 55% of the rainfall may leave as run-off. Urban run-off, by itself, is often sufficient to pollute coastal waters. For example, despite a billion-dollar sewage treatment plant, urban run-off from Long Island brings enough bacterial contamination to the shore that some New York shellfish beds will have to remain closed permanently. Sources of the contamination include street litter, household refuse, automobile drippings, pet wastes, spillage from bulk chemical storage, and exposed dirt piles (see chapter 20).

SALT WATER INTRUSION. Groundwater resources face growing pressure from coastal communities as shore-based aquifers are pumped for industrial and domestic water use. In southern California, Florida, New Orleans, and Long Island, the aquifer has already been overpumped and contaminated by salt water intrusion (see chapter 10).

Oil

Oil spills are dramatic, visible, and locally important. Despite enormous quantities of oil spilled onto coasts worldwide, however, spills represent a relatively minor concern compared with the more chronic problems outlined above

(Mitchell 1993; Shaw 1992). Wave action scours even badly coated shores, and migration from nearby shore areas enables biotic communities to recover relatively rapidly. Prince William Sound has shown remarkable recovery from the apparently devastating 1989 Exxon Valdez spill. Bald eagle nesting has recovered and invertebrate and plant recolonization has begun. On the other hand, long-term disruptions of salmon runs have been noted, and otter, seals, and birds probably will not return in abundance until shellfish populations recover.

Oil pollution, in combination with other contaminants, has created a larger problem in the Mediterranean Sea. Particularly vulnerable because of low rainfall, nutrients, and species diversity, the Mediterranean is also especially heavily used. It is a major shipping and transport route and, even after the establishment of a Regional Oil Center in 1976 to reduce oil pollution, one-fifth of the world's oil spills in the 1980s occurred in the Mediterranean Sea (Platte 1995).

Despite the experiences of the Mediterranean, on a global level, intentional discharge of oil during tanker operations remains a much larger source of oil pollution than spills. Although intentional discharge has decreased by approximately 90% since 1970, its volume still far surpasses the volume of oil spills. Other important sources of oil, constituting approximately half of the total anthropogenic input of oil to marine systems, are land-based sources such as municipal and industrial wastes and urban run-off (Windom 1992). While not as dramatic as spills, these inputs represent a chronic level of contamination and may present a greater long-term threat to coastal communities and water quality than the dramatic but relatively rare spills.

Dredging, Filling, Dams, and Impoundments

While dredging, filling, dams, and impoundments have different goals, their effects on the coastal zone are similar in that each activity alters either flow of fresh water into the coastal zone or flushing action of tides and waves. Thus, they act to isolate coastal areas and increase the potential for concentration of pollutants. Sedimentation is also a common result of any construction activity and may affect the ecosystem through decreased water clarity (which affects productivity), as well as through smothering organisms and their potential food sources. These structures also alter the flow of sediment to coastal areas, and can contribute to severe coastal erosion, loss of coastal dunes, and changes in other coastal habitats.

Radioactivity

Radioactivity is another area of concern that is of minor importance now but has the potential to grow into a significant issue in the future. Radioactive

contamination from atmospheric sources (above-ground testing and accidents), ship collisions, and oceanic disposal of radioactive waste may all pose
significant localized water quality hazards in the future.

Focus on Estuaries

Water quality issues in estuaries are particularly compelling because of the extraordinary productivity and sensitivity of these aquatic ecosystems. In addition, some of the highest human population densities occur around estuaries,
and, as discussed earlier, estuaries provide breeding and feeding grounds for the
vast majority of commercial fish. Estuarine conditions are dependent upon a
specific combination of saline–fresh water flow, depth, and clarity that management must protect. These conditions, illustrated in Fig. 11.2, include:

- Confinement: This property provides shelter by protecting the estuary
  from wave action, enabling plants to root and shellfish larvae to attach and
  permitting the retention of suspended life and nutrients.

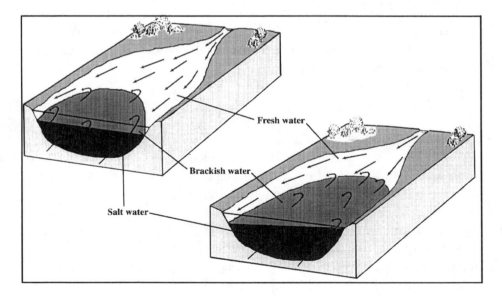

Fig. 11.2 Cross section of a typical estuary. Estuaries exhibit both vertical and horizontal gradients
that affect water quality. Salt water enters the estuary in a wedge along the bottom, mixes with fresh
water, and exits at the surface as brackish water. The degree of mixing is controlled in large part by
the flow of fresh water into the estuary. High fresh water flow will create a steep salt–fresh water
gradient and will prevent salt water from intruding far into the estuary. Low fresh water flows enable
salt water to intrude toward the river mouth. Coriolis (spin) forces of the earth frequently cause a
deflection of the salt water wedge. (Modified from Stowe 1987)

- Depth: Shallowness permits light to penetrate to plants over much of the bottom, allowing the growth of marsh plants and tide flat biota, improving flushing, and discouraging the incursion of oceanic predators (which avoid shallow waters).
- Salinity: Fresh water flow dilutes salt water and promotes an especially rich and varied biota. It also deters oceanic predators and creates a beneficial stratified flow of water.
- Circulation and tides: Fresh water outflow, tidal patterns, and salinity gradients together create a beneficial system of water movement and transport for suspended life (particularly effective when stratified). Tidal energy provides an important driving force; tidal flow transports nutrients and suspended life, and dilutes and flushes wastes. Tidal rhythm acts as an important regulator of feeding, breeding, and other functions.
- Nutrient storage and cycling: Estuarine plants—in particular, submerged grasses—have a high capacity for nutrient storage. Estuarine physical conditions also promote the retention and rapid cycling of nutrients.

Studies in North Carolina indicate that the greatest pollution control is achieved by giving priority to sources in and near the estuary (Phillips 1991). This statement does not imply that coastal water resource managers should ignore watersheds that feed estuaries. Inland sources can have significant influences on coastal water quality but, according to Phillips (1991), heavily concentrated coastal zone development has a greater impact than upstream sources. Phillips suggests that a watershed-based approach to water quality management is necessary for protecting coastal water quality, although the relative importance of inland pollution sources on the zone is often overestimated.

In contrast, studies of the Chesapeake Bay have tended to emphasize the importance of nonlocal sources in degradation of the bay. In particular, Correll *et al.* (1992) note that atmospheric deposition and nonpoint run-off are the major sources affecting water quality in the Chesapeake Bay. This bay is, however, an unusual estuary in that its drainage basin is enormous, spanning approximately 64,000 square miles and draining parts of six densely populated east coast states and the District of Columbia.

Implications for Management Strategies

- Prevent significant alteration of fresh water, and nutrient inflow.
- Avoid significant reduction of natural oxygen concentrations.
- Maintain natural vegetation.
- Avoid discharge of pathogens, toxicities, and suspended solids.
- Avoid reducing or blocking water flow into or out of estuaries.

- Maintain natural salinity regimes.
- Maintain the natural light and temperature environment.

Coral Reefs

Coral reefs are among the most biologically productive, diverse, and aesthetically pleasing communities in the world. Their practical values include fish
production and protection from waves along tropical coastlines. Reefs are also
easily damaged, however, and are increasingly threatened. Because corals generally grow slowly and their growth depends on the viability of conditions for
symbiotic coralline algae, reefs are slow to recover from damage. Even under
optimum conditions, reef recovery takes several decades. Yet, reefs such as
those along the atoll pictured in Fig. 11.3 are intimately connected to and influenced by shoreline features. Consequently, reefs suffer from a variety of anthropogenic impacts, as discussed below.

Sedimentation

Dredging and sedimentation from erosion have a severe effect on coral reefs by
reducing the light available to coralline algae, thus inhibiting coral growth. A
sufficiently thick blanket of sediment may kill expanses of coral within days.
Sedimentation also affects recolonization because the coral larvae and other
reef invertebrates cannot settle on soft, shifting sediment. In Kaneahoe Bay
Oahu, for example, sedimentation has been so great that water depth has decreased more than 6 feet because of poorly regulated land clearing in the adjacent watershed and resulting sedimentation (Laws 1993).

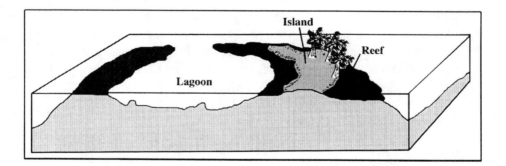

Fig. 11.3 Cross section of an atoll showing the restricted water flow into and out of a coral reef la-
goon. The restricted water flow, shallowness of the reefs, and their proximity to land make coral
reefs highly susceptible to land use impacts. (Redrawn from Stowe 1987)

Sewage

Respiratory rates of tropical organisms are higher than those of temperate species, although dissolved oxygen levels in tropical waters are lower. Consequently, most tropical organisms live closer to their lower oxygen limits. Sewage represents an enormous hazard to these communities because oxygen-consuming substances in sewage have a proportionately larger effect on these tropical systems. In addition, dissolved nutrient levels are generally much lower in tropical waters. While introduction of sewage in temperate systems may raise nutrient levels by 30% in a coral reef system, the same nutrient input may raise nutrient levels by 1000%. Sewage pollution reduces the diversity of species of corals and other reef organisms, as well as overgrowth of algae.

Fresh Water Run-off

Another result of poor land management is accelerated run-off of fresh water, which has lowered coastal salinities in some areas enough to kill shallow reef communities within hours after a major storm.

Other Effects

Pollutants that affect nearly all coastal systems also affect coral reef systems. Thermal pollution from industries alters reef communities; heavy metals (e.g., copper from desalination and power plants) are toxic to coral reefs. Chlorine (used to treat sewage outfalls) and oil spills in which oil comes into contact with coral have major impacts. Each of these pollutants is locally important, although their effects are minimal over the extent of global reef systems.

Implications for Management Strategies

- Improve land use practices that cause sedimentation (e.g., contour farming, buffer strips).
- Create coastal wetlands that can filter nutrients and trap sediment.
- Reduce fresh water run-off (buffer strips, limit channelization, control logging practices).
- Educate coastal communities about the effects of land-based activities on reef communities.
- Improve waste treatment and disposal methods to limit exposure of reefs to sewage.
- Create and enforce effluent standards for industries and coastal communities.

Coastal Management Institutions

Although many countries have adopted regulations concerning shipping and ocean dumping, few have adequately addressed the complex of land-based and diffuse sources that affect coastal zones. Many countries within the Organization for Economic Cooperation and Development (OECD) have recently acknowledged the inadequacy of previous regulations and have begun developing comprehensive coastal zone management plans. Translating this general awareness into management plans and priorities can be a slow process, and is often hindered by low public concern and existing sectoral planning structures (Hildebrand and Norrena 1992). Because coastal zones cross traditional institutional lines and because pollution in these areas is still not widely acknowledged, coastal zone management continues to be an irregular part of each country's water resources and water quality planning. The need for integrated coastal management to encompass all sources, scales, and planning levels makes it a difficult endeavor. The integrated matrix presented in Fig. 11.4 gives some sense of this complexity. Below is a brief survey of some management approaches used throughout the world.

New Zealand

New Zealand has taken a strong institutional stance in response to problems of coastal zone management. In a 1991 law, the Resource Management Act effectively replaced more than 20 major statutes concerned with land, air, and water resources, pollution, and the coasts. The act is distinctive because it promotes sustainable management of natural and physical resources by focusing decision making on "intended outcomes" rather than regulation of resource uses. It also specifies the functional responsibilities of planning and resource allocation agencies and sectoral agencies as well as detailing national, regional, and local level responsibilities. It requires that all levels of government, indigenous people, special interest groups, and the public become involved in initial consultations to develop a credible policy statement. Regional Councils have been given 2 years to prepare regional coastal policy statements. Both coastal management and management of inland waters are now the direct responsibility of these Regional Councils: one agency, therefore, has the responsibility and accountability for developing a comprehensive overview of the management of coastal land, coastal waters, and inland waters, with more detailed planning carried out by subagencies. This model is likely to be emulated in many other countries in the next 25 years.

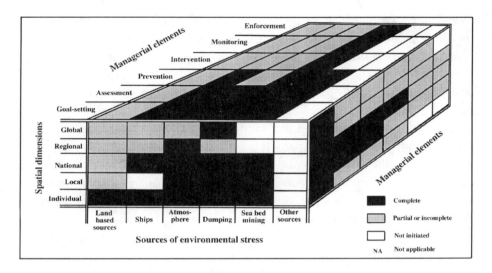

Fig. 11.4 A comprehensive marine management program requires simultaneous attention to several management concerns, operating at a variety of spatial scales and reflecting a variety of environmental stresses. This matrix illustrates the status of comprehensive marine management planning in Canada. (After Cote 1992)

The United States

In 1972, Congress passed the U.S. Coastal Zone Management Act (CZMA), which was amended in 1976. Under the CZMA, states have full legal authority to manage and protect their coastal resources such as the coastal preserve pictured in Fig. 11.5. The federal government offers financial and technical assistance as an inducement to states. The federal government is also committed to acting consistently with state-level programs once they have been reviewed by federal agencies and approved by the Secretary of Commerce.

Mexico

Mexico has historically been an inland-oriented state. Despite the fact that the country includes coasts on both the Atlantic and Pacific Oceans, less than 13% of Mexico's population lives in the 126 coastal municipalities (Merino 1987). Population density in coastal states is half that of the inland ones and the primary economic centers are all far away from the coast. In addition, the defined area of the coasts under public management is only 20 m wide, a zone too narrow for any effective controls. Coastal zone management in Mexico is further complicated by the nation's heavy reliance on offshore oil drilling; because oil production constitutes a vital sector of the economy, the Mexican government is unlikely to impose regulations or monitoring efforts (Merino 1987).

Japan

The coastal areas of the island nation of Japan play a vital role in its economy and its national self image. Japan's relative coastline is many times longer than those of most other countries. For example, England has 28.4 m of coastline per square kilometer, the United States 6.0 m/km^2, and Japan 91.1 m/km^2 (Someya *et al.* 1992). In addition, Japan's coastal cities, towns, and villages cover 31% of the country's entire land area and include 52% of the population. Coastal zones are also vital areas of economic activity, housing the majority of industrial capacity and wholesale trade.

Japan has a three-tiered approach to coastal management. The national government sets policies and standards, but implementation plans are made by prefectural governors. Local governments take responsibility for initiating port

Fig. 11.5 Hawaii's rich marine habitat is being managed through a network of coastal preserves. While the immediate aim of these preserves is protection of fish and invertebrates, the protection must extend to the adjacent land such as these palm forests. (Photo by L. Vanderklein)

and harbor projects, although these projects are loosely coordinated by the national government.

Korea

Korea's coastline has been estimated at between 1300 and 6200 km, depending on how accurately one measures its bays and inlets (Hong 1991). Korea's population and development is heavily concentrated along the coast. The coastal urbanization rate is higher than that of the entire country and density of coastal zone populations is 1.4 times greater than that of the country as a whole. Furthermore, the gross regional product of the coastal zone accounts for more than one-third of GNP. According to Hong (1991), coastal zone resource management was a simple issue with few functional linkages in Korea until the 1960s. Since the early 1970s, major social and environmental changes such as rapid industrialization, increased energy needs, growing population density, increased foreign trade, and technological advances have complicated coastal zone management. In response, the Korean government has created more than 12 institutions, each with subordinate agencies, specifically to address major problems in coastal zones. Sectoral in nature, these organizations set their own policies and priorities, neither conforming to, nor taking responsibility for, creating and enforcing a national master plan. More influential ministries (i.e., Construction, Commerce and Industry, Agriculture and Fisheries, Home Affairs, Energy and Resources, Transportation, Science and Technology) exercise a great deal of discretion in application of rules and regulations.

In addition to the large number of competing agencies, Korea has more than 54 laws relating to coastal zone management. Mechanisms for resolving interministerial conflicts over management of these resources and environments vary widely from law to law. The government also operates 17 committees and commissions under various ministries for resolving interministerial conflicts or improving efficiency of coastal zone management policies. Although there has been official recognition that these laws and agencies must be coordinated to achieve effective management, the structure remains entrenched.

India

India's vast coastline is a critical resource. It contains more than 100 major ports and more than 200 fishing harbors and anchors an Exclusive Economic Zone nearly two-thirds as large as India's entire land area (Nayak *et al.* 1992). Despite the importance of the coastal zone for national development, India has not yet identified a coastal zone management policy and its implementation strategy. This fact is particularly intriguing considering that India has enacted comprehensive environmental protection laws to control water and air pollution. The Central Beach Erosion Board, which was constituted in 1971 as a research and advisory board, has proposed establishment of a coastal zone management authority in every maritime state of India. That proposal has not been implemented.

Summary

- Coastal water bodies are fragile ecosystems that integrate the biogeochemistry of the adjacent land but are subject to the physical forces of the sea.
- Numerous coastal zone resources are heavily impacted by local and regional scale land use practices. In contrast to most other water resources, most coastal zone impacts are local-scale rather than watershed-scale. A significant exception involves the impact of global change, which causes major changes in coral reef ecosystems.
- Offshore disposal of municipal wastes represents a significant and controversial coastal zone management issue. Untreated sewage sludge is pumped into deep water off the coast of many large cities. In some cases, impacts are highly localized (i.e., limited to several hundred square meters of deep benthos). In other cases, many kilometers of coastline and many very valuable resources are heavily damaged.
- The institutional and policy environment for coastal zone management is highly fragmented in most countries. This condition poses a constraint to effective management in most areas of the world.

For Further Reading

Hanson H, Lindh G. Coastal erosion: an escalating environmental threat. *Ambio* 1993; 22:188–195.

Hildebrand LP, Norrena EJ. Approaches and progress toward effective integrated coastal zone management. *Marine Pollution Bulletin* 1992;25:94–97.

Kudo A, Miyahara S. Predicted restoration of the surrounding marine environment after an artificial mercury decontamination at Minimata Bay, Japan—economic values for natural and artificial processes. *Water Science and Technology* 1993;25:141–148.

Lundin CG, Linden O. Coastal ecosystems: attempts to manage a threatened resource. *Ambio* 1993;22:468–473.

Machiwa JF. Anthropogenic pollution in the Dar es Salaam harbor area, Tanzania. *Marine Pollution Bulletin* 1992;24:562–567.

Matthews GJ. International law and policy on marine environmental protection and management: trends and prospects. *Marine Pollution Bulletin* 1992;25:70–73.

12

Lakes and Water Quality Impacts

Overview

Lakes contain an enormous proportion of the earth's available fresh water. They represent a vital source of domestic water, but are also popular destinations for recreation, important for transportation, essential for fishing industries, and critical for the maintenance of bird and wildlife populations.

One distinguishing characteristic of lakes is that the volume of water in the basin is proportionally greater than its annual inflow and outflow. Consequently, the water in lakes develops unique chemical and physical characteristics such as temperature and density stratification. The larger a lake's volume, the greater its ability to assimilate pollutants. Larger volume also means longer residence time, so these lakes also respond less quickly to improvements in water quality. Lakes that are deep enough to stratify may generate in a surface layer that is oxygen-rich but nutrient-limited; their lower depths, on the other hand, are generally oxygen-poor, nutrient sinks. Seasonal mixing of the stratified layers becomes a critical regulator of water chemistry and productivity. Biological productivity of lakes differs from that of rivers in that lakes will support the growth of a suspended photosynthetic community (i.e., phytoplankton) as well as allowing submerged plants to survive at greater depths than in turbid rivers. From a water quality perspective, volume, residence time, stratification, and water clarity are critical properties that define the potential uses and values of a lake.

Nonpoint pollution is generally the major concern in lake water quality management, and eutrophication commonly results from nonpoint nutrient input. Other important impacts include acidification, the introduction of exotic species, sedimentation, changes in lake level, and contamination by toxic compounds.

Introduction

Globally, fresh water lakes account for only a fraction of a percent of global fresh water reserves (0.26%). Because lake water is so accessible, however, it constitutes an enormous percentage of available fresh water. In the United States, nearly 70% of the water used by the largest utilities comes from lakes (or reservoirs). Similar percentages are found wherever lakes are a prominent feature of the landscape. The Great Lakes alone serve as the domestic water supply for more than 40 million U.S. and Canadian citizens (NRC 1992). Lakes can also serve industrial water needs, provide seemingly unlimited recreational opportunities, fulfill irrigation requirements, serve as transportation routes, and facilitate waste disposal. Although these intentional uses place important water quality stresses on lakes, the greatest impact upon lake waters comes from the large volumes of sediment and chemical inputs (i.e., nonpoint source pollution) produced by surrounding land use activities.

In many countries, management of lake water quality issues is changing due to the increased awareness of nonpoint source pollutants as well as to shifting societal values that have grown to recognize the ecological values of lakes. These new management considerations include the effect of lakes on local climates, on streamflow regulation, and recharge of groundwater aquifers, in addition to their provision of wildlife habitat and contribution to landscape diversity. Weighing the dominant values of any lake ecosystem is extraordinarily difficult, however. To homeowners, lake quality might be considered poor if the lake falls below a certain level in the summer or floods property in spring. Maintaining such strictly defined lake levels may, in turn, have a negative impact on near-shore spawning grounds of important commercial and recreational fisheries. Other conflicts in values are often less obvious. For example, the fertilizers used by homeowners to maintain a green lawn down to the shoreline inadvertently contributes to nutrient enrichment, growth of undesirable weed beds, and eutrophication with its algal mats, summer stench, and changes in aquatic community. Eutrophication, in turn, makes sitting on the green lake-side lawn significantly less enjoyable.

One important characteristic of lakes is that large sections of shoreline are frequently accessible. With the exception of steep-sided mountain lakes, development commonly encircles a lake and uses are divided into recognized, and sometimes mandated, zones. For example, swimmers occupy near-shore areas, boaters occupy offshore zones, domestic intake pipes occur along the shoreline, and sewage outfall is piped to deep shore areas. Use pressures in less developed countries are similarly intense, but generally occur with less subdivision into use zones—washing clothes, obtaining drinking water, waste disposal, and fishing all occur where the shoreline is most easily accessible.

Inadequate data preclude global estimates of how many lakes are degraded by anthropogenic activities and the sources of their impacts. Even within the United States, estimates of impacts and their sources vary according to the extent of lake area surveyed and the year of the study. According to 1992 U.S. EPA data, the dominant land uses of concern to U.S. lake managers are:

- Agriculture (affecting 56%)
- Habitat and hydrology modification (36%)
- Storm sewers and urban run-off (24%)
- Municipal point sources (21%)
- Other wastewater disposal (16%)

Other sources such as industrial processes, construction, resource extraction, silviculture, and combined sewer flow affect smaller areas of U.S. lakes, although local impacts may be quite pronounced. The causes of water quality impairment are clearly related to the dominant land uses. In decreasing order of importance (on a lake area basis), these causes in U.S. lakes include (EPA 1992):

- Metals (47%, but data are questioned due to suspected sampling errors)
- Nutrients (40%)
- Organic enrichment (24%)
- Siltation (22%)
- Priority organic chemicals (20%)
- Suspended solids (16%)

Data such as these provide useful frames of reference but are far from exact. Year-to-year differences in categories may vary the estimates of primary concerns, sampling errors can bias data, and different percentages of assessed lake areas can yield different estimates of degradation. Such averages can also lead to false security over the magnitude of certain problems. For example, acidification, flow alteration, and the introduction of exotic species are all critical problems in specific regions and specific lakes, even though they are not listed in the above rankings.

Lake Hydrology from a Water Quality Perspective

Globally, lakes vary tremendously in terms of size, depth, age, and salinity and are affected by temperature, seasonality, and exposure to different weather regimes. Each of these features influences the details of a lake's hydrology, and consequently its biology and susceptibility to water quality degradation. Despite all of these differences, lakes do have a few common elements that distinguish them from rivers or other dynamic water bodies:

- Large volumes that dilute pollutants and other inputs.
- Their static place in the landscape, which enables these same pollutants and inputs to accumulate over time.
- Stratification of water layers and different patterns of mixing.
- Physiographic irregularity that subdivides lakes into differentially susceptible bays and inlets.

As discussed in the following sections, the interplay of depth, volume, and latitude are critical elements in water quality dimensions of lakes.

Thermal Stratification and Chemical Mixing

With the exception of extremely shallow lakes, lakes usually stratify into distinct layers as illustrated in Fig. 12.1. The surface layer, or epilimnion, is well-lit and oxygen-rich because of gas exchange with the surface and aeration through wind and waves. Consequently, it is also the most productive layer in a biological sense. Productivity in the epilimnion is often nutrient-limited, however, because nutrients from decomposing material filter to lower layers and concentrate in sediments or bottom waters.

Below the epilimnion is the metalimnion (or thermocline), a layer whose size and location varies tremendously with the size, depth, and location of a lake. The metalimnion is a transition zone in deeper lakes where temperatures and light levels decrease; it may not even form in shallow lakes. Where it does occur, its size varies from only a few meters deep in small lakes to nearly 30 m thick in large lakes (Schindler 1991).

If a lake is deep enough, it may also form a hypolimnion, a cold, deep layer that supports little photosynthetic activity but does serve as a site for benthic decomposers and as a repository for nutrients that percolate down from the productive upper layers. Productivity of the hypolimnion is minimal not only because little light filters to its depths, but also because the hypolimnion is oxygen-poor.

This tripartite model of thermal stratification provides a useful conceptual approach to lake hydrology. It is not a universal phenomenon, however. The model is most commonly found in temperate latitudes where significant seasonal changes in temperature cause stratification. In maritime, tropical, alpine, and Arctic regions, lakes are frequently only weakly stratified.

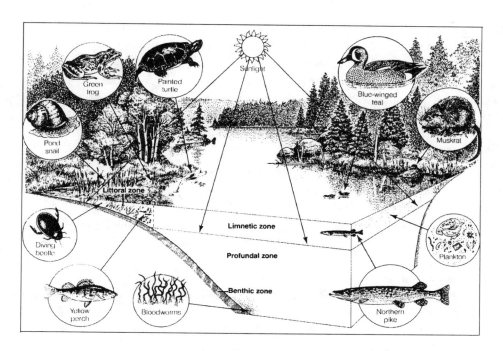

Fig. 12.1 Lakes may be divided into many vertical and horizontal zones. Vertically, lakes are frequently stratified by temperature and chemical gradients. The epilimnion is the shallow, productive zone; the metalimnion is a thermal transition zone to the deep, often oxygen-poor but nutrient-rich hypolimnion. Other zones are defined by the proximity to the shore. Littoral areas, for example, are highly productive, shallow, near-shore areas. While the zones are flexible in time and space, they represent important characteristics that define sensitivity to water quality degradation and improvement. (After Miller 1992)

Whether the lake is weakly or strongly stratified, mixing the layers constitutes a critical feature of lake productivity and pollutant cycling. Because upper lake levels are the sites of maximum photosynthetic activity and lower levels are the sites of nutrient accumulation, destratification enables the oxygen-rich waters from the epilimnion to mix with oxygen-poor waters of the hypolimnion. This process brings oxygen to the benthic communities and thus facilitates decomposition and other metabolic activities. Similarly, the nutrient-rich waters of the hypolimnion are allowed to mix with surface waters, enabling surface waters to remain productive. Mixing can occur through two primary processes. First, horizontal movement of the upper layer (due to wind) can cause vertical mixing. Second, lakes may mix as seasonal temperatures equalize surface and deep water temperatures, eliminating stratification. Frequently, temperate lakes undergo mixing in spring and fall, a feature which is termed dimictic. Lakes may be monomictic (mixing only once a year) if annual seasonal temperature variations are small. In contrast, shallow and tropical lakes are often polymictic, mixing frequently in summer in response to windy periods or diurnal temperature changes (Fig. 12.2).

Vertical mixing (also termed lake turnover) is vitally important to lake productivity. In small to medium-size lakes, most chemical exchanges occur in a horizontal direction and result from exchanges with sediments at the same depth, rather than vertically. According to Schindler (1991):

> The operation of mixing processes may, for practical purposes, be visualized
> as a series of rapidly moving horizontal conveyor belts, transporting sub
> stances between the water column and the sediment surface at the same
> depth in the lake. The relative vertical movement is small, perhaps no larger
> in relative magnitude than the vibrating of a conveyor belt. [p. 93]

Persistent winds can result in horizontal and vertical mixing and restratification. Steady, strong winds can create diagonal stratification in lakes, with warm waters piled against the far shore (Fig. 12.3). Similar to upwelling in coastal zones, the persistent winds in a lake "push" surface waters away from one shore, causing deeper waters to rise to the surface.

Water Renewal Time

Water renewal time is one of the most important characteristics of water quality risk in a lake. This measure is calculated by dividing the volume of a water body by volume of the outflow. It represents the time it takes to replace the water in a lake with "new" water. Although simple to calculate, renewal time results from many factors such as rainfall, catchment area relative to lake volume, and evaporative loss. In addition, Schindler (1991) notes that "when the input of a chemical is changed, but the water input remains constant, the chemical concentration will change at a logarithmic rate dictated by the rate of water renewal" [p. 99]. For managers, the implication is that improved conditions cannot be achieved in a single renewal cycle, but over the course of many renewals. Consequently, although renewal times range from several hours for small lakes formed on large rivers to hundreds of years in the Great Lakes (Schindler 1991), expectations for the time to degradation or improvement must allow for multiple renewal times.

Renewal time is of central importance in water quality management. Lakes with long renewal times have a large "buffering" capacity against pollution. The largest lakes, such as Lake Baikal and Lake Superior, act as environmental sinks. They are slow to show impacts, but similarly slow to recover to a nonpolluted

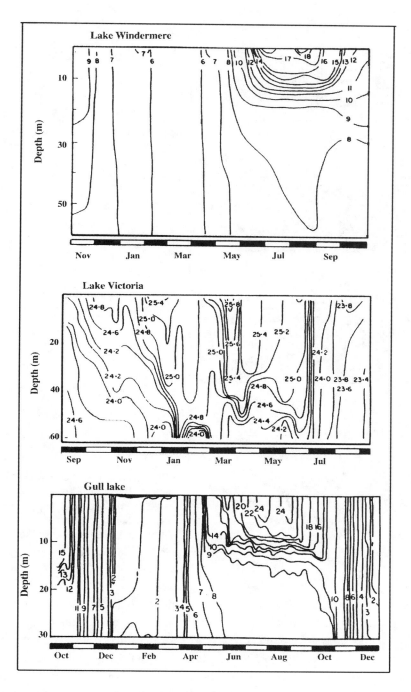

Fig. 12.2 Depth time diagrams display changes in water chemistry of a lake both through the year and among depths. Lake Windermere is stratified in the summer and has uniform top-to-bottom temperatures in the winter. Because the climate is maritime, water temperatures in the lake vary little among seasons. Patterns in Lake Victoria vary even less through the year but show small areas of stratification in deep waters in January and February. Gull Lake, Michigan, is strongly stratified in winter and water temperature varies significantly throughout the year. (After Moss 1991)

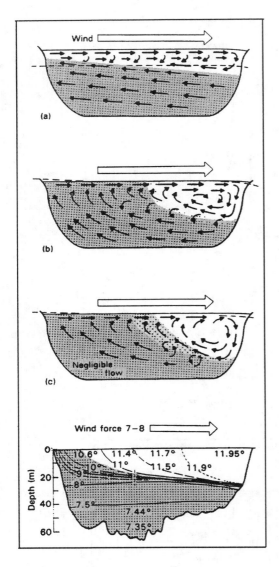

Fig. 12.3 Wind is the principal force causing lake mixing. Winds blowing consistently from one direction cause uneven mixing and significant differences in water quality around a lake basin. (After Schindler 1991)

state. In contrast, smaller lakes are degraded more quickly but also recover more quickly once the source of pollution is reduced or eliminated. Rapid flow-through lakes such as reservoirs most closely resemble slow rivers. Lakes with closed basins, such as Mono Lake in California, have renewal times of thousands of years because evaporative loss is their only outflow. Thus, chemicals become concentrated in these lakes, which then become saline.

As critical as renewal time is, it may not speak to the elimination of contaminants from a lake. Many compounds are fixed to sediments so recontamination of water may occur continually, even though the water itself is renewed frequently. For example, in the Yellow Sea, which resembles a large lake, the sediments retain heavy metals and the sediment load of metals is 1000 times more concentrated than the heavy metal load in the water (Platte 1995).

Chemical Concentrations

Another determinant of lake water quality is related to biological use of certain chemicals. "Biologically conservative" chemicals are those not used in any significant proportion by plants and animals. Consequently, their concentrations are fairly easily predicted given inputs, water volume, and solubility.

In contrast, "biologically nonconservative" chemicals are substances used by plants and animals. Their concentrations in the water column are difficult to predict, and their effects can appear to be disproportionate to the actual concentration. Phosphorus and nitrogen are two examples of such chemicals (Schindler 1991). Availability of these compounds is an important limiting factor in the growth of aquatic plants. When their concentrations increase—for example, through agricultural inputs or sewage outfalls—these nutrients are taken up in large concentrations and increase plant productivity. As discussed in chapter 21 on eutrophication, increased productivity can lead to severe deoxygenation of waters through accumulation of decomposing material. Thus, the real water quality impacts of increased phosphorus and nitrogen lie in the chain reaction they begin and, in the final analysis, in deoxygenation of waters. The implication for water quality management is that biologically nonconservative chemicals can lead to numerous unexpected and dramatic impacts.

Chemical reactions between elements can also have notable and unanticipated effects on water quality. For example, oxidation and reduction processes alter the forms of certain compounds such as sulfate. Sulfate-reducing bacteria in sediments and hypolimnion remove oxygen from sulfate, thus storing the sulfate in a reduced form. This sulfate retention has been identified as an important process in improving the acid-buffering capacity of acidified lakes (Schindler 1991). The rate of sulfate reduction does not prevent lake acidification, but it does modify, or buffer, the process and improve recovery time.

Lake Physiography and the Expression of Hydrology

Lake hydrology is a critical dimension in water quality studies that is expressed differently depending upon physiography of the lake itself. For example, lake depth is a critical determinant of the degree of stratification, productivity, and susceptibility to eutrophication and oxygen depletion. Similarly, the occurrence and shape of basins and shorelines can effectively subdivide a lake into smaller and often shallower lakes. Thus, shoreline features control local and whole-lake stratification, productivity, and water movement. For example, Lake Tanganyika has thousands of bays that act like distinct, small lakes. Difficulties with low oxygen levels (and associated fish kills or migration) eutrophication, concentration of pollutants, and sedimentation may not affect the entire lake, but will affect the shoreline bays where most human activity takes place. In contrast, Lake Baikal has a largely undivided coastline and consequently constitutes one large undifferentiated basin.

Spatial effects are also evident in the unequal concentration of pollutants in biologically pivotal littoral (shallow water) zone. The littoral zone of a lake supports the majority of productivity, complex fisheries, and breeding grounds for most fish and many insects. Deep water lake trout, for example, usually lay eggs in the littoral zone. Even if only a small area of a lake is affected, if this area is concentrated in the littoral zone, then the entire ecosystem can be severely affected.

Although lake waters can mix, as described earlier, this mixing is often ineffective and incomplete, with pollutants often remaining concentrated near the source of impact. Consider the city of Cleveland, which uses Lake Erie for both potable water and sewage outfall (discussed in Laws 1993). Overall, the bacterial counts in the lake are low enough that the water is potable. To ensure that this situation continues, intake pipes are located 5 km from shore. In contrast, sewage outfall pipes discharge to near-shore areas, and these areas (which are most accessible and desirable for swimming, shoreline fishing, and boating and for wildlife) are highly polluted by bacteria.

The shallowness or depth of a lake also influences water quality risk. Despite similar surface areas, the dynamics of Lake Baikal, with a maximum depth of 1700 m, is far different from that of Tongle Sap (Cambodia), whose maximum depth is 12 m (Gleick 1993). In shallow lakes, the mixed layer may reach to (or near) the bottom during much of the year. Consequently, nutrient limitations do not occur: nutrient recycling is efficient and productivity levels are high. Both productivity and shallowness carry their own risks from a water quality point of view, however. Shallow lakes are more susceptible to eutrophication, and have less buffering capacity against stresses such as temperature, accumulation of toxic compounds, and effects of sedimentation. They also depend heavily on wind mixing to maintain high oxygen ratios. Several days of calm weather can lead to serious problems with oxygen depletion. Dramatic drops in oxygen level may even be experienced at night when photosynthetic production of oxygen stops.

In contrast, very deep lakes are generally less productive because the mixed layer may extend to great depths and production will be light-limited. When the mixed layer is very deep, phytoplankton are distributed within a thick layer but only those that happen to be nearest the surface, where light is adequate, can photosynthesize efficiently. Additionally, nutrient recycling in very deep lakes is an inefficient process. Even during lake turnover, the very bottom layers may never become mixed.

Biological Communities in Lakes

Because lakes are static and their waters have long residence times, their biology is different from that of streams. Within lakes are found suspended and littoral communities, vertically stratified species, and benthic insects that settle in bottom sediment. Fig. 12.4 illustrates these spatial–temporal differences for selected biophysical and biological components of a lake. Additionally, most changes in a lake are predictable on an annual basis; lakes are not disturbance-adapted in the same way as rivers and streams. Hydrologic and chemical stratification and seasonal variations in temperature, visibility, and lake level create special conditions that influence the response of lake-adapted species to water quality impacts.

The response of a lake community to water quality stress is determined by many factors. Some factors relate to the nature of the stress itself, such as its duration, magnitude, toxicity, and synergistic effects of combined stresses. How a biological community is affected by stress also is determined by the composition of the affected community. In particular, the presence of diverse species and the replication within the community of functional roles represent critical elements in the susceptibility of a community to stress (see chapter 16). In a simplified example, if a community has one predator and prey, elimination of either species would lead to a collapse of the entire "system." In contrast, the community as a whole would be buffered by the presence of multiple species capable of filling similar roles. Consequently, the natural biotic diversity of lakes is an important concept in determining lake susceptibility to water quality degradation.

The diversity of life in a lake results from many factors; age, latitude and altitude, diversity of the surrounding terrestrial habitat, and temperature are the most important determinants. Rates of speciation can also play important roles in isolated cases, as discussed by Schindler (1991). For example, Lake Victoria is a relatively young lake (less than 14,000 years) but developed an enormous diversity of one family of fish, the cichlids. More commonly, older lakes develop enormous species diversity. For example, Lake Baikal, an ancient lake in Russia, evolved in isolation from similar habitats in other lakes. Consequently, it has developed a diverse fauna that encompasses more that 700 endemic species of animals. Other examples include the deep and ancient rift

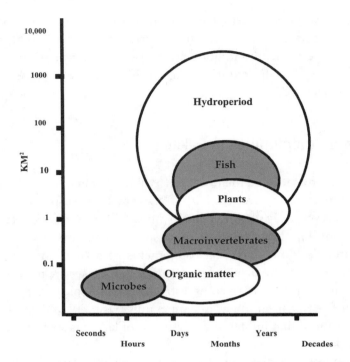

Fig. 12.4 Lake ecosystems are relatively stable through time, but lake communities often differ among bays or in pelagic versus littoral areas. Organisms that are more wide-ranging and are more often associated with open water habitats vary more in space and time than do organisms that are less free-ranging. For example, fish are more variable than insects.

valley lakes such as Lake Tanganyika, with 247 endemic species, and Lake Malawi with 500 endemic species. In contrast, the younger, but far more extensive, Laurentian Great Lakes contain far fewer than 170 fish species.

Very young lakes, especially those at high altitudes and latitudes that have only recently emerged from glaciers (i.e., are less that 10,000 years old), may contain only a few tens or hundreds of species. Schindler (1991) cites the example of lakes along the Canadian Shield, many of which contain fewer than five species of benthic crustaceans and rarely more than a few fish species. Similarly low numbers in lakes can be related to small size and/or species-specific limiting chemical compositions (e.g., low-silica waters support few species of diatoms, while low-calcium waters will support few molluscs).

Important Water Quality Stresses

Water quality impacts in lakes, rivers, and groundwater are intricately interrelated because each system, although under separate controls, feeds the others.

Consequently, the source of a lake's water is an important consideration when evaluating water quality impacts. Seepage lakes, which are fed primarily from rainfall, will differ significantly from lakes fed by rivers. Similarly, the quality of a lake will differ depending on the watershed drained by the feeder river.

Where rivers or streams flow into lakes, the effect of land use activities surrounding the river will affect flow into the river, the amount of sediment flow into the lake, and concentrations of pollutants that the lake must then assimilate (or dilute). Laws (1993) details a study of Lake Jackson, Florida, a highly segmented lake whose northern basin is dominated by a forested watershed and whose southern basin is dominated by an intensely urbanized watershed. Important water quality variations within the lake resulted from differences between these two watersheds. Streams discharging from the urban watershed, as compared with the forested watershed, carried one-third more total run-off, more than 500 times the suspended solids, and two and a half times the inorganic nitrogen into their respective bays.

Three of the best-known water quality impacts on lakes are eutrophication, acidification, and introduction of exotic species. The source of each varies widely, indicating the broad ways in which humans influence lake water quality. For example, eutrophication generally results from nutrient enrichment primarily from nonpoint sources such as agricultural run-off. Acidification results from long-range transfer of particulates in the air, and ultimately from industrial air pollution. Introduction of exotics may occur either intentionally (attempting to establish a desired fishery) or accidentally (e.g., the release of zebra mussels from Europe in the bilge water of a boat) but in either case stems from use of lakes for recreation and transportation. Despite these differences in origin, the end-results are similar for the biotic community: diversity is reduced, dominance by tolerant species is increased, and food webs are simplified. In addition, each of these may change the appearance of the affected lake and consequently affect certain recreational values. Because of the prevalence and notoriety of these problems, each is covered in a separate chapter (see chapters 21, 22, and 24).

Although these three factors receive the most "press coverage," lake water quality is also commonly affected in other important ways. Throughout history, the earliest and most severe impacts of water quality degradation have been related to pathogens. While treatment of drinking water keeps pathogenic risks smaller in developed countries, pathogens remain an enormous problem throughout the world (see chapter 17). Other significant effects on lake water quality include water level regulation, sedimentation, and contamination by toxics. As exemplified by Lake Srinagar (Fig. 12.5), however, polluted lakes do not always *look* polluted. Each of these topics is discussed below.

Fig. 12.5 While some water quality problems are obvious, others, such as pathogens, heavy metal pollution, and invasion by exotic species, remain largely hidden from view. Consequently, local communities sometimes find it difficult to accept that a given lake has developed serious water quality problems. (Photo by D. Vanderklein)

Regulation of Lake Level and Water Quality Effects

Regulation of lake level is a common practice for lakes in developed areas. To protect lakeshore property, swimming areas, and other recreational facilities, dams are frequently built at the outflow of a lake. While water levels may be naturally low in summer (or dry season), these dams serve to maintain an artificially high, "desirable" water level. Similarly, in spring (or wet season) when flood danger becomes a problem, significant amounts of water can be released, again with the goal of maintaining desirable water levels.

In the long run, water stabilization creates a new set of undesirable effects. Lake aquatic communities are adapted to fluctuating water levels and some aspects of their function depend on that fluctuation. In particular, near-shore sediments are oxidized by fluctuating water levels and plant and animal diversity is promoted by the changing environment (NRC 1992). When water levels remain constant, sediments become oxygen-poor (anoxic) and unable to support spawning areas or fish populations. In other cases, near-shore areas are converted to dry uplands. The lack of seasonal flushing can also lead to concentration of pollutants along the shore area or bay of entry.

Use of feeder rivers may also affect water levels for upstream water needs. Rather than maintaining water levels, however, tapping feeder streams lowers the lake level, increases salinity, and leads to loss of wetlands. The Aral Sea example has been discussed previously, but is hardly the only lake to suffer from decreased water levels. Mono Lake, California, is one of numerous lakes that have also been compromised by diversion of stream flow for urban and agricultural uses.

Sedimentation

In discussing the effects of sedimentation on rivers, it was noted that sedimentation can clog the feeding mechanisms of filter feeders, smother (rock) attached invertebrates, and decrease the efficiency of predators through decreased visibility. The effects of sedimentation are somewhat different in lakes, however, because a soft sediment substrate is more common and the biological community tends to be more mobile. One important problem linked to suspended sediment loads in lakes is a decrease in light penetration. Macrophyte productivity may decline in consequence, leading to a concurrent reduction in fish habitats. Fisheries may also be affected through sedimentation of spawning areas and smothering of egg masses. Sediment build-ups can also reduce the lake depth, making it less viable as a storage reservoir, reducing the quality of drinking water (and increasing the need for filtration), increasing its susceptibility to eutrophication, and reducing the desirability of the lake for recreational uses.

Another effect of sedimentation can be a change in the nature of the phytoplankton population. For instance, the phytoplankton species blue–green algae (cyanobacteria) contain gas-filled bubbles that enable them to float on the water's surface. In times of increased sedimentation, decreased water clarity may severely reduce available light to the majority of the phytoplankton community. Consequently, production by cyanobacteria will dominate and their floating surface mats (scum) can dramatically reduce the aesthetic appeal of a lake. In addition, cyanobacteria are less palatable to the zooplankton that feeds on phytoplankton, and changes in their feeding patterns can cascade throughout the lake food web.

Toxic Contamination

Water quality of lakes (and slow-moving rivers) is not always related to the water itself. Concentration of inputs such as metals and organochlorines in sediments can continue to cause quality problems even after water has "turned over" in a lake or river. For example, DDT is still found in lakes of the lower Mississippi, even though the substance was banned in the United States more than 20 years ago. Toxic chemicals may derive from point sources (such as inadequate disposal or industrial spills) or nonpoint sources such as deposition from the atmosphere or run-off.

Mercury contamination is one example of an element whose presence may be attributable to point sources as well as to long-range atmospheric transport and burning of fossil fuels by power plants. Toxic in high concentrations, mercury biomagnifies and consequently poses the greatest threat to top-level carnivores and the predators that eat them (e.g., eagles, humans). Significant efforts have been made in reducing point sources of such elements. For example, Meybeck *et al.* (1989) describe the process of identifying, and tracing mercury pollution in the Great Lakes. In Lake Erie, a chloralkali plant on Lake St. Clair (the western basin of Lake Erie) was responsible for significant mercury pollution. Elimination of the point discharge in 1970 led to a rapid decline in the mercury content of Lake St. Clair (and Lake Erie) fish—from more than 3 mg/km in 1970 to approximately 0.5 mg/kg in 1980. Despite reductions in point sources, long-range aerial transport is still responsible for significant mercury pollution in many areas and favored sport fish such as walleye continue to have particularly high "body burdens" of mercury. As of 1992, 21 states had issued fish consumption advisories because of mercury contamination.

The management of other compounds, such as PCBs, is also complicated by the fact that they may enter a lake system through any source: point, nonpoint, or atmospheric deposition. The National Research Council (1992) documented numerous points of entry for toxic compounds in U.S. lakes. For example, within the Great Lakes, PCBs have entered through the point source of improper disposal (Waukegan Harbor, Lake Michigan) as well as the nonpoint widespread use of chemicals from the 1920s to 1970s (especially Lakes Erie, Ontario, and Michigan). In Lake Superior, which is subject to only minor point and nonpoint sources of PCBs, PCB concentrations in some lake trout have still been high enough to warrant a health advisory. Atmospheric transport of PCBs from hundreds and possibly thousands of miles is significant enough to have a major impact on aquatic health. PCB contamination has also persisted despite the banning of the substance in 1979. In Lake Erie, for example, concentrations in most species have decreased, but remain above the target level of 100 ppb. Notably, concentrations in lake trout have increased dramatically from 1984 to 1987 (although the International Joint Commission for the Great Lakes notes that sample size for lake trout in these years was limited). Metals such as arsenic, cadmium, chromium, copper, lead, mercury, manganese, and zinc have also

been identified in lake sediments, as have cyanide, grease, oil, hexachloroben-
zene (HCB), and dioxins (Mason 1991).

Lake Water Quality Management

Managing lake water quality requires a delicate balance between reducing (or
eliminating) the inputs of toxics, nutrients, or other stress agents and changing
hydrological relationships to improve the affected ecosystem. It also necessitates a delicate balance in forging relationships with other resource managers.
For example, fisheries and water quality professionals can work at cross purposes in lake management. In Medical Lake, Washington, water quality managers were trying to reduce nuisance algal blooms. Fisheries managers
inadvertently aggravated the problem by stocking the lake with fish (NRC
1992), whose feeding caused a decline in herbivorous zooplankton that were essential in keeping algal blooms under control.

The list of techniques used in lake management is long, and includes strategies for both reducing inputs and improving the entire lake ecosystem. The
National Research Council (1992) includes the following recommendations
among its list of viable techniques:

Input Reduction

- Diversion: Rechanneling effluent flow around lakes and farther downstream. Although rarely implemented, diversion has been used with partial
  success in Madison, Wisconsin, and Seattle, Washington.
- Land disposal: Disposing of treated municipal wastewater on agricultural
  lands or golf courses. Both disposal methods are becoming more common
  in warm climates as a means to simultaneously conserve water and divert
  nutrients.
- Product modification: Including banning manufacture, sale, and distribution. Product bans have been successfully imposed on chemicals such as
  DDT, PCBs, and phosphate detergents, but residual effects from earlier use,
  and some atmospheric transport from countries without similar bans, continue to complicate management.
- Wastewater treatment: While wastewater treatment has been used successfully to treat bacterial contamination, more advanced treatment (tertiary
  treatment) can provide greater removal of phosphorus as is required for
  wastewater entering the Great Lakes.

- Interception of nonpoint sources: Improved agriculture and forestry practices and improved urban drainage patterns can "intercept" nonpoint sources, channeling them to settlement basins, slowing run-off, and improving uptake by vegetation.
- Dilution and flushing: Increasing flow through a lake is one approach to decrease phosphorus sedimentation. Although effective, it is an expensive and rarely used process.

Ecosystem Improvement

- Alum treatment: Reduces phosphorus release from lake sediments by binding the phosphorus through use of aluminum salts; at low pH, aluminum is toxic.
- Sediment skimming: Removes the sediment surface layer where most of the phosphorus is concentrated.
- Sediment oxidation: Injecting calcium nitrate into sediments to catalyze a series of reactions that will oxidize sediments and bind phosphorus to sediment iron.
- Deep-water discharge: Siphoning or discharging phosphorus-rich hypolimnetic waters to reduce phosphorus loading.
- Biomanipulation: Modifying fish populations to control algal blooms and promote desired food web relationships.
- Artificial circulation: Reduces summer stratification and promotes habitat expansion, reduces soluble compounds such as iron and hydrogen sulfide, and reduces algal mats and growth of blue–green algae.
- Biocides: Algicides, herbicides, and piscicides can eliminate or reduce blue–green algae populations.
- Biocontrol agents: Grass carp can be used to control macrophyte growth; predatory fish can control exotics; herbivorous insects can control exotic weeds.
- Drawdown/sediment desiccation: Macrophyte growth can be controlled by occasional drawdowns.
- Bioharvesting: Macrophytes in a eutrophic lake can be mechanically harvested.
- Aeration: Winter fish kills in shallow, productive northern lakes can be prevented by aeration.
- Dredging: Removing sediment loads and toxics bound in sediment is one approach to management.
- Liming: Treatment with lime can buffer acidified lakes.

These remedies must be exercised with caution, since each technique has side effects that may negate the overall improvement. Besides these strategies, efforts aimed at changing land use practices can have an enormous impact on lake (and stream) water quality. For example, zoning requirements that limit shoreline vegetation removal and use grasses instead of some curb and gutter drainage, setback restrictions from shoreline construction, and on-site storm water retention devices have all been used effectively to reduce impacts of shoreline development. Some techniques for reducing impacts of urban run-off include the diversion of run-off to wetlands, retention and detention basins, street sweeping and vacuuming, and public education to reduce lawn fertilizers and litter accumulation. Similarly, many land use improvement efforts have been launched on agricultural lands to reduce their run-off to lakes. Some of these programs include contour farming, winter cover crops, no tillage, detention ponds, improved timing, measure, and type of fertilizer applications, crop rotations, and storage of manure.

Some of these water management options affect large areas (e.g., statewide phosphate bans) and control significant amounts of nutrient loading; others, such as best management practices, work in smaller, cumulative ways. In contrast, impacts such as acidification or invasion by exotics leave the manager with far fewer options. For example, in-lake methods to control acidification are limited to liming; reducing concentration may be improved by flushing or dilution. The only long-term solution is a reduction of the industrial and fossil-fuel-burning processes that lead to acid precipitation. The need for integrated management and cooperation between different management sectors is enormous.

Summary

- In the past, lakes have been viewed as static ecosystems. In fact, they are quite dynamic, although their timeframes of response are longer than those of streams, rivers, or wetlands. The volume of incoming water per unit time compared with lake volume defines the turnover, or residence time. That variable is the single most illustrative classifier of lake response to water quality impacts.
- Lake ecosystems are driven by autochthonous, or within-system, properties like carbon and energy flows. Allochthonous, or outside-the-system, properties such as watershed impacts are less important. As such, most management concerns in lakes are quite local, and are managed at the bay, shoreline, or whole-lake level.

- Water in lake ecosystems regularly mixes due to wind and hydrology. Most such mixing is horizontal. In many lakes, strong vertical stratification occurs at certain seasons, further restricting vertical mixing.
- Major lake management concerns include eutrophication, acidification, and exotic species introductions. Management strategies include controlling inputs (e.g., of nutrients, exotic propagules, or other water chemistry alterations) and manipulation of the lake ecosystem (e.g., by physical, biological, or hydrologic changes).

For Further Reading

Laws EA. *Aquatic pollution: an introductory text,* 2nd ed. New York: John Wiley & Sons, 1993.

Mason CF. *Biology of freshwater pollution,* 2nd ed. New York: Longman Scientific and Technical, 1991.

Meybeck M, Chapman D, Helmer R, eds. *Global freshwater quality: a first assessment.* Oxford: Basil Blackwell Ltd, 1989.

National Research Council. *Restoration of aquatic ecosystems: science, technology and public policy.* Washington, DC: National Academy Press, 1992.

Platte AE. *Worldwatch* 1995;6:10–19.

Schindler DW. Lakes and oceans as functional wholes. In: Barnes RS, Mann KH, eds. *Fundamentals of aquatic ecology,* 2nd ed. Oxford: Blackwell Scientific Publications, 1991.

13

Wetlands: Productive, Vital, Cleansing, and Threatened

Overview

Historically, wetlands have been valued only for the space they occupy when drained or filled. Throughout the 1980s, however, the importance of wetlands became much more widely accepted, and they are now recognized as a vital feature of the landscape. The greater valuation placed on wetlands comes from several sources: (1) an increasing emphasis on biological diversity; (2) a recognition that wetlands can "cleanse" waters of many pollutants; (3) an increased regard for wildlife and other noncommodity uses of natural resources; and (4) a recognition that wetlands act as a valuable buffer between other land forms. For example, riverine wetlands reduce channel erosion, reduce the intensity and frequency of floods, and can "filter" many effects of urban run-off, thereby reducing adverse impacts on adjacent streams or lakes.

A wetland is difficult to define (a problem that has frustrated policy makers), but generally occurs in a depression or along a shoreline and represents an area in which the soil is at least occasionally waterlogged. These areas support dense, often reedy vegetation, along with rich algae and epiphytic growth. Over small spatial areas, significant changes in water level often result in a great diversity of plant (and animal) species. Consequently, wetlands are a vital source of habitat diversity, providing cover and food for birds, amphibians, and other wildlife. They are also among the most productive aquatic communities. Their development is generally influenced by water level, nutrient status, and natural disturbances. Water quality management of wetlands is dominated by three issues: (1) maintaining wetland hydrology (e.g., drainage, fluctuating discharge and recharge); (2) using wetlands as waste treatment alternatives; and (3) ensuring wetland productivity (e.g., wetland reconstruction and maintenance).

Introduction

Wetlands are a unique and enigmatic water resource. Long considered simply wasted space, the positive social and ecological values of wetlands have been widely recognized only since the 1980s. Consequently, wetland "management" has historically been dominated by efforts to drain wetlands for agricultural and urban expansion. Wetlands meet few of the traditional and familiar values ascribed to aquatic systems: they serve no transportation use, have little utility for drinking water, provide no swimming and little fishing use, and cannot support irrigation or industrial demands. In addition, wetlands are difficult to traverse, support large populations of biting insects (e.g., mosquitoes), and support vegetation and wildlife that have little utility for eating or cultivating. The resultant dark social image of these spaces is conveyed by terms such as "bog" and "swamp" that are laden with negative connotations. Throughout folklore and modern tales, swamps, marshes, and bogs are used as an allegory for fear, while these spaces are used in modern language to convey images of stress and immobility ("bogged down," "swamped under," "mired"). The combination of these factors has led to the rush to drain and fill wetland rather than to conserve them.

Drain and fill operations have been extremely effective in ridding the landscape of wetlands. In the United States, more than 50% of the original wetlands in the lower 48 states have been lost to agricultural or development activities. More dramatic, however, is the concentration of the losses. Heavy wetland losses have occurred along coastal areas of the southeast United States and along the greater Mississippi River drainage basin (Fig. 13.1). State-by-state totals provide equally compelling evidence that wetland loss has been dramatic: California has less than 10% of its wetlands left; Ohio and Iowa are not far behind with 90% and 89% losses, respectively. Overall, 10 states have drained more than 70% of their wetlands (EPA 1992). The extensive draining is reflective of a U.S. national policy that, until the 1970s, was specifically devoted to facilitating the draining and filling of wetlands. The earliest of these laws, the Swamp Lands Acts of 1849, 1850, and 1860, ceded federally owned wetlands along the Mississippi River to their respective states. States, in turn, sold the land to private individuals at discount prices. In a truly ironic twist, the wetlands were developed, thereby subjecting these new developments to floods; individuals sought flood protection from state and federal agencies and currently the federal government is paying "enormous sums" of money to buy back those lands for conservation and flood protection (Mitsch 1993).

Wetland destruction is not limited to the United States—rather, it is a global phenomenon. Between 1950 and 1980, the Netherlands and the former Federal Republic of Germany "reclaimed" more than 50% of their wetlands,

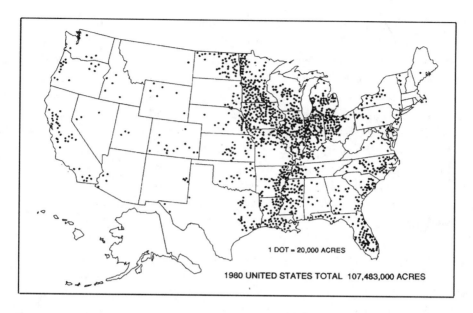

1 DOT = 20,000 ACRES

1980 UNITED STATES TOTAL 107,483,000 ACRES

Fig. 13.1 Wetlands in the eastern United States have been drained extensively since 1870, especially along coastal plains and riverine systems such as the greater Mississippi drainage basin. Each dot equals 8000 ha (20,000 acres) of drained wetlands. (After Mitsch 1993)

while other countries such as Sweden lost fully one-fourth of their wetlands (World Resources 1992). Countries of Asia and Africa have lost even more of their vital coastal wetlands. For example, more than 70% of the mangrove swamps in the Philippines have been lost through reclamation or degradation through pollution and aquaculture development (see chapter 11). Inland wetlands have also been severely affected as riverine wetlands throughout the world have been converted to agriculture land or dried up through the creation of levies and dikes. In Nigeria, more than 300 km² of wetlands have been reclaimed along one river alone (the Hadejia). Beyond sheer losses of area, many more hectares of wetlands have been degraded by hydrological changes, introduced species, and pollution stress.

The Challenges of Wetland Management and Valuation

Despite the historical disregard for wetlands, the newly recognized values placed on these environments have far-reaching importance. The list of positive ecological and social values includes the vital function that wetlands play in bird and wildlife habitat. Some estimates claim that more than one-third of the endangered or threatened species in the United States depend on wetlands for some part of their life cycle, even though wetlands constitute only one-third of the nation's land (NRC 1992). Among the animals that depend on wetlands are migratory waterfowl such as bald eagles, ospreys, hawks, egrets, herons, kingfishers, and other shore, marsh, and passerine birds such as ducks, geese, and swans. According to the NRC (1992), more than half of the waterfowl in the United States nest in the prairie pothole wetlands of the Midwest. Many other birds are facultative users of wetlands. Mammals that depend on wetlands include the muskrat, beaver, and swamp rabbit; others such as the otter, mink, raccoon, bobcat, meadow mouse, moose, bear, and deer use wetlands as feeding areas (NRC 1992). Wetlands are also critical habitats for many species of reptiles and amphibians.

Other critical values of wetlands include their roles in nitrogen cycling, groundwater recharge, flood prevention, and shoreline stabilization. Their quiet waters and productive vegetation are also responsible for the settling of sediment, nutrient removal, toxic material removal, and support of fish and shellfish breeding areas. The diversity of values, variety of wetland types, and size and distribution of wetlands all contribute to the management challenges.

The challenges of wetland management are also greatly complicated by the fact that their positive social values are so distinct from those of more familiar water resource values. Social values of rivers, lakes, and coastal zones tend to cluster around water's role in meeting physical needs such as drinking, agriculture, fish, and transportation, and its role as a source of recreation through swimming, boating, or recreational fishing. The values ascribed to wetlands cluster at significantly different poles, however. One set of wetland values is extremely pragmatic: they can ameliorate, in a natural and inexpensive way, some of the waste problems humans have created. The other set is primarily aesthetic and ecological: wetlands increase wildlife habitat and landscape-scale biodiversity. Both sets represent relatively new and somewhat unstable values on which to base policy and management decisions.

Wetland management is also complicated by the fact that wetlands' greatest values are cumulative ones. With the exception of particularly large wetlands such as the Florida Everglades, the ecological or social value of any single wetland tends to be small. Instead, the presence of many wetlands in the landscape confers value for biodiversity and wildlife habitat. This understanding has led to the popularity of the "no net loss" policy, which states that any wetland can be developed, as long as a compensatory, replacement wetland is established. Creating a viable wetland has proved far more difficult than once thought, from both an engineering and biological standpoint. While "no net loss" makes an easy policy rule, the concept has generally become discredited among wetland biologists, even as the patchwork of viable wetlands continues to be degraded.

Yet another challenge in wetland management is the problem of wetland ownership. In contrast to rivers and lakes, which are commonly recognized as public resources, many wetlands are small and are located within individually owned properties. Almost 75% of remaining U.S. wetlands are found on private lands (NRC 1992). In contrast, the environmental, social, and economic benefits of wetlands are primarily public values—individual owners receive little direct benefit from their wetlands. Conflicts between managing land for the "public good" and rights of private property owners to manage their land as they see fit are age-old and often intractable. Nevertheless, they represent a relatively new dimension for water quality management.

Lastly, management of wetlands is complicated by the presence of many different kinds of natural wetlands and the relative infancy of much of wetland science. An overview of wetland ecology as it is currently understood, and emerging issues in wetland science, are presented below.

What Is a Wetland?

There is no clear definition of where a wetland begins and ends. The primary defining characteristic of a wetland is that it contains waterlogged soils during at least some part of the year. Vernal pools found in the western United States represent one extreme of wetlands. These wetlands are shallow and ephemeral, drying up for large portions of the year during the warm summers and dry autumns. Their wet period, during the cool winter and spring, typically spans only 3 or 4 months. Coastal salt water and coastal fresh water marshes represent the other extreme—permanently flooded wetlands that experience little seasonal variation in depth. Between these two points lie important wetland types such as the prairie potholes of the upper Midwest, which occur in glacial depressions and vary over a cycle of 5 to 25 years (Mitsch 1993). Prairie potholes may dry up during the long summer of a dry year or stay saturated through the year, depending on precipitation patterns. Typical hydroperiods for these wetlands are pictured in Fig. 13.2. Wetlands may also be fed by, and serve to recharge, groundwater or they may occur along the banks of rivers. In all cases, their form and temporal variability reflect their source of water and its replacement time.

As the importance of wetlands continues to emerge, scientific understanding of wetland ecology is growing exponentially. Until relatively recently, hydroperiod was considered the primary factor regulating productivity of all wetlands. As indicated throughout this chapter, new studies are refining that

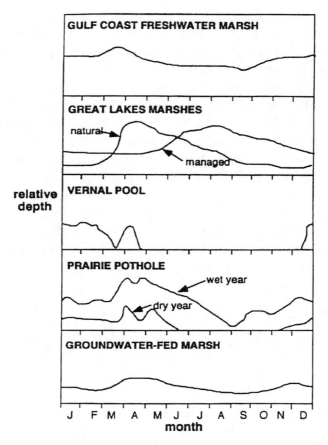

Fig. 13.2 Wetland hydroperiod controls many of the features and functional roles of wetlands. The variations in hydroperiod are enormous and range from coastal wetlands whose water level is nearly constant throughout the year to vernal pools that may be dry for more months than they are wet. (From Mitsch 1993)

understanding and pointing to hydroperiod as only one mechanism controlling the extreme variability in wetland ecology. Other newly recognized mechanisms include the placement of a wetland in the landscape, nutrient limitations, autogenic processes, and the source of water for different wetlands. Inherent in studies of wetland variability is the increased awareness of spatial and temporal scales in wetland ecology and management. These issues are discussed, for example, in a 1992 study in which Mitsch attempted to define appropriate scales of study for different wetland parameters, demonstrating that controls on wetland form and function operate at widely different spatial scales.

An important emerging concept in wetland science is that wetlands have different functions depending on their position in the landscape. Detenbeck *et al.* (1993), for example, noted a close correlation between adjacent land use and wetland characteristics. Looking at a different set of variables, Brinson (1993) suggests that a wetland's location along a stream greatly influences its role in nutrient regulation of that stream. In particular, he indicates that lower-order (e.g., smaller) streams yield more value for water quality improvement of non-point pollution than high-order streams, and notes that this finding is antithetical to the current U.S. wetlands policy that allows headwater wetlands to be filled. Looking for measures of wetlands that relate more closely to function than area, Brinson also suggests that the length of riparian wetland is "a better index of potential for enhancing water quality than area" [p. 68]. Brinson's work also points to upstream/downstream differences in function, especially for water quality. For example, his review of wetland studies catalogues enormous variations in the nutrient flow patterns of different wetland types. Thus, although some wetlands fit the familiar model of nutrient traps or filters, with high nutrient inflow and low outflow, other wetlands have little effect on nutrient levels and still others oxidize compounds and produce higher nutrient outflows than inflows.

Finally, Brinson draws attention to the importance of different sources of water in different wetlands and suggests new terminology that would describe the source of wetland water and, consequently, the susceptibility to contamination from incoming water. Similar new classification schemes are emerging from other studies, such as Lloyd *et al.*'s (1993) study of wetlands in the United Kingdom. As with Brinson's simpler scheme, Lloyd *et al.* display an increased sensitivity to the origin and flow of waters into wetlands rather than focusing on the timing of hydroperiod per se. These suggested classification schemes are compared in Figs. 13.3a and 13.3b. These models contrast dramatically with Cowardin's (1979) early classification of wetlands by geographical location and hydroperiod. Cowardin's model, however, is still widely used today. While new terminology has yet to be adopted, these works are just several among many indicating the rapidly expanding knowledge of wetland systems, and their role in water quality management.

Productivity and Nutrient Cycling

The productivity of wetlands varies according to nutrient inputs. At one extreme are northern (ombrotrophic) bogs that are fed primarily by precipitation and consequently are nutrient-poor. In these systems, the high pH and oxygen-poor (anoxic) soils slow down the decomposition of organic matter,

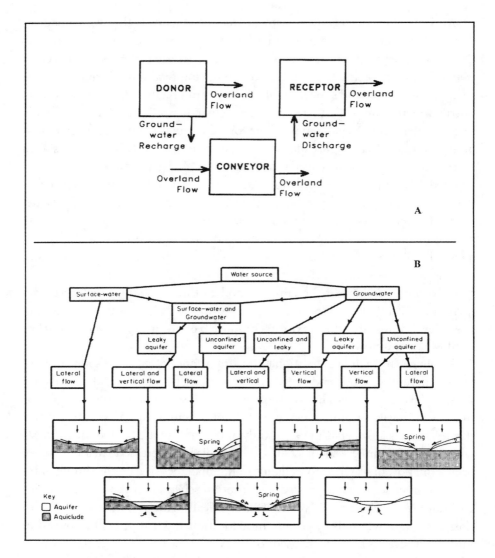

Fig. 13.3 As wetland science evolves, new terminologies and classifications for wetlands are being developed that reflect landscape-scale hydrologic movement and the functional role of wetlands in the landscape rather than the traditional wetlands' classification based on soils, hydroperiod, and plant community assemblages (not pictured). (13.3a after Brisson 1993; 13.3b after Lloyd 1993)

leading to the formation and accumulation of peat. In contrast, eutrophic tidal wetlands are highly productive and are flushed with new nutrients with the tide (once or twice a day). Although decomposition is not inhibited in these wetlands, organic matter still accumulates because production outpaces decomposition. The productivity of other fresh water wetlands is related to the rate and source of feeding waters. For example, groundwater flow is an important source of nutrients and dissolved oxygen in many inland marshes.

The accumulation of organic matter in wetlands is of critical importance to water quality because it is through this accumulation that carbon and nutrients are sequestered and "removed" from the water. Yet, as implied above, wetlands vary tremendously in both their ability to sequester nutrients and in the processes that lead to that sequestering. Some sense of the variability of wetlands can be gained from Mitsch (1993):

> "Organic matter can vary from a very high content (75 percent) as is found in floating freshwater marshes in coastal Louisiana to a low content (10–30 percent) in marshes fed by inorganic sediments from agricultural watersheds or open to organic export. Concentrations of total (as distinguished from available) nutrients are reflections of the kinds of sediments in the marsh. Mineral sediments are often associated with high phosphorus content, for example, whereas total nitrogen is closely correlated to organic content." [p. 341]

Wetland Plants

Early theories on wetland vegetation held that wetlands were continually evolving to terrestrial forms—that is, accreting organic matter until they were no longer saturated, and succeeding to a terrestrial forested climax community. Accumulation of peat occurs only in anoxic sediment, however. Under oxidized conditions, peat subsides, so formation of a dry habitat from peat is unlikely unless the water table is lowered. Recent studies, including analysis of pollen records, support the idea that most wetlands are stable systems operating under strong environmental influences. Rather than following patterns of succession over time, they respond to long-term changes in lake level or changes in a river channel that alter the flux and source of water (Mitsch 1993).

An important characteristic of wetland plants is the vertical zonation caused by hydrological conditions. Wetlands exhibit sharp and predictable zones at the land–water interface. For example, in the typical wetland profile pictured in Fig. 13.4, the periodically flooded bank contains lowland grasses; the waterlogged, shallow water zone contains sedges (*Carex*) and arrowhead (*Sagittaria*); deepening waters support reedy plants (emergent macrophytes) such as cattails (*Typha*) and bulrush (*Scirpus*); and the deeper water supports floating leaved and submersed plants such as water lilies (*Nymphaea*), pond weeds

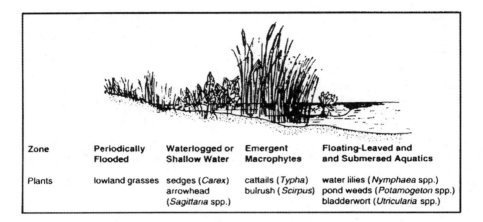

Zone	Periodically Flooded	Waterlogged or Shallow Water	Emergent Macrophytes	Floating-Leaved and and Submersed Aquatics
Plants	lowland grasses	sedges (*Carex*) arrowhead (*Sagittaria* spp.)	cattails (*Typha*) bulrush (*Scirpus*)	water lilies (*Nymphaea* spp.) pond weeds (*Potamogeton* spp.) bladderwort (*Utricularia* spp.)

Fig. 13.4 Plant communities in wetlands are dependent on water and soil characteristics and thus grow in narrow zones defined by those characteristics. (After Mitsch 1992)

(*Potamogeton*), and bladderwort (*Utricularia*). This zonation results from hydrologic and salinity gradients caused by the elevational changes (however minor) across a wetland.

Although environmental (allogenic) forces exert an extremely strong influence on the formation and functional processes of wetlands, the growth and establishment of wetland plants can also undergo significant autogenic development. For example, as described by Mitsch (1993), wetland productivity causes the sediment to become increasingly organic and, therefore, increasingly light. In some instances, the entire sediment mat becomes buoyant enough to float. As a result, the sediment is always saturated, but because it floats, it never becomes flooded and is no longer responsive to flooding stresses.

Wetlands are unique in that the vegetative community contains characteristics of both very young and very mature communities. Productivity is very high, as in immature ecosystems, but is detrital-based and supports the complex food webs typical of mature systems. The spatial zonation is typical of a mature system, but strong allogenic influences are more typical of "young" systems. Other mixed characteristics derive from the life cycles of wetland consumers. The cycles are usually short (a characteristic of immature systems) but extremely complex (a characteristic of mature systems).

Although water renewal rates of wetlands can vary from once per year in a northern bog to 7500 times per year in a riparian wetland, and the amount of nitrogen delivered to wetlands varies with similar intensity (e.g., five orders of magnitude), wetlands are more similar than these numbers would indicate. According to Mitsch (1993), total stored biomass ranges only from 40 to 60 km/m^2, and soil nitrogen varies only threefold, from about 500 to 1500 g/m^2.

Net primary production varies somewhat more (from 600 to 2000 g/m² per year), but still does not reflect the orders-of-magnitude differences that might be expected. Mitsch proposes that these close values are related to the nitrogen mineralization rate, which is controlled by temperature and hydroperiod. Except in northern bogs, it is thought that temperatures during the growing season are similar enough that mineralization rates are also similar, causing nitrogen supply rates to be relatively close in different systems. Because productivity in most wetlands is limited by nitrogen rates, similarities in nitrogen mineralization could explain primary productivity rates, even though other important processes are variable.

Wetlands and Water Quality

Wetlands are unique in that their relationship to water quality is twofold: they are negatively impacted by certain pollutants and land use activities but are also efficient at removing some pollutants from the water, thus improving water quality. Water quality relationships in wetlands are characterized by extensive interaction between water and vegetation, relationships of water and soil, and temporal variations in hydrology (Fig. 13.5). Alterations that affect water quality and wetland function can be either biological, chemical, or physical, or, in the case of urban development, may be all three (Fig. 13.6).

In Mitsch's (1993) review of wetland management, he notes that overall little work has been done on cataloguing the diverse effects of pollutants on the life history of wetlands. Several pollutants, however, are known to lead to extensive degradation of wetland vegetation. These contaminants include oil spills, acid drainage from mines, and sulfates. Other pollutants may have less effect on wetland vegetation but devastating impacts on the wetland wildlife. For example, selenium poisoning of the Kesterson Wildlife Sanctuary in California led to widespread death and deformities among bird populations. Mitsch also reports that other inputs, such as phosphorus, may improve the success of invading species such as *Typha*, resulting in changes in species composition and overall wetland functioning (e.g., patters of nutrient cycling).

The use of wetlands as a water quality filter has received considerably more attention. Because wetlands retain sediments and "trap" nutrients through increased productivity, they have been used in concert with waste treatment efforts. Early enthusiasm for utilizing wetlands in this way came from studies that demonstrated wetlands at Houghton Lake (Michigan) removed 70% of the ammonia nitrogen, 99% of nitrite and nitrate nitrogen, and 95% of total dissolved phosphorus as the water moved through the wetland (Mitsch 1993). One

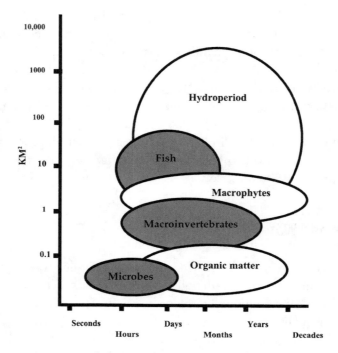

Fig. 13.5 Wetlands are comparatively uniform through space and time (compare with Figs. 9.6 and 12.4). Hydroperiods of small prairie wetlands are more variable than are Gulf Coast swamps, but all vary less than do streams. Animal communities in these systems are relatively constant but plants vary on longer-term cycles (e.g., decades).

of the most celebrated waste treatment wetlands is found in Arcata, California. The city of Arcata channels primary treated wastes into a 63 ha settling pond/wetland system. Water in the wetland has approximately a two-month renewal time. During those months, the water is cleansed by natural processes to a quality higher than that of the receiving bay. Other equally successful projects have been widely documented, including some that remove important pathogens and some metals from the water.

As with any manipulation of a natural system, these studies must be viewed with site-specific caution. Arcata's success is partially related to the size of the receiving wetland compared with the relatively small city of 15,000 inhabitants. Larger waste treatment demands can overload the processing abilities of the receiving wetlands. As indicated above, wetlands vary tremendously; consequently, their ability to filter nutrients and pathogens, and their ability to remain relatively unaffected by the changed nutrient regime, will vary im-

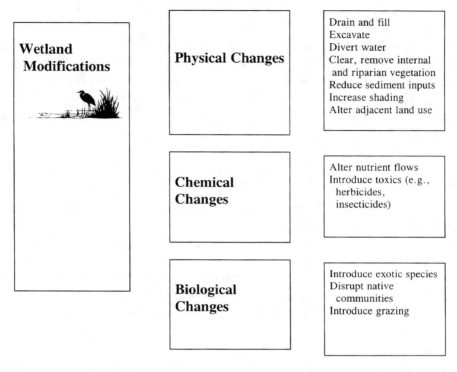

Fig. 13.6 Wetlands are altered in many ways, by numerous anthropogenic activities. Each type of alteration has a distinct suite of impacts on the wetland ecosystem.

mensely. The effectiveness of many waste-receiving wetlands has been shown to decrease after several years. In particular, wetlands that do not accumulate significant peat organic layers or that do not have associated woody vegetation (like cypress trees) may become saturated for those inputs and no longer serve as effective sinks. Additionally, the nature of the vegetation itself usually changes under increased nutrient regimes. Even in unimpacted conditions, wetlands with higher nutrient levels tend to be less diverse than those with low nutrient levels. The same trend has been shown in waste-receiving wetlands.

Water Quality and the Function of Wetlands in the Landscape

Other significant water quality issues revolve around the role wetlands play in determining the water quality of the surrounding landscape. Draining and filling of wetlands have effects far beyond the altered surface area. For example, riverine wetlands serve important buffer functions, storing excess water in times of flood and providing significant bank protection. Flood protection value of wetlands is the estimated extent of property damage if wetlands were removed from the river. Along the Charles River in Boston these values have been estimated at $17 million per year, or $2000 per acre of wetland (Turner 1991). These same riverine wetlands also trap sediment and nutrients and process nitrogen in soils, providing an added water quality benefit.

Upland wetlands also have an important water quality role. In particular, they aid in retarding erosion, slowing water flow rates, and reducing in-stream sedimentation. Consequently, the destruction of these wetlands increases downstream water quality degradation.

An emerging area of study in wetland water quality is the effect of cumulative impacts on wetland function and quality. As Johnston (1994) describes, loss of wetlands has been correlated with increases in stream nitrogen content. Nevertheless, "the cumulative loss of wetland function at the landscape scale may or may not be linearly proportional to area lost" [p. 53]. Johnston cites studies demonstrating that the initial losses of wetland area have smaller effects on wetland function than later losses; for example, losses from watersheds containing 10% to 50% wetlands have little effect on flood flow, but losses from watersheds containing less than 10% have a large effect on flood flow. A similar "10% threshold" was found in the susceptibility of wetlands to increased loadings of suspended solids. Johnston provides further evidence that loss of connectivity between wetlands through landscape fragmentation can significantly affect the cumulative function in nutrient processing. Thus, "wetland impacts that seem minor on an individual basis may become major when considered collectively over time and space" [p. 54] (Johnston 1994). These findings have significant implications in wetland management, but more research is needed on this issue.

Wetland Management and Restoration

One of the most important outcomes of wetland research of the past few decades has been the recognition that the best management for preserving their ecological value involves retaining wetlands in their natural state. Altering water levels, hydroperiod, species type, and nutrient input all affect the functioning of the wetland and interfere with its primary value. Where wetlands have already been altered, several options exist for recreating former conditions. These techniques relate to restoring an appropriate hydrology, eliminating contaminants, and reestablishing native species.

Several techniques are used to reestablish wetland hydrology. Diverted rivers may be redirected to flow back into original river channels and their marshes. Removing flood structures is also used to reestablish hydrology in dammed and diked rivers, streams, lakes, and wetlands. In some cases, such as in the prairie pothole region of the upper Midwest, entire drainage systems were established to facilitate the drying of wetlands (tile drainage system); recovery of many of these wetlands ultimately depends on the destruction of these drainage systems. Other options include reestablishing topography in modified landscapes (e.g., removing material from filled wetlands).

Recovery options can be extremely expensive. Estimates for restoration of one small (260 acres) urban wetland in Los Angeles have reached as high as $50 million (NRC 1992). While this estimate is unusually high, the costs of land purchases, excavation, road alterations, sewers or drainage systems, pumps, and other devices often invite questions about the value of any single wetland.

Water Quality Equivalency of Created Versus Natural Wetland

Wetlands are unique in that they are the only natural resource that has been treated as both expendable and recreatable. Humans have modified, often drastically, the habitats of forests, rivers, and lakes, but these modifications are generally acknowledged as unequivalent approximations of the original. In forestry, specific nomenclature has been developed for different forests based on their logging history and the distribution of species and ages within the forest. The term "channel" is used to connote a section of river that has been redirected or whose bank has been extensively modified. The theory behind much of wetlands policy, in contrast, is that one wetland can be filled as long as a compensatory wetland is built elsewhere. Implicit in that approach is a poor understanding of wetlands as a natural, complex, and local phenomenon. Also incorporated in that approach is a reliance on structural replication of a wetland (a depression with waterlogged soils and plants) instead of a functional replication that would include such characteristics as sustainability, low susceptibility to invasion by exotics, productivity comparable to natural systems, ability to retain nutrients, and biotic interactions similar to those of reference systems.

The success of created wetlands is difficult to quantify because of poor follow-up and monitoring. Although most projects note the acreage dredged to establish a new wetland, and may report on the number of seedlings established in the created wetland, the long-term success of these projects is of the greatest environmental and water quality concern. A tremendous range of success and failure has occurred in establishing viable wetlands. Among the few wetlands that have been monitored for at least several years, some have lost initially successful plantations to insect outbreaks, while others have been thwarted by continued development activities. In still other studies, vegetation was successfully established in the created wetland. In some cases, the established species are similar to those found in other wetlands in the area; in others, they diverge and reflect different species compositions. In almost all cases, vegetation in created wetlands is lower in density than that noted in natural wetlands. Created wetlands begin with fewer canopy strata and fewer trophic levels with the expectation that these will evolve into a fully functional wetland. The failure of many such wetlands to evolve as hoped indicates the lack of knowledge about keystone species, functional processes, and soil dynamics (NRC 1992; see also chapters 14 and 15).

The question of vegetation establishment is only the first part of wetland creation. While establishment of a full canopy may take only several years, it is not just the existence of vegetation—but its structure and architecture—that are important from a wildlife habitat point of view. Development of different height profiles in the canopy might take 10 times longer than the establishment of vegetation per se. For water quality concerns, the way in which a wetland cycles and stores carbon and nutrients is critically important. Some estimates expect that this function will build up only over a period of decades.

Because such a large percentage of wetlands were destroyed before their values were recognized, and because scientific knowledge of wetland structure, function, and diversity is incomplete, there are both significant opportunities and important risks in wetland management. It is undoubtedly one arena that will continue to be a subject of intense policy debate and scientific research.

Wetlands Around the World—A Footnote

This chapter has focused on U.S. wetlands primarily because most of the studies in this emerging field have been conducted in the United States. The problems of wetland management, however, are by no means confined to U.S. soils. In a review of international experiences with wetland management, Hollis (1990) demonstrates that draining, hydrologic alteration, and pollution of wetlands is a universal phenomenon. His descriptions include detrimental effects on rare species in Greece, social disruption in Nigeria, and a long list of failed wetland management projects from Canada to Tunisia and Ghana. Hollis suggests that wetland destruction is furthered on a global basis by international subsidies that advance large-scale agricultural projects, by narrowly focused local-scale planning that fragments wetlands, and by sectoral approaches to planning that ignore the interconnected ecology and function of wetlands and divide wetlands into planning projects for different disciplines.

One successful international agreement, the Ramsar convention, seeks to protect wetlands important to the international migration of waterfowl. The

Ramsar convention has been used to identify, and target for conservation, wetlands from Africa to Northern Europe. Although the specifics vary, it is still a global truth that wetland value is only beginning to be recognized and that effective policy and sustainable wetlands' management is not yet a reality.

Summary

- Wetlands are diverse, highly productive bodies of shallow, standing water. Some wetlands are "windows in the groundwater"—places where the water table is exposed to the surface. Others are flow-through systems along rivers or at the mouths of rivers or estuaries. Still others are isolated bodies of water and saturated soil. All of these forms represent biologically diverse interfaces between the terrestrial and aquatic environments.
- Wetlands act as efficient filters, removing a wide range of materials (e.g., organic matter, sediment, nutrients) from flow-through waters. They represent highly important migration corridors and habitat islands in the terrestrial and semi-aquatic landscape and provide valuable habitats. Although they are known to serve major nutrient and biodiversity functions, their overall role in landscape function remains poorly understood.
- Between 30% and 75% of wetlands around the world have been lost to drainage and filling, with the specific percentage varying among countries and wetland types. Coastal wetlands have been the most heavily impacted.
- Management concerns in wetlands are dominated by issues of drainage, other changes in hydrology, and water chemistry. They are widely used as waste treatment systems but effects of such use on natural wetland properties are poorly understood.
- Restoration of wetlands is becoming a major management action, although its effectiveness is hampered by inadequate knowledge of wetland and landscape ecology. We know how to create an area hydrologically equivalent to a drained wetland. Very little agreement has been reached, however, on what constitutes functional or ecological equivalence between natural and created wetlands.

For Further Reading

Brinson MM. Changes in the functioning of wetlands along environmental gradients. *Wetlands* 1993;13:65–74.
Cowardin LM, Carter V, Golet FC, La Roe ET. *Classification of wetlands and deep water habitats of the United States.* Office of Biological Services: Fish and Wildlife Service:

U.S. Department of the Interior. Washington, DC: U.S. Government Printing Office, 1979; GPO 024-010-00524-6.

Johnston CA. Cumulative impacts to wetlands. *Wetlands* 1994;14:49–55.

Lloyd JW, Tellam JH, Rukin N, Lerner DN. Wetland vulnerability in East Anglia: a possible conceptual framework and generalized approach. *J Environ Mgt* 1993;37:87–102.

Mitsch WJ. *Wetlands*, 2nd ed. Van Nostrand Reinhold, 1993.

14

Structuring Management Goals by Ecological Level

Overview

Previous chapters have addressed three principles that form the foundation of water quality management: (1) human values and goals define decision making (chapters 1–6); (2) the ecology and biophysical characteristics of water resources define "what is possible" and "what is at risk" (chapters 8–13); and (3) ecological processes operate at different spatial and temporal scales (chapters 6–13). Chapters in this section integrate these three principles and address the effectiveness and necessity of incorporating "scale planning" into management. They present a central tenet of water quality management—that sustainable management depends on focusing attention on the appropriate ecological level to meet human goals. The appropriate ecological level, however, is not always (or often) the apparent one. Although many products for which we manage are discrete (e.g., game fish, drinking water), the overall approach must not be limited to this level. Instead, it must consider community and ecosystem dynamics that regulate the more discrete levels. Section 4 (chapters 17–24) builds upon the importance of scale planning to address water quality issues at their different spatial and temporal scales of relevance.

Management that incorporates scale planning gains much from a study of hierarchy theory and ecological levels, the conceptual tools used in matching goals to ecological realities. Current ecological theory regards ecosystems as *nested sets of open biological, physical, and chemical systems operating on different temporal and spatial scales,* not as "closed" cycles as previous theory proposed. In this approach, some interactions such as those between predator and prey are recognized to be short-term and oscillatory or cyclic. These relationships represent components of large complex systems such as oceans, the atmosphere, or large forests that are more homeostatic and stable. These hierarchical differences offer important guidelines for choosing management levels (discussed in this chapter), evaluating ecosystem stress (chapter 15), and establishing management plans (chapter 16).

Introduction: Scale-Dependent Questions

The key to effective management is learning to ask the right questions. As discussed in the chapters of section 1, defining goal statements is a critical first step: what are the societal values associated with given water uses? If goal statements are out of sync with the biophysical nature of a water resource, the goal statement must be reevaluated to fit the ecological and biophysical contexts. (These contexts were the focus of the chapters in section 2.) Effective management begins when biophysical and ecological reality is integrated with value statements. The degree to which this integration is successful depends, in large part, on understanding the interplay of temporal and spatial scales in ecology. Goals, for example, may state that a given river should be managed for trout fishing. Yet, implementing that management goal depends not only on whether that river can support trout, but on managing water chemistry, prey species, and other ecological conditions that control viable trout populations. Thus, success depends on managing at a higher ecological level than the trout population to affect the variables that control population-scale processes.

This chapter outlines hierarchy theory, one of the most useful tools in conceptualizing different spatial and temporal scales and suggests applications of hierarchy theory to management practice. Hierarchy theory is increasingly accepted as a model for ecological processes, in part because it is so flexible. Hierarchical levels may be continuous or discontinuous, nested or nonnested; the theory does not force observations into a given structure, but rather provides a framework for the way scientists think about the processes they are observing.

Hierarchies and Scales, the Theory Behind Practice

Hierarchy theory outlines the relationship of different spatial and temporal scales to one another and serves as a useful tool for focusing management questions at the appropriate ecological scale. The concept that ecosystems are hierarchically organized is not new. People have always known that medium-size fish eat little fish, and that bigger fish eat medium-size fish. What is new in ecological theory is the recognition of the full extent of ecosystem hierarchies and the conceptual and managerial power behind them. Simple observations about natural phenomena yield new insights when seen as examples of larger ecosystem organization. For example, the observation that butterflies have shorter lives than swallows, which have shorter lives than eagles, is a shorthand way of expressing important temporal hierarchies in nature. In addition to temporal hierarchies, there are trophic hierarchies among species, size hierarchies within and among species, and spatial hierarchies of territory size and distribution.

Hierarchy theory has been applied in myriad disciplines to help organize thinking about interactive and complex systems. Its most familiar application involves administration and political systems. Each neighborhood is its own unit, with its own local economy, local decision-making boards, and local issues such as noise, crime, or pets. Its spatial boundaries define the neighborhood as does its identifiable internal flow of resources. Local ordinances and decisions, however, are made within the context and frameworks of city government. On one level, city government represents the compilation of local neighborhood governing boards. At the same time, city government has its own unique structures. Cities operate within the framework of state government, and states function within the framework of the federal hierarchical "unit." At each level, the governing units constitute and, in turn, are constrained by levels above and below them. Goods and services cross from one level to the next, but the flow within any given level is distinct. It is also true that at each successive level, the flow of resources appears to be relatively contained within the higher sociopolitical borders.

Hierarchical structures are found everywhere from politics to social and natural systems. They can be described generically as a series of self-regulating, open units interacting to form larger self-regulating, open units (Muller 1992). At each higher level, units appear more closed and processes within them operate on longer time scales and over larger distances. In ecology, the multiplicity of interacting levels complicates hierarchy theory. The power of hierarchy theory for natural resource management lies in its ability to allow for just that complexity. As summarized by Fox (1992): ". . . the theory leads us away from the naive mistake of searching for a fundamental hierarchy or level of analysis. The theory suggests that we must consider different ways of structuring the data we collect and choose the hierarchy and level of interest according to the problem at hand" [p. 292].

Ecological Hierarchies

Traditional hierarchies of business or politics tend to be strictly vertical (e.g., the office worker answers to the manager, who answers, in turn, to the vice president). Hierarchies in ecology may be represented by trophic levels of predator and prey. In this system, the lowest levels of a hierarchy define biotic potential of a system, while the highest levels coordinate and constrain lower levels. For example, prey reproductive rates define the population potential for higher-level predators: the oyster population, for example, regulates the otter population. In turn, predators constrain biotic potential of prey. Where ecological hierarchies deviate from many other hierarchical systems is in the presence of multiple relationships nested within each level. Fig. 14.1 illustrates the interplay of multidimensional hierarchies for one study area of the Arctic.

For example, differences in size segregation, productivity, and species distribution can all exist within a given hierarchical level. In practice, they reflect effects of environmental conditions. For example, the space or light available for plants is an environmental constraint that determines plant growth, productivity, and, ultimately, an individual's survival. Integrating hierarchy theory with earlier theories of species distribution has, in fact, produced a more flexible portrayal of the relationships between species within a landscape, allowing simultaneously for species that have narrow ranges, broad ranges, overlapping ranges, and discontinuous ranges (Collins *et al.* 1993). These environmentally controlled subsets of hierarchies are critical in ecosystem management because they represent the redundancy for any given ecosystem component. As demonstrated below (and in chapter 15), redundancy is one of the key features regulating the ability of an ecosystem to remain stable despite environmental or management pressures.

An example of an hierarchical model was developed by Frissel *et al.* (1986) for stream systems. In their model, the stream incorporates, on successively lower levels, stream segment, reach, pool/riffle, and microhabitat subsystems. Each level of the hierarchy occurs at a different spatio-temporal scale and is controlled by different forces. As Frissel *et al.* (1986) describe, "Geologic events of low frequency and high magnitude cause fundamental evolutionary changes in stream and segment systems, while relatively high-frequency, low-magnitude geomorphic events can change the potential capacities of reaches, pool/riffle systems and microhabitats" [p. 201]. They also note that the hierarchies are spatially nested, with higher-level systems forming the environment of lower-level ones.

Hierarchy theory concepts can be summarized in five "rules of thumb" for defining hierarchies (adapted from Muller 1992 and Fox 1992):

1. The spatial extent of higher levels is broader than the extent of lower levels. To distinguish a level, one must identify spatial scales.
2. Higher, coarser levels change more slowly than lower, finer-scale levels. Significant changes require longer periods at higher levels.

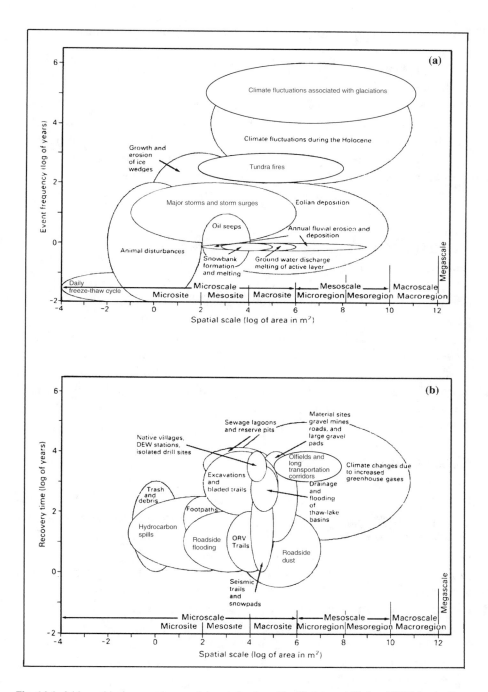

Fig. 14.1 A hierarchical ecosystem model was developed by Walker and Walker (1991) for Arctic ecosystems. (a) Represents spatial and temporal scales of natural disturbances. (b) Represents spatial and temporal scales of recovery from anthropogenic disturbances. In this example, anthropogenic disturbances are more compressed in both time and space. Walker and Walker's graphic representation of their work is an elegant illustration of both the utility of visualizing ecosystems from hierarchical viewpoints and the complexity of these hierarchies. (Modified from Walker and Walker 1991)

3. Higher levels control lower levels by dictating physical, chemical, and biological limits.
4. Higher levels usually contain lower levels (i.e., nested hierarchies).
5. Higher-level phenomenon will more accurately predict effects at lower levels than lower levels will predict effects on higher ones. (For example, a national or regional phenomenon may predict local conditions, but it is more difficult to aggregate local conditions into a national or regional phenomenon.) [See, however, Lawton's (1991) proposal that in some cases, simple models of characteristics of individual organisms can generate understanding of both population dynamics and species' distributions.]

Flows Among Hierarchies: What Characterizes a Healthy Ecosystem?

In water quality management, as in any natural resource management, characterizing specific ecological hierarchies is less important than understanding the way energy and matter flow among them. Hierarchies in a healthy, stable ecosystem interact on many levels but important points of vulnerability exist in the flow pathways among hierarchies that determine an ecosystem's susceptibility to degradation and opportunity for recovery. Four of these points are particularly important: ecosystem feedback loops, homeostasis, sustainability, and redundancy.

Feedback Loops

Feedback loops are links from one system that regulate processes in another system. A classic example is the release of nutrients through decomposition that are "fed back" to phytoplankton and enable high productivity. The rate of decomposition determines the potential productivity of the system, but actual productivity may be controlled by constraints such as thermal stratification that prevents the nutrients from reaching phytoplankton (see chapter 12). Nutrient cycles are examples of positive feedback loops (i.e., processes that maintain a given process).

Ecological processes are also controlled by negative feedback loops that constrain a given process. Cyclical population crashes, for example, often result from negative feedback loops. In many populations, high population density facilitates the spread of devastating diseases that would normally be held in check by dispersed populations. The disease causes the population to decrease to a lower level at which disease transmission is checked. Another example is increased predation on a rapidly rising population, which in turn causes the population to decrease (Engel 1990).

Ecosystems are in a constant state of flux as their components respond to climate variances, natural disturbances, immigration, emigration, mutations, and natural population cycles. True to their hierarchical nature, however, while lower levels (e.g., populations) respond quickly to these fluxes, feedback loops at the community level ensure that community-level changes occur much more slowly. Ecosystem-level changes take place even more slowly as feedback loops at the ecosystem level buffer fluxes from lower levels. Consequently, managers may examine energy transfer throughout hierarchical levels to determine the "health" of, or damage to, an ecosystem (see chapter 15). Inadequate buffering of fluctuating energy transfer, or communities and ecosystems that experience rapid change, can indicate instability and potential for more serious impacts.

Homeostasis

Exchange of energy within an ecosystem is relatively constant (homeostatic), despite fluctuations at population and community levels. An ecosystem operates in a way somewhat analogous to a thermostat with a mechanical set-point, or a brain that maintains a constant body temperature based on feedback from the body and the environment (Kelly and Harwell 1990). An important difference exists between feedback systems in ecosystems and these mechanical or physiological analogies, however. Human physiological functions have evolved to maximize reproductive success of the individual organism. In other words, a single "goal" has driven the evolution of these interactions. In contrast, no single goal has driven development of ecosystem functions and feedback loops. Instead, they have evolved because of interactions between interdependent populations and their abiotic environments.

Sustainability

A related characteristic of a "healthy" ecosystem is that it is self-sustaining, at least at a scale of hundreds to thousands of years. Cycles of production, consumption, and decomposition within the system serve as recycling loops such that, without significant disruptive forces, the system will remain relatively intact indefinitely. Of course, stochastic, disruptive forces—from storms to global climate changes—are a natural part of the global ecosystem. Ecosystems do not, in fact, remain intact indefinitely. One of the greatest challenges facing management is distinguishing the effects of "natural" fluctuations and disruptions from anthropogenic ones, and determining whether disturbances are inherent to, or disruptive of, ecosystem sustainability.

A familiar terrestrial example of disturbance and sustainability comes from the field of fire ecology and management. For hundreds of years, fires were seen as disruptive and fire control was viewed as an essential part of natural resource management. After a long history of fire control, people now realize that many species and qualities we prize in ecosystems depend upon periodic fires. For example, oak regeneration and native prairies both require periodic burns. Currently, managers are charged with the difficult task of allowing, and planning for, a "natural level" of burns while suppressing fires that occur too frequently, or at too high an intensity. Even more challenging, they must try to tell the difference between the two conditions.

Redundancy

One way that ecosystems maintain homeostasis is through feedback loops; another method is through redundancy. In a healthy ecosystem, many species fill different but similar roles. This practice provides ecosystem stability even if a catastrophe strikes individual species. Thus, if one bottom feeder is removed, several will remain to fill the ecosystem function.

An important concept in hierarchy theory and ecological practice is that successively higher hierarchical levels filter or dampen the fluctuations of energy and matter received from lower levels. For example, a given invertebrate population may fluctuate widely, depending on predation pressure, local disturbances, and food source. At the higher level of the whole lake, variations in individual populations are dampened, or averaged out, and invertebrate populations are stable. A similar example was presented in chapter 7: the distribution of fish species on any given atoll is regulated by stochastic events and is not predictable. On a regional hierarchical level, however, these differences are dampened out and the distribution of fish species is extremely predictable.

This hierarchical model suggests that higher levels of organization become progressively more stable as local or short-term fluctuations in productivity or species composition are factored out. This stability depends in large part on redundancy in each ecological level. In the example given in chapter 7, if a regional reef community consists of only two atolls, then wide fluctuations in species distribution in any given atoll would have a large influence on the predictability of fish at the regional reef level. A lack of redundancy in any given ecological characteristic creates a vulnerable link. A disruption to these vulnerable elements can cause a chain reaction of instability that radiates up and down the ecosystem hierarchy. When internal regulations break down so that effects radiate to a whole-ecosystem level, ecosystem function and stability are seriously impaired. This process leads to one major unanswered question in natural resource management: what happens when enough ecosystems are destabilized that regional or global hierarchies can no longer buffer changes?

Hierarchies from a Management Perspective

Natural resource managers seek to maintain appropriate feedback loops, homeostasis, sustainability, and redundancy. These characteristics are relatively unquantifiable, however, and hierarchy theory is largely a way of organizing thoughts about the complex relationships in ecology (May 1989). Managers require, instead, a more tangible system of variables that will indicate whether those four pillars of ecological hierarchies are being maintained. These tangibles are found in a series of functional and structural attributes of ecosystems, such as changes in the flow of energy across ecosystem boundaries or the balance of autotrophs to heterotrophs. Some of these usable attributes, which can be broken down into features of ecosystem "structure" and "function," are described below.

Ecosystems operate on two broad levels: kinds and numbers of things found in an ecosystem define its *structure*, while the rates at which processes occur describe ecosystem *functions*. Structural and functional attributes have been likened to economic concepts of goods and services because they make up those qualities that society finds valuable (e.g., tree biomass production, water flow regulation, biodiversity maintenance). These components are also manipulated when managing ecosystems. Intentionally or otherwise, changing structure and function alters the quantity or quality of "service" provided by that ecosystem. For example, damming the Snake River enables energy production and reliable agricultural production in thousands of hectares previously dependent upon highly variable rainfall. At the same time, increased cropping lowers soil fertility and forces farmers to plant different crops and add different nutrient subsidies. Similarly, a new fishery is created in the reservoir impounded behind the dam, but indigenous fisheries for anadromous species are lost. At each point choices and ramifications exist because each management action changes the very goods and services desired from the ecosystem.

Rate and Function Hierarchies

Management decisions must reflect an understanding of rates and scales controlling ecosystem behavior. Among these factors, four functional properties are particularly important and represent tangible clues to whether relationships among ecological hierarchies are being maintained: patterns at ecosystem boundaries, nutrient and energy transfer, metabolic intensity, and the stage of development.

PATTERNS AT ECOSYSTEM BOUNDARIES. Ecosystem borders or boundaries are often called an ecotone, or ecological edge. Actual ecosystem boundaries are approximate, but are often best defined by landscape features. Landscape boundaries can be anticipated, for example, at the terrestrial–aquatic edge, with a meandering stream, or at the wetland where the stream feeds into a lake. Ecosystem boundaries can, therefore, be defined at a variety of scales.

The term boundary is misleading—ecosystem boundaries are not inviolate. The distribution of many plants and animals extends through adjacent ecosystems, and management must account for these effects. For example, the manager concerned with a salmon population must consider more levels in management planning than the manager of a trout population because salmon migrate among widely differing ecosystems and serve important ecological roles in each system. Changes in salmon populations thus radiate to a coarser spatial scale than do changes in trout populations.

Boundaries operate as highly dynamic transition zones with high energy transfer rates. While the greatest exchange of energy and resources occurs *within* each ecosystem, all matter exchanged among systems must cross these relatively narrow boundaries. Consequently, boundaries represent areas of comparatively rapid change and high diversity. Fig. 14.2, for example, shows that plant communities change rapidly at zone borders, as does the rate of nitrogen fixation. Ramifications of a management action can be much greater if they cross boundaries than if actions are contained within a single ecosystem. Consequently, borders serve as an important monitoring point for ecosystem response to stress and management actions (see Naiman *et al.* 1988 for a review of the importance of boundaries in aquatic landscapes).

NUTRIENTS, ENERGY, AND MATTER. As mentioned earlier, ecosystems are open. Nutrients, energy, and matter all cross ecosystem boundaries because cycles regulating these flows operate globally. The movement is difficult to measure, however, and its importance is typically underestimated. Water quality managers commonly overlook the importance of properties such as nutrients and sediment.

The Aswan Dam in Egypt is an often-cited example of this oversight. The dam was primarily created for power production and, secondarily, to expand the area of irrigated land along the Nile. Unreliable rainfall has plagued farming throughout Africa. The most common solution to this problem has been the construction of dams that can buffer rainfall irregularities and, theoretically,

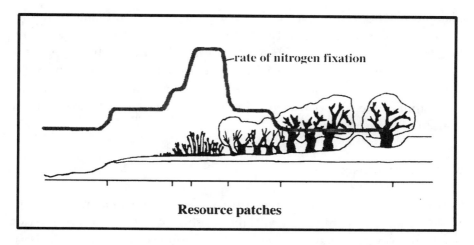

Fig. 14.2 Process rates, represented here by relative rates of nitrogen fixation, change abruptly at ecosystem or ecological zone boundaries. Process rates change even at smaller scales of resource patches that are found within any given ecosystem. (Modified from Naiman *et al.* 1988)

greatly increase agricultural production. From an economic and flood prevention point of view, the dam has been extremely successful. Biswas (1994) states categorically that Egypt would face economic collapse without the dam. From an ecological point of view, however, it has been less successful. Early planners did not consider the relationship between annual flooding and the provision of the nutrient-rich silt that enabled crop production. Because the dam blocks the silt, the land has lost productivity, a cycle that necessitates increasing use of chemical inputs. Had early managers raised their planning scale to model inputs of nutrients into the system, they may have anticipated this result and designed a system that met both economic and ecological requirements.

METABOLIC INTENSITY. The balance between heterotrophs (consumers, including decomposers) and autotrophs (green plants or primary producers) serves as an important indicator of ecosystem metabolic rates. Metabolic intensity is an important principle of ecosystem behavior. A higher rate (i.e., higher respiration or higher nutrient or energy consumption rates) will require the system to maintain greater input and output fluxes to sustain the rates, leading to more open (leaky) boundaries. Similarly, ecosystems consist of coupled synthesis (production) and degradation (decomposition) processes that allow a persistent, recognizable biological assemblage to persist through time.

STAGE OF DEVELOPMENT. Developmental stage is another ecosystem attribute that warrants consideration. Ecosystem behaviors differ according to the stage of development. Younger systems require more input from, and export more to, adjacent systems; older systems, in contrast, are more self-contained with smaller cross-boundary fluxes. This factor becomes vital when considering recovery of systems from disturbance. For example, streams that experience widespread fish and invertebrate kills after toxic spills or pesticide run-off depend upon inputs from tributaries for recolonization of the affected stream. Recovery time is often correlated with distance from the source of impact relative to location of feeder streams.

Structural Hierarchies

Ecosystem structures are usually described either by population metrics, which group organisms according to spatial relationships, or by trophic level metrics, which group organisms according to levels in the food chain. While the two parameters overlap, important distinctions exist. From a management context, these distinctions relate to specification of management goals. Ecosystem services are usually thought of, and phrased in terms of, populations and communities. Ecosystem function, on the other hand, more closely follows trophic levels. Thus, the water quality manager must be conversant with both constructs.

FROM INDIVIDUALS TO ECOSYSTEMS. At the smallest practical level, ecosystems are composed of individuals. (Clearly, many levels exist below that of the individual, from enzymes to cells to tissues. These are not generally considered in water quality management, however.) All individuals of a particular species living in a given area represent a population—whether it be a population of lake trout, a population of river snails, or a population of duckweed in a pond. A community consists of all populations of all species living in a particular area—that is, the plant community, the tree or shrub community, or the entire community of plants and animals within a square kilometer or a watershed. The community and its abiotic environment make up the ecosystem. Thus, ecosystems comprise complex mixtures of individuals of many species (i.e., several populations), interacting to form one or more communities, which in turn interact with the abiotic environment.

Humans value populations, communities, and ecosystems for different reasons. Maintaining viable populations of specific species may represent an important goal for both recreation and commercial purposes; species-specific differences make management of individual populations quite challenging. At the next higher scale, communities are often valued for their rarity or for historical, biological, or recreational value. At yet another scale, ecosystems are valued for multiple goals, and their management consists of the balancing act of managing discrete populations or community units and their abiotic surroundings. By its nature, water quality management is an ecosystem-level science.

TROPHIC LEVELS: A CONCEPTUAL TOOL. Energy flow patterns throughout an ecosystem, or trophic levels, were mentioned earlier as an example of a vertically organized hierarchical system. In the traditional model of a trophic system, green plants or primary producers occupy the lowest level. Primary producers use photosynthesis to convert sunlight and raw materials (e.g., nutrients, minerals) into organic matter. Primary consumers eat green plants, secondary consumers eat primary consumers, and decomposers break down dead organic matter and release the raw materials (minerals, nutrients, CO_2) that plants later reuse.

The natural distribution of species among trophic levels provides important information to water quality managers about anthropogenic impacts at different hierarchies. The primary production rate in an ecosystem usually determines upper limits on the size of the animal population. Thus, significant changes in plant community size (although not necessarily in species composition) will bring repercussions for animal-carrying capacity and the exchange of energy and material along trophic pathways. Animals near the top of the food web in a community will have the lowest productivity rates and (usually) smallest total biomass and numbers of individuals. These populations usually consist of small numbers of large animals such as elephants, tigers, or, in an aquatic context, crocodiles or predatory fish such as muskellunge or perch. Many top-level consumers have low reproductive rates and long lifespans that render them especially susceptible to disturbances. Their success is ensured only if the ecosystem remains stable for relatively long periods. The well-documented sensitivity of larger organisms to a multitude of stresses has led to the suggestion that biotic-size spectra be used as an indicator of ecosystem health (Rapport 1990).

The familiar image of orderly trophic levels has inspired significant debate. Oksanen (1991), for example, details the disagreements in the ecological community over the utility and application of trophic levels as many scientists begin to question the theory. Others maintain that trophic-level concepts are viable representations of ecological structures. Pimm *et al.* (1991), for example, note that "recently discovered widespread patterns in the shapes of webs ... indicate that webs are orderly and intelligible and have foreseeable consequences for the dynamics of communities" [p. 669]. One example of the changing awareness of trophic dynamics comes from recognition of the increased importance of the microbial community in food webs. Traditional thought downplayed the significance of bacteria and protozoal activities in this linear food chain. More recently the importance of microorganisms in aquatic food webs has gained critical recognition (Stone and Weisburd 1992). These organisms control major fluxes of energy and nutrients, in some cases diverting more than 50% of photosynthetic production into microbial loops and

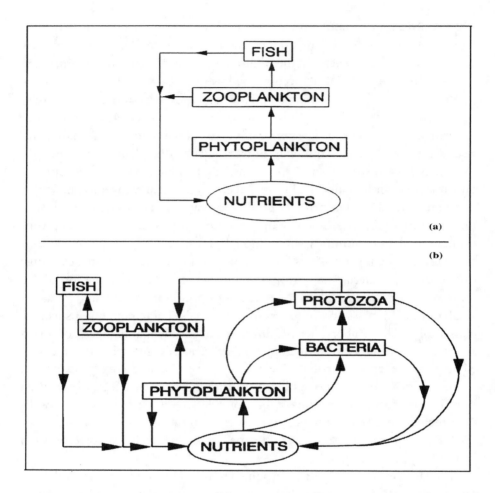

Fig. 14.3 The classic model of trophic pathways is a simple and linear arrangement with only one large feedback loop (a). At this degree of simplification, however, the vertical food chain model has little value for ecosystem management or science. Recent models present more complex (albeit still simplified) representations of trophic pathways that include a larger role for protozoa and bacteria and more complex feedback loops among trophic levels (b). (After Stone and Weisburd 1992)

rapidly remineralizing primary production. Stone and Weisburd (1992) use this information to develop a nonlinear refinement of classic trophic-level relationships (Fig. 14.3). As highlighted by both new research and ecological debate, ecosystem descriptors such as trophic level are conceptual tools and valuable abstractions, not discrete truths. Their analytical power is significant, as long as natural resource managers recognize that they are "reasonable abstractions" (Oksanen 1991).

TOLERANCE RANGES. A third structural delineation of organisms in an eco-system—tolerance ranges—is difficult to characterize. Tolerance ranges of species represent a more multidimensional set of hierarchies in which organisms occupy space in an ecosystem according to many characteristics, including water temperature, salinity, shelter, food sources, and territory size. The relevance of tolerance ranges and their different hierarchical axes is that disturbances affect organisms differently because each species has a specific range of tolerable conditions. The breadth of abiotic and biotic conditions within a system (i.e., the degree to which conditions approach the tolerance level of each species) determines the species composition in any ecosystem; changes in those ranges, in turn, affect different organisms in potentially many ways.

Changes in water temperature, for example, may cause different reactions in different species. For species already near their thermal maximum, increased water temperature may prove lethal because it is above the species' tolerance range for that characteristic. Even nonlethal effects may cause secondary changes that affect population, community, and ecosystem-level properties. For example, elevated temperature in one section of a lake may prompt one species to migrate to cooler waters. Overlapping territories of two previously asynchronous species could lead to food competition or increased predation. Each species has an optimal range for each condition and, beyond that, a range of tolerance. Relative densities or number of individuals of each species occurring in an area will be determined by the degree to which conditions approach the optimum for that species.

Matching Management Scale to Goals: Finding the Right "Ecolevel"

Previous sections in this chapter have reviewed hierarchy theory and different examples of ecological hierarchies to illustrate some of the scale-dependent contexts within which management decisions must be made. Water quality managers must understand the ways in which energy and materials flow among hierarchies so that they can ascertain the scale of ultimate impact for each water quality concern—that is, the scale at which hierarchical processes fully dampen lower-level effects. Previous sections have also reviewed some of the structural and functional attributes of ecosystems that reflect the healthy (or interrupted) flow of energy and materials across hierarchical levels within an ecosystem. These latter factors show that each ecosystem holds "clues" or indicators of how impacts may radiate through different hierarchies.

This section examines these previously discussed concepts from the point of view of water quality management decision making. Given the above theories, how does a water quality manager find and influence the appropriate ecological scale to fulfill management goals? These questions are addressed in familiar contexts: managing for discrete populations, for communities, and for ecosystem-level values.

Managing at the Population Level

As indicated earlier, managers are frequently called upon to manage for discrete populations. Fisheries management, for example, is based upon population management. Each species and population fills a unique ecological role. Even when managing for discrete populations, the water quality manager must act at or strive to influence the appropriate ecological scale to fulfill management goals and to be aware of secondary effects of achieving those goals. Using accurate information on a population's life history characteristics and functional niche, a manager can begin to identify appropriate ecological scale by asking two questions:

1. Does redundancy exist in this population's ecological role?
2. Is the population naturally cyclic?

REDUNDANCY. Redundancy will tell the water quality manager how far-reaching the effects of population management will be. High redundancy confers the largest number of management options because it implies that the ecosystem can more effectively buffer effects of management. On the other hand, low redundancy indicates that the scale of management must be raised because it implies that adequate buffers are not in place and effects of management will radiate to higher (and lower) ecolevels.

Establishing that a desired product or service has low redundancy is, however, simply part of the picture. Managers must then determine the nature of radiating effects, a process that can be answered only by understanding the functional role of the desired product in its ecosystem. If a given species serves as a critical food resource for a given predator, decreasing its population will cause the predator's population to decline as well. Alternatively, decreasing a given species population may lead to an explosion of its prey species. For example, in the marine intertidal on the coast of Washington, United States, the starfish *Pisaster* regulates and dominates much of the ecological community. In the presence of the starfish, mussels are integrated as a part of the community, with moderate mussel population densities and high species diversity (the community has 15 species). Without the starfish, the total number of species drops to eight. In addition, individual mussels grow very large and attain high population densities, thus greatly reducing species diversity. Consequently, actions to control the starfish, a nuisance population, have widespread negative effects throughout the ecosystem. In this case, low redundancy requires an ecosystem scale of management; removing the starfish from the community will affect overall community and ecosystem dynamics.

As in the previous example, many questions of redundancy can be reduced to considerations of predator–prey relationships. Does the target population have sufficient redundancy in predators and prey such that its addition or removal will not have widespread repercussions? Consider the predatory fish *Cichla ocellaris*, which was introduced to Gatun Lake, Panama, to increase fish production. The goal was to reduce population density and increase average size of fishes in the lake. The new species bred rapidly, spread throughout the lake, and within a short time had devastated the native fish community. Changes in the fish community altered bird populations, zooplankton density, and phytoplankton composition, and caused a resurgence of mosquito-borne malaria.

Because management acts on ecosystems by changing the number and types of species as well as by altering energy flow patterns, managing a natural system means learning to anticipate possible chain reactions throughout these hierarchies. The manager must find the highest level to which effects will radiate and the point at which small-scale changes are fully buffered, and then attempt to lower the scale of detectable (or acceptable) effect. For example, a new wastewater treatment plant is unlikely to affect regional water quality, but could easily influence water quality of a whole lake. Management faces two critical questions: Can the effluent be treated or discharged in such a way that the level of detectable effect is lowered below the level of the whole lake? Can effects be limited to a smaller spatial and temporal extent? Failing to plan according to ecological levels threatens to set up a chain effect of destabilizing decisions. The ability to plan "ecologically" increases the probability of sustainability.

POPULATION CYCLES. Understanding and accounting for population cycles represents a temporal-scale dynamic (e.g., patience) that is often difficult to incorporate into water quality management. Population levels of many species, aquatic and terrestrial alike, are frequently cyclic and may fluctuate over a period of many years. Inadequate information on ecological and population cycles also has often lain at the heart of poor management plans.

An important development in population-level management came in the 1970s with the increased prominence of "catastrophe theory" (Kay 1991). This theory derives from the cyclic nature of population increases and crashes that characterize so many predator–prey relationships. While many examples exist, the terrestrial example of rabbit and fox populations is perhaps the best known: as rabbit populations increase, they support increasingly large fox populations. Eventually, the increases become unstable and the rabbit population crashes. This event is followed by a crash of the fox population. While fox numbers are low, the rabbit population rebounds. Thus, the cycle is complete and continuous. Instead of seeing these events as examples of an imbalance in ecosystems, catastrophe theory suggests that they are balance regulators—that one level may, in fact, "need" to be stressed to maintain effective feedback loops. Catastrophe theory allows natural resource managers to view some population disequilibriums as "normal" processes. A population analog of stress-dependent ecosystems such as flood plains, catastrophe theory postulates that some populations are stress-dependent. As such, it introduces a new tool, as well as new complexity, into natural resource management.

Managing at the Community Level

Community-level management is usually implemented either for management of biological diversity or for populations, which, by their nature, must be understood and managed at a higher scale. Community management focuses on specific communities of plants and animals and aims to ensure that they are maintained productively. In many cases, this approach remains experimental, operating midway between "product" management for specific populations of given fish or resources and management for whole-ecosystem integrity. Effective management depends on the same questions, however: Does community-level redundancy exist such that changes at the community level will not affect the whole ecosystem? Are there natural cycles that must be considered?

Managing at the Ecosystem Level

In all but a few cases, an ecosystem perspective and an understanding of impacts at the ecosystem level are critical for understanding and managing resources at all levels. The probability that a water quality management strategy will be sustainable depends largely on the relationship between the management practice and ecosystem function. Even population-level goals must consider the relationship of that population to its environment to sustain the initial objective (e.g., fish production). In most cases, management considers multiple products and must recognize that those products result from intrasystem interactions (e.g., water control, nutrient cycling, temperature regulation).

The Limitation of "One Decision at a Time"

Economist Alfred Khan has proposed the idea of "the tyranny of small decisions" to describe how a society frequently makes many small decisions that appear consistent with larger goals and policies but taken en masse result in actions that are contrary to desired managerial goals. Ecologist W. E. Odum has applied Khan's "small decision" ideas to environmental management. For example, a poll taken on the U.S. East Coast showed that a very high proportion of the public (greater than 75%) favored preservation of coastal wetlands and dunes. During a 20-year period, however, 85% of the coastal wetlands and dunes along this 2000-mile coastline were destroyed. That destruction occurred through thousands of small decisions (e.g., to build one more shopping center, one more recreational home) rather than one large decision to sacrifice coastal resources in favor of development.

Our attention to (or our intellectual imprisonment by) small decisions has allowed us to develop management strategies in great detail for small problems. In other words, we know how to stabilize a sand dune but seem unable to address larger problems higher in the hierarchy. Instead of stabilizing a thousand sand dunes, society should be trying to control the onward creep of civilization onto the coastal flood plain.

Human society seems to be severely limited in its abilities to move up the hierarchy from one level to any of the higher levels above. We have difficulty extrapolating from studies on individual species, or even whole communities, to field-scale ecosystem conditions because (1) ecosystem components behave differently when they are isolated from the overall system, and (2) because compensatory mechanisms in the field may attenuate (or accentuate) responses of a given component to some stressor. While our attempts to extrapolate from one level of the hierarchy to another remain rudimentary, advances in ecological theory in combination with accumulated experience have provided significant progress in the ability of water quality managers to tune their management to the appropriate ecological levels.

Summary

- Water quality management consists of managing water bodies and their adjacent terrestrial and atmospheric systems in such a way that the highest-priority human values are protected. To accomplish that goal, we must recognize that aquatic systems are open, integrative ecosystems that remain relatively homeostatic over long time periods. We must then manage for discrete values (e.g., a fish population, a drinking water supply) in the context of whole ecosystem structure and function.
- Management at the ecosystem level can be effective only if we understand the hierarchical nature of ecosystems. Hierarchy theory allows us to predict certain properties that characterize healthy ecosystems. For example:

    Homeostasis: Energy and other cycles in healthy ecosystems are relatively constant through time.

    Sustainability: At the scale of hundreds of years, ecosystem properties will be self-perpetuating.

    Redundancy: Each flow of carbon, nutrients, or energy for each unit of information has several parallel paths in a healthy ecosystem. This property allows systems to remain homeostatic and self-perpetuating.
- Most of those proprieties are most easily measured at ecosystem or patch boundaries, where properties and processes operate in a state of flux.

For Further Reading

Kelly JR, Harwell MA. Indicators of ecosystem recovery. *Environ Mgt* 1990;14:527–545.

Naiman RJ, Decamps H, Pastor J, Johnston CA. The potential importance of boundaries to fluvial ecosystems. *J N Am Benth Soc* 1988;7:289–306.

Oksanen L. Trophic levels and trophic dynamics: a consensus emerging? *Trends in Ecology and Evolution* 1991;6:58–60.

Paul-Wostl C. The hierarchical organization of the aquatic ecosystem: an outline of how reductionism and holism may be reconciled. *Ecological Modeling* 1993;66:81–100.

Rapport DJ. Challenges in the detection and diagnosis of pathological change in aquatic ecosystems. *J Great Lakes Research* 1990;16:609–618.

Walker DA, Walker MD. History and pattern of disturbance in Alaskan arctic terrestrial ecosystems: a hierarchical approach to analyzing landscape change. *J Applied Ecology* 1991;28:244–276.

15

Responses to Stress at the Ecosystem, Community, Population, and Individual Levels

Overview

The previous chapter reviewed several approaches to conceptualizing and evaluating healthy ecosystems, and examined ways in which water quality managers could use that information to choose appropriate ecological levels in managing for different goals. It also stressed the importance of the patterns of energy flow and matter transfer in healthy ecosystems by discussing flow among ecological hierarchies, differences between structural and functional attributes, and key concepts such as redundancy. This chapter reviews the ways in which stressors cause changes in those same structures, functions, and hierarchies. Because we depend on ecosystems for goods and services, the way in which these systems respond to stress is of paramount importance. Whether the stress consists of climate change, pollution, or changes in land use, the root question remains the same: what will these changes mean for our use of these systems, and how can we manage appropriately?

Ecosystems are complex, and each system responds in a unique manner. Nevertheless, the hierarchical nature and homeostatic mechanisms built into ecosystem organization provide a few reliable indicators of stress. Specifically, ecosystem structure represents the first level of system-wide response to the introduction of a stressor, with stress-sensitive species suffering reductions in numbers, increased dominance by a few species, and reduced diversity overall.

In cases of severe or prolonged stress, system energetics will also respond, most notably through increased respiration rates, increased nutrient export, and increased "leakiness" across ecosystem boundaries. Through monitoring and evaluation, a manager can use these characteristics as indicators of stress in a given system. They can then be called upon to support policy discussions and changes.

New Perspectives on Response to Stress

In many ways, response to stress is what most people envision when they hear the phrase "water quality." What are the effects of agricultural run-off? of acid rain? of a paper mill upstream? of increased demand? of logging? of climate change or drought? Before these specific questions can be addressed in detail (chapters 17–24), it is critical to review the ways in which structural and functional attributes and hierarchical structures are influenced by stressors. While many of the concepts discussed in this chapter were introduced in chapter 14, here they are presented in more depth and with a focus on how the attributes of a healthy ecosystem change when subjected to stress. This information has fundamental importance for water quality managers because it provides the basis for understanding the impacts of specific anthropogenic and natural stresses. As described in chapter 16, it also provides the basis for understanding how similar impacts will have different effects on water resources, depending on regional attributes of geology, soil chemistry, and geography.

Despite the wealth of studies tracing the effects of different stressors on water quality, no ecosystem stress cookbook exists to guide water quality managers. The ingredients are far too variable; the effects on one system cannot necessarily be translated to another system, nor are the effects of all stresses the same. The nature, degree, source and duration of a given stress all influence responses, as do functional and structural properties of the affected ecosystem.

Despite this broad caveat, a growing body of literature demonstrates that predictable trends can be identified in ecosystem response to stress. Authors who believe these generalities are useful and valid generally hold a large-scale view of ecosystems—acknowledging small-scale differences but focusing, instead, on large-scale similarities. Fundamental papers contributing to the generality theory include Odum (1985, 1992), Schindler (1990), Perry et al. (1987), Harris et al. (1985), and Rapport (1989). Other authors focus their attention on the wide variability in ecology and ecological responses to stress. Concentrating on small-scale patterns and processes, these authors believe that the differences among systems are more acute than the similarities. They contend that current models do not adequately account for the complexities that ecosystems exhibit in their response to, and recovery from, stress. Authors representing this view include Cairns (1990), Kay (1991), and Kelly and Harwell (1990).

Because water quality managers must be sensitive to local conditions, little danger exists that they will overlook local ecological idiosyncrasies in favor of a generalized large-scale model. Significant danger does emerge, however, in retaining a local-level focus and not having the tools or perspectives to recognize large-scale impacts and common responses among similar systems. A water quality manager's perspective must be large-scale. Political, financial, and time constraints all mandate that the manager make informed decisions from general principles, corroborated, if possible, by local studies. Consequently, the suggestion that ecosystems have predictable, generalized responses to stress is of great importance. Even in an unfamiliar setting, a manager can use these generalizations to make early and reliable assessments of ecosystem stability, sustainable production, and risks to users or to the ecosystem itself. Generalizations do not allow definitive prediction of how an individual ecosystem will respond, but they do permit *the anticipation* of likely responses.

What Is Stress?

The term "ecosystem stress" is verbal shorthand for describing the cumulative effect of an individual organisms' response to stress. Stress, like other components of natural systems, occurs at different scales and hierarchical levels. Thus, while ecosystems per se do not respond to stress, their assemblages of individuals, populations, and communities do. At an individual or species level, these responses vary widely. As these impacts are aggregated to higher hierarchical levels, the variations become less extreme and general responses become more predictable. Other ecological hierarchies are also differentially sensitive expressions of stress response. For example, within individual ecosystems, the species that suffer from stress vary dramatically. In contrast, at an ecosystem level, the species that suffers is often less important than the functional role of that species. At this level, responses of functional roles are more predictable and important than the species that fill those roles.

Stress is used throughout this book as an all-encompassing term to describe significant shifts in ecosystem attributes or characteristics. Each stress prompts adjustments within the biological community, ranging from changes in an individual organism's physiology to large-scale shifts in resource patterns. From an ecological perspective, the term "stress," which connotes shifts that lower ecosystem productivity, differs from the term "subsidy," which connotes shifts that benefit an ecosystem's productivity. From a water quality perspective, these distinctions may not exist, as any perturbation—whether "positive" or "negative"—can cause a shift in the valuation of water resources. For example, the addition of nutrients such as phosphorus to a system represents a subsidy that increases ecosystem productivity. Both biologically and socially, however, the increased productivity affects the goods and services derived from a water body; from a management perspective, it can then be termed a stress.

Which Responds First: Structure or Function?

At the whole-ecosystem level, structure and function are not as closely linked as once thought. In fact, significant structural changes may be observed without parallel observations of functional change. The principles of redundancy and homeostasis outlined in chapter 14 explain much of the looseness in this link; as long as sufficient redundancy exists, the functional effects of increases or decreases of individual species will be buffered. Thus, structural attributes will change in response to any given stress, but the functional roles may remain unaffected (Table 15.1). This concept is one of the first important principles of ecosystem response to stress: *functional attributes (production, respiration, nutrient cycling) are more robust than structural qualities (species composition or species diversity).*

Examined more closely, this statement appears to be contradictory. In order for species composition or diversity to change, individuals must die or multiply. In other words, the real scale of first response must occur at the physiological level, with changes in individual functional qualities such as respiration and production (Parsons 1991). While acknowledging that physiological functional changes are the root cause of changes in population and community structure, we are currently concerned with larger-scale, whole-ecosystem processes. At a whole-ecosystem level, functional redundancy and homeostasis cause functional qualities such as ecosystem rates of respiration and production to respond more slowly than component community structures. Clearly, however, the structural and functional roles are tightly interwoven and one defines the behavior of the other.

First Response: Structural Properties

Community structure reflects the sum of species-specific responses to stressors on individual and population levels. In all, managers can incorporate six structural attributes into monitoring and evaluation plans to detect stress.

LIFESPAN AND REPRODUCTIVE DECREASE. One of the most basic stress responses is the decline in lifespans and reproductive rates due to increased respiration and

Table 15.1 Odum (1985) predicted that ecosystems under stress would show a variety of specific responses. In the Experimental Lakes Area in Canada, scientists conducted a large number of whole-ecosystem manipulations, including acidification and artificial eutrophication. This table contrasts Odum's predictions with results from acidification and eutrophication experiments. (Modified from Schindler 1990)

Response (After Odum 1985)	Acidification Lakes 223, 302S	Eutrophication Lakes 227, 226NE
Community respiration	Decreased in winter in whole lake; increased in periphyton	No change
P/R ratio	Increased	Increased
P/B ratio	No change	No change
Excess primary production	Increased	Increased
Nutrient cycling rate	Minor changes in carbon, nitrogen	Increased P, N, C cycling
Loss of nutrients	Small increases in N and S losses	Higher losses of P, N, C
Proportion of r-strategists	Increased for zooplankton; decreased for fish	Increased for zooplankton
Organism size	Decreased for chironomids, phytoplankton, zooplankton; increased for fish	Decreased for zooplankton; increased for phytoplankton
Organism lifespan	Decreased for fish, benthic crustaceans	Decreased for crustaceans, zooplankton
Food chain length	Shorter	No change
Species diversity	Decreased	Decreased
System openness	No change in inputs; increased nitrogen losses	Increased inputs and outputs of N and C
Efficiency of resource use	Decreased efficiency for NO_3, NH_4, allochthonous matter	No change
Resistance of functional properties	Generally more resistant	Generally more resistant

metabolic rate (Schindler 1990; Odum 1990). Species will also be more likely to spend longer periods in resistant stages such as eggs, resistant spores, and pupa. These changes represent the cornerstone of most of the community modifications that follow, but, from a managerial standpoint, their importance varies. In some circumstances, the reduction in fecundity and lifespan is important as it provides a reliable early indicator of stress and can be used to initiate policy discussions on stress reduction. In other cases, it has specific utility. For example, where a specific desired population may be threatened, these reductions provide a way to measure the degree of response and may serve as the basis for new management rules, such as catch limits. Kolasa and Pickett (1992) caution that a decline in a given species or component does not necessarily indicate ecosystem stress. Ecological change is ongoing and, as discussed in chapter 14, populations are often cyclic over long periods. Water quality managers must be acutely aware of the temporal and spatial scales involved in these changes to categorize indicators of stress and natural fluctuations correctly.

Species at the lowest and highest end of the trophic spectrums are commonly held to be most susceptible to stress (see, however, Schindler 1990 and chapter 22). At the lowest extreme, species occupy narrow specialized niches that make them sensitive to small environmental fluctuations. At the highest extreme, many predators are susceptible to stress because their limited numbers of progeny mean fewer chances for the appearance of successful, resistant variations. Multiple stresses in the Great Lakes, for example, have eliminated many large benthic fish species (Rapport 1989). For some types of stress, such as pollutants, higher-trophic-level species may suffer from the accumulation of pollutants in their tissues (*bioaccumulation*). The most famous example of this phenomenon was the accumulation of DDT in the tissues of predatory birds discussed in chapter 5. Another frequently cited example of the disappearance of high-level trophic species is the decrease of fish production, especially of higher predators, in acidified lakes (Schindler 1990). Other compounds such as PCBs are now causing selective decline of large species in the Black Sea and elsewhere (Rapport 1989), while disease and heavy metals are causing similar declines in countries such as Chile, South Africa, and Romania.

CHANGED INTERACTIONS. As the numbers of species change, the types of species will often change as well. For example, stress and its associated impacts on organism physiology reduce the resistance of many individuals to disease and parasitism. Consequently, stress favors parasites and diseases so that negative interactions will increase and positive interactions (e.g., mutualism and symbiosis) will decrease (Perry *et al.* 1987; Rapport 1990). Resource scarcity, a special case of stress, may increase the incidence of mutualism; this aspect of stress response has received little study to date, however.

DECREASED DIVERSITY. Diversity, a dynamic measure of ecosystem integrity, is a broad term that may be broken down into numerous categories. For example, species richness is the most commonly cited measure of diversity in ecological studies, while genetic diversity is often of paramount importance for conservation biology. Habitat diversity is another measure of ecosystems that has major implications for sustainable landscape-level processes. While each focuses on a different question, the three metrics are interrelated. Habitat diversity is frequently correlated with species diversity, while genetic diversity of a given species may more closely reflect the stability of environmental conditions in a given area.

   In stressed ecosystems, the disappearance of individual stress-sensitive species, along with increased disease and parasitism, generally leads to a decrease in species richness and a related increase in dominance by a few species (Fig. 15.1). This change is critical because decreased diversity usually means decreased redundancy and, therefore, a decrease in self-regulatory capability. The less redundancy, the greater the chances that subsequent or more prolonged stress will eliminate occupants of key niches. At its most severe, this process can spur a chain reaction where ecolevels lose their ability to buffer fluctuations and ecosystem stability is threatened. Species diversity is perhaps the most well-documented indicator of environmental effect and has been incorporated into numerous biotic indices (Rapport 1989; Karr 1991; Odum 1990; see also chapter 6).

   Diversity does not always decrease, however, and its characteristics are best considered on a variety of spatial and temporal scales. Although not the norm, some stresses actually increase habitat diversity. In particular, frequent, low-level disturbances increase diversity through the creation of multiple niches in time and space. In these cases, species diversity may also increase. Temporal

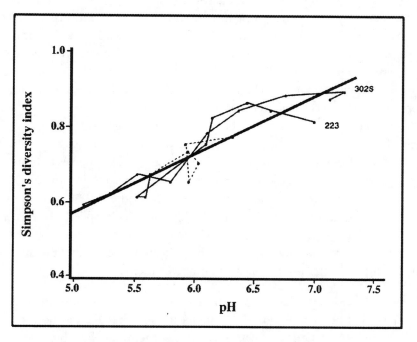

Fig. 15.1 The decline of phytoplankton diversity with increasing acidification is a characteristic response of ecosystems to stress. The heavy line represents the average response of two lakes (numbers 223 and 302S) studied in Canada. (From Schindler 1990)

scales may be relevant in a consideration of diversity as well. For example, shortly after a disturbance, diversity may increase as "opportunistic" species appear before native assemblages are completely disrupted. Because of the different spatial and temporal dynamics in diversity, water quality managers must ensure that measures of diversity are considered against a reference system and account for both natural fluctuations in diversity and relic effects from a stressed ecosystem in flux (Kelly and Harwell 1990; Rapport 1989).

INCREASED PATCHINESS. Another correlated effect of the loss of individuals and species is an increase in spatial heterogeneity (patchiness). Through the elimination of some individuals and colonization by other species, stressed ecosystems exhibit increased unevenness. Temporal heterogeneity is also characteristic of some stressed ecosystems. Consequently, while diversity of species at any one site will be generally lower in a stressed system, the distribution of species is more likely to be patchy. Spatial diversity is most pronounced when stresses are intermediate in intensity and occurrence.

"SUCCESSIONAL" STAGE SETBACK. The concept of succession in stream and lake communities was briefly described in chapters 9 and 12. Although the term itself may be misleading, predictable patterns of changes in communities and community dynamics occur over different spatial and temporal scales. Thus, stream communities, for example, change from upstream to downstream, from the bank to the channel, and from depth to surface. Under stress, ecosystem character frequently reverts to conditions representative of those found earlier in time or space. Frequently a setback to an earlier successional stage reduces diversity, but not necessarily redundancy.

A corollary to "successional setback" involves changes in the proportion of species following different reproductive strategies. In 1985, Odum hypothesized that stress would result in an increase in r-strategists (species that produce numerous young but have limited involvement in the rearing process). This idea has been borne out in numerous studies. Schindler (1990), for example, notes that both eutrophic and acidified lakes included a species shift in zooplankton toward more r-strategists. Schindler does note, however, that fish species in acidified lakes were increasingly k-selected (species that produce only a few young at greater individual cost and investment).

ORGANISM SIZE CHANGES. Early predictions of ecosystem response to stress hypothesized that organisms would be smaller under stressed conditions (Odum 1985). In fact, this trend has been shown for some organisms, but not for all. For example, phytoplankton appear to respond with increased size while zooplankton conform to the predictions and exhibit smaller sizes. Fish size has been reported to increase and decrease, depending on the ecosystem and the nature of the disturbance. For a water quality manager, recording changes in organism size may indicate stress, but the direction of the changes may be species-specific, and the degree of stress will require further investigation.

Second Response: Ecosystem Function

If structural changes are pronounced enough, they will affect the ecosystem buffering capacity and result in changes in function at the ecosystem level. According to Odum (1992), "When the stress produces detectable ecosystem-level effects, the health and survival of the whole system is in jeopardy" [p. 542]. The loss of ecosystem stability may be temporary and the ecosystem may "settle" into a different stable level, but the instability may also lead to a cycle of unpredictable oscillations. From a managerial standpoint, this unsettled situation may mean a loss of desired goods and services. It may also—and more typically—mean that management agencies will spend enormous sums of money and further inputs to try to keep the goods and services flowing at expected levels from a system no longer structured to provide those resources.

ENERGETICS. Energetics refers to the cycling of carbon through a system—that is, the cycle of photosynthesis, respiration, growth, and decomposition. Stressed ecosystems are less energy-efficient. In particular, respiration rates increase so that a smaller unit of biomass is supported by each unit of energy consumed and by each unit of energy respired. Decomposition also generally decreases, and a greater reliance on energy from outside the system occurs. In addition, excess and higher exports of primary production or biomass are likely. Perry and Troelstrup (1988) reported that aquatic systems stressed by insecticide application were characterized by an increase in drifting organic matter and invertebrates.

These qualities do not always hold true for each component of an ecosystem. For example, periphyton community respiration has been shown to increase in acidified lakes, but not in eutrophic lakes (Schindler 1990). In addition, Schindler (1990) reports that acidified lakes exhibited no change in exported or unused primary production, while both increased in eutrophic lakes. (This example may demonstrate a case in which the effects of a subsidy differ from those of a stress.) Pesticide application has been shown to reduce overall decomposition at the same time that it dampens the physiological activity of individual decomposers. Acidification, on the other hand, appears to have a more species-specific effect on decomposition as illustrated in Fig. 15.2 (Perry *et al.* 1987).

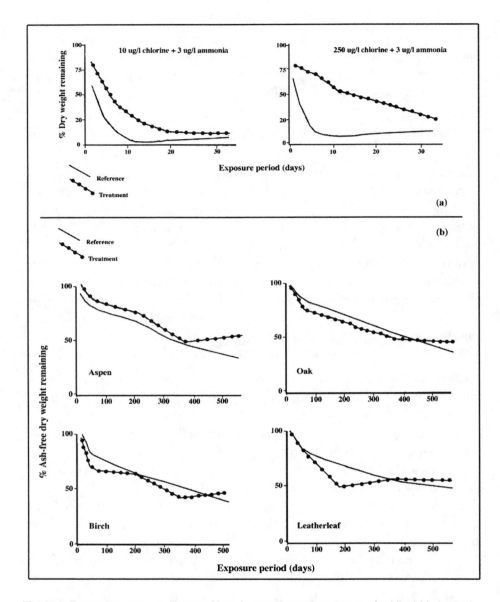

Fig. 15.2 Ecosystem stress, as illustrated here by experimental treatments of acidity (chlorine and ammonia), varies with the concentration of the stressor (a) and with species-specific responses to a given stress (b). Experimental acidification of Little Rock Lake, Wisconsin, affected decomposition of four common species differently, decreasing the decomposition rate of birch and oak leaves but increasing it for aspen. Leatherleaf exhibited still another response, with its decomposition rate varying over time, relative to a reference condition. (Data from Perry *et al.* 1987)

NUTRIENTS. Another measure of "integrity" involves nutrient use and cycling. Stressed ecosystems manage nutrients less efficiently, exchanging more nutrients across ecosystem boundaries, increasing nutrient turnover within ecosystems, and increasing the horizontal dimension of nutrient cycling. Consequently, the standing crop of nutrients decreases with stress while nutrient export increases. For example, Cs, Al, Ca, Mg, and K increase in streams after forest harvest (see chapter 18). Nutrient losses may indicate system malfunctions such as imbalances in the coordination of cycles for different nutrients, impairment in biological activity, or simply supply of one or more nutrients in excess of maximum ecosystem uptake (Schindler 1990).

Exceptions to this pattern have been identified, however. Acidification of terrestrial ecosystems causes phosphorus to bind to aluminum in acidified soils. Consequently, phosphorus input to lakes is often reduced from acid-stressed terrestrial ecosystems (rather than increased). While this exception may occur as the result of a geochemical interaction (rather than biological mechanisms), it illustrates that these principles are simply general indications, with many notable system-specific exceptions.

TEMPORAL TRENDS. Variability over time is another measure of a stressed ecosystem. While it warrants its own discussion, temporal variation is really a result of changes in the interactions between community structure and function, and the interruption of buffering capacities. Decreased diversity in stressed systems leads to reduced redundancy because the network contains fewer duplicative links. In such a simplified system, stochastic variations in the environment (e.g., temperature or precipitation) will cause wide swings in ecosystem function as each of the species in the system represents a greater proportion of the total system. For example, a significant disruption of nutrient cycling by one species will bring a major change in cycling in the entire system. Stable systems are able to dampen these stress–response oscillations, while unstable systems will experience accelerated oscillations through time.

Recovery and Temporal Trends

The structural and functional responses described above may cause a variety of shifts in the whole ecosystem, with each having very different ramifications for ecosystem recovery and for the changes in goods and services that the system can produce. Kay (1991) presents several examples of paths that ecosystems can take after being stressed: (1) the ecosystem does not move from its original path;

(2) the ecosystem moves from its original path, but then returns to it; and (3) the system moves permanently from its original pathway, either collapsing, establishing a new but similar structure, or establishing a completely new set of structures. These alternate pathways are presented schematically in Fig. 15.3. The recovery pathway followed by an ecosystem depends on a wide range of factors, including the severity and nature of the stress, its duration and timing, subsequent disturbances, sensitivity of key species or processes, redundancy, the proximity and mobility of sources of recolonizing organisms, residual effects of the stresses, and the actions of management agencies (Kelly and Harwell 1990; Cairns 1990).

Kay's first model, in which an ecosystem is not fundamentally changed from the original, might be most typical for minor and small-scale disturbances such as a well-managed forest harvest. Ecosystems that change dramatically but return to their former structure and function may be more typical of one-time-only disturbances such as oil spills. While full recovery in these systems may require 10 or more years, the oil does not persist in large amounts; once it dissipates, numerous adjoining bays and shorelines are available to serve as a source of recolonizing organisms.

Kay (1990) presents the example of exotics in the Great Lakes as a model for an ecosystem that establishes a new, but similar structure. As discussed in chap-

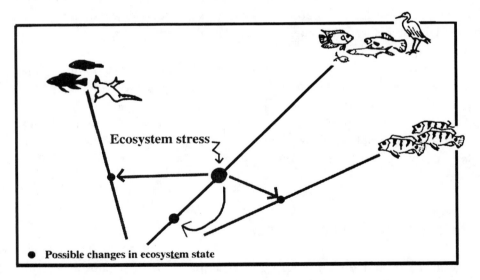

Ecosystem stress

● Possible changes in ecosystem state

Fig. 15.3 Possible pathways for change of ecosystems, where lines represent approximate paths for development of species assemblages and ecological relationships. Ecosystem stress may cause an ecosystem to suffer a temporary setback, but a return to its previous composition and functional processes, or it may cause an ecosystem to recover to a variety of new, but stable compositions along different ecological pathways. (Modified from Kay 1991)

ter 24, exotics are a unique biological pollutant that, once established, may alter previous food webs and trophic dynamics significantly but usually not completely. Thus, some species from the original ecosystem continue to be represented but the overall structure of the ecosystem is significantly altered. In most cases, even if the exotic were successfully eliminated, the cascade of changes in the food web would prevent recovery to the former structure.

Kay's last model is the ecosystem that recovers to a completely new structure and function. Severe acidification of mountain lakes, for example, may eliminate virtually the entire aquatic community. If the lake is isolated from sources of recolonization, even recovery from an acidified state is unlikely to produce an ecosystem resembling the former one. The ecosystem may still recover to a different stable state, however, and it may have desirable values. Society's and management's goals must be restructured when ecosystems end up on a different evolutionary "branch" after being subject to stress.

Stress, Society, and Management

Because stress affects ecosystem function and because societies rely on those functions, ecosystem stress reflects on society's ability to use and benefit from an ecosystem. From fisheries to flood control, societal uses evolve under expected conditions of supply and quality. The management challenge is to continue to provide expected products and services. Consequently, the ability to recognize stress and anticipate how an ecosystem will respond to it, or recover from it, is a vital part of effective and timely management.

When Does Change Become a Stress?

In an applied field such as water quality management, ecosystem stress and water quality stress are not necessarily synonymous. An ecosystem may respond to environmental changes, but if that response does not conflict with management goals, little or no action is required by the water quality manager. To revisit an example presented in chapter 1, if management goals for a lake center on swimming and boating goals, ecosystem responses to acidification would be of little to no consequence. On the other hand, if management goals center on fishery production, then fish kills or changes in species composition would present major management challenges.

Water quality managers may also have a different perspective when broad social goals conflict with ecologically sustainable water resources management. They might, for example, evaluate the success of a hydroelectric dam differently than politicians or economists. Such differences of opinion may be especially acute in cases when management changes the ecology of stress-adapted systems.

Indicators of Stress

Despite characteristic ecosystem-level responses to stress, effects are often difficult to detect. Responses by individual species may, for example, cancel each other out or remain masked as they translate from the individual to the system level. In such a case, biological indicators have a vital role to play: they act as warning signs or clues that components of the ecosystem—and hence the system itself—is under stress, even before large-scale changes in the system become evident.

Water quality professionals use two groups of indicators to detect stress. The first group of general indicators include measures such as changes in community dynamics, lowered diversity, and disappearance of predators. The other group, called biotic indicators, differs in that they are less generic but, within a given ecosystem, are more exact. In all cases, reliance on an indicator requires foreknowledge of the aquatic system and the awareness that indicators cannot substitute for site-specific accuracy. Extrapolating from one system to another having different geology, climate, and water chemistry will lead to erroneous conclusions.

Biotic Stress

The coal miners' canary is perhaps the most famous example of an indicator species. Because canaries are particularly sensitive to poisonous gases, coal miners used them to provide early warnings of gas leaks. All ecosystems possess indicator species for different conditions: these species are the least tolerant to variations in some factor, whether it is a level of a given pollutant, temperature fluctuations, drought, or another influence. Biotic indicators in aquatic systems may be specific fish species, mussels, plants, or any other species that has been established as sensitive to water quality changes. While many indicators accurately reflect ecosystem integrity, only a few provide adequate early warning of pathological change (Rapport 1990). Some of the most successful early warning indicators incorporate suites of species that fill different functional roles in an ecosystem. Concentrating on any given species may not indicate ecosystem stress, but analysis of response guilds may prove more effective. For example, Brooks and Croonquist (1990) suggest that avian response guilds are particularly sensitive and integrative indicators of riparian wetland disturbances.

Another kind of biotic indicator is known as a *keystone species* or *keystone process*. Keystone species or processes regulate critical ecosystem interactions. For example, some water columns have extremely low concentrations of inorganic nutrients, which might suggest that phytoplankton growth should be limited by nutrient concentrations. Because of the rapid remineralization by microbes of much of the organically bound nutrients, however, enough nutrients are recycled to maintain near-maximal phytoplankton growth rates. In this context, the microbial feedback loop that continually resupplies the system with nutrients may be considered a keystone process.

In coral reefs, grazing on algae by fish and urchins is another example of a keystone process. Grazing stimulates algal productivity, causing growth rates to be two to ten times greater where natural levels of grazing are found than where grazing is experimentally eliminated. In this case, the keystone process may be interrupted by overharvesting several species; harvesting of any given species may not be a significant concern for managers, but if enough grazers are eliminated, system productivity will fail. Through monitoring the keystone process, managers may determine acceptable limits of harvest or product removal.

Ecosystem Stress and Recovery to a New State: An Example

The 20-year history of Belews Lake, North Carolina, provides an interesting example of whole-ecosystem change in response to stress. The lake was formed in 1973 after impoundment by the Duke Power Company. In 1975, the lake contained a regionally typical mixture of 24 species of fish, including both piscivores and planktivores. In 1976–1977, selenium poisoning led to the rapid elimination of most of the fish species, including all piscivores. Selenium concentrations were not high enough to affect zooplankton communities, but the fish were affected through bioaccumulation. A follow-up survey in 1983 found only 10 species represented in the lake—all of them planktivores—and at densities twice those found in 1975. This example illustrates several key responses to ecosystem stress:

- An initial response through decreased reproduction and lifespan (i.e., death).
- A selective effect on high trophic levels, through bioaccumulation of toxicants.
- Deceased diversity and increased dominance of selected species.

Interestingly, this example does not end with the recovery of the planktivorous fish. Because the recovering fish community was dramatically different than the original community, a new stress was added to Belews Lake—predation pressure. Although unaffected by the original selenium poisoning, zooplankton communities have since been altered by the predation of the new fish community. In turn, zooplankton diversity has been reduced from 18 species to 14, an overall reduction has occurred in zooplankton size (due to size-selective grazing), and the dominance structure of the remaining zooplankton has shifted.

Belews Lake provides an excellent illustration of the chain of effects set in motion by the presence of many stressors: changes brought about by accidental chemical poisoning led to new stresses unrelated to the original pollution, but certainly as dramatic for the structure of the ecosystem. This case also illustrates that recovery can lead to a stable ecosystem, albeit one that may differ significantly from the original ecosystem. Depending on the popularity of the remaining fish, local communities may regard these changes as positive.

Summary

- Although ecosystems are complex, some degree of generality can be discerned in the ways that systems respond to stress. Structural (kinds and numbers) variables respond more quickly than functional (rate or process) variables.
- Six classes of variables are consistent in ecosystems altered by stress:
  A decrease in lifespans and reproduction
  Different species–species interactions, often reducing positive interactions like symbiosis
  Decreased diversity at many levels
  Increased spatial and temporal patchiness
  A reset to an earlier successional stage
  A reduction in the organism's average size
- After structural variables have been altered, functional variables like nutrient, carbon, and energy cycles are modified.
- Several responses can be useful in water quality monitoring programs striving to evaluate ecosystem-level management strategies. These responses include integrative variables like energy flux or decomposition as well as changes in individual species—particularly significant or keystone species.

For Further Reading

Harris HJ, Sager PE, Richman S, Harris VA, Yarbrough CJ. Coupling ecosystem science with management: a Great Lakes perspective from Green Bay, Lake Michigan, USA. *Environ Mgt* 1987;11:619–625.

Odum EP. Great ideas in ecology for the 1990s. *BioScience* 1992;42:542–545.

Odum EP. Trends expected in stressed ecosystems. *BioScience* 1985;35:419–422.

Perry JA, Troelstrup NH. Whole ecosystem manipulation: a productive avenue for test system research? *Environ Toxic Chem* 1988;7:941–951.

Rapport DJ. Evaluating ecosystem health. *J Aquatic Ecosystem Health* 1992;1:15–24.

Schindler D. Experimental perturbations of whole lakes as tests of hypotheses concerning ecosystem structure and function. *Oikos* 1990;57:25–41.

16

Regionalization in Natural Resource Management: Ecoregions

Overview

Water quality managers have long recognized that water quality varies in response to land use practices, soil types, vegetation types, and climate. Nonetheless, the descriptive and predictive value of this awareness went largely unrecognized until the 1980s. Since that time, the process of regionalization—that is, stratifying the landscape according to similar biophysical characteristics—has emerged as an increasingly powerful tool. The theory of Ecoregion (i.e., ecological region) delineation states that natural water quality characteristics of lakes or streams within a single Ecoregion will be more similar than the characteristics between Ecoregions. Qualities such as water clarity, phosphorus loading, and residence time (all critical elements of water quality) vary according to regional patterns in geomorphology, soil type, and climate. Consequently, attainable water quality standards in one region are different from those of another region.

Regionalization is a powerful tool that enables managers to clarify and compare water resources, establish reasonable standards, predict effects of practices and controls, locate monitoring sites, and extrapolate to larger areas. Notably, Ecoregions may be defined at any scale. At a national scale, for example, five to six Ecoregions have been defined across the United States. Within these areas, each state may define its own Ecoregions based on finer-scale maps of the same or similar variables. While the Ecoregion concept is becoming an important tool in the United States and in selected European countries such as the Netherlands, it has yet to be adopted widely in other countries, which continue to rely on the more linearly defined river basin management approach.

Introduction

Water quality exists in a landscape framework; neither normal nor impacted conditions of water resources can be separated from controlling influences of the surrounding landscape. In chapter 12, for example, basin geomorphology was described as a critical control of lake stratification and mixing, both of which are mediators of lake water quality. Similarly, in chapter 9, channel morphology was seen to define characteristics of river flow and community development. Likewise, groundwater, wetlands, and coastal water resources are first, and foremost, land-based features.

Naturalists, scientists, and managers have always been obliquely aware of the importance of landscape features for water quality. Tropical systems have been contrasted with temperate ones, montane lakes have been distinguished from lowland lakes, and Arctic conditions are recognized as being unique for their harsh climate and relative youth. Until the early 1980s, however, these broad regional descriptors were just that—scientific caveats explaining why results of one scientific study might not be replicated in another system. An enormous gap remained between scientific awareness of regional differences in ecological characteristics and any relevance of that awareness for management and policy formation. That gap began to close in the early 1980s as the predictive possibilities of regional differences became evident. A twist of logic and perspective enabled the recognition that, instead of seeing landscape differences as features that confounded relatively clear issues, those differences could be used as tools to facilitate water quality science and management.

This chapter (and chapters 14 and 15) trace ways in which the remarkable developments of ecological theory in the 1980s have created an enormous potential for change in the applied ecological fields of natural resource management and, in particular, water quality management. In all scientific fields, it takes time for the full import of the implications and critiques of new theories to emerge. For example, as the awareness of scale issues developed in the field of ecology, the discipline first appeared to forestall any attempts to create general theories and ecological comparisons. Any analysis at one scale seemed to be invalidated at other scales. Over time, however, important principles of scale differences have emerged that have brought enormous implications for management. This chapter traces one of the most important of these developments—the ability to use landscape-level variation as a tool for creating appropriate expectations and enforceable standards. In concert with increased awareness of stress responses and management at the appropriate ecological level, an Ecoregion focus provides the opportunity for water quality management to become predictive and anticipatory, rather than strictly reactive.

Regionalization Concepts

Regionalization is an effort to understand and classify landscapes using spatial patterns of relevant environmental variables. As noted above, regional variation is not a new concept. The process of regionalization, however, uses the concept in a new way. Instead of beginning with the result (e.g., water quality) and noting that it varies according to biophysical parameters such as land use, soils, and vegetation, regionalization begins with the variables themselves and views them as the cause of natural differences in water quality. Landscape features consequently become a predictive asset rather than a confounding irritation. It follows logically that natural conditions of lakes in an area characterized by highly erodible, nutrient-rich soils would differ from those of lakes in areas of granitic rock or sand.

Water quality management cannot be effective unless it is defined within a biophysical framework. Until the Ecoregion concept was developed, few adequate tools were available to facilitate assessment of natural, biophysical conditions. Previous analytical or management delineations relied on political or hydrological groupings of lakes, streams, and other resources, but these groupings do not relate closely enough (and some not at all) to natural, attainable conditions. Once attainable conditions are determined, the process of setting realistic goals and standards, and the process of monitoring change and responses to stress, are improved dramatically.

Regionalization is primarily a process of defining "like-enough" regions in which processes can be expected to be similar. Consequently, properly conducted regionalization helps separate "noise" among scales from real patterns. Ecoregions themselves may be delineated according to a variety of different scales. On a very broad scale, for example, the United States may be divided into five or six Ecoregions. A finer-scale division developed by Omernik (1987) divided the country into 76 Ecoregions (Fig. 16.1). Even Omernik's Ecoregions, however, are too broad for management action by many state agencies. Sufficient ecological variation is always present so that each region can be further subdivided into relevant subregions. Thus, the same logical process may reveal three or four Ecoregions within a small or relatively homogenous state (e.g.,

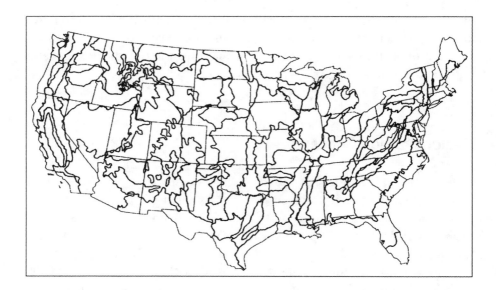

Fig. 16.1 Delineation of Ecoregions depends on the goals of the planner. The aquatic regions of the United States, developed by Omernick in 1987, provide appropriate detail for regional planning efforts, and illustrate the extraordinary variety of water quality conditions across the country. Omernick's regional delineations, however, are too detailed for many national-level planning efforts and insufficiently detailed for some aspects of state planning. Consequently, coarser-scale Ecoregions can be developed for national-level planning efforts and finer-scale Ecoregions can be developed for state- or local-level planning. (Modified from Omernick 1987)

four in Kansas), or more than 10 Ecoregions within larger, more diverse states (e.g., 11 in Oregon) (NRC 1992).

Early Efforts

One of the earliest, successful compilations of "ecoregional" maps was developed in 1972. *Land Resource Regions and Major Land Resource Areas of the United States* (Austin 1972) was developed to provide a geographic basis for making national and regional-level management decisions about agricultural concerns. According to Gallant *et al.* (1989), the effort sought to inventory and determine research needs, extrapolate information from site-specific research to other areas, and spatially organize resource conservation programs. Despite difficulties inherent in the original scheme, including the unreliability of available information on soil types, its utility and appropriateness were obvious and it was quickly adapted for other natural resource concerns.

Bailey (1976) developed a second widely used Ecoregion model to provide a hierarchical geographic framework for the U.S. Department of Agriculture Forest Service. *Ecoregions of the Unites States* integrated climate, soil, vegetation, and land-surface patterns in compiling an ecoregional view of U.S. forest areas. As noted by Gallant *et al.* (1989), however, the characteristics that are useful for delineating forest areas are not necessarily applicable to aquatic systems. In addition, Bailey's model considered relatively few parameters in setting regional boundaries, which made the classification incomplete. This void created a need for a different approach to Ecoregion delineation for water quality management.

Definition of aquatic Ecoregions began in 1980 with maps of alkalinity designed to predict effects of acid rain. While useful for that narrow purpose, these maps were too specific for other water quality management needs. In the mid-1980s, scientists from the U.S. Environmental Protection Agency (EPA) (e.g., J. Omernik, D. P. Larsen, R. Hughes) began to develop a more generic ecoregional mapping process for water quality. The EPA team aimed to classify streams for more effective management of water resources, synthesizing mapped patterns of land-surface form, land use, potential natural vegetation, and soils.

Development of *Ecoregions of the Conterminous United States* (Omernik 1987) was begun as a joint effort between the EPA and the states of Arkansas, Ohio, and Minnesota to establish and evaluate regional management frameworks. After testing and validation, these state-level maps were "scaled up" to a national level, and the ecological variables used in the state efforts were mapped at a national level. Integration of ecologically relevant variables into a single map of "Ecoregions" of the United States has served as an important conceptual development for ecologists and natural resource managers. State-level delineations, however, continue to be the cornerstone of much water quality management because they are more appropriate for local and regional management decisions and actions.

Ecoregional Variables

Many variables combine to give an Ecoregion its peculiar character. Importantly, different processes are paramount in different Ecoregions. Consequently, defining Ecoregions is an integrative process using many different physiographic and biological variables. As Gallant *et al.* (1989) explain, the chosen variables may either cause regional variations (e.g., climate, mineral availability, physiography) or integrate causal factors (e.g., soils, vegetation, land use).

The utility of causal factors is readily apparent. Climate, for example, determines maximum and minimum water temperatures, the potential for seasonal mixing of stratified layers, relative abundance and seasonality of precipitation, and the length of ice and ice-free seasons. Mineral availability determines important characteristics of water chemistry, including salinity, phosphorus levels, and acidity. Physiography determines other important variables such as water discharge of a stream, depth and mixing characteristics of a lake, and seepage and flow rates of groundwater.

Maps of causal factors alone, however, often do not adequately indicate how these factors interact to determine natural water quality parameters. In some areas, for example, mineral availability may outweigh other factors in determining the potential utility or value of a water resource. In other areas, seasonal temperature is dominant; in still other areas, mixing by wind dominates. Because single-feature maps cannot indicate subtle interactions between causal factors, researchers turn to integrator maps. Soil maps, for example, indicate the interaction of geology, climate, vegetation, and topography. As a result, they can point out important influences on water quality, such as where water resources are likely to receive inflow of nutrient-rich sediment or where water inflow emanates primarily from surface run-off.

Other indicator maps use familiar frameworks in unusual contexts. For example, land use maps are not employed to define Ecoregions per se, but rather serve as indicators, or artifacts of spatial changes in natural environmental characteristics and resource quality (Gallant *et al.* 1989). Land use provides key subtleties missed when measuring only climate, topography, and physiography—subtleties that imply important consequences for water resource character and sensitivity to degradation. Gallant *et al.* also report that spatial patterns of past and current land use have a major influence on water quality sensitivity and that land use maps provide important indicators of those spatial patterns. Finally, land use maps are more accurate than other component maps, and have been developed at far finer scales. Thus, they confer great utility in ecoregional determination if used in conjunction with other component maps. The risk with land use maps is that alterations to natural landforms (e.g., from urbanization and development) can subsume and mask natural potential. In those cases, land use maps receive comparatively less scrutiny and component maps carry more weight.

Delineating Ecoregions

Two important issues surround the delineation of Ecoregions. One is the choice of scale—that is, the size of each Ecoregion. The other issue is the delineation of boundaries.

Omernik's 1987 Ecoregion map shows a large range in Ecoregion size. The Willamette Valley Ecoregion is small (15,000 km²), while the Southern Plains

Ecoregion, a large relatively homogenous area, covers 330,000 km^2; the average is approximately 130,000 km^2. Subregional characterizations are used to address smaller variations within large ones. Within the Appalachian Mountains, for example, small drainage basins of only 500 km^2 contain a patchwork of forested mountains and agricultural valleys. Choosing the actual size of a region or subregion is part art and part science. As described by Gallant *et al.* (1989), "Delineating actual regional boundaries involved an iterative process of both map overlay and qualitative analysis of the relative accuracy and level of generality of each component map. . . . Since each component map was compiled at a different level of generality, with varying levels of accuracy relative to true locations of the characteristics represented, as well as to the source material used in map compilation, the usefulness of the map alignments for drawing Ecoregion boundaries varied" [p. 23]. In the final analysis, map boundaries represent "mental averaging" of the boundaries of each component map. This mental averaging is illustrated in Fig. 16.2 for Nebraska's Sand Hill Ecoregion.

Delineating boundaries is the process of implying sharp changes in gradual ecological gradations. Drawing a line indicates changes in spatial patterns of some variable. Actual boundaries are never as abrupt as lines on a map imply; ecoregional maps are only approximations of field conditions. Because these maps are used to set policy and standards, map-drawn boundaries inevitably take on a larger-than-life significance, potentially leading to sudden changes in legal standards and regulations that actually cover an area of gradually changing biophysical conditions.

Ecoregion Validation

The concept of water quality Ecoregions has been tested and validated in many diverse areas of the United States, including several streams in Arkansas, Colorado, Kansas, Minnesota, Ohio, and Oregon, and lakes of Michigan, Minnesota, Ohio, and Wisconsin (NRC 1992). Typical findings include the following:

- Strong differences among Ecoregions in nutrient and major ion concentrations in Ohio. Complexity and health of fish assemblages also varied with Ecoregion (see the case study later in this chapter).
- Strong differences in total phosphorus (and phosphorus cycling), clarity, and water residence time among Minnesota Ecoregions (Table 16.1).
- Strong differences in ionic strength, total phosphorus and nitrate/nitrite–nitrogen, and clear differences in fish assemblages among lakes in Colorado Ecoregions.

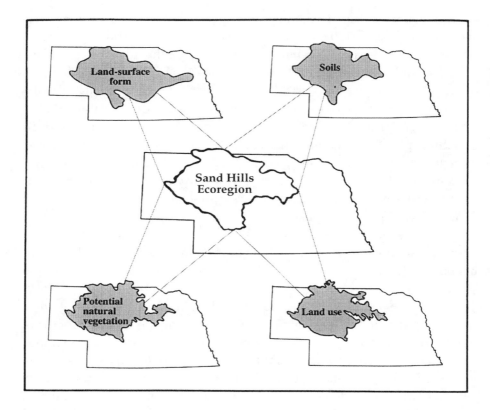

Fig. 16.2 Delineating Ecoregions involves a mental averaging of boundaries of component maps. In this figure, the boundaries of Nebraska's Sand Hills Ecoregion are seen to be a composite of the component maps of surface morphology (land-surface form), soil characteristics, potential natural vegetation, and land use. Land surface form has less influence over the final Ecoregion than potential natural variation and soils. Because Ecoregion delineation is a judgment of the planner, surface morphology may weigh more heavily in the determination of other Ecoregions where topography is more variable. Land use may be equally important in other heavily modified areas. (Modified from Gallant *et al.* 1989)

Ecoregions Versus River Basins

"River basin management" is sill a more familiar term in water resources management than is "ecoregional management." Outside the United States, ecoregional delineations are rare. A notable exception is the Netherlands, which has used a two-level hierarchical delineation of Ecoregions for planning since 1988 (Klijn *et al.* 1995). An ecoregional framework has also been developed for Romania and the Upper Seine in France, but its purpose is descriptive rather than managerial. Throughout much of Europe (e.g., France, England, Germany),

Table 16.1 Minnesota has four major Ecoregions, which contrast strongly in their physiographic and water resource characteristics. For example, the Northern Lakes and Forests region is more than 66% forested and contains 30 large lakes. The Northern Great Plains is 73% cultivated and has only eight large lakes. (Ecoregions: NCHF = North Central Hardwood Forest, NLF = Northern Lakes and Forests, NGP = Northern Great Plains, WCBP = Western Corn Belt Plains.)

Variable	Units	NCHF	NLF	NGP	WCBP
Number of lakes		36	30	8	11
Land use					
Cultivated	%	34.8	1.8	73.0	60.6
Pasture	%	18.0	3.9	9.2	5.9
Urban	%	0.7	0.0	2.0	1.5
Residential	%	6.4	4.8	0.4	9.9
Forested	%	16.4	66.2	0.0	7.0
Wetland	%	2.5	2.1	0.6	1.2
Water	%	20.9	20.9	14.4	13.6
Watershed area	ha	4670	2140	2464	756
Lake area	ha	364	318	218	107
Mean lake depth	m	6.6	6.3	1.6	2.5
Total phosphorus	μg/L	33	21.6	156	98
Chlorophyll-a	μg/L	14	6	61	67
Secchi depth	m	2.5	3.5	0.6	2.9
Total P load	kg/yr	1004	305	1943	590
Inflow P	μg/L	183	358	5666	564
Areal P load	kg m^{-2}/ yr^{-1}	276	96	891	551
Outflow volume	km^3/yr	6.2	5.3	0.9	1.0
Water residence	yr	9.3	5.0	36.2	4.8
Stream total P	μg/L	148	52	1500	570

water quality continues to be managed through a river basin approach. In many ways, river basin management has become synonymous with water resources management: it is accepted that basins share similar properties and concerns and represent a logical framework from which to manage. The dominance of river basin management reflects the historical dominance of the linear, up-stream–downstream orientation of hydrologists and managers. It is less suited to management of nonlinear lakes, groundwater, or wetlands.

In addition, the assumption that basins share similar properties is not always borne out. River basins are often linked only by the water that flows through them. Large areas may be characterized by different geological formation, seasonality may change with attitudinal gradients, vegetation will reflect climate and geology/soils, and land use can vary considerably along river basins. Although a river basin may offer a logical framework for water supply management, for water quality management it is less applicable. The river basin

formation is not necessarily a factor controlling the sensitivity of water resources to degradation, and the variety of uses and conditions along a large river
basin necessitate widely different standards, approaches, and scalar planning.

In many ways, it is not surprising that the United States is leading the development of a new approach to natural resources management. The phenom-
enal diversity of the U.S. landscape and the scale of its water resources requires
a different approach than that taken in smaller countries that must contend
with comparatively less diversity. River basin planning in the United States takes
on enormously complex dimensions simply because of the sheer size and diversity of these water resources. The Tennessee Valley Authority, for example, is
one of the showcases of river basin planning, as well as a quintessential example of its complexities. The Tennessee River is huge; according to Newson
(1992), it covers an area representing 80% of the area encompassed by England
and Wales, with an average discharge 24 times that of the Thames or 70% that
of the Nile. These figures appear even more daunting when the Tennessee is
seen as only one of many tributaries of the Ohio River, which is itself one of
many tributaries of the Mississippi River.

Ecoregion Case Studies

Ohio

To those unfamiliar with the Midwestern region of the United States, Ohio
would seem to represent a relatively homogenous state: there is no high relief,
all of the forests are hardwood varieties, and cropland dominates much of the
state. Because of these apparent similarities, Ohio makes a good case study for
aquatic Ecoregions, allowing us to ask if the tool can identify important but perhaps subtle patterns in water quality.

The ecoregional delineation began in Ohio, as it does in all cases, with a
"weeding out" of heavily impacted sites and the selection of least-disturbed
watersheds of various sizes. Streams were then sampled over a 16-month
period for numerous variables, including alkalinity, total organic carbon, nitrogen, and phosphorus (Larsen *et al.* 1988). The results were used to develop
a clear delineation of five Ecoregions within the state (Fig. 16.3). Phosphorus,
total ionic strength, and conductivity, for example, all grade from highest
values in the northwest corner of the state (Huron/Erie Ecoregion) to the
southeastern corner (Western Allegheny Plateau). In addition to these clear gradations, the Ecoregions reflect spatial differences. For example, total phosphorus values are relatively homogenous within the Huron/Erie Ecoregion and

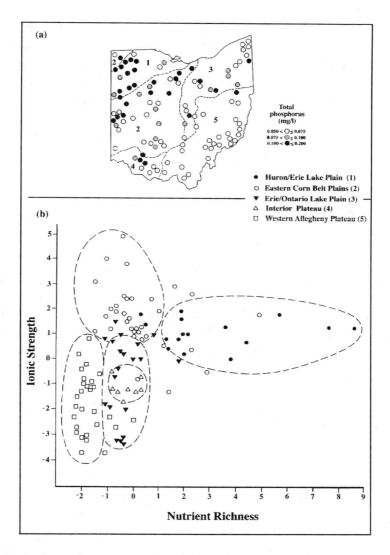

Fig. 16.3 Water quality in Ohio varies across the state and has led to the delineation of five Ecoregions. Shown are gradations of total phosphorus in lakes across the state (a), and clustering of water chemistry characteristics among the five Ecoregions (b). See text for further discussion. (After Larsen *et al.* 1988)

the Western Allegheny Plateau; the same values are, however, extremely variable in the Eastern Corn Belt Plains. This finding underscores the fact that Ecoregions do not describe finite differences between regions. For planning and management, it is just as valid to consider that one region of a state has highly variable conditions as to identify that other regions are more homogenous. On a larger scale, the variable regions of Ohio are perhaps more accurately viewed as extended border areas between the areas of high and low phosphorus, conductivity, and ionic strength.

The Ecoregion concept has been used creatively in Ohio to choose "best candidate" reservoirs for eutrophication restoration efforts. Traditionally, lakes and reservoirs slated for improvement are those representing a worst-case scenario. Improving these resources can lead to enormous outlays of money, time, and effort, but the choice of "worst case" has traditionally seemed self-evident. In contrast to this approach, researchers in Ohio evaluated the "potential" trophic state of selected reservoirs after restoration (as estimated from ecoregional averages) against their actual state. As illustrated in Fig. 16.4, this framework allowed managers and policy makers to visualize the degree of possible improvements for each reservoir. As described by the NRC (1992): "Lake 1 [Fig. 16.4] is hypereutrophic, and its water quality is among Ohio's poorest. It is located in an Ecoregion with rich humic soils in which the principal land use is agricultural. Its attainable trophic state does not differ significantly from its current state" [p. 103]. Such a framework enables managers to understand that, although expensive improvements could be made, they would endure for only the short term. The lake would quickly refill with silt and return to its current trophic state. Under traditional modes of selecting sites for restoration, Lake 1 would be a high priority. The ecoregional framework has enabled managers to evaluate whether real improvements are possible and whether funds could be better allocated to sites with more long-term improvement potential.

Texas

The state of Texas exhibits a very high diversity among its aquatic ecosystems. It is home to several large rivers and more than 3700 small to medium-size streams. The state has regulatory authority over water quality management in those water bodies, and is required to set standards and effluent limits for dischargers. Texas operates on the assumption that biocriteria are more representative of water quality conditions than are water chemistry values. Biocriteria are, therefore, used to establish site-specific effluent limits for each discharge. A traditional approach to developing site-specific standards has been to use an upstream–downstream model where effluent conditions are not allowed to force the downstream condition outside the norm of the upstream. In many Texas streams, this model is inappropriate because multiple effluents along a reach or wide variations in stream flow (e.g., where the effluent may constitute more than 90% of the downstream volume) preclude reasonable comparisons. As a solution, Texas has used a regional

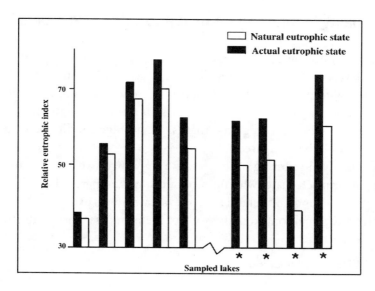

Fig. 16.4 The difference between actual and potential trophic states in Ohio lakes has been used to facilitate planning of restoration activities. Traditional restoration decision making chooses the "worst" condition for improvement. An ecoregional framework, on the other hand, allows planners to balance the costs and effort against potential improvement. In the first group of lakes, for example, the current degree of eutrophication is only slightly higher than the natural potential state. The starred lakes are those that show the greatest difference between current, actual state of eutrophication and the potential, natural state. These lakes have the greatest potential for successful, cost-effective remediation efforts even though they are not always the most heavily eutrophied. (Data from NRC 1992)

approach to developing and implementing biocriteria for management (Hornig *et al.* 1995).

Fish, invertebrates, and habitat represent various components of the ecosystem; all three are, therefore, used in development of Texas biocriteria. Limits were developed for only 11 of the state's 12 EPA Aquatic Ecoregions due to limited numbers of discharges in the twelfth area. In each Ecoregion, scientists selected the "most typical area" and sampled least impacted streams in those areas. They analyzed 1–11 replicate sites in each Ecoregion. Three discrete indices were used for classification: a modified fish Index of Biotic Integrity (see chapter 6), an invertebrate index based on diversity of pollution-intolerant forms, and a quantitative habitat quality index (HQI). All three indices were calculated for each site, and statistical analyses were then performed to identify patterns. As a result, the state developed one integrated index and three discrete index values for each site. Those values represent the "expected" or background condition for stream reaches in a given region. Effluents are then permitted, regulated, and monitored against that background (Hornig *et al.* 1995).

Water Quality Monitoring from an Ecoregional Perspective

Ecoregion monitoring centers around delineation of "most typical" areas in which the levels of chosen variables are most clearly distinct from levels in other Ecoregions. Within this most typical area, reference sites are identified from among the least impacted locations. These sites represent "best attainable water quality" for a given Ecoregion—that is, the average conditions with which other sites should be compared.

Water quality monitoring is, by definition, a comparative science. Monitoring can compare impacted sites with unimpacted sites, compare sites with objective values such as water quality standards, or evaluate trends over time. Monitoring within an ecoregional context makes each of these processes more effective.

Impacted Versus Unimpacted Sites

Detecting impact depends on detecting differences in hydrology, physics, chemistry, and biology. Consequently, it is imperative that "natural" variations in these variables be factored out of any determination of impact. In this context, an ecoregional framework increases the probability that the resulting comparison will be meaningful and defensible because sites are grouped based on their biogeochemical similarities. Furthermore, if reference sites are chosen from the least-impacted, most typical area, the reference condition will typify "expected, unimpacted" conditions, providing the most meaningful comparison.

Comparison with an Objective Standard

Water quality laws and regulations are designed to protect the "value" of a water resource. Thus, objective standards are established based on designated uses. If designated uses (and thus standards) do not correspond with those relevant to a given Ecoregion, they will often prove unrealistic.

Before the advent of Ecoregions, states typically adopted statewide standards for any given variable. For example, based on reference to a trout fishery, a state might have adopted a minimum acceptable dissolved oxygen standard of 5.0 mg/L for all waters. In contrast, healthy non-trout fisheries flourish under a variety of oxygen regimes. In southwestern Minnesota, for example, water bodies often exceed 16 mg/L during summer days, falling to 1.5 mg/L on summer nights. These small shallow lakes house a diverse community of warm-water fishes. Some stream trout populations along Minnesota's North Shore might be damaged or eliminated if oxygen fell below 7.0 mg/L. Thus, one statewide oxygen level will neither protect all communities nor adequately guide management.

Standards and monitoring designs can be calibrated to ecoregional conditions such that the minimum acceptable level accurately reflects the values society has established for a given water body. In southwestern Minnesota, the oxygen standard might be expressed as "no more than 8 hours below 1.5 mg/L." In Minnesota's North Shore streams, the oxygen standard might be expressed as "no lower than 7.0 mg/L." Those standards cannot be established or enforced differentially for every lake or stream reach. Thus, an ecoregional delineation allows generality at a level lower than that for the entire state.

Trends ThroughTime

Long-term monitoring has emerged as an increasingly important concern as society becomes more aware of, and scientists understand more about, cumulative impacts of pollutants and as we ask whether management activities are achieving their desired goals. Unfortunately, long-term monitoring at a large series of sites is usually not feasible because of budget limitations. Instead, managers sample several locations representative of temporal trends by choosing newly impacted sites and sites with longer histories of impact and management. The more similar the "representative" sites are, the more useful this technique becomes, because differences in soil chemistry, climate, or other variables might mask or accentuate impacts and invalidate the estimate of temporal trends. Thus, choosing sites within an ecoregional framework helps to ensure that sites are similar and that differences in levels of impact can be assumed to represent temporal trends.

Summary

- Natural water quality variations are controlled by a series of broad-scale variables such as climate, soil, geology, and land use. We can integrate and then map those variables, developing a map of "Ecoregions." Such regions are being used in the United States to guide management more effectively than geopolitical or strict hydrologic delineations. Although they have been developed in the Netherlands and parts of France and Romania on an exemplary basis, Ecoregions are not yet used in water quality management outside the United States.
- An aquatic Ecoregion can be divided into a least typical area and a most typical area (i.e., the area where the defining variables are most

homogenous and consistent). Waters in a most typical area that are least impacted can be used to generate a suite of characteristics describing the "best attainable water quality." Management actions may be judged against this reference or standard.

- Several states in the United States were early leaders in using Ecoregions to guide water quality management. Ohio, Maine, and Texas have developed biocriteria for streams. Minnesota manages its lakes on an ecoregional basis. Most other states are beginning to develop monitoring or management strategies based on an Ecoregion framework.

For Further Reading

Klijn F, De Waal RW, Oude Voshaar JH. Ecoregions and ecodistricts: ecological regionalization for the Netherlands' environmental policy. *Environ Mgt* 1995;19:797–813.

Larsen DP, Dudley DR, Hughes RM. A regional approach for assessing attainable surface water quality: an Ohio case study. *J Soil Water Conservation* 1988;43:171–176.

Newson M. *Water, land and development: river basin systems and their sustainable management.* London: Routledge Press, 1992.

Omernik JM, Rohm CM, Lillie RA, Mesner N. Usefulness of natural regions for lake management: analysis of variation among lakes in northwestern Wisconsin, USA. *Environ Mgt* 1991;15:281–293.

Wilson CB, Walker WW Jr. Development of lake assessment methods based upon the aquatic ecoregion concept. *Lake and Reservoir Management* 1989;5:11–22.

17

Effects of Land Use on Water Quality

Overview

Many land use activities affect water quality in similar ways, despite the varied origin of impacts. For example, deforestation may increase sediment loading to a stream bed, but sediment loading may also occur through agricultural run-off or through urban run-off via storm drains. Similarly, water temperature may be changed by forestry activities that reduce stream-bed shading or by high-temperature effluent from power plants. Ten common and critical water quality impacts result from different land uses: changes in suspended sediment load, organic matter and biological (and chemical) oxygen demand, bacteria and viruses, nutrient loads, temperature, heavy metals, toxins such as pesticides and herbicides, acidification (pH), salinization, and changes in the water flow itself. This chapter introduces and discusses these 10 major changes; each factor is also discussed in some detail in chapters 18–24 as well.

In brief: suspended sediment affects water quality by reducing water clarity and photosynthetic potential of aquatic plants as well as by clogging fish gills and smothering eggs and macroinvertebrates. Biological and chemical oxygen demand from organic material can reduce the available oxygen in the water to the point that it can no longer support aquatic biota. Bacteria and viruses, heavy metals, toxins, salinization, and acidification may all have effects that are either directly toxic or sublethal but still damaging. Nutrient loads can lead to overproduction of vegetation, reduced water clarity, and, ultimately, deoxygenated waters. Finally, changes in the water volume may modify the concentration of pollutants, the ability of plants and animals to establish viable communities, and susceptibility to temperature changes, eutrophication, salinization, and other impacts. More than any single impact, the synergistic effects among impacts serve to define and characterize water quality conditions.

Introduction

Virtually every land use has the potential to change water quality. Actual effects vary enormously depending on the volume of a pollutant discharged into the water, its concentration, rate of delivery, water body renewal rate, and the nature of the specific stressor. Subsequent chapters discuss specific land uses and ways in which they affect water quality. From the perspective of the aquatic community, however, the source of an impact is usually irrelevant. If, for example, stream temperature is increased by the introduction of industrial effluent or from increased light resulting from deforestation, effects on the aquatic community will be relatively similar.

This chapter focuses on the similarities among impacts and reviews the chemical, biological, and physical effects of those stresses. Included in this discussion are the 10 most common, significant impacts on water quality: changes in suspended sediment load, organic matter and biological oxygen demand, bacteria and viruses, nutrient loads, temperature, heavy metals, toxins such as pesticides and herbicides, acidification, salinization, and changes in the water volume itself. Importantly not only the effect that these stresses have on water quality, but also synergistic effects among them, define both water quality impacts and the scope of management activities. These synergistic effects are discussed briefly at the conclusion of this chapter, and in more detail in subsequent chapters.

Suspended Sediment

Extraordinary variation is observed in natural sedimentation rates on a worldwide basis. Novotny and Olem (1994) note measurements ranging from less than 1 metric ton/km²/yr for some rivers in Poland and Australia to more than 10,000 metric tons/km²/yr for some of the tributaries of the Yellow River in China and some rivers in Kenya. For 14 major rivers, Gleick (1993) provides figures ranging from 11 metric tons/km²/yr for the Zaire to 1620 metric tons/km²/yr for the Haihe, yielding corresponding sediment concentrations of 34 mg/L for the Zaire and 40,500 mg/L for the Haihe.

Differences in baseline sediment load constitute an important reference point for managers, indicating the important soil and climate properties that must be accepted as givens. Baseline sediment loads also define the goods and services available from a water body and the composition of the aquatic community. Although a wide range of sediment loads are carried by rivers and lakes, biological communities and human populations are peculiarly adapted to the bed loads of their local rivers and lakes. Consequently, when land uses increase or decrease baseline sediment load, these changes affect the functioning of the biological community, local perceptions of water quality, and potential changes in goods and services. Documented changes following human activity range from nearly eliminating the entire sediment yield of a river after dam construction (Postel 1995) to 4500-fold increases in sediment yield following deforestation and cultivation in highly erodible areas of Brazil. More typical estimates are 2- to 50-fold increases in sediment yield following agricultural, forestry, or construction activities (Novotny and Olem 1993).

Suspended sediment comprises any material (e.g., silt, clay, organic debris such as leaves) held in suspension in the water column of a lake or stream. It plays a key role in aquatic ecology. For example, some species, such as sardines, require the high nutrient levels associated with high sediment loads. In areas where sediment loads have been eliminated, such as the Nile River delta in the Mediterranean, sardine populations have declined by more than 80% (Postel 1995).

In most of the world, severe water quality impacts are more commonly associated with increases of sediment load than with decreases. The principal result of increased sedimentation is a change in the physical environment of the river or lake, which affects the biotic communities in diverse ways. Suspended sediment reduces light infiltration and, therefore, may lead to reduced plant growth (i.e., reduced primary productivity) and diversity. Other direct effects on biota include clogging and abrading gills of fish and macroinvertebrates and covering eggs and spawn with a smothering layer of silt. Although not directly toxic, suspended sediment serves as a medium to which toxic compounds can bind. Some heavy metals and organochlorines are typically carried into aquatic systems in this way. Consequently, increased sediment loads can increase the risk of severe health risks to both aquatic biota and humans.

Increases in sedimentation also influence water use and valuation. High sediment loads decrease the aesthetic values of a water resource by lowering visibility, one of the critical factors in judging water quality. Highly turbid water can also increase water delivery costs because filtration is required to remove sediment from drinking water supplies. Another important concern found around the world is the accumulation of sediments behind dams and levees.

The accumulating sediment diminishes water volume and can lead to changes in thermal regime, stratification, and nutrient cycling. Each decade in the United States, approximately 3% of water storage capacity is lost through sediment accumulation in reservoirs in the contiguous 48 states (Gleick 1993). In areas of normally high sedimentation, deposition of sediments behind dams and flood control barriers creates an even more serious problem; for example, the Nizamsagar reservoir in India has lost 60% of its capacity in only 40 years (Newson 1992).

The largest concentrations of suspended sediment are generally found in running waters, where the moving water erodes (scours) stream channels and prevents the particles from settling. Concentrations of suspended material are highly dependent upon stream velocity and generally increase during periods of high stream discharge. In contrast, waters moving over a slow-moving flood plain will deposit sediment and carry reduced volumes of suspended sediment downstream.

Determinants of Sediment Delivery

Natural variations in sediment production primarily result from the interaction of three factors: erosivity of different soils, the magnitude and severity of storms, and topography (with steep slopes showing greater erosive potential). Sediment potential can be categorized in different ways based on each of these factors. For example, Newson (1992) summarizes two different classifications of sediment yield. The first indicates the importance of climate:

	metric tons/km²/yr
Tropical, rainy	71.5
Dry	169.0
Humid, Mediterranean	714.4
Humid, cool	46.5

The second roughly indicates the importance of soil type and erodibility:

	metric tons/km²/yr
North America	97
South America	63
Africa	27
Australia	33
Europe	35
Asia	600

In all cases, however, it is important to recognize that the broad regional differences in sediment deposition rates often mask significant local variation due to topographic extremes, local climate patterns, and local geology. In general, high rainfall, steep slopes, and poor vegetative cover are the principal contributors to accelerated rates of erosion.

Land use also controls sediment yield through changes to vegetative cover and slope. Each such change can have a large impact on increased sediment yield to surface waters. Baker (1993) demonstrates differences between sediment loads and associated phosphorus loads associated with agricultural, forested, and urbanized watersheds feeding Lake Erie. General findings include the following: point sources dominate in urban watersheds; nitrate levels are highest in the agricultural watersheds (the results of tile drainage systems), lower in urbanized watersheds, and lowest in forested watersheds; residue from historical phosphorus inputs to soil cause elevated phosphorus levels in the sediment of agricultural watersheds; and extensive annual variability occurs in the amount of total phosphorus exported from the agricultural watersheds (compared with steady, but decreasing, point source inputs), with the greatest amount of phosphorus export occurring during winter and spring storm events.

Sources of sedimentation include agricultural fields, pastures and livestock feed lots, logged hillsides, degraded streambanks, road and building construction sites, and urban storm sewers. In the United States, for example, more than 40% of increased sedimentation is attributed to agricultural activities (Gleick 1993). In much of the developing world, poorly managed and uncontrolled forest cutting contributes to significant sedimentation, while sediment production in many parts of Africa is increased by the reduction of grass cover through overgrazing. Urban areas are also becoming recognized as important sources of sediment load (see chapter 20). Each of these land use changes is discussed in more detail in the subsequent focus chapters on water quality in agricultural, forested, and urban landscapes.

Organic Matter and Dissolved Oxygen

The largest quantities of any pollutant discharged from point sources into water courses involve organic matter. When the organic material enters the water, microbes colonize and begin decomposing it. In the process, these microbes consume oxygen in proportion to the amount of organic material present. Thus, large releases of organic material can result in severe depletion of oxygen near and downstream of the discharge. This oxygen depletion is a cornerstone process that leads to fish kills and reduced abundance and diversity of aquatic life. The demand on oxygen is most commonly referred to as biological oxygen demand (BOD) (see chapter 2), a term that represents the most readily decomposed material. It may also be represented by the broader term chemical oxygen demand (COD), which incorporates more-resistant material, or by sediment oxygen demand (SOD), which is the oxygen demand specifically associated with material in and on the sediment.

Dissolved oxygen is essential for survival of most aquatic animals, just as atmospheric oxygen is essential for most terrestrial animals. In aquatic ecosystems, oxygen is often in short supply. Besides its significance for higher organisms such as insects and fish, oxygen has often been termed the Master Variable because of the cascading effects that oxygen concentrations have on water chemistry and biology. These effects range from influencing the spatial distribution of fish populations to controlling the form of toxic trace metals such as mercury.

Surface waters exchange oxygen with the atmosphere: oxygen-poor waters will absorb oxygen from the air, while extremely oxygen-rich waters may lose oxygen to the air. Because of the air exchange, surface waters are usually oxygen-rich. The actual concentration of oxygen that a body of water can hold varies with temperature, salinity, and atmospheric pressure. Waters in balance with the atmosphere at a given temperature and pressure are called fully saturated. Land use changes that increase stream or lake temperatures have an important impact on oxygen-carrying capacity of water because they may alter the saturation level as well as the actual oxygen concentration.

Streams and lakes are not fully saturated at all times. Considerable spatial and temporal variation exists in oxygen saturation, with aeration, photosynthetic activity, and rates of respiration all affecting the oxygen level in water. For example, surface waters in a lake frequently can be supersaturated (to 120% saturation) due to rapid photosynthetic production of oxygen. Deeper waters often remain well below the saturation level (or even anoxic, with 0% saturation) due to consumption of oxygen by microbial respiration and the fact that these deep waters are not open to the atmosphere. Waters that are very low in oxygen have a very limited ability to support aquatic biota.

Low levels of dissolved oxygen can either affect an entire water body (e.g., eutrophication) or a portion of the resource, such as an area downstream of a point source. Point source discharges that are high in BOD or COD create an "oxygen sag" effect below the effluent as microbial consumption of the waste depletes the oxygen levels of waters. Downstream

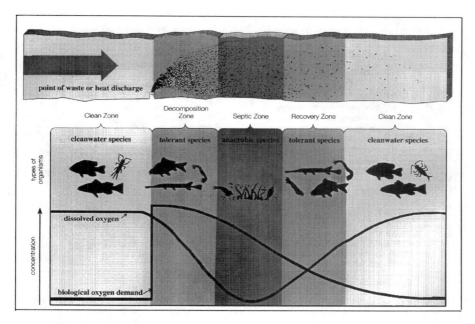

Fig. 17.1 Oxygen sag is the term given to the decrease in oxygen below a point source of organic matter. The oxygen decrease results from metabolic activity of decompositional microbes. A gradual reoxygenation occurs as surface waters take up oxygen from the air. Low-oxygen waters support fewer species of fish and other aquatic organisms, and poor-quality organisms such as sludge worms. (Modified after Miller 1992)

of the point source, however, reaeration of the water through continuous oxygen exchange with the atmosphere allows oxygen levels to rebound (Fig. 17.1). A cumulative effect on oxygen levels may occur when outfalls are located close together.

Waste streams also contain inorganic plant nutrients such as nitrogen and phosphorus that stimulate primary productivity, indirectly affecting oxygen concentrations. Increased primary productivity results in increased dissolved oxygen during the day. In contrast, night-time respiration by plants, algae, and heterotrophic microorganisms feeding on organic matter cause oxygen declines. This diel cycle—in which increased primary productivity leads indirectly to decreased oxygen availability and in some cases anoxic conditions—serves as the primary mechanism of oxygen depletion in eutrophic waters. This process is discussed in more detail in chapter 21.

Sources of organic matter include agricultural fields and pastures, landscaped urban areas, municipal sewers, logged areas, and chemical manufacturing and other industrial processes.

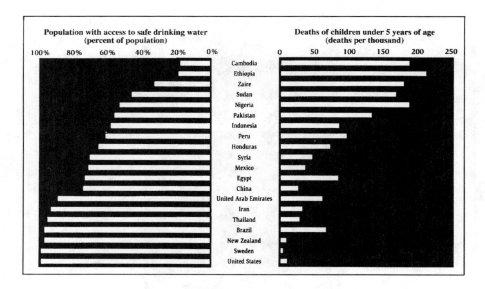

Fig. 17.2 Worldwide, water quality is closely correlated with child survival: waters high in disease-causing agents cause enormous personal and social costs. (Modified from Engelman and LeRoy 1993; courtesy of Population Action International)

Bacteria and Viruses

Bacteria, parasites, and viruses in water represent one of the most serious global health hazards, as water quality is clearly linked with child survival (Fig. 17.2). Not all water-related diseases are spread in the same way, however. More than 250 million people worldwide suffer from water-borne diseases—that is, diseases transmitted by water-borne fecal parasites, bacteria, and viruses. Important agents of these water-borne diseases include cholera, *Salmonella,* amoeba, *Giardia,* typhoid and paratyphoid, giant roundworm, hookworm, and whipworm. Another 200 million people suffer from water-based diseases in which water is the habitat for infected snails or crustacea; contact with these organisms spreads the diseases. A third category, called "water-related" diseases, includes malaria, yellow fever, sleeping sickness, and river blindness. These conditions are transmitted by insects that breed or live near water and affect by far the largest number of people globally. According to Nash (1993), malaria accounts for 20% to 30% of all childhood deaths, and more than 2 billion people (40% of the world's population) are at risk for contracting malaria. Other important water-related diseases such as sleeping sickness (*Trypanosomiasis*) affect 25,000 people per year, but more than 50 million people are at risk for these illnesses in 36 countries.

Sources of disease agents include raw and partially treated sewage, animal wastes, and nonpoint sources of human waste such as urban run-off. Control devices that reduce water velocity or increase water temperatures exacerbate effects.

Water-borne Diseases

Water-borne diseases are transmitted through drinking water contaminated with human or animal excreta. Consequently, these diseases are associated with poor sanitation, urban sewage, feedlot contamination, wildlife stream crossings, or human use of a common water source for sanitation and water supply purposes. Fecal coliform (a measure of *Escherichia coli* bacteria) is used to determine bacteriological contamination of water; drinking water standards for WHO, the EC, and the U.S. EPA all specify acceptable levels as zero colonies per 100 mL (Gleick 1993). If water is treated and disinfected before consumption, fecal coliform can be eliminated from the potable water supply. This strategy is employed in much of Europe, where fecal coliform rates in surface waters are often high, but impacts on public health are low because nearly all municipal water supplies are adequately treated and disinfected.

Much less is understood about viruses and protozoans. Generally, they are more resistant to treatment than bacteria but their distribution is localized. Yet, it is unusual for them to create a problem in treated public water supplies. A major exception occurred in Milwaukee, Wisconsin, in 1993, when more than 200,000 people became ill from the protozoan *Cyptosporidium* that contaminated the city's water supply. Viruses and bacteria also have important impacts on aquatic communities, reducing survival and reproduction in some fish, shellfish, and associated species. Consequently, contamination reduces availability of these species for human consumption.

Water-based Diseases

Water-based diseases such as *Schistosomiasis* and *Dracunculiasis* are widespread, although their distributions are changing. The occurrence of *Dracunculiasis* (guinea worm disease) decreased dramatically in the 1980s. Currently, approximately 3 million people are infected, but the occurrence is limited to 12 countries in Africa and South Asia. The occurrence of *Schistosomiasis* is increasing globally, however. Currently, more than 400 million people are at risk from *Schistosomiasis* in over 70 countries (Nash 1993). In numerous cases, dramatic increases in the occurrence of *Schistosomiasis* have been seen following irrigation and dam construction (Nash 1993). For example, in the Nile delta, construction of the Aswan dam resulted in infection rates of nearly 100%; similarly, after the construction of the Akosombo dam in Ghana, West Africa, infection rates in children rose from 10% to more than 90%.

Water-related Diseases

Water-related diseases such as malaria, yellow fever, sleeping sickness, and river blindness are often not considered a problem caused directly by water quality. They are clearly related to water quality, however, because the construction of irrigation channels, dams, and flood control mechanisms has created slower-moving, warmer waters more suitable for the insect hosts. Incidence of malaria tends to be heavily concentrated in villages within a region, and regions within a country, indicating that malarial infection is related to local water conditions.

Nutrients

While nutrients are critical for all biological communities, high nutrient loads in a water body can result in eutrophication, algal blooms, decreased oxygen, and increased organic loading. These changes can lead to reduced diversity and increased growth of large plants, release of toxins from sediments, reduced diversity in vertebrate and invertebrate communities, and fish kills. Nitrogen and phosphorus are the two most problematic nutrients. High nutrient concentrations in some effluents have led to eutrophication in reaches of many rivers in central Europe, a problem previously thought to be restricted to lakes and reservoirs. Increased nutrient loading may also limit availability of food resources to humans because of population declines of fish, shellfish, and associated species. Lastly, increased nutrient input frequently impairs recreational uses of a water body due to algal blooms, macrophyte growth, odors, turbidity, and dermatological parasites that flourish in nutrient-rich waters.

The combination of agricultural run-off, detergent waste, and human sewage from 81 million people in the Danube drainage basin, for example, has resulted in approximately 340,000 tons of nitrogen and 60,000 tons of phosphorus carried to the Black Sea. These numbers have increased sixfold and fourfold, respectively, in 25 years (Platt 1995). These nutrients routinely cause major eutrophication problems in the Danube River.

Sources of nutrients include agricultural fields, pastures, livestock feed lots, landscaped urban areas, raw and treated sewage discharges, and industrial discharges.

Nitrogen

Nitrogen is a critical and naturally occurring element that is essential for amino acids, the building blocks of proteins, and thus also essential for many metabolic pathways. Nitrogen is abundant in the atmosphere and dissolves in water as the gas N_2. For photosynthetic production, nitrogen must be in the form of nitrate (NO_3^-) or ammonium (NH_4^+). A simplified diagram of the nitrogen cycle is presented in Fig. 17.3. Increased nitrogen (and consequently nitrate) levels in surface waters and groundwater can create several problems for human populators. Nitrates represent a serious concern in drinking water, where they pose a risk of reduced oxygen-carrying capacity in infant blood; they also generate carcinogenic nitrosamines (Nash 1993). In marine environments, naturally low nitrogen levels limit primary productivity. When nitrogen levels are increased, proportional increases in primary productivity can result in eutrophication. Nitrogen may be limiting in tropical fresh waters but rarely limits growth in temperate fresh waters.

Run-off from the landscape, atmospheric deposition, and nitrogen fixation by aquatic microorganisms are the major sources of dissolved inorganic nitrogen in aquatic ecosystems. Two major groups of organisms accomplish nitrogen fixation (i.e., reduction of N_2 gas to NH_4^+): blue–green algae and free-living heterotrophic microorganisms. Microbial reactions accomplish further conversion of NH_4^+ to NO_3^- (e.g., oxidation of NH_4^+ to NO_2^- [nitrite]) by bacteria such as *Nitrosomonas*, and then further oxidation to NO_3^- by groups such as *Nitrobacter*). In fresh water, blue–green algae can overcome nitrogen deficiency through nitrogen fixation. Hence, nitrogen is rarely limiting in fresh water.

As illustrated in Fig. 17.4, many land use practices increase delivery of nitrogen to aquatic ecosystems, including industrial and municipal waste, agricultural fertilizers, manure spread on agricultural lands, and forestry operations (Fig. 17.4). Most forestry impacts on nitrogen cycling can be attributed to mineralization of organic nitrogen in the forest floor, followed by nitrification of NH_4^+ produced by mineralization; this process results in increased NO_3^+ levels. More important anthropogenic sources of nitrogen are chemical fertilizer applications on agricultural land and the spreading of manure on agricultural land. Europe is experiencing increasing problems with nitrogen overfertilization and the proliferation of nitrate in area waters. Nitrate loadings in European waters have also grown because of atmospheric deposition, primarily from automobiles. According to the World Resources Institute (1992), approximately 25% of the EC population is drinking water that has nitrate loadings that exceed the EC recommended maximum level (25 mg/L). In Germany, the Netherlands, France, and the United Kingdom, water supplies are particularly threatened by increasing nitrate levels.

Fig. 17.3 A simplified diagram of the aquatic nitrogen cycle. Nitrogen enters lakes and streams from the atmosphere, where it is in the form of nitrogen gas (N_2). At the water's edge, nitrogen-fixing bacteria begin the conversion process to biologically available nitrogen through conversion to ammonia (NH_4) or its intermediate (NH_3). Ammonia may then either be oxidized by bacteria and converted to nitrites (NO_2) and nitrates (NO_3), or reconverted back to nitrogen gas by bacteria in anaerobic conditions. Nitrites and nitrates are available to various components of the aquatic food chain and re-released as ammonia in waste products. Land uses, including agriculture and logging, frequently increase the flow of nitrates to area waters.

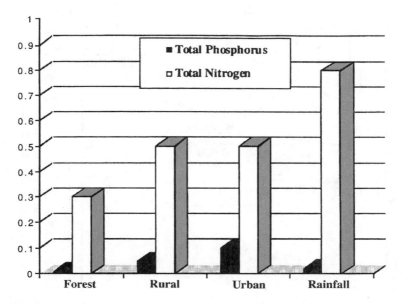

Fig. 17.4 Quantities of nitrogen and phosphorus derived from various land use practices in the United States. (Data from Mason 1993)

Phosphorus

Phosphorus is an essential element of ATP, the primary energy carrier in higher organisms. The major natural source of phosphorus is rock weathering. Although it is found in a variety of forms in aquatic ecosystems, particulate phosphorus accounts for 70% of total phosphorus found in fresh waters. Fertilizers, detergents, and sewage are the largest anthropogenic sources of phosphorus in waterways and provide a readily available form of this nutrient. Phosphorus cycles through aquatic ecosystems but, in contrast to nitrogen, no biological processes exist for phosphorus "fixation." Instead, phosphorus is often bound to iron; when these particles settle into deeper waters, anoxic conditions enable phosphorus to dissolve from the iron. Return of released phosphorus to the oxygenated, productive lake stratum is not an efficient process and often depends on seasonal mixing of waters.

Unlike nitrogen compounds, phosphorus does not carry apparent human health risks, but phosphorus is a critical driving variable in eutrophication of fresh waters. Primary production in fresh waters is generally limited by low phosphorus levels (Laws 1993). In phosphorus-limited lakes and rivers, phosphorus input results in a proportional increase in primary production.

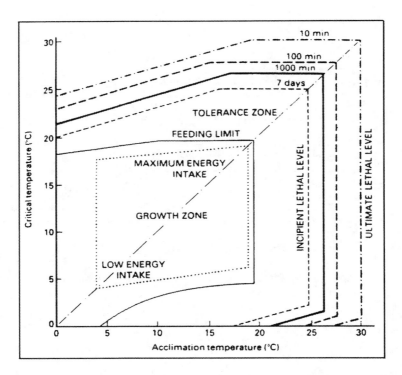

Fig. 17.5 Aquatic organisms exhibit a range of temperature tolerances for different physiological activities. These values are typically represented in a temperature tolerance polygon, as seen here for brown trout. The diagram indicates that certain critical physiological functions, such as growth and feeding, cease before temperatures become directly lethal. This figure also shows that acclimation can extend the temperature tolerances of certain fish. (After Hellawell 1986)

Eventually, overproduction can lead to algal blooms, anoxic waters, and reduced biotic diversity (see chapter 22).

Temperature

Increased water temperature can cause dramatic, local effects, whether from power plant effluents or changes in stream or lake-side vegetation. Water temperature is critical because it regulates metabolic activities; beyond certain species-specific limits, breeding stops, feeding stops, and high temperatures eventually prove fatal (Fig. 17.5). Because respiration rates increase in high-temperature environments, some fish may be physiologically stressed and become more susceptible to diseases and other disturbances. In fish, temperature effects range from direct impacts on eggs and fry, which are significantly less temperature-tolerant than older fish, to behavioral changes such as changes in migration and spawning, to alterations in competitive ability.

An important issue in thermal pollution is the mobility of the potentially affected organism. Some fish are attracted to warm effluents in cold seasons; they may, therefore, be adversely affected by plant closings and sudden changes in water temperature. More often, however, mobile organisms such as fish can avoid areas of increased temperature. Fish may also acclimatize to a certain level of increased temperatures (Mason 1991).

The effects of thermal discharge are more important for less mobile or immobile organisms such as plants and macroinvertebrates. Increased temperatures, for example, usually increase the growth of attached algae, cyanobacteria, and tube worms, but overall species diversity declines in heated waters (Mason 1991). Higher stream temperatures can also accelerate the development of aquatic insect larvae, leading to early adult emergence. Earlier emergence reduces population viability because early emergers fail to find mates, freeze, or are food-limited.

Other impacts commonly attributed to increased temperature include the elimination of cold-water fish and shellfish species, reduced dissolved oxygen due to increased plant growth, and increased decomposition rates.

Sources of increased temperature include urban landscapes, unshaded streams, impounded waters, reduced discharges from dams, and discharges from power plants and industrial facilities.

Heavy Metals

Although metals are natural constituents of soil and water, in recent decades the worldwide production and use of metals have expanded dramatically, as has the associated problem of aquatic pollution. For example, the heavily polluted Yellow River (Huang He) carried 751 tons of cadmium, mercury, lead, zinc, arsenic, and chromium into the Bohai Sea in 1989 (Platt 1995). The majority of these compounds settle to sediment surfaces, where they concentrate in crustacea and other bottom-feeding animals. Other dramatic examples of heavy metal poisoning come from heavily industrialized sections of Eastern and Central Europe. Table 17.1 lists exemplary concentrations of selected metals in unpolluted and polluted waters.

Incidents such as Jintsu River cadmium contamination, which resulted in more than 100 deaths from Itai-Itai disease, mercury poisoning of Minamata Bay, and similar episodes in Sweden and Canada provide examples of indirect

Table 17.1 Natural sea water contains lower concentrations of most metals than does typical fresh waters. Coastal areas subject to significant anthropogenic impact often have very high concentrations of metals in sediments. (Data from Laws 1993)

Location	Mercury	Arsenic	Copper	Zinc	Lead
Sediment					
Oahu					
Pearl Harbor					
Upper West Loch	0.25	11	57	80	50
Ala Wai Boat Harbor	58	0.16	0.007	0.0014	0.005
Australia					
Halifax Bay			7.0	31	16
Cleveland Bay			6.2	30	15
Water					
Average river water	0.7	1.7	1.5	30	0.1
Average sea water	0.001	1.5	0.1	0.1	0.003

poisoning of humans, that is, poisoning in which contaminants in water become concentrated in foodstuffs. Acute poisoning of humans by metal-contaminated drinking water is rare because of the large doses required. The primary health concern posed by metals is either indirect poisoning, particularly through formation of organic metal complexes in foodstuffs, or long-term chronic effects (as is the case for lead) (World Resources 1992).

Many metals that are not particularly toxic to humans are, nonetheless, highly toxic to aquatic life. Among these substances are copper, silver, selenium, zinc, and chromium. The toxicity of metals to aquatic plants and animals is controlled by their oxidation state and their ability to form particular complexes in the environment, which is largely determined by prevailing water chemistry (e.g., pH, salinity, temperature).

Lead remains the most prevalent heavy metal contaminant on a world-wide basis. Present-day levels of lead in industrialized countries have been estimated to be two to three orders of magnitude higher than those of the preindustrial age. While no other metal has the same prevalence as lead, several are widespread in aquatic environments. For example, levels of cadmium have increased, aluminum has been mobilized on a regional scale in soils and water affected by acid precipitation, and mercury contamination has emerged as a problem of regional significance. From a human health perspective, the metals of greatest concern are lead, mercury, arsenic, and cadmium.

Sources of heavy metals include domestic effluent (the largest source), geologic weathering, ore and mineral mining and processing, use of metals and metal components, leaching of metals from garbage, solid waste dumps, and animal and human excretions, mobilization through acidification of waters, and excessive irrigation.

Lead

Lead is one of the most toxic metals found in aquatic systems. It is particularly toxic to young children, posing hazards that include kidney damage, metabolic interference, central and peripheral nervous system toxicity, and depressed biosynthesis of protein, nerve, and red-blood-cell formation. Although lead does not biomagnify in the food web, it can affect early life stages, particularly by causing developmental deformities. The detrimental effects of lead (Pb) appear to decrease with increasing alkalinity; health requirements indicate that concentrations of Pb in water should not exceed 5 μg L^{-1} at alkalinities of 20 mg L^{-1} and 25 μg L^{-1} at alkalinities of approximately 80 mg L^{-1}.

Mercury

Mercury is also highly toxic to humans and unique among the metals because it is consistently biomagnified within the aquatic food chain. In its inorganic form, mercury may cause kidney damage and ulceration (Nash 1993). In surface waters, however, mercury is frequently converted to its more toxic organic form, methyl mercury. Ingestion of methyl mercury affects the central nervous system and can cause death even at relatively low doses. In fish, both diet and age influence the relative concentration of mercury. Trout in Lake Ontario, for example, have mean Hg concentrations of about 0.13 μg g^{-1}, a biomagnification factor of up to 100 times. While levels of Hg in whole fish higher than 0.5 μg g^{-1} exceed consumption guidelines for humans, little field evidence has been gathered to show that fish health has been significantly impaired by this contaminant. In particular, walleye, with recorded mercury concentrations up to 20 μg g^{-1} (among the highest ever recorded in fresh water fish), exhibited no apparent effects from these high concentrations. Severe health impacts can occur in humans dependent on these fish in their diet.

Cadmium

Cadmium is used for a wide variety of industrial activities: electroplating, batteries, pigments, as a stabilizer for plastics, and even in some pesticides, phosphate fertilizers, smoke detectors and luminescent dials. It is also an impurity in fossil fuels and in zinc compounds, making it an element that is both ubiquitous and difficult to control. Worldwide production has climbed from nearly nothing in 1930 to more than 20,000 metric tonnes in 1992 (Laws 1993).

Cadmium pollution of waters generally results from industrial discharge, seepage from settling ponds, and volatilization of cadmium from fertilized agricultural lands. Effects can be locally severe, as cadmium has a long half-life and accumulates for decades in kidney and liver tissues. Severe cadmium poisoning affects bone strength. The most infamous case of cadmium poisoning occurred in the 1950s in Japan (Itai-itai disease) in an area heavily impacted by a mining operation that produced zinc, lead, and cadmium.

Other Metals

Copper uptake occurs largely at the gill surface. Small fish, therefore, have a greater potential for copper uptake per unit weight of body than larger fish. Copper in mussel tissue, for example, decreases as fish size increases. Copper can have extensive sublethal effects, and safe concentrations in water have been set at 5 μg L^{-1} for aquatic life.

Other metals that pose chronic health threats include aluminum, which may be a confounding factor in Alzheimer's disease; chromium, which is associated with dermatitis, pulmonary congestion, and nephritis; and organotin complexes that are neurotoxic. High aluminum concentrations also interfere with many metabolic processes in fish, and the mobilization of this metal in acidic waters is the primary mechanism for fish mortality in many acidified lakes and rivers.

Toxic Chemicals

The category of toxic chemicals includes the spiraling numbers of pesticides and herbicides used globally to improve agricultural productivity and urban aesthetics. Species are variably sensitive to toxic chemicals, but effects usually include reduced growth and survivability of fish eggs and young as well as various fish diseases. In humans, toxins may increase the risk of rectal, bladder, and colon cancer. Toxins may also reduce the availability and healthfulness of fish, shellfish, and associated species (Nash 1993). Toxins such as DDT have been responsible for massive fish kills and the near-extinction of large, water-dependent predatory birds such as the bald eagle and osprey.

Like heavy metals, synthetic organic compounds such as PCBs and certain pesticides concentrate in higher levels of the food chain. While stringent health regulations minimize human exposure, many forms of aquatic biota cannot avoid long-term exposure to low levels of organic contaminants and may concentrate such materials as a result of bioaccumulation. Reproductive failures and deformities have been reported in a wide range of aquatic fauna. For example, toxic effects are thought to be the primary cause of recent ecological changes in the Lake Ontario basin.

The effects of organic contaminants probably are more prevalent and more significant than the effects of metal contamination. Effects of toxics on ecosystem components include the following:

- Low concentrations of PCBs causing possible suppression of photosynthesis in the smallest sizes of phytoplankton.
- Sensitivity of benthic invertebrates to contaminants, including deformities and population decrease.
- Increases in fish tumors, particularly in benthic feeding fish; possible elimination of sensitive fish such as the slimy sculpin.
- Decreased viability of embryo and larval fish stages (especially of Piscivores) and damaged swim bladder function and inflation (particularly from PCBs).
- Deformities in turtles, dead and deformed embryos of snapping turtles, susceptibility of mink and otter to PCBs and dioxin (leading to reproductive failure), and impaired reproduction by several species of fish-eating birds (gulls, terns, double-breasted cormorants).

Sources of toxic compounds include urban and agricultural run-off, municipal and industrial discharges, and landfill leachate.

Acids

Acidification is a severe water quality problem in some regions. Reduced water pH and reduced alkalinity lead to elimination of acid-sensitive aquatic organisms; reduced species diversity and increased predominance by a few species; a shortened food chain; increased water clarity; release of trace metals from soils, rocks, and metal surfaces such as water pipes; reduced availability of fish, shellfish, and associated species; increased aluminum concentrations; and increased risk of mobilization of lead and other heavy metals from drinking water systems. Acidification is discussed in more detail in chapter 22.

Sources of acids include atmospheric deposition, road run-off, industrial discharges, sludge and discharges from sewage treatment plants, creation of reservoirs, acidic mine effluent, and degradation of plant material.

Salinization

Salinization of fresh water is a common result of many human activities. At high levels, salinity is toxic to fresh water life. It reduces the availability of drinking water supplies, soil and agricultural field productivity, and availability of fish, shellfish, and associated species. In coastal areas, salinization is often generated by overextraction of groundwater; in agricultural areas, it results from heavily irrigated crops and the concentration of salts as the water evaporates; in industrial areas, it can be produced by waste discharges. Other significant sources of salinization are leaching from oil fields, underground injection of brines, and road de-icing salts. See chapter 19 (agriculture) and 20 (urbanization) for more information.

Changes in Water Volume

Humans rarely leave water flowing in a natural state. From water withdrawal to channelization, dams, and creation of impervious surfaces, the history of humans is largely a history of changing water courses. Slowing and reducing water flow through damming and other flow barriers creates serious water quality impacts, even as it reduces aeration, potentially increases stream temperature, and concentrates pollutants in a smaller volume of water. Similarly, increases in water flow have resulted from straightening water courses, reducing wetland area and creating impervious urban surfaces that channel rain into receiving waters (see chapter 20).

Other impacts on water flow may come through water withdrawal. Human water use has increased more than 35-fold over the past three centuries, an increase that far exceeds the accompanying increase in world population. In 1850, for example, world population stood at nearly 1200 million people; by the year 2000, it had reached an estimated 6000 million. At the same time, per capita water availability decreased from 33,300 m³ per year in 1850 to 8500 m³ per year today (Shiklomonov 1993). In recent decades, water withdrawals have been increasing about 4% to 8% per year, with most of that increase occurring in the developing world. Water use is stabilizing in industrialized countries, where the rate of the increase of withdrawals is expected to decline to 2% to 3% annually in the 1990s. Annual per capita water use varies widely, from 1692 m^3 in North America to 726 m^3 in Europe, 526 m^3 in Asia, 476 m^3 in South America, and 244 m^3 in Africa (Gleick 1993).

While withdrawals are typically considered an issue of quantity rather than quality, the two are tightly intertwined, with withdrawals increasing pollutant concentrations in remaining water and prompting society to consider more distant sources for adequate water. Withdrawals may be from either groundwater, surface waters, or both, depending upon local resources. Land subsidence is a problem peculiar to groundwater withdrawals, with major cities such as Mexico City and Venice sinking several centimeters per year as they deplete their groundwater. Depending upon the supply system used, water quality impacts may result from source water contamination or from changes to intermediate areas' distribution systems (e.g., through introduction of exotics, temperature increases, pollution transport, water seepage, and salinization).

Synergistic Effects

Unless severe, a single impact will often not cause adverse long-term effects on water quality. For example, coral reef communities can recover from short-term increases in sediment load, whether it originates from urban run-off or from forestry activities. A multiplicity of impacts commonly causes too many stresses on a biological community, however, with synergistic effects being the most problematic. Consequently, coral reefs near urban areas may also be impacted by increased fresh water flushed to sea by impervious urban surfaces. As illustrated in Kanehoe Bay, Hawaii, combined stresses of reduced salinity and heavy siltation can have enormous impacts, although, by itself, neither impact would be lethal (Laws 1993).

A great deal of water quality management relates to managing the combined effects of two or more stresses. Documenting toxicological synergistic effects is difficult, however, and many effects have only site-specific value. Consequently, most water quality criteria and standards reflect maximum permissible concentrations of single pollutants (Table 17.2; chapter 5). It is the responsibility of the astute manager to consider and test for synergistic effects.

Table 17.2 Comparison of water quality standards for different compounds discussed in this chapter. (Data from Gleick 1993)

Metal	Maximum Permissible Concentration (μg/L)
Mercury	0.144
Lead	5
Cadmium	10
Selenium	10
Thallium	13
Nickel	13.4
Silver	50
Manganese	50
Chromium	50
Iron	300
Barium	1000

Water Quality Monitoring and Assessment

Each of the impact chapters that follows (chapters 18–24) evaluates how various management practices affect water quality. Clearly, water quality monitoring and assessment is a central element of proper management. Society uses legal and economic policies and incentives to ensure that certain practices are conducted or to ensure that practices are implemented in certain ways. Managers use "compliance monitoring" to determine whether a certain practice is being used properly. Water quality monitoring allows the manager to address two questions:

1. When is (are) a management practice (or series of practices) considered in place, and is the resource condition such that one or more designated uses are protected?
2. Are the uses designated for a water body the appropriate ones, in light of costs and benefits?

The first question deals with issues such as proper implementation of a selected management practice, cumulative effects of a series of practices on a water body, and unanticipated impacts from one or more practices. The second question really asks if the goals established for a water body are appropriate. In some cases, management practices may be properly implemented and impacts contained within anticipated bounds, but the desired uses are not being met. Monitoring programs help to answer those

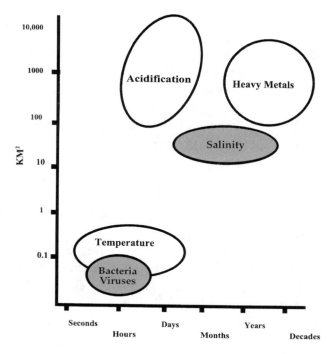

Fig. 17.6 Spatial and temporal patterns of five water quality parameters. Temperature and bacteria are more limited in their spatial scale and temporal endurance, but their effects can be dramatic within those spans of time and space. Acidification is placed in a midrange for temporal scales because many ecosystems will recover toward previous conditions when the stress is removed. In contrast, heavy metals can persist in the sediments for decades.

two central questions and to form strategies for changing either practices or goals.

As described elsewhere, and illustrated in Fig. 17.6, all natural resource management questions are scale-dependent. When examining a natural resource or water quality management question, it is critical to understand the spatial and temporal (and ecological) scale of the problem and to ensure that any resource management strategy is presented at the appropriate scales. Therefore, in addressing water quality monitoring issues for any given class of management practice (e.g., forestry or agriculture), we must understand the variables commonly affected by the practice, the ecological impacts of any changes, and the spatial and temporal scales upon which those changes occur. Those elements then define our monitoring design and the analyses of our monitoring results. In each of the impact chapters, we address these questions in a separate monitoring section to provide guidance about understanding and measuring effects of different water quality management practices.

Summary

- Water quality impacts from land uses are highly site-specific in detail, but can be generalized to 10 major classes of expected change. Those classes are: changes in suspended sediment load, increased organic matter and biological–chemical oxygen demand, bacteria and viruses, nutrient loads, temperature, heavy metals, toxics such as pesticides and herbicides, acidification and other changes in pH, salinization, and changes in the hydrograph.
- Several of these impacts act as physical agents affecting stream organisms. For example, heat denatures proteins in plants and animals, while sediment smothers or abrades living organisms.
- Other impacts such as toxics have chemical–physical effects on the physiology of aquatic organisms.
- A third class of impact is indirect, in which stressors affect one trophic level directly (e.g., reducing bacterial and fungal populations, thereby reducing decomposition rates) and affect upper trophic levels indirectly (e.g., by reducing food availability throughout the detrital food chain).
- The most significant effects of land use on water quality, and the ones where our management ability is least well developed, are cumulative effects and synergistic interactions among various stressors.

For Further Reading

Gleick P. *Water in crisis: a guide to the world's freshwater resources.* Oxford: Oxford University Press, 1993.

Laws E. *Aquatic pollution: an introductory text,* 2nd ed. New York: Wiley and Sons, 1993.

Mason C. *Biology of freshwater pollution,* 2nd ed. London: Longman Scientific and Technical, 1991.

Novotny V, Olem H. *Water quality: prevention, identification and management of diffuse pollution.* New York: Van Nostrand Reinhold, 1994.

18

Management of Water Quality in a Forested Landscape

Overview

Nearly every forest management activity has the potential to have a significant impact on local water quality. By their very nature, forestry practices alter soil and vegetative features that play an important part in regulating water quality. Forestry activities have been linked to changes in water yield, ground water recharge, stream temperature, run-off, sediment loads, nutrient balance, and pollution from pesticides and herbicides. In many developing regions, restrictions have been placed on where forests can be cut, but few constraints have been instituted on how that cutting takes place. Consequently, the end-results can be quite severe. In most developed regions, awareness of impacts has led to development of "best management practices," or activity guidelines that reduce the impact of selected activities on the aquatic environment. For example, buffer strips along stream banks help keep stream temperatures constant, reduce sediment loading and erosion, and are vital in aiding establishment of new vegetation.

The water quality risk that forest management poses is difficult to judge, because it depends on the nature of the landscape (steep watersheds are most sensitive to disturbance), the intensity of forestry activities, postharvest land use, and inherent characteristics of the soil and water. Generally, well-managed forestry activities may result in temporary changes in stream flow and dissolved nutrient levels but these changes will remain within normal bounds for an aquatic ecosystem. Consequently, communities can completely recover from such stresses. In contrast, slow ecosystem recovery is most likely to be associated with chronic forest harvest occurring over large areas and, on rare occasions, with lethal doses of short-lived pesticides or chronic doses of persistent toxicants.

Introduction

The connection between forestry activities and changes in run-off and sedimentation has been recognized for thousands of years. Writings from ancient Greece bemoan the denuded lands and massive erosion that resulted from widespread deforestation. These observations remained unquantified, however, until this century. During the last 50 years, controlled studies have documented a wide range of effects from large-scale deforestation. More encouraging is the progress that has been made in documenting ways in which many of those effects can be minimized. Today, forestry activities with the potential to have large-scale impacts usually bring only minor effects in most developed regions. For example, according to 1992 estimates, forestry practices have a negative but minor impact on 7% of the assessed U.S. waterways (less than 1% of the assessed waterways are seriously impacted by forestry activities) (U.S. EPA 1992).

Frequently, people associate the term forestry with logging activities only. Throughout history and in many parts of the world, unregulated logging, with its attendant environmental consequences, has dominated forestry. Managed forestry actually encompasses a much broader territory. Throughout the life of a forest, the vegetation, soil, and water flow change continuously. Forest managers both respond to and direct these changes by controlling species composition, density, age distribution, and growth of understory vegetation. The myriad ways in which the vegetation at a site can be manipulated create almost endless variations in the degree of positive and negative environmental and water quality impacts.

Some of the typical water quality impacts from forestry activities (managed or unmanaged) include changes in hydrology, sedimentation and erosion, increased stream temperature, and altered site nutrient status. Each of these conditions, in turn, affects in-stream community composition and function. As with all natural resource management, however, societal goals and needs determine whether these effects are seen as negative or positive. If a logged catchment resulted in increased water yield and availability (for human use) in an arid area, it would be construed as a positive impact. In fact, logging is sometimes used intentionally to increase water yields. In other cases, the increase of yield following logging may create problems with flooding or a raised water table (with increased likelihood for contamination). In either case, the manager's most important tool is the ability to predict and control likely impacts.

A Silvicultural Overview

Forest stands are managed to meet a wide variety of goals. Timber production is the most visible of these goals, but others include the maintenance of landscape biodiversity, the preservation of wildlife habitat, recreational uses such as hiking and hunting, and any combination of recreation, timber revenue, aesthetic enjoyment, and landscape values. More recently, forest management has been used to complement water conservation efforts—e.g., cutting vegetation to increase yield in arid areas and planting forests to reduce water yield in vulnerable areas. Although less common, forest management has also been used to improve water quality (e.g., in agricultural catchments). More frequently, changes in water quality represent inadvertent side effects of other silvicultural goals. (Silviculture is the practice of managing forest composition, growth, and productivity.)

From a water quality perspective, the most important silvicultural activities are the harvesting regimes, the creation and design of access roads, and the use of herbicides and pesticides to control growth. Recent evidence suggests that afforestation efforts also belong in this list: afforestation may have unexpectedly large consequences for in-stream chemistry and water quality. Such effects are traceable not only to logging but also to an array of forestry activities that influence soil, water flow, soil and water chemistry, and vegetation. In this equation, topography and soil characteristics define an area's sensitivity to water quality degradation, while the choice of machinery, harvesting design (including road construction and the use of herbicides or insecticides), and severity and size of an operation determine the expression of impacts on the landscape. Management difficulties come, not in understanding potential effects of an operation, but in predicting them.

Harvesting may occur in a wide array of intensities and areas. The most dramatic harvesting regime (and the most significant for water quality impacts) is the clear-cut, or removal of all trees in a given area. When trees are removed in a clear-cut, roots no longer retain the soil and the flow of water is no longer slowed or taken up. Consequently, these sites become susceptible to increased flow, erosion, and leaching of nutrients. Recovery to preharvesting conditions will depend on the speed with which new vegetation is established, as well as on the species composition of the new vegetation.

Other harvesting regimes are more selective, ranging from harvesting in small patches to removing only occasional, individual trees. In general, water quality impacts reflect the proportion of trees removed, so selective harvesting results in fewer water quality impacts than does clear-cutting. The spatial area harvested will also affect water quality, as will the proximity of removals to a body of water. Almost more important than the harvesting activities themselves are the design, construction, and maintenance of logging roads, which has been shown to be among the primary causes of water quality degradation in harvested areas (see below).

The species composition of vegetation established after a forest harvest also influences water quality. For example, revegetating a native hardwood forest with conifers can increase stream acidity and aluminum concentrations, affecting macroinvertebrate and fish community structure (Omerod et al. 1993). In another example, Smith (1992) found that stream-side afforestation (with pine) of a pastured catchment increased nitrogen, phophrus, and sediment levels, at least in part because of the loss of stream wetlands. While previous studies have established the basic associations between deforestation and increased yield and sedimentation, the sensitivity of forestry and water quality management is increasing as studies such as these contribute to a more fine-tuned understanding of how forest composition affects water quality.

Deforestation in Developing Countries

The most severe effects of forestry come from unregulated cutting in which the harvested area extends all the way to river banks, covers large areas, and is not replanted. While this model is no longer practiced in most developed countries, unregulated deforestation and its associated severe impacts are still prevalent in many developing countries. The World Resources Institute (1992) reports a quadrupled rate of deforestation in Indonesia since 1970, with an annual loss of 1 million hectares of rainforests. The report also notes that Thailand's forested land area has been reduced by half since 1960, and continues to disappear at a high rate despite a logging ban that went into effect in 1989. In Chile, more than half of the forests designated as protected and three-fourths of those designated for sustained yield have already been cleared. Similar figures are evident from numerous developing countries (Table 18.1).

Despite these dramatic numbers, deforestation-related water quality degradation is difficult to pinpoint in developing countries. Deforestation, by itself is rarely a problem. Instead, deforestation occurs as a symptom of, and its impacts are greatly exacerbated by, population and development pressures. La Plaza, Puerto Rico, exemplifies the complex and interconnected nature of these problems. In La Plaza, forest managers responded to increased urban water demand by deforesting upland catchments to increase water yields. Poor planning, however, has led to heavy siltation of reservoirs and streams, a significant decrease in water storage capacity, and degraded water quality. Simultaneously, urban sprawl is resulting in the paving of previously forested, groundwater recharge

Table 18.1 Deforestation and reforestation rates in various countries

Country	Total Forest Area[1] (000 ha)		Average Annual Deforestation (000 ha)		Average Annual Reforestation (000 ha)
	1980	1990	1980	1990	1981–1985
Africa					
Sudan	146,438	43,266	504	482	13
Zaire	207,312	113,335	370	732	0
Madagascar	20,966	16,092	156	135	12
North America					
United States	288,076	257,000	159	1894[2]	1775
Canada	436,400	435,000[3]	N/A	857[3]	720
Mexico	134,009	48,741	1000	678	22
Central America					
Costa Rica	2041	1104	42	50	0
South America					
Chile	16,917	N/A	50	20	74
Asia					
Bangladesh	1370	1104	8	38	17
Nepal	2480	5103	84	54	4
Thailand	17,089	13,491	235	515	24
Indonesia	160,073	118,299	1000	1212	131
China	155,510	149,000	N/A	N/A	378
Europe					
United Kingdom	2178	2207[2]	N/A	246[4]	40
Bulgaria	3800	3386[2]	N/A	88[4]	50
USSR	929,600	754,958	N/A	22,300[4]	4540
Oceania					
Australia	105,900	145,613[2]	N/A	156[2]	62

[1]All data from FAO 1990 unless otherwise noted.
[2]*The Forest Resources of the Temperate Zones.* Forest Resources Assessment, vol. 1. United Nations 1992; UN-ECE/FAO 1990.
[3]Compendium of Canadian Forestry Statistics. National Forestry Database. Ottawa: Canadian Council of Forest Ministers, 1994.
[4]Data for these countries represent 1980–1990 averages, based on UN-ECE/FAO 1990.

zones, so groundwater levels are dropping and drinking water quantity and quality have reached a critical point (NPR 1995).

Deforestation concerns in other areas typically revolve around the increased flow and possible links with increased downstream flooding. Perhaps the best-known example of this condition is the long-debated relationship between forest clearing in the highlands of Nepal and increased flood regimes in lowland Bangladesh. Historically, Bangladesh's major floods occurred approximately every 50 years (WRI 1993). Since the 1970s, however, that interval has narrowed to every 4 years, and it was commonly acknowledged that upland deforestation was responsible. Recent studies indicate that failed flood control efforts in Bangladesh (channelization), destruction of native mangrove swamps that provided flood control, and habitation of marginal areas may explain more of Bangladesh's plight than upland deforestation. In other areas along the Gangetic plain, which should be equally affected by upland deforestation, flooding has not increased (Hofer 1993).

While the connection with downstream flooding is still a matter of debate, large-scale deforestation is clearly affecting mountain streams through increased sedimentation and stream scouring. The seasonality of intense rainfall in much of Asia and Africa is likely to make these impacts a seasonal, "pulsing" of inputs, rather than a continuous problem. Mild, tropical climates where soils are not adversely affected by deforestation may alleviate the duration of effects through quick regeneration of stabilizing vegetation.

In developing countries, deforestation impacts are often dramatic, but few controlled studies have been conducted in these environments. As a result, conjecture, debate, and trial and error tend to characterize planning in these regions. Until more accurate ecosystem-specific studies become available, the best tool (albeit a poor one) remains extrapolation from studies in other areas. The rest of this chapter will concentrate on temperate forests because most conclusive research has been conducted in these areas and because these studies have demonstrated the effectiveness of simple modifications to forestry practices—applicable globally—that significantly reduce water quality impacts.

Wetland Forestry and Wetland Function

Early awareness of forestry impacts stemmed from dramatic logging efforts in the Pacific Northwest of the United States and in parts of the mountainous New England states. Only recently, however, has a distinction been drawn between the sorts of effects noted in high-slope forestry and those noted in flat and wetland forestry. Important differences between the systems revolve around the slower water flow and wetter soils of forested wetlands. These topographic differences result in less potential for erosion, sedimentation, and high flow, as detailed below, but create a unique set of impacts involving the wetland function itself.

Forested wetlands make up slightly more than half (50.2%) of all wetlands in the conterminous United States. Of these forests, approximately 95% comprise bottomland or other hardwood species that are desirable for forest products (Wal-bridge and Lockaby 1994). In addition to logging these existing forests, many wet-lands are being used as sites for forest plantation establishment. In North Carolina, for example, forest plantation establishment has altered 52% of the state's wet-lands (while agricultural conversion has resulted in a 42% loss) (Richardson 1994). One of the most immediate effects of plantation establishment is the re-duction of flow within a wetland (Richardson noted a 14% reduction). Forestry activities may also reduce flow through draining of the wetlands to remove vege-tation. The impacts of reduced flow can include the loss of flood control and flood storage functions, loss or lessening of sediment and erosion control functions, and modifications in a wetland's ability to "cleanse" nutrients and pollutants.

Currently, the most common harvesting method in forested wetlands is clear-cut followed by natural regeneration. Removing the felled trees leads to impor-tant silvicultural choices. Wet sites may be drained and roads constructed, trees may be removed without drainage, or trees may be removed by air (e.g., heli-copter). Aerial removal, the most expensive choice, carries fewer side effects for soil compaction and leads to faster natural regeneration and fewer alterations in the composition of the regenerated community. In contrast, removal of trees by ground machinery (skidders, feller-bunchers, tractors) compacts the soil, thus decreasing hydraulic conductivity and oxidation–reduction potential. Walbridge and Lockaby (1994) have indicated how much is unknown about the long-term impacts of these sorts of changes on water quality and the wetlands:

> The direct effect of these changes on N and P retention have not been exam-
> ined. Lower redox potentials should favor increased rates of denitrification,
> given sufficient amounts of labile carbon and available nitrate. Increased soil
> bulk density and reduced hydraulic conductivity might reduce phosphate ad-
> sorption by soil minerals by reducing infiltration, although adsorption effi-
> ciency could increase due to increased contact time between water and
> sediments. [p. 14]

Draining a wetland for forestry and clearing the vegetation can lead to a significant increase in decomposition rate, which results in a loss of a wetland's ability to accumulate organic matter (Fig. 18.1). Partial drainage can double the decomposition rate of some wetlands; more complete drainage can raise the decomposition rate by up to five times. Each increase translates into a loss of wetland function and productivity: drained wetlands may accumulate less than 0.25 mm/yr of peat whereas, worldwide, peat accretion rates average 1–2 mm/yr and some wetlands with an extended hydroperiod may add as much as

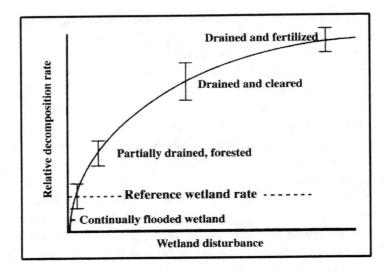

Fig. 18.1 Forestry activities in forested wetlands typically include draining the wetland and may include clearing and fertilizing the forested area. Each of these activities can increase wetland decomposition rates and result in a loss of certain vital ecosystem functions. (Redrawn from Richardson 1994)

4–7 mm/yr. Notably, the U.S. Clean Water Act requirements and evaluations required by each state do not include measures of wetland function (e.g., biogeochemical cycles or decomposition rates); thus, function has no legal standing.

Traditional concerns such as sedimentation and water yield are not significant issues in water quality impacts from wetland forestry. Studies of wetland forestry are only just beginning to indicate that other long-term concerns related to wetland function may have important implications for the affected wetland resource and its associated "filtered" water.

Upland Forestry: Impacts on Water Quality

Forest practices occur on a template of topographic conditions. For wetland forests, that template is hydroperiod, with effects including changes in decomposition and nutrient cycling. In upland forests, however, the most important template is slope and soil erodibility. Consequently, the greatest areas of concern in the impacts of upland forestry on water quality stem from erosion potential, siltation pressure, and changes in overland flow.

Changes in forest composition affect the amount and velocity of water draining a watershed, impact of water on the soil (e.g., sedimentation and erosion rates), and the ability of light to penetrate to the soil, streams, or lakes in the watershed. Additionally, many forestry activities are accompanied by applications of insecticides and herbicides. These toxins may affect water quality directly or through prolonging periods of high flow and sedimentation. The more frequently observed effects are discussed below.

Changes in Hydrology

Forestry activities may cause a series of changes in the hydrology of a watershed, each of which has associated implications for water quality. Trees take up enormous quantities of water, so removal of trees during logging operations can result in significant changes in site water status. Specifically, tree removal generally results in an elevated water table and decreased water infiltration capacity and a resultant increase in surface discharge. Each of these effects varies enormously with individual site characteristics, logging intensity, and tree removal systems.

The more dramatic effects of forestry on hydroperiod involve short-term water yield. Timber harvest, mechanical site preparation, and herbicide applications can all increase water yields by reducing the vegetational water use on a site. Effects on stream flow are largely related to the amount of basal area removed, although an effect is not usually detectable if less than approximately 20% of the basal area is removed. Clear-cuts or other large-scale removals of basal area can result in as much as a 30% to 40% increase in water flow. Fig. 18.2 illustrates how variable hydroperiod effects can be, depending on harvest frequency, intensity, topography and affected species.

In addition to the degree of harvesting, the configuration and timing of the cut influence short-term water yield. For example, cutting a single block—especially one on the lowest portion of the catchment—will result in large increases in water yield, while cutting a series of strips or harvesting individual trees will produce smaller increases (Hornbeck *et al.* 1993).

While the short-term changes in water yield described above can be dramatic, long-term influence on stream flow is generally mild, approaching the predisturbance level in 3 to 5 years as vegetation regrows. Elevated flows are common after this period, but usually remain within the ranges associated with normal variation. If vegetative regrowth is controlled by herbicide applications, or further silvicultural activities (such as thinning) take place, the return to preharvesting flow may take decades. Soil compaction during harvesting may also prolong increased annual yields.

The species composition of regrowth also has long-term effects on water yield and nutrient dynamics of a watershed. For example, in the Hubbard Brook experiments, pin cherry and birch replaced the former birch, beech, and maple

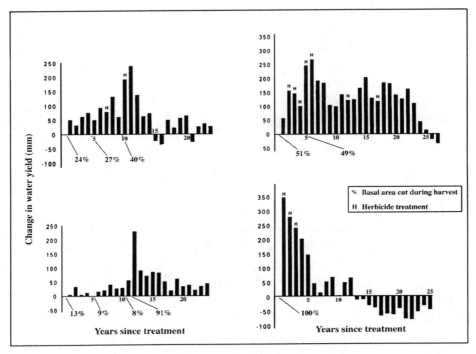

Fig. 18.2 Examples of increased water yield following different harvesting regimes. Percentages denote the amount of basal area cut during harvest treatments; H denotes herbicide application to cut portions of catchments. Data are from Fernow Experimental Forest, Hubbard Brook, and Leading Ridge Watershed Research Unit. (Excerpted and modified from Hornbeck *et al.* 1993)

forest. Both replacement species have greater transpiration rates than the previous ones and caused decreases in water yield relative to preharvest conditions. A similar effect is noted following conversion of a hardwood forest to conifers. In addition, areas with faster regrowth will experience a more rapid return to preharvest levels than will more slowly regenerating forests such as some high-montane western forests.

Siltation/Sedimentation

Sedimentation and associated increases in turbidity can be among the most dramatic and harmful water quality impacts from forestry activities. Both natural sedimentation rates and responses to forest harvest are extremely variable, however, making generalizations difficult. Demonstrated sediment increases range from 10% to more than 1000% for clear-cut forests (Binkley and Brown 1993). When variable degrees of harvest and postharvest vegetative control are factored in, sediment rate increases of more than 40-fold have been recorded; when significant road construction is added to the equation, sediment yield may increase even further. Sediment production varies dramatically depending on slope, soil properties, and harvesting intensity.

Except on steep slopes, sediment yields from undisturbed forests and plantations are low, generating in-stream sediment concentrations generally below 10 mg/L. (Average annual concentrations in excess of 20 mg/L are not, however, uncommon.) Sediment yields from thinned or selectively harvested forests are similarly low. Sediment production increases as site disturbance increases, but is a relatively short-term phenomenon that decreases as vegetation regrows and annual water yield diminishes.

One of the most important elements in sediment production is not the harvest itself, but the construction and maintenance of access roads. In montane forests, harvests facilitated by the construction of roads may produce more than 30 times the sediment of those harvested by helicopter; sediment production from forest roads has been measured at 50 to 90 tons ha/yr (Grayson et al. 1993). Typically, road use continues long after the forest has regenerated, in particular for recreation, and can extend the period of increased sedimentation indefinitely.

Related to the problem of forest roads and harvesting is the persistent localized impact on streams from stream crossings and increased landslide risk. The greatest potential for landslides occurs 3 to 10 years after harvest, when roots of harvested trees have lost their holding potential but new vegetation is not firmly established. By preventing rooting of new vegetation, herbicide application prolongs the period of landslide risk.

Temperature

The temperature of water in forest streams controls many factors, including species distributions and concentrations of dissolved oxygen. As early as 1958, a major study in western Oregon (Alsea Watershed Study) documented effects of timber harvesting on the temperature of small, forest streams and contributed to increased awareness of the effects of logging. Because harvesting and thinning involve the removal of vegetation, previously shaded forest streams may be exposed to increased direct sunlight. Ensuing temperature increases and fluctuations can prove lethal to some stream flora and fauna. Increased stream temperature may also generate increased autochthonous production and increased fish biomass. For example, Bisson and Sedell (1984; from Binkley and Brown 1993) noted a 50% increase in total fish biomass following the removal of forest stream canopies in Oregon and Washington. To some species, tem-

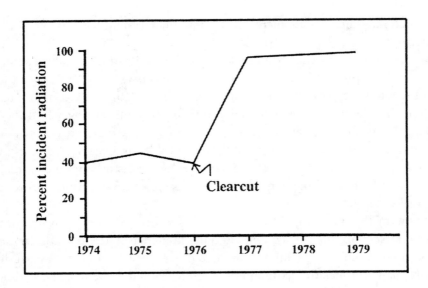

Fig. 18.3a Clear-cuts that do not leave a buffer strip around a stream dramatically increase the amount of light reaching that stream. This situation is illustrated in the Carnation Creek watershed that was clear-cut in 1976. (Adapted from Shortread and Stockner 1982)

perature increases reflect degraded water quality; to others, it would clearly be an improvement.

Clear-cuts and clearings can lead to a 50- to 1500-fold increase in radiant energy reaching the water surface and can raise stream temperature by 3 to 15°C, depending on the size of the stream and the intensity of the harvest (Fig. 18.3). Even more relevant are seasonal effects. Studies in deciduous forests of the eastern United States have found relatively small changes (no more than +/− 2°) in winter stream temperatures due to forest harvesting. Effects are heightened during summer, however, when solar radiation is more intense and more direct; increases range from 3 to 10°C in eastern deciduous forests and from 1 to 8°C in the Pacific Northwest.

Woody Debris

It is easy for branches and tree trunks cut during a harvest to end up in a stream, but less simple to determine whether these inputs degrade or im-prove water quality. The impacts of this debris vary with the amount of slash and size of the stream. In some cases, the added debris acts as a temporary sediment trap that inevitably fails, flooding the downstream reaches and

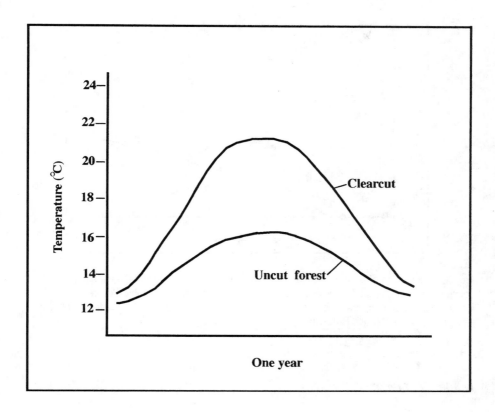

Fig. 18.3b A direct result of increased light penetration is increased stream temperatures in streams running through clear-cut areas. Illustrated here is a generalized mean monthly maximum temperature in uncut and clear-cut areas. Daily fluctuations (not illustrated) are also increased, and exhibit a sharp mid-day peak in clear-cut areas. Values are typical of West Coast ecosystems.

scouring that portion of the stream bed. Unstable debris dams can also restrict fish migration. Other links have been found between large debris accumulations and depressions in dissolved oxygen; Binkley and Brown (1993) report several studies that recorded dissolved oxygen concentrations as low as 0 mg/L because of logging debris.

Nonetheless, natural wood accumulations play a vital role in stream ecology, in some cases reducing channel scour (by slowing stream flow), reducing macroinvertebrate drift, supplying refugia, and increasing primary and secondary productivity. In general, it appears that natural wood accumulations should remain in the stream, while artificial increases due to forest harvest will usually prove detrimental to water quality.

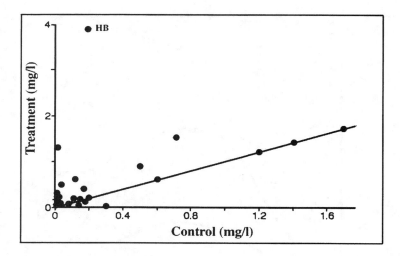

Fig. 18.4 Response of stream water annual nitrate-N concentration to forest harvesting. Data points represent different experimental watersheds; HB represents the highly elevated nitrate-N found at the Hubbard Brook experiment and exists as a notable outlier compared to other studies. (Modified from Binkley and Brown 1993)

Nutrient Alterations

Forestry operations may result in a variety of impacts on nutrient dynamics of streams and associated lakes, but these impacts tend to be minor and short-term. For example, changes in phosphate concentration have been demonstrated only in isolated cases such as high-intensity slash burns. Increases in nitrate–N concentrations after harvest are more common, but only one study (Hubbard Brook; Fig. 18.4) has demonstrated large-scale postharvest increases in nitrate. As before, experimental forests have shown tremendous variations in nutrient impacts. Harvesting severity appears to play a role, as does the slope, choice of tree-removal system, choice of site preparation technique (e.g., mechanical versus chemical site preparation), forest type, and characteristics of replacement vegetation. In general, nitrate–N levels from streams in central hardwood and conifer forests appear to be relatively unresponsive to forestry activities, but northern hardwood forests show an increase if more than 70% of the watershed is harvested (Binkley and Brown 1993). Except in isolated storm pulses, increases in nitrate typically remain well below the drinking water standard of 10 mg/L, and most are less than 0.5 mg/L. Increased nutrient levels frequently return to near-base conditions after several years and the regrowth of native or forest vegetation.

It is significant, however, that fertilization raises nitrate–N levels and that in several experiments these levels have exceeded drinking water standards. Again, forest type appears to be an important factor regulating response. In the nitrogen-poor forests of the Pacific Northwest, fertilization rates increase but generally remain below 5 mg/L. Eastern forests appear to be more susceptible to large increases, with several experiments in Pennsylvania's Fernow experimental forests showing increases well above 10 mg/L.

Pesticide and Herbicide Residues

Although pesticide contamination of ground and surface waters is an important national concern in the United States, agricultural inputs remain the primary source of these organochlorines. Forestry herbicide use represents only 10% of agricultural use. The EPA's 1992 water quality report does not even list silvicultural activities as a factor in groundwater contamination. Use of herbicides and insecticides in forestry present such a low pollution risk because each block of treated land tends to be relatively small (2–4 hectares is a typical size of a southern forested wetland) and the chemicals are applied only rarely in a typical rotation (once or twice in rotations of 25 to 75 years). In addition, application rates are generally low (< 2 kg/ha) in forestry activities. According to Neary et al. (1993), the greatest potential hazard to groundwater comes not from regular forestry operations but from spills that occur during the transport, storage, application, or cleanup of chemicals.

Regional, confined groundwater aquifers are not likely to be affected by normal use of silvicultural herbicides. Surface, unconfined aquifers in the immediate vicinity of herbicide application zones have the most potential for some contamination. Residues from pellet applications of chemicals such as picloram and haxazinone become apparent in groundwater within days to months following a forestry application, and may persist for upward of a year. Concentrations above acceptable standards are rare, however, and generally result from improper application.

Indirect effects from use of pesticides in forested watersheds are of mixed benefit. While some studies have documented a temporary increase in nitrate (nitrogen) losses, others have documented a reduction in sedimentation. Because sedimentation is the largest concern in water quality effects from forestry management, this reduction constitutes an important trade-off if other contamination risks are low. Other indirect effects include temporal changes in terrestrial invertebrate abundance, reduced plant diversity, and changes in particulate organic matter transport in streams.

Streamside Management: Goals and Tools

Riparian zones are natural areas of erosion, sedimentation, and continuously changing stream position and flow. Erosion, siltation, and stream channel movement cannot be, nor should they be, eliminated. Stream and lake ecosystems depend, in part, on fluctuations in water level and influx of new sediment. The goal in forest management is to make the inevitable changes better anticipated and more controlled.

Three primary strategies have been developed to control the vast majority of silvicultural impacts on water quality: (1) use of buffer zones; (2) restrictions on slope for roads and skid trails; and (3) maintenance guidelines to ensure that long-term impacts are minimized. Strategies are implemented through forestry "best management practices" (BMPs), as described below, and detailed in the Minnesota GEIS case study given later in this chapter.

Buffer Strips

Buffer strips (also known as buffer zones) are an elegant concept in forestry. Simple, yet highly effective, a buffer zone is a strip of unharvested, untreated vegetation left along stream banks, roads, or other sensitive areas. Buffer strips offer a wide array of benefits, including reducing siltation and stream bank erosion, reducing peak flows, reducing input of herbicides and pesticides into the streams and lakes, and providing sufficient shade so that stream or river biota are not affected by temperature and light increases. Over-reliance on buffer strips will not address important water quality problems, however. The effectiveness of buffer strips is still dependent on a well-planned harvest that results in slow, uniform water run-off (Barling and Moore 1994). Further, Barling and Moore (1994) note that buffer strips are better filters of sediment than of nutrients, so buffer strips should be a secondary measure after reducing the source of pollution. In addition, Omerod et al. (1993) demonstrate that the choice of species in the buffer strip is important, especially if a plantation consists of non-native species.

Slope Restrictions

One approach used to regulate both water flow and erosion/sedimentation is to reduce the gradient of roads and skid trails. Typical guidelines state that skid trails should be at no more than a 20% grade over no more than a 300-

foot distance, and roads should not exceed a 10% grade for more than 300 feet. In addition, most regulations call for log landings to be on level slopes and stable ground.

Maintenance Guidelines

Several (e.g., 3 to 10) years are required for regrowth of ground cover vegetation that will effectively capture sediment and stabilize soil. Consequently, roads and skid trails must be constructed such that they do not increase the potential for water quality effects in the interim between harvesting (or other practices) and regrowth of vegetation. Regulations commonly specify that ruts must be filled to reduce gullying, that culverts are maintained as a permanent feature of roads, and that waterbars be constructed and maintained to reduce erosion potential. Regulations may also call for reseeding of roads with clover or other low ground cover.

Monitoring Considerations

Monitoring is an extremely important part of ensuring that best managment practices are used in an effective manner. Intensive watershed-scale monitoring of stream quality is cost-intensive, with estimates running as high as $100,000 to assess silvicultural impacts on water quality in a watershed. As a result, many states are turning to indirect measures of silvicultural impacts. These include state agency resource commitments, the number of public complaints and enforcement actions, user recognition of and attitudes toward BMPs, water quality and soil erosion surveys, and computer modeling.

Variables Impacted

Monitoring evaluates the most commonly impacted variables: sediment, water volume, energy, and organic matter. Sediment is delivered from the disturbed landscape to the stream or lake. Smaller particles are held in suspension in moving water and are assessed as mg/L or kg/day suspended sediment. Heavier particles are deposited, then moved downstream in extreme hydrologic events. These larger particles represent a major, but rarely quantified component of the sediment load, called bed load. Disturbance to the landscape allows water to be held on the land less effectively and to reach the stream channel more quickly. Thus, a stream in a harvested landscape will

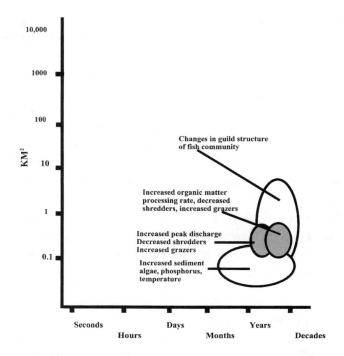

Fig. 18.5 Spatial and temporal effects of forestry activities on aquatic communities and water quality characteristics. Steeper slopes will tend to increase the spatial range of effect and higher mean air temperature will tend to decrease the temporal impact.

be more "flashy." That is, peak flow (maximum m³/s) will occur more quickly after a storm event or during snow melt, and the peak will be higher but will subside more quickly than in "natural" landscapes. Total discharge (m³/ha/yr) is most often increased by forest harvest, but results usually depend on the principal forest species involved. Nitrate-N export (as kg/ha or mg/L) is usually increased by forest harvest. Finally, harvest of the riparian zone allows more light to reach the water surface (increasing temperature) and decreases the quantity of incoming organic matter (as kg/yr leaf litter) (Fig. 18.5).

Ecological Impacts

The principal ecological impacts evaluated in monitoring efforts include increases in algal populations, increases in grazing invertebrates, and decreases in shredders. Fish communities may be affected by altered food sources or by reducing fry survival through reduced dissolved oxygen.

Spatial Scale

Monitoring generally concentrates in the most heavily impacted spatial scale. In forest harvest, this scale is often concentrated in first- to third-order streams (usually those less than 10 m in width) and forested wetlands. Principal impacts are within 0.5–1.0 km of a harvest site or within 1–3 km of a road.

Temporal Scale

Forest impact monitoring must evaluate both short-term and long-term changes. Peak discharge, suspended sediment load, nitrate-N, and temperature are primarily short-term concerns (e.g., < 3 years). Changes in total water volume, in light entering the riparian zone, and in the invertebrate communities are medium-term issues (e.g., 3–10 years). Changes in riparian energy dynamics and in some fish populations are still moderate but longer-term (e.g., 5–25 years).

Caveats

Most data on water quality impacts of forest harvest have been collected near 35–50° N latitude; the spatial and temporal frameworks described above are both dominated by that view. Certain influences will cause relatively predictable changes in the patterns discussed earlier. Most notably, increased average temperature (e.g., in more southern climates) will usually reduce the timeframe of impact. Increased topographic relief (e.g., in mountainous terrain such as in the Rockies, Alps, or Cascades) will increase the spatial scale (i.e., impacts will be detectable over larger areas).

Science, Understanding, and Policy

A review of the extensive literature on the effects of forestry on streams and water quality reveals a gradual change in focus, from a concentration on physical impacts such as increased water yield and erosion, to a focus on the biological impacts of forestry activities. In the 1960s and 1970s, research catalogued the enormous increases in sediment and water yield that resulted from many large-scale forest clearings. Throughout the 1980s, adoption and testing of BMPs led to a series of summary reports that demonstrated how practices such as buffer strips and road maintenance could keep impacts within the range of normal variation (Fig. 18.6). Most recently, studies published since early 1992 appear to suggest that, even though protective measures such as buffer zones reduce

Fig. 18.6 Sedimentation and run-off from poorly designed roads, especially in areas of erodible soils, are among the largest water quality impacts from forestry activities. (Photo by J. Perry)

damage, some long-term impacts on in-stream productivity and functioning have been overlooked in previous studies that concentrated on the physical impacts. These studies reflect not only the refinement of knowledge in this variable and complex field, but also an alteration in values. Some of those value shifts can be inferred from the following highlights.

A Historical Example: Deforestation Around Lake Ontario

The Lake Ontario region experienced quick and severe deforestation in the 1800s as "unlimited" supplies of white pine were harvested and used for railroad ties and construction of expanding cities such as Chicago. While effects of this deforestation were not quantified at the time, significant changes in species compositions, sedimentation rates, and chemical changes in the lake are preserved in lake sediment. Using both sediment records and historical accounts, effects of deforestation on the Lake Ontario ecosystem can be recreated. For example, deforestation caused significant changes in peak flow and spring melt, resulting in increased run-off rates, with greater extremes in high and low flow conditions. In addition, loss of forest canopies caused summer surface water temperatures to increase, and created more rapid rates of spring heating and fall cooling (Sly 1991).

Decreased shading and associated temperature rises in streams and nearshore areas most likely affected fish communities. Sediment records demonstrate changes in both aquatic plant and associated fish communities dating from the time of deforestation. In addition, inputs of leaf and terrigenous materials changed, probably modifying levels of dissolved organic carbon and composition of invertebrate communities. The Atlantic salmon was one species that did not survive these combined effects. Lake Ontario represents the southern limit of the salmon's range, and deforestation presumably raised stream temperatures above critical thresholds for survival of this species.

Deforestation did not occur without other important land use changes. Cutting timber was only part of the process. Transportation and processing of logs also affected streams and lakes, as did the growth of roads and human habitation. Specifically, effects of deforestation in the area were closely linked to development of mills and creation of dams. Dams likely had an important impact on stream temperatures, as they reduced flow rates and would have allowed greater temperature increases and presumably contributed to lowered dissolved oxygen levels and increased nutrient levels (Sly 1991).

A Model for the Future: Minnesota's GEIS

Environmental impact statements are used to assess impacts of proposed projects on local environmental resources. Several states are using a new approach in environmental management, called a generic environmental impact study (GEIS), to improve management planning. By posing alternate scenarios and studying presumed impacts of these scenarios on a wide array of environmental conditions, a GEIS can be an effective planning tool. In the early 1990s, Minnesota completed a large-scale GEIS studying the potential impact of three future timber harvest levels on state resources (i.e., 0%, 50%, and 100% increases over 50 years). Results are assisting state planners in taking a long-range, proactive view of resource planning. Among the reports produced for the study, is the analysis of effects of the three harvesting intensities on water quality, a 300-page document.

The study concludes that, except for the highest-intensity harvests, timber harvest practices are unlikely to affect larger water bodies. The most likely detectable impact is slightly increased annual run-off. Impacts become increasingly apparent higher upstream. In a relatively small area, the GEIS study predicts that during peak snow melt, stream flow will double. In headwater streams, increases are likely in stream-dissolved ions, followed by increases in lake nitrogen. The study also notes that ensuring use of BMPs is critical. Even where BMPs are implemented, small watersheds will experience localized impacts, including increased nutrient and sediment loads, and changed structure and functional rates of aquatic communities. Where BMPs are not used (as is still the case among half-private forest ownerships), affected streams will experience increased sediment loads and light levels, with their attendant changes in primary productivity and temperature.

None of the three alternate harvesting intensities would yield effects detectable at the Ecoregion level. They predict, however, that local effects will be evident from all levels of harvesting. The following BMPs are listed as ways to minimize adverse impacts from harvesting.

Water Volume Changes

- To prevent storm flow peaks and potential increased flooding, harvesting plans should consider the watershed area affected. For third-order or larger watersheds, forest area should not be reduced by large increments (20% to 30% of area) within any 10-year period.
- To reduce peak discharges, avoid over-bank flow, and minimize stream bank erosion and sedimentation, harvesting plans should incorporate improved road culvert designs along with improvements in design of other water conveyance systems.
- To reduce leaching, extensive clearing of large, contiguous areas of lowland conifers (peatlands) should be avoided where: (1) a small mineral soil upland component exists in the watershed, and (2) a receiving stream sensitive to reductions in stream flow during dry periods exists. Clear-cuts of 30 to 50 acres in most such watersheds will have little effect on low flows.

Water Quality BMPs

- Document possible threats to water quality associated with all timber sales, including potential access problems.

- Establish a minimum 25-foot filter strip along all intermittent and permanent streams, lakes, rivers, and wetlands. Wider strips should be used as slope, slope length, and soil erodibility increase. Width of a particular filter strip along streams, lakes, rivers, and wetlands will vary with the percent slope and slope length, as well as with soil erodibility; the general rule is 25 feet plus 2 feet for each 1% rise in slope between the stream and the soil disturbance, including roads.
- Establish a minimum 25-foot filter strip along all temporary and permanent roads near water bodies. Roadways should not parallel a water body within the 25-foot limit. Roadway placement in areas of higher slope and soil erodibility should reflect the greater erosive potential of these areas, with roadways being located a greater distance from water. Proper road design minimizes road slope to grades of 1% to 2% where feasible; grades of more than 10% should be avoided, unless care is taken to provide drainage and avoid discharge directly from road surfaces to water bodies.
- Prevent unmitigated crossing of all permanent streams at any season and of streams large enough to have open water during winter. BMPs include recommendations for angle of crossing (i.e., perpendicular to stream), construction material (nontoxic), drainage allowance (to minimize damming and provide for fish migration), approved bridges (permanent structures), and coordination with other land use patterns.
- Prevent use of soil in drainage and stream crossings, including temporary ones. Prevent use of winter roads during spring thaws, particularly in wetland areas.
- Prevent direct drainage of diverted water into lakes, streams, or wetlands. Water should be drained into a filter strip of appropriate width for the slope and distance to water; roadway drainage constitutes a serious concern in reducing effects of harvest activity on water quality. Grade rolls or dips, open-top culverts, cross drains, and lead-off ditches all reduce the sediment-carrying capacity of roadway run-off by reducing velocity.
- To reduce erosion, attain full implementation of the BMP recommendations about excavation and drainage components of roadway construction. These recommendations include careful placement of debris in a manner not impeding water flow, careful shaping and stabilization of borrow pits, installation of drainage structures on roads as soon as possible, armoring culvert inlets and outlets to reduce bank and channel erosion, roadway surfacing in high slope areas, and stabilization of exposed soil surfaces with grasses or sod.
- Place barriers across inactive roads, along with signs stating road closure. Establish water bars in areas of greater than 5% road grade. Conduct follow-up inspections to ensure compliance with BMPs.

- Improve education of recreational users of forested lands to sensitize individuals to potential impacts on water quality and the role they can play to prevent water quality degradation. Postharvest recreational vehicle activity on permanent and temporary roads can extend the period of disturbance far beyond the time of tree removal.

Summary

- Forest harvest involves removing standing vegetation from patches in a landscape. The probability that forest harvest will have a negative water quality impact is highly scale-dependent.
- Most forest harvest that occurs within tens of meters to a few hundred meters from a water body will have a measurable, local-scale impact. That impact most commonly will involve changes in stream flow (i.e., increased total run-off volume and a more flashy hydrograph) and increases in sediment load. Those changes will be quite local in scale, often extending less than 500 m downstream.
- The larger the percentage of a watershed that is subject to harvest, the higher the probability that a water quality impact will occur. As percent harvested area increases, changes in hydrology increase first, followed by changes in nutrients, and finally by changes in other physical variables such as sediment load. As percent harvest increases, the effects become detectable further downstream.
- Biological effects from forest harvest are less common than are physical–chemical impacts. The first biological impact is usually local increase in primary producers, followed by species and guild changes in macroinvertebrates.
- Impacts on the fish population are typically a product of changes in temperature, but rarely an impact of other physical–chemical or biological changes.
- Stream systems are the most likely to respond to forest harvest. Rivers and wetlands may be expected to show localized impacts. Forest harvest effects on lakes are relatively rare.
- Best management practices (BMPs) have been widely developed for forest harvest. If harvest is conducted following BMP guidelines, impacts can be expected to be highly localized and short-term.

For Further Reading

Barling RD, Moore ID. Role of buffer strips in management of waterway pollution—a review. *Environ Mgt* 1994;18:543–558.

Binckley D, Brown T. Forest practices as nonpoint sources of pollution in North America. *Water Resources Bulletin* 1993;29:729–740.

Hofer T. Himalayan deforestation, changing river discharge and increasing floods—myth or reality. *Mountain Research and Development* 1993;13:213–233.

Hornbeck JW, Adams MB, Corbett ES, Verry ES, Lynch JA. Long-term impacts of forest treatments on water yield: a summary for northeastern USA. *J Hydrology* 1993; 150:323–344.

Richardson CJ. Ecological functions and human values in wetlands: a framework for assessing forestry impacts. *Wetlands* 1994;14:1–9.

19

Management of Water Quality in an Agricultural Landscape

Overview

Agricultural uses account for approximately 70% of annual global water withdrawals and result in some of the most significant water quality impacts around the world. Because of the tremendous volume of water used for irrigation, water quality in many areas is affected by water drawdown and salinization. The effects of agricultural practices extend beyond the sheer volume of water used, however, to include problems of run-off from agriculture land. Fertilizers, pesticides, and other inputs leach or run off into area waters and contribute to eutrophication and toxic effects. Significant decreases in water quality also come from large volumes of sediment run-off from agriculture fields, a by-product of soil-disturbing cropping methods. Feedlots and wastes from livestock farms are significant contributors to biological oxygen demand (BOD), eutrophication, and toxic inputs.

The area of land devoted to agriculture will certainly continue to grow. Given that, management can take three directions: (1) reducing water demand; (2) changing agricultural practices to reduce land and water impacts; and (3) managing the edge of water bodies to reduce impacts. Reduced demand can be effected through efficient water management systems such as drip irrigation, through improvements in the integrity of irrigation canals to reduce seepage, or through using more drought-tolerant crops. Reduced land and water impacts can be achieved through techniques such as soil conservation practices, integrated pest management, and use of cover crops. Use of improved crops and integrated agriculture may achieve similar gains in less developed nations.

Planted buffer zones, seepage ponds, and wetland construction are other techniques that can be employed along the water edges to moderate the impacts of agricultural run-off.

Introduction

Of all human uses of water, agriculture accounts for the most significant water quality impacts. That dubious distinction is gained in part because of the sheer volume of global agricultural production and in part because of the water-intensive nature of growing crops. While approximately 1500 million hectares of land are devoted to agriculture, it takes about 1000 kg of water to grow 1 kg of grain and 2000 kg to grow 1 kg of rice. In the United States, approximately 620,000 gallons of water are required to grow one year's worth of food for one adult. Maintaining plants and livestock requires approximately 2500 km^3 of fresh water annually, or almost 70% of the total global withdrawals.

Agriculture exerts an uneven influence on the water demand of different continents, with agricultural water use being especially high in Africa, Asia, and South America. Asia uses 86% of its water for agriculture, mainly for irrigation; Egypt uses 98% of its water for agriculture (Gleick 1993). Even more importantly, these numbers are not static. Although some estimates maintain that total irrigated area will increase nearly 20% between 1993 and 2000, trends cited by Postel (1993) indicate that the rate of increase of irrigated land is now only about 1% per year. Rising costs of irrigation projects, environmental concerns and criticism, and low commodity prices have all contributed to the reduced growth rate of irrigated land. Although irrigated area will continue to increase, as will the pressure on water resources, per capita irrigated area has been decreasing since the late 1970s (Postel 1993).

The amount of water used for irrigation has increased fivefold since 1900 and currently stands at slightly less than 250,000 cubic hectares (Gleick 1993). In many parts of the world, those volumes of water come only from aquifers that recharge slowly. Severe drawdowns are being experienced that raise the cost of further water extraction and result in land subsidence and infrastructure damage to roads and buildings. In addition, drawdown reduces the volume of water available to recharge lakes and streams and can severely deplete water bodies as well as the concentration of minerals and salts in those water bodies.

On average, crop use accounts for about 25% of the water withdrawn for agriculture. Approximately 50% of agricultural waters never reach the crop. Instead, they are lost through unlined canals and leaky pipes and may raise local water tables, resulting in both waterlogging and salinization of affected land. In some areas, irrigation efficiency is even lower; Postel (1993) notes that major irrigation projects in Asia lose fully 70% of the water to leaks. Water not lost to leaks and not taken up by the plants returns to streams and lakes as wastewater, carrying with it both physical and chemical run-off from farms. Many agricultural practices disturb the soil surface and create significant losses of topsoil during rains. This sediment accumulates in streams and lakes, causing severe local impacts to biotic communities. Other impacts result from run-off containing pesticides, fertilizers, and herbicides.

The combined effects of large withdrawals, run-off, and leaching of agricultural chemicals makes agriculture the single largest contributor to decreased water quality. In the United States alone, the EPA reports that agricultural non-point sources are responsible for 72% of river water quality degradation, 56% of lake degradation, and 43% of degradation in estuaries. Agriculture represents the largest source of contamination for streams and lakes, and the third largest source for estuaries (EPA 1992).

Extensive studies of agricultural effects on water quality have revealed some common principles, as detailed below. These studies also detail extreme variability in impacts, however. Stream size and flow, parent geology, and rainfall in combination with different crops, tilling practices, water sources, and agricultural techniques all create dramatic differences in water quality impacts. Predicting and responding to these forces represents a major challenge for natural resource managers. Complicating the decision-making process is the propensity for one management decision to create an economic or water quality trade-off in another area. For example, Baker (1993) suggests that increasing nitrate concentrations in the Lake Erie watershed may be a trade-off associated with conservation tillage, an erosion and phosphorus control measure. Conservation tillage increases the amount of water percolating into the soil and thus can increase the concentrations of nitrate, a soluble (not sediment-bound) chemical. To date, the risks from increased phosphorus input have outweighed those of increased nitrate levels. As new information emerges and nitrate levels continue to rise, the trade-off may become unacceptable.

Irrigated Agriculture: Salinization and Waterlogging

Although barely 20% of the globe's cultivated land is irrigated, those lands account for more than one-third of total global agricultural production. They also account for the majority of agricultural water use. Irrigation is extremely water-intensive. In Asia, where irrigation supports large rice crops, 82%

of total water withdrawals are used for irrigation waters. In the United States, irrigation accounts for 40% of withdrawals, while in Europe the figure stands at 30% (Gleick 1993). Not surprisingly, severe water overdrafts are becoming increasingly common. For example, the southern portions of the Ogallala aquifer, which stretches from South Dakota to Texas, have been decreased by almost 25% due to overpumping for irrigated agriculture. Postel (1993) cites other cases: "Groundwater levels are falling up to a meter per year in parts of northern China, and heavy pumping in portions of the southern Indian state of Tamil Nadu reportedly dropped water levels by as much as 25–39 m in a decade . . . In Saudi Arabia . . . during the latter half of the 1980s, ground water pumping exceeded estimated recharge more than five-fold" [p. 59]. In addition, waterlogged and salted lands, declining and contaminated aquifers, shrinking lakes and inland seas, and destruction of aquatic habitats create a high environmental price for irrigated agriculture. Mounting concern about this damage is making large new water projects increasingly unacceptable.

Most of the water quality impacts of irrigation revolve around poor water management and heavy price subsidies that remove the incentive for efficient use of water. In one common scenario, unlined canals and overwatered fields prevent adequate drainage. Consequently, seepage causes the water table to rise. Eventually, the root zone becomes waterlogged, starving plants of oxygen and inhibiting their growth. In another common scenario that is most frequently observed in dry climates, evaporation of irrigation waters near the soil surface leaves behind a layer of salt that reduces crop yields and eventually kills the crops. In the case of salinization, the quality of the irrigation water is less important than the sheer volume of water used and evaporated. Even the best irrigation supplies typically have concentrations of 200–500 parts per million total dissolved solids. (U.S. drinking water standards are 500 ppm.) Applying 10,000 m^3 of water to a hectare per year—a fairly typical irrigation rate—thus adds between 2 and 5 tons of salt to the soil annually.

In the United States, an estimated 20% to 25% of all irrigated land (some 4 million ha) suffers from salinization; globally salinity now seriously affects productivity on about 20 to 30 million ha (about 7%) of the world's irrigated land. Approximately 1 to 1.5 million ha each year are newly affected. Salinization is a persistent problem in the plains of China, the Indian subcontinent, Central Asia, the Aral-Caspian lowlands of the NIS, southeastern Europe, the Middle East, North and West Africa, and Australia. In other words, salinity is a problem wherever irrigation is practiced on a large scale. Half of all affected irrigated cropland is located in South Asia, with Pakistan and India being the most seriously affected countries. According to various estimates, 30% to 60% of Pakistan's 15 million ha of irrigated cropland suffers from salinity and waterlogging. In India, the figure is nearly 25% and in Egypt nearly 30%. In Australia's arid Murray-Darling river basin, salinity is so severe that a salinity bank has been instituted to control irrigation return flows.

In addition to salinization and waterlogging, extensive irrigation may facilitate mobilization of toxins in the soil. For example, in the western United States, death, deformities, and reproductive failure in fish, birds, and other wildlife have been linked to high concentrations of selenium in water. Selenium is a naturally occurring element that is highly poisonous in large concentrations and has been leached out of the soil by large quantities of irrigation water. Since 1985, intensive investigations throughout California have found lethal or potentially hazardous selenium concentrations at 22 different wildlife sites, including Kesterson National Wildlife Refuge (see chapter 3).

Agriculture and Nonpoint Pollution

Management efforts are complicated by the diffuse nature of agricultural impacts. Rather than controlling output from easily identified outfall pipes, agricultural waste comes from many thousands of individual farms dispersed over countless miles of rivers, streams, and lakes. The impacts of any single farm on local water quality may be low, but the cumulative impact of many different farms draining into different reaches of the same river or lake can prove both severe and difficult to control. Inflow from the heavily agriculture-impacted Minnesota River has been termed the single largest source of pollution to the whole of the Mississippi River. Yet each farm along the Minnesota contributes only one small part of that total pollutant load. Motivating and enabling individual farmers to change land use practices is a significantly harder task than regulating outfall from urban and industrial sources.

Sedimentation is the most significant water quality impact from agriculture for two reasons: sediment causes direct harm to biota and habitats, and sediment is a carrier for chemicals and nutrients. When sedimentation rates are lowered, rates of input of most pesticides, herbicides, and nutrients also decline. This fact explains why a significant proportion of water quality legislation in the United States is housed in farming bills that promote soil conservation.

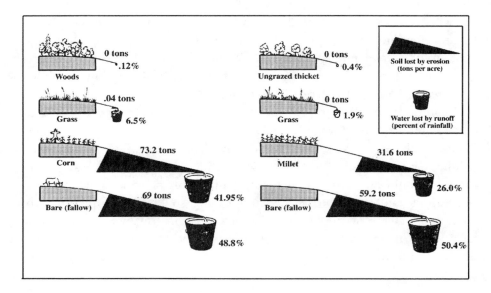

Fig. 19.1 Different cropping patterns yield significantly different rates of soil loss by erosion (i.e., sedimentation to streams and lakes). Meadows and uncut woods have low sedimentation rates and high uptake of water. Water run-off and soil loss increase in proportion to the amount of time a field lies bare (e.g., either while fallow or while new crops are germinating and growing). (Modified from Rapp *et al.* 1972)

Sediment

Sediment constitutes the largest volume of all nonpoint source pollutants in U.S. surface waters, and the largest portion of that sediment arises from cropland erosion (Fig.19.1). Yearly discharge from agricultural land to waterways in the United States is estimated at 1079 million metric tons of sediments and 477 million metric tons of total dissolved solids. Sediment accumulations fill drainage ways, culverts, and stream and lake beds, resulting in restricted flow, habitat degradation, navigational difficulties, and reduced productivity. Sedimentation rates in the lower Mississippi River may exceed 8 cm per year. The regions most affected by sedimentation are associated with row crop agriculture such as production of corn and soybeans. Wheat-producing areas (Great Plains and Northwest) that are prone to both wind and water erosion also represent major problem areas.

In lakes, suspended sediments constrain aquatic primary productivity, especially where suspension limits light penetration. In addition to burying fish habitat, plants, and aquatic insects, sediment accumulations are sinks for persistent pesticide residues and metals such as DDT, arsenic, and mercury. Sinks

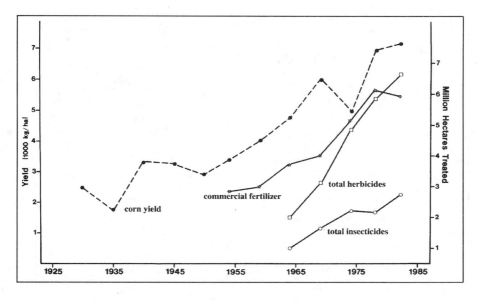

Fig. 19.2 The relationship between increasing corn yield and increases in chemical applications of fertilizer, herbicides, and insecticides in Iowa. The lure of higher crop yields carries associated water quality costs not previously taken into consideration. (After Lighthall and Roberts 1988)

add to long-term contamination because they permit low-level reentry of pollutants over long periods.

Nutrients

Between 1951 and 1972, the rate of fertilizer use per acre in U.S. croplands increased 7% per year (Fig. 19.2). Today, U.S. agriculture discharges 1.16 million metric tons of phosphorus and 4.65 million metric tons of nitrogen into surface waters annually (Cooper and Lipe 1992). Enriched run-off from cropland is the primary mechanism responsible for increased productivity in waters and the proliferation of eutrophic conditions in water bodies. This enriched run-off remains the major seasonal nutrient input into fresh waters, with the greatest potential for leaching occurring during fallow periods (Fig. 19.3).

Many parts of Europe are also experiencing profound difficulties managing nitrate from agriculture, and the use of fertilizers is increasing in many parts of Asia and Africa. Notably, fertilizers are not the only agricultural source of nutrients—failing animal waste treatment facilities are a significant, if more local, problem that has been compared in magnitude to major urban point sources of nutrients.

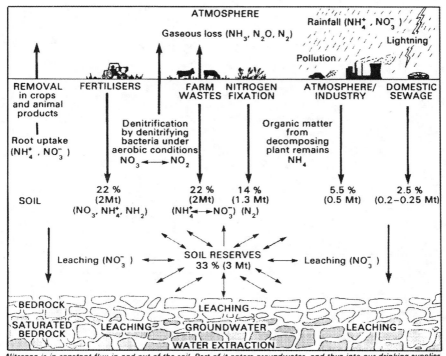

Nitrogen is in constant flux in and out of the soil. Part of it enters groundwater, and thus into our drinking supplies.
(Figures come from French research.)

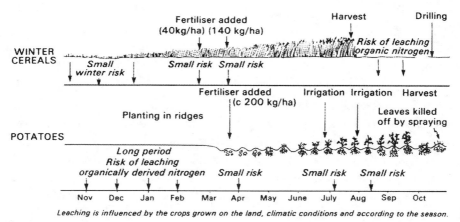

Leaching is influenced by the crops grown on the land, climatic conditions and according to the season.

Fig. 19.3 The nitrogen cycle is altered dramatically when land is under cultivation and subject to domestic and industrial development. Farm wastes and fertilizers contribute the largest loading of nitrogen leachates to area waters, while the extended fallow periods for some crops (such as potatoes) create a large potential for nitrogen leaching. (After Newson 1992)

Nitrates and phosphorus represent the two major concerns in nutrient management of waters. Of the two, phosphorus is generally considered the greater risk to water quality in temperate fresh waters. Most fresh water plants are phosphorus-limited, so increased phosphorus is the primary link to eutrophication in streams and lakes. In contrast, nitrates are more limiting to the growth of marine plants, so minimizing nitrate concentrations in estuaries and coastal zones is of paramount importance. Nitrate levels also create more concern than phosphorus for human health (see chapter 17).

Globally, enormous variation exists in both baseline nitrate levels and nitrate levels in degraded systems. Interestingly, a wide variation is also present in recommended standards, with WHO and the U.S. EPA recommending drinking water limits of 11 mg/L and 10 mg/L, respectively; the EC limit is far higher, at 50 mg/L. A commonly quoted, though conservative, estimate for baseline natural nitrate concentration in pristine rivers is 0.1 mg/L (Meybeck 1982). Pollution can boost the nitrate concentrations up by many factors of 10. The global median for non-European rivers is 0.25 mg/L, while the European average is 4.5 mg/L (Nash 1993). Background nitrate levels have increased since the 1940s in almost all developed countries, and standards are exceeded increasingly often. Nitrate pollution is also increasing in developing countries, although high nitrate concentrations are more often associated with poor sanitation and waste disposal options, than with agriculture, in those areas. For example, Jacks and Sharma (1983) studied nitrate pollution in southern India and found that village wells had higher nitrate levels (ranging up to 1500 mg/L) than irrigation wells.

The 1992 EPA Water Quality Assessment reports that more states reported impairments of lakes due to nutrients than any other single pollutant. Logan (1993) suggests that this trend will continue, although the nutrient of greatest concern will shift from phosphorus to nitrate. Current best management practices (BMPs) are primarily erosion control measures. They significantly reduce sediment-bound pollutants, such as sediment-bound phosphorus, but are largely ineffective on nitrate and phosphate (and pesticides) leaching. Reduced use of phosphorus fertilizer across the United States and implementation of BMPs have led to some regional declines in phosphorus loads attributable to nonpoint sources (Logan 1993). Logan's suggestion that nitrate is little impacted by BMPs is supported by the increasing proportion of nitrate-contaminated wells and surface waters. According to the U.S. EPA (1992), 49 states have reported significant problems with nitrate contamination of groundwater. The 10 mg/L drinking water standard has been exceeded in more than 4% of sampled wells on a national basis, and far higher levels have been discovered in many agricultural areas (e.g., 15% of wells in the Delmarva Peninsula of Delaware, Maryland, and Virginia).

Phosphorus, in its many forms, and nitrates follow different pathways that influence the susceptibility of a water body to contamination. The fact that nitrate is water-soluble rather than sediment-bound makes nitrate "pulses" less likely—except in Karst systems—than pulses of sediment-bound phosphorus. In addition, the fact that phosphorus is sediment-bound changes the impact of phosphorus loading in different size watersheds. For example, small watersheds are more "flashy," and the times of peak sediment transport account for the majority of total sediment transport. In larger watersheds, peak concentrations are lower but last longer.

Herbicides

From 1971 to 1976, the numbers of hectares of U.S. soybeans, corn, and wheat treated by herbicides doubled; the area continued to increase through the 1980s. Approximately three-fourths of all preemergent herbicides used in the United States are applied to row crops in the "corn belt" of the Upper Midwest (Thurman *et al.* 1991). That same area is highly vulnerable to soil erosion, allowing herbicides to be carried to surface waters via this overland flow. Herbicides may also fall into surface waters when applied by plane (crop dusting) or may leach into groundwaters; thus, it is not surprising that herbicides have been detected widely in ground and surface waters of the midwest. An emerging problem with herbicides is the proliferation of new types of chemicals. New generations of herbicides are being produced in the industrialized nations at such a rapid pace that current water quality monitoring is inadequate to assess actual or potential water quality degradation.

Application of herbicides (and consequently their occurrence in surface and ground waters) is usually a seasonal phenomenon, becoming most intense in late spring and early summer before crops are fully established. These seasonal pulses may be an order of magnitude larger than preplanting levels and may occur at a time when many native plants and animals are at vulnerable stages (e.g., eggs and fry are more sensitive to toxins than are adult fish). In addition, surface water contaminated by spring run-off may contribute significantly to groundwater contamination. Rapid increases in stream flow in spring may reverse flow downward into adjoining alluvial aquifers. Contaminated surface water may also enter alluvial aquifers by recharge from flood water and upland run-off. In their study of the Midwestern corn belt, Thurman *et al.* (1991) found that several herbicides exceeded U.S. drinking water standards during spring flush. In particular, 52% of sites exceeded the 3 μg/L standard for atrazine, 32% exceeded the 2 μg/L standard for alachlor, and 7% exceeded the 1 μg/L standard for simazine.

Herbicides may persist in the soil for years, degrading continuously but very slowly. Degradation products may be carried by run-off into surface waters or leached into groundwater. Desethylatrazine, a degradation product of atrazine, is one of the most persistent and mobile of these products. In fact, it has been used as an indicator of nonpoint source pollution of groundwater and may be used as a tracer of groundwater movement into rivers.

The four major herbicides used in the United States, in decreasing order of stability, are atrazine, metachlor, alachlor, and cyanazine. Atrazine is the most frequently detected herbicide, occurring in up to 90% of preplanting samples of rivers and 80% of harvest-time samples in rivers.

Non-agricultural threats in the United Kingdom come primarily from weed control on road and rail systems, where herbicides banned for agricultural purposes are permitted for transportation clearance. Some evidence suggests that these uses have increased rates of leaching because road and rail beds are more permeable, allowing herbicides to pass through the soil profile more readily.

Insecticides

Producers applied about 400 million pounds of pesticides to the 10 major field crops in the United States in 1988 (Cooper and Lipe 1992). Through run-off and leaching, the more persistent of these chemicals find their way to surface water and groundwater resources. In fact, more than 100 different pesticides have been detected in groundwaters. While approximately one-third of these chemicals are from point sources (e.g., industrial pollution), the remaining two-thirds are from agricultural nonpoint sources.

Agricultural pesticides—primarily organochlorine insecticides—are a serious water quality problem. The early classes of these insecticides included chemicals such as DDT. Although DDT was banned in the 1970s in the United States, residues are still widely detectable. In southern states, the concentration of DDT remained sufficiently high to accumulate in fish tissue 10 years after it was banned. Newer organochlorines have significantly lower persistence (days or weeks instead of years) but sublethal effects on aquatic life and water quality may be subtle and not well defined. In addition, acute toxicity of newer compounds poses a potentially serious threat to aquatic ecosystems if they are flushed into streams and lakes soon after application.

While many of the most persistent and ecologically damaging pesticides are banned in the United States and Europe, many developing countries continue to use extremely hazardous and persistent organochlorine compounds. For example, India uses DDT in cotton production and in malaria control; the compound is also widely available in stores throughout South East Asia. DDT and other compounds restricted in industrialized nations continue to be used in Latin America, Africa, and other regions as well. Use is not only ongoing, but is increasing: pesticide use in Africa nearly doubled in a 5-year period in the mid-1980s. Latin American imports of pesticides increased by almost 50% between 1973 and 1985.

Insecticide management is complicated because each compound degrades at a different rate, binds with soils differently, and dissolves at a different rate. Consequently, insecticide decisions need to be based on soil properties, crop type, the target insect, and the solubility, binding properties, and degrading properties (half-life) of each insecticide. In addition, monitoring is complicated by high costs and spatial and temporal variability. The fact that lag times between application of a compound and its appearance in groundwater may be on the order of years creates unique problems for a manager attempting to demonstrate links between contamination and a given agricultural practice.

Europe has responded aggressively to pesticide contamination of groundwater. In the Po Valley of northern Italy, widespread irrigated cultivation of maize and rice in an area of permeable gravel aquifers led to regional contamination of shallow aquifers used for drinking water. This finding prompted a nearly Europe-wide prohibition of specific pesticides for agricultural usage in vulnerable areas. Other countries such as Denmark, France, Germany, and the Netherlands have also severely restricted or banned pesticides such as atrazine that are considered a particular threat to potable water supplies. In Germany and the Netherlands, a wide range of the more mobile pesticides are commonly banned in Water Protection Zones.

Landscape Alteration

Just as irrigation is vital in areas with inadequate soil water or rainfall, too much soil moisture can also preclude agriculture. Areas with high water tables or poor percolation require improved drainage to ensure good crop production. Skaggs et al. (1994) note that in many parts of the United States and Canada, artificial drainage is required for more than 50% of all cropland. Many of these lands are drained wetlands. The U.S. EPA (1992) reports that from the mid-1950s to the mid-1970s, agriculture was responsible for 87% of the loss of wetlands in the nation. In the following decade, agriculture was responsible for 54% of the loss (conversion), for a total in three decades of more than 750,000 acres.

The water quality consequences of improved drainage are tremendously variable and site-specific. Evaluation of impacts also depends on the reference condition. For example, when compared with uncleared land, improved drainage and initiation of agriculture results in increased peak run-off, increased loss of sediment, and nutrient loss. When compared with surrounding nondrained agricultural areas, however, the same site might have lower peak outflows and sediment losses. In the first case, drainage would indicate degraded water quality, while in the second case, it could be interpreted as improving water quality. Skaggs (1994) also notes a temporal effect in which run-off and leaching rates change as the drainage system ages and agriculture becomes more established at a site.

In addition to these variables, the method employed to drain a site can influence water quality. Channelization, for example, increases peak run-off and sediment loss. Subsurface drainage systems such as tile drains generally have less run-off, lower peak outflow rates, and lower sedimentation rates than those depending on surface drainage (Skaggs 1994). Consequently, losses of sediment-bound pollutants such as phosphorus, organic nitrogen, and some pesticides will decrease in subsurface drainage systems, but losses of mobile pesticides (e.g., atrazine) and nitrate will increase.

Because the most substantial water quality problem from subsurface drainage is increased nitrate levels, efforts have been made to promote denitrification. Most "controlled drainage" efforts rely on providing a short period of anaerobic conditions, created by flooding fields before releasing water and by raising the water level in ditches during winter months to promote anaerobic conditions. Such efforts improve nitrate leaching by an average of 45% (10 kg/ha) (Skaggs 1994).

Livestock Agriculture: Point Sources of Pollution

Agricultural effects on fresh waters are not confined to croplands. Livestock production also accounts for serious water quality impacts. In Wales, for example, more than 90% of serious pollution incidents are related to the discharge of silage liquor and cow slurry into streams. Livestock wastes cause a variety of problems, ranging from deoxygenation of waters to eutrophication to bacterial contamination.

Oxygen Demand

Untreated cow slurry is, in terms of biological oxygen demand, 40 to 80 times stronger than untreated municipal sewage. Overloading storage lagoons with excess water is a significant cause of the great majority of slurry pollution incidents in Wales. In the United States, the EPA has estimated that pollution from faulty livestock waste storage facilities now rivals urban point sources.

Silage production and the by-product of silage liquor also has a severe water quality impact. Silage liquor itself carries up to 200 times the biological oxygen demand of untreated domestic sewage, as well as having a high ammonia content. Wilting of grass prior to ensiling can virtually eliminate silage liquor, but weather frequently prevents efforts at ensiling. Pollution from silage liquor typically results from inadequate storage of liquor or failures in the base or walls of the silo or of the liquor collection system.

Nutrient Enrichment

Nutrient enrichment of waters from livestock wastes is a serious problem throughout Europe, particularly in the Netherlands. In the United States, livestock-related, nonpoint source problems are concentrated in Washington, New York, Pennsylvania, Vermont, California, the Southeast coast, and the area draining into Chesapeake Bay. One of the important practices contributing to nutrient-enriched waters is the manuring of fields with livestock waste. While an efficient waste management technique, current agriculture is usually practiced with high animal densities. With less land per cow or pig, manuring results in a heavy leaching of nutrients into surface and ground waters. Manuring tends to be a locally severe problem that follows cultural patterns of land management; for example, it is particularly severe in the Netherlands as well as in areas of the eastern United States farmed by people who trace their lineage to the Netherlands (e.g., the Pennsylvania Dutch).

Coliform Bacteria

Coliform contamination of rural and urban surface waters has created a major environmental and public health concern for decades. Specific concerns from agriculture have centered on water supplies that receive direct run-off from pastures, feedlots, and land disposal areas. Sources of contamination are variable and complicated, however (Fig. 19.4). Both soil and stream bottom sediments act as reservoirs for coliforms. Wildlife and livestock contributions are difficult to quantify, but fecal coliform contamination at wildlife refuges may be as high as those at cattle crossings (Cooper and Lipe 1992).

Fig. 19.4 In many areas, livestock roam freely and contribute to fecal contamination of drinking water. (Photo by D. Vanderklein)

Remediation of Agricultural Effects

Remediation of agricultural effects is as individual as are agricultural practices, and mostly a matter of common sense (Table 19.1). In developed nations, technology offers many options for careful and accurate application of inputs based on site, microsite, and individual plant needs. In most of the developing world, the best hope lies in choice of crops, improved cropping patterns, and other large-scale approaches. In between lie medium-scale approaches involving improvement of irrigation efficiency, improvement and/or choice of drought-resistant and low-nutrient-input crops, and either improved inputs (i.e., pesticides and herbicides with lower environmental impact) or a cost structure that includes "true costs" of input agriculture (i.e., not subsidizing agricultural input costs).

Controlling Sedimentation

The most erosion-vulnerable period for cultivated fields is during seedbed preparation and planting. As much as 85% of total sediment loss may occur

Table 19.1 Classification of conservation practices and agricultural best management practices (BMPs) by environmental objective, pollutant type, and medium impacted. (Modified from Logan 1993)

Practice	Primary Objective	Pollutant Type	Medium Impacted	
			Surface Water	Groundwater
Structural				
Terraces, ditches	①	❶❸	●	■○
Grass waterways	①	❶❸	●	■
Tile drains	④	❶❸❷	●○	●
Irrigation	④①	❷	●	●
Sediment basins	②③	❶❸❷	●	○
Filter strips	②③	❷❸❹	●	●
Cultural				
Conservation tillage	①②③	❶❸	●	●
Contour cropping	①②	❶	●	●
Strip cropping	①②	❶	●	●
Crop rotation	①	❶	●	●
Stream bank protection	①	❶❸❷	●	●
Management				
Integrated pest management	③	N/A	●	●
Animal waste management	②③	❷❸	●	●
Fertilizer management	②③	❷❸	●	●
Pesticide management	③	N/A	●	●
Irrigation management	④③	❷❸	●	●

● Positive impact   ❶ Erosion, sediment   ① Erosion control
○ Adverse impact   ❷ Nutrients   ② Eutrophication
■ Variable or no impact   ❸ Pesticides   ③ Overall water quality
                           ❹ BOD$_5$ ④ Salinity

during this critical stage, before crop canopy and roots develop. Methods for reducing sedimentation include the following:

- On highly erodible land, cultivate only with protective soil and water conservation practices.
- Use conservation tillage, cover crops, grassed waterways, terraces, filter strips, water and sediment control basins, and contour farming.
- Use no-till or cover crops. (No-till systems reduce annual soil loss by nearly 99% compared with conventional tillage systems. Winter crop cover can reduce rainfall run-off by half and soil loss by 75%.)
- Build farm ponds and water/sediment retention structures. (Farm ponds can remove as much as 75% of annual sediment inflow from mixed-

pasture and row-crop watersheds. Sediment traps have the additional benefit that sediment can be excavated periodically and returned to farm fields.)
- Use field outlet or overfall pipes to eliminate lateral stream headcuts, another source of sediment.

Controlling Nutrient Pollution

Many of the sediment control techniques outlined above will also reduce nutrient pollution. In addition, both simple and highly technological approaches may be used to ensure that only needed fertilizers are applied and that, once applied, they are filtered out of the soil and water before reaching lakes and rivers. Efficiency in soil testing and fertilizer application alone can reduce potential nutrient discharges by 94%.

Some of the simpler, low-technology options for nutrient management include the following:

- Agroforestry, or utilizing tree crops with agronomic crops, is being used in many areas as a way to reduce soil erosion and promote soil productivity.
- Mixed rotation cropping and mixed crops, especially when mixed with nitrogen-yielding legumes, are time-honored ways of reducing the need for fertilizers and maintaining soil productivity.
- Use of slow-release nitrogen fertilizers and split applications can lower nitrogen losses.
- In addition to reducing soil erosion, implementation of terraces, crop rotations, and other conservation tillage practices can result in more efficient nutrient uptake by crops, thereby reducing nitrogen and phosphorus losses. No-till systems may have less than 10% of the nutrient losses found in tilled systems. Without tilling, however, farmers may need to increase herbicide use.
- Use of impoundments can trap nutrients.
- Use of constructed wetlands will lower nutrient, biological oxygen demand, and coliform bacteria levels.
- Composting (can be either a high- or low-tech option) can replace nutrients and lower water needs of crops. (The choice of compost material is critical in controlling the nutrient leaching as well as possible contamination of compost by heavy metals or other toxicants.)
- Improved animal nutrition lowers the amount of animal waste generated.

Exemplary technical solutions rely on computer technology to evaluate actual soil nutrient levels and plant nutrient needs. For example, crop-specific tissue tests, calibrated manure spreaders, GIS profiles of farms, and integrated soil testing and nitrogen applicator systems are all techniques that utilize site- and often plant-specific information to control nutrient applications.

Controlling Insecticide Pollution

As with herbicides, practices that reduce sedimentation also reduce run-off of other pesticides. In addition, farmers may implement these practices:

- Integrated pest management can substantially reduce the amount of pesticides needed and offer less environmentally hazardous alternatives.
- Continued education and recertification of both private and commercial applicators will ensure proper application, mixing, loading, transport, storage, handling, and container and waste disposal.
- Continued education can keep farmers up-to-date about risks, benefits, and alternatives to pesticide use.

Reducing Coliform Bacteria

Reductions in coliform bacteria can be achieved through the same techniques used to reduce sedimentation, nutrients, and pesticides. In addition, the following practices may be implemented:

- Constructed wetlands may help filter bacteria before wastes are discharged into area waters.
- Limiting livestock access to streams can reduce bacterial contamination.
- Structural improvement and proper maintenance of animal waste systems is of vital importance for limiting coliform bacterial pollution.

Improving Irrigation

Improving efficiency of water delivery systems is one of the most effective methods of reducing negative effects from irrigation. Techniques include lining canals to reduce water loss and implementing drip irrigation or other low-water systems. In the American West, improving irrigation efficiency (which averages 55%) by just 10% would double the amount of water available for urban residences and businesses. A 10% increase in water efficiency in parts of Pakistan, where present efficiency is 40%, would release enough water to irrigate another 2 million ha (Postel 1993).

Pricing reforms have also been widely recommended as an effective means of improving irrigation. In most of the world, water use is heavily subsidized, leading to inefficient and environmentally damaging overuse. In the western United States, for example, the federal government spends more than $530 million each year on irrigation supplies, while collecting only $58 million in water charges. Mexico subsidizes agriculture by supplying irrigation water at 10% of what it costs to produce the water.

Providing irrigation usually involves large-scale projects with large-scale water quality effects. Environmental managers are encouraging the use of smaller-scale, decentralized projects to provide water without transporting it over extensive distances and affecting enormous areas of land. Some of these smaller-scale approaches include improving techniques to trap moisture at each site, using controlled floods to provide irrigation water, restoring deforested watersheds, using recycled wastewater (known as "greywater"), and utilizing more salt- and drought-tolerant crops.

Monitoring and Assessment

Variables Impacted

Monitoring for agricultural impacts must consider the mode of transport of the different variables. Sediment is delivered from numerous agricultural practices, including irrigation return flow, row crop agriculture, and overgrazing. Most agricultural sediment is relatively fine-grained and is measured as suspended matter. Phosphorus is the principal nutrient responsible for agricultural water quality impacts; it typically moves while being adsorbed to sediment particles. Adsorbed phosphorus is not biologically active, however. Therefore, water quality managers usually measure dissolved forms such as ortho-phosphate and soluble reactive phosphorus (SRP) as well. Nitrogen moves through shallow soils in dissolved forms (i.e., primarily nitrate) and is a common agricultural contaminant in surface and groundwaters. Particulate nitrogen is less common and moves at a much finer spatial scale. Pesticides of many kinds move from agricultural landscapes to the water. In general, pesticides are very difficult (i.e., expensive) to measure and, therefore, are sampled relatively rarely. A final variable often associated with agricultural management is water volume, because water withdrawal is the principal agricultural use. Stream dewatering is widespread and relatively easily assessed. In some arid climates, salinization (measured as increases in sodium and relative decreases in magnesium and calcium) presents a massive water quality problem resulting from agriculture.

Ecological Impacts

Ecological monitoring efforts include increases in algal production due to nutrient supplements, major changes in invertebrates toward a more simplified, stress-tolerant community based on algal grazing, and a change in fish communities away from salmonids toward cyprinids, cyprinodontids, and catostomids.

Spatial Scale

Agricultural management typically affects second- to fifth-order streams (usually those between 5 and 20 m in width). Agriculture affects thousands of hectares of wetlands each year, primarily through drainage but also through contaminated run-off. Agricultural impacts to wetlands and lakes are hydrologically determined and often exceed distances of 20 m from shore. In many cases, of course, this area represents the entire water body. Agricultural impacts to streams are often detected several kilometers downstream, and cumulative effects may extend far beyond that point (Fig. 19.5).

Temporal Scale

Changes in phosphorus and nitrogen are associated with hydrologic events and will continue to occur as long as the land is managed for agricultural use (e.g., for centuries in parts of Asia). Sediment loads are usually associated with newly disturbed lands and are seen for 3–5 years after disturbance. Some pesticides are highly toxic to aquatic life but are relatively short-lived in the environment (e.g., days to months), making them especially difficult to detect. Others are more easily sampled and traced, and have longer lives (e.g., several years to a few decades), but are much less toxic. Impacts to fish and invertebrate populations are essentially permanent (i.e., persisting as long as the land is maintained in agriculture).

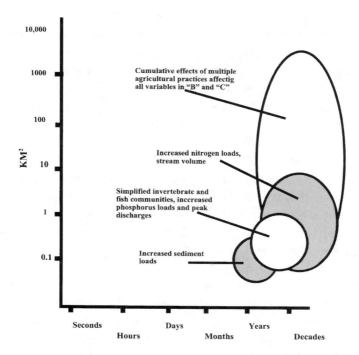

Fig. 19.5 Spatial and temporal effects of agricultural activities on water quality and the aquatic community. Rather than short-duration pulse events, agricultural activities typically affect small to mid-range areas but the effects endure for years or decades.

Historical Example: Agriculture's Impacts on Lake Ontario

The agricultural history of the Great Lakes region began in the early 1800s with clearing of vast forests of white pine for use in shipping, railroads, and building (see chapter 18 for a review of the effects this forest clearing had on Lake Ontario water quality). Agriculture moved into the cleared land almost as a secondary activity. Eventually, the American Civil War created a viable market for Ontario's products, particularly barley. Between 1861 and 1864, production of barley became the Great Lakes' primary economic activity (Sly 1991).

These changes can be learned from existing historical records, but they are also recorded in Lake Ontario sediment. Clay mineralogy, particle size, and phosphorus concentrations in sediments increased during the period of deforestation and early European agriculture. The barley years are reflected in a sustained peak in phosphorus and changes in sediment size and composition between 1850 and 1890.

Erosion is also evident in sediment cores. Reconstructing the lake's history through its erosion indices illustrates important trends in agriculture in the Lake Ontario basin. For example, coarser granules of suspended load, which indicate significant erosion, began to appear in the early 1800s, at the same time that widespread deforestation was occurring and the first mills and dams were being constructed (Sly 1991). This erosion peak was followed by a more stable time in which erosion rate was high but constant, indicating that a permanent change in land use had occurred. A trend to finer sediments indicates reduced erosion rates since about 1900.

Sly describes the utility of using diatom communities detectable in sediment cores as another avenue for agricultural history. According to the diatom-based historical reconstruction, Lake Ontario was characterized by clear waters and limited nutrient supply from a stable, mature landscape, with low algal productivity and high macrophyte productivity. The earliest European interventions apparently caused only minor changes in total algal abundance but appreciable changes in composition of the flora. Sediment cores can even reveal effects of culturally distinct land use patterns on the same land: the diatom flora, for example, differed during the period of French settlement from that seen after the arrival of British Loyalists and the consequent expansion of settlements and rise of the square timber trade.

Impacts from modification of the lake's watershed were relatively sudden. The entire transition from oligotrophy to eutrophy occurred in only about 20 years (1830 to 1850). Large increases in rate of sedimentation and nutrient concentration suggest that fire, poor agricultural practices, and deforestation acted in concert to create significant impacts on the lake. Between 1860 and about 1900, nutrient loading to the lake increased, presumably as a result of vegetation recovery and restabilization of the landscape. That period also marks the approximate end of the period in which agriculture dominated the water quality impacts on Lake Ontario. The more recent history of the lake has been influenced more by urban development than by agriculture (Sly 1991).

Agriculture in the European Community

Agriculture in the European Community is the primary cause of rising nutrient and pesticide concentrations in ground and surface waters throughout Europe. These increases have caused grave concern and significant policy shifts in the past decade, in large part because the EC depends on groundwater for 70%

of its drinking water (Cartwright 1991). Eutrophication of surface waters is also a source of major concern (see also the EC case study in chapter 26).

EC potable water standards for pesticides are precautionary and controversial because they are not based on toxicology and because monitoring low concentrations is very difficult. Some agricultural pesticides considered particularly hazardous come under the control of the EC Dangerous Substances Directive (Cartwright 1991). The EC Groundwater Directive also regulates water quality, and requires that particularly hazardous substances be prevented from entering groundwater, while the introduction of less hazardous materials is limited "so as to avoid pollution." Some national standards established to protect aquatic communities have even more strict criteria for insecticides that are toxic to aquatic life but less toxic to humans.

Controls on diffuse inputs are incorporated in a proposal that would require each EC Member State to establish "vulnerable zones" (i.e., surface waters with the potential for eutrophication and ground or surface waters used as sources of drinking water if they exceed or are likely to exceed 50 mg NO_3/L). Member States would have to impose restrictions on diffuse inputs of nutrients within these zones (e.g., limiting livestock density and manure spreading; establishing maximum application rates for chemical fertilizers based on crop uptake and residual nitrogen in the soil; developing rules to control timing and location of spreading of both manure and chemical fertilizers; crop rotation, crop cover, and irrigation improvement). Set-asides, reforestation, diversification, training, education, and research are also potential avenues for improvement.

In the short term, high nitrate levels are resolved by blending, source replacement, or treatment. Reducing nitrate is the only viable long-term solution, however. Denmark, the Netherlands, and Germany have all imposed national or state-wide restrictions on manure spreading. These restrictions—although uniform—do not consider local factors such as soil type, denitrification, mineralization, or crop requirements. In lieu of prohibitively expensive monitoring, research organizations throughout the EC are developing predictive models relating nitrate leaching to land use, with the goal of refining manure spreading guidelines to local conditions.

Water Protection Zones are used throughout Europe to varying degrees to control land use around groundwater abstraction points and to reduce nitrate leaching to the aquifer. Both legislated restrictions and voluntary agreements between farmers and water companies are employed to control diffuse inputs of nitrate. In the former (e.g., Baden Wurttemberg, Germany), suitable application rates for different crops are published to achieve desired soil nitrate concentrations. Reductions achieved by farmers are monitored by state authorities. A tax on drinking water provides funds for compensation of farmers for the yield lost due to decreased fertilizer usage. In another area of Germany, North Rhine–Westphalia, local farmers have entered into a voluntary agreement with the water company. The water company purchased particularly vulnerable areas of the catchment for conversion to unfertilized grassland. Farmers within this catchment restrict the rate and timing of fertilizer application according to advice based on soil nitrate levels.

Summary

- Agricultural impacts on water quality are very widespread and nearly ubiquitous. Agriculture claims nearly 70% of all water used in the world each year. The uses represented in production of food for humans include direct consumption (i.e., water consumed by a plant or animal and either incorporated or evaporated/transpired/respired), indirect impacts (e.g., irrigation return flows that pass the plant and return contaminated to the water resource), and ancillary or processing waters.
- Agricultural impacts are notably difficult to manage because they emanate from practices that result in direct benefit to human communities. As such, a powerful constituency exists in each society that suggests that water quality (or other environmental impacts) are appropriate uses for the world's ecosystems.
- Best management practices are not nearly as widely developed or as available for agriculture as they are for forestry. Some practices, such as ploughing with the contour or terracing, have been used for centuries. In most cases, those practices evolved to protect the landscape (e.g., prevention of gullies) rather than to protect a water resource. Therefore, they are often not fully effective in water quality management.
- Cumulative and watershed-scale effects are water quality impacts frequently seen in an agricultural landscape. The most common effects are changes in sediment load, hydrology, and nutrients.

For Further Reading

Cartwright N, Clark L, Bird P. The impact of agriculture on water quality. *Outlook on Agriculture.* New York: Pergamon Press, 1991.

Jacks G, Sharma VP. Nitrogen circulation and nitrate in groundwater in an agricultural catchment in southern India. *Environ Geol* 1983;5:61–64.

Logan T. Agricultural best management practices for water pollution control: current issues. *Agriculture, Ecosystems and Environment* 1993;46:223–231.

Skaggs R, Breve M, Gilliam J. Hydrologic and water quality impacts of agricultural drainage. *Critical Reviews in Environmental Science and Technology* 1993;24:1–32.

Sly PG. The effects of land use and cultural development on the Lake Ontario ecosystem since 1750. *Hydrobiologia* 1991;213:1–75.

Thurman E, Goolsby D, Meyer M, Kolpin D. Herbicides in surface waters of the midwestern United States: the effect of spring flush. *Environ Sci Tech* 1991;25:1794–1796.

20

Management of Water Quality in an Urban Landscape

Overview

Urban development introduces a complex array of water quality impacts that range from localized impacts, such as those from sewage outfalls, to landscape-level impacts, such as those caused by the construction of dams to meet urban water and energy needs. In addition to broad differences in scale, urban development introduces vastly different types of impacts—from industry-based toxins to increased flow from paved or hardened-soil surfaces. Although each impact results from a different component of urban development (e.g., paving, industrial growth, waste removal needs), management must consider urban problems as an interrelated whole to achieve adequate water quality protection.

One of the major problems faced by any urban society is waste disposal and its associated pathogens. Urban waste streams contain more than human disease agents, however. Run-off from paved streets includes lawn chemicals and fertilizers, waste oil from cars, road salt, and other substances. Household waste streams frequently include corrosive cleaning agents, paints, and other persistent compounds. Wastes from the industrial sector contain ever greater numbers of toxic pollutants. Processing wastes is one of the most significant water quality problems in existence, and has been a constant companion to human settlement throughout history.

Water course modification is another major impact on water quality in urban landscapes, and just as old. Levees for flood control, dredging and filling in to create more land, and river channelization to improve land for building or agriculture are all common practices that influence water flow and pollution loading in urban watersheds. Dams are also an important side effect of urban development. Because dams block the flow of rivers and create reservoirs, virtually every aspect of water quality changes following dam construction. Effects of any specific dam depend upon the geology and hydrology of the area, as well as on dam size, elevation in the dam from which waters are withdrawn, and water level management.

Introduction

The problem of urbanization is a problem of concentration and additive effects. When populations concentrate in cities and towns, the amount of water needed for drinking, industrial uses, waste disposal, energy generation, and other purposes multiplies exponentially. Increased use brings about many attendant stresses, such as water withdrawal and conflicting valuations of water quality. In addition, water quality impacts multiply as the human population expands, bringing with it a concomitant increase in waste; likewise, more diverse uses of the water result in more diverse kinds of waste (Fig. 20.1). The need for source water reliability also creates critical water quality impacts. When people settle in one location, the requirements increase for steady and predictable sources of water. Consequently, existing water courses are frequently modified to reduce flood potential or to increase year-round reliability (e.g., via dam construction and channelization).

An important problem with urbanization is that it often outpaces infrastructure development. In 1950, only one city in the developing world had a population in excess of 4 million; in 1980, 16 such cities exceeded that mark. Projections suggest that more than 60 of these cities will exist by the year 2000 (Meybeck *et al.* 1989). Recent unprecedented levels of migration and high urban birth rates have contributed to record population increases in many of the cities in Africa, Latin America, and Asia—regions that are now growing too fast to plan and develop adequate water and sanitation systems. An estimated 90% of population growth will occur in urban areas in the future, yet infrastructures are already severely overloaded, existing water supply and sewerage systems are overtaxed, and the ever-greater volumes of industrial wastewater discharges go largely untreated. Provision of both fresh water and adequate sanitation lag behind urban growth, and the magnitude of this problem is expected to double every 10 years in large urban centers in the developing world (Fig. 20.2).

The search for safe and adequate water supplies is often the beginning of a series of chain reactions that pit technological developments against environ-

Fig. 20.1 Urbanization causes multiple stresses on local and regional waterways and its impacts are frequently difficult to control. Some communities, such as Madison, Wisconsin, have initiated an active program of citizen education as well as infrastructure changes such as frequent street cleaning to reduce the flow of wastes into storm drains that flow untreated to area lakes. (Photo by T. Walz)

ment resilience. Water quality conditions in urban landscapes are rarely the result of a single isolated source or impact. Mexico City, for example, suffers from the combined problems of water supply, sewerage, and overpumping of the nearby Mexico Valley aquifer that supplies more than 80% of its water supply. Bangkok, Thailand, suffers from similar combined difficulties: with essentially no municipal waste treatment disposal system, Bangkok discharges 10,000 metric tons of raw sewage and municipal wastes daily into the city's rivers and canals. At the same time, overexploitation of the groundwater is causing the city to sink two to four inches per year, leading to cracked pavements, broken sewer and water pipes, sea water intrusion, and flooding. Along the lower Rhone, in France, a two-century process of urbanization began with levees for flood protection and progressed to dikes for improved navigation, hydroelectric plants, and channelization to provide cooling water for power plants, concentrated urban areas, and the spread of petrochemical plants (Fruget 1992). The result in the Rhone—and in many other rivers—is low morphological and biotic diversity and a biota tolerant of pollution.

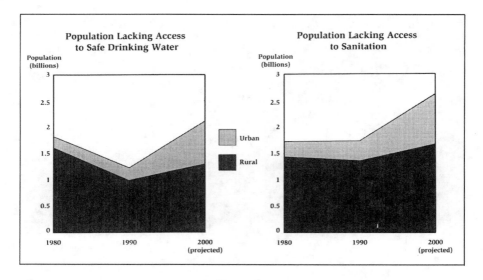

Fig. 20.2 Until approximately 1990, improvements in supply of fresh water and sanitation kept pace with increases in global population. Population increases since 1990, however, have outpaced improvements. Current projections suggest that the number of people without a safe water supply or sanitation options will increase dramatically through the 21st century. The problem is likely to be especially acute in urban areas. (Redrawn from Engelman and LeRoy 1993; courtesy of Population Action International)

The examples of Mexico City, Bangkok, and the Rhone demonstrate that urbanization creates a complex set of water supply and water quality issues, affecting both surface water and groundwater sources and spanning a broad range of temporal and spatial scales. Water quality impacts of urbanization also extend to the conversion of prime farm land to agricultural production. Charbonneau and Kondolf (1993) suggest that one of the most important water quality "footprints" from urbanization is the displacement of agricultural and other activities onto marginal lands that are more erodible and less fertile, requiring greater application of fertilizers and other inputs. This displacement is a universal phenomenon, with effects felt from California to the Kathmandu Valley of Nepal.

Of these impacts, the use of water to carry human waste is the most universally important health and water quality issue. Of increasing importance in developed areas, however, is the influence of street wastes in the waste stream. Other effects of urbanization, such as flood control structures and dams, transform rivers into homogeneous water courses with less habitat and species diversity, less variable flow rates, and lowered assimilative capacity. These effects are intricately linked to the demands of building an urban infrastructure. Waste management (including urban run-off), industrial contamination, and physical modification of watercourses are each discussed later in this chapter.

Urban Waste Streams: Sewage and Street Run-off

Pathogens

It has been estimated that half the world's population suffers from water-related diseases, contracted by drinking and bathing in polluted waters. Most of those affected are poor, and almost all live in developing countries, where epidemics of water-borne diseases are on the rise. Although virtually all water-related disease epidemics begin in water-disadvantaged communities, each year contaminated water from sewer overflows or breakdowns in waste treatment plants force beach closings, contaminate shellfish beds, and cause temporary prohibitions on drinking city tapwater. Urbanization in the developing world has created a serious health and water quality problem.

Despite international efforts, infrastructure development on a global basis continues to trail behind increases in population growth. Without adequate sanitation, communities remain susceptible to epidemic outbreaks of more than 30 diseases carried in human waste, including cholera, typhoid, schistosomiasis, and diarrheal diseases. Diarrheal diseases alone kill more than 4 million people each year. Historical outbreaks of diseases such as bubonic plague in the Middle Ages, which killed fully one-third of the population of affected areas, are just as attributable to poor sanitation as modern-day outbreaks of cholera in Peru. Traditionally, cities and towns (if they had sewers) channeled sewage to rivers and lakes, where the large volumes of water would effectively diffuse the input of waste. Today, however, towns and cities are home to enormous populations, and the concentration of human populations has overwhelmed the assimilative capacity of waterways.

Adequate sanitation, although imperative, has not always been a high priority. Ignorance, cultural perceptions, and technological and economic barriers have all contributed to ineffective sanitation systems. Currently, barely 65% of the world's urban population has access to adequate sanitation. While that figure represents a jump of 8% from 1980 to 1990, and service for an additional 300 million people in urban areas, nearly 400 million urban residents still have no safe way to dispose of waste (Meybeck 1989). Notably, the urban population constitutes only one-third of the world's population, and more than 1.5 billion people in rural areas remain without sanitation systems.

Fig. 20.3 Poorer sections of expanding urban areas are significantly less likely to receive adequate provision of potable water, or adequate development of sewer or waste disposal options. These densely populated "outer reaches" of urban areas rely heavily on the assimilative capacity of local riverways, but high population densities have outstripped the assimilative capacity of most of these rivers. (Photo by J. Perry)

Among the rural population, only 20% has adequate sanitation. It may be even more significant that during the 1980s (the U.N. Water Decade), 80% of the investment in water and sanitation went to the wealthiest 20% of the population (Fig. 20.3).

Urban Run-off

Accomplishing adequate sewage treatment begins with its collection in sewer systems (the term sewage refers to the raw liquid waste collected from houses, buildings, and industries). At its simplest, sewer systems are open gutters that run down streets. They may empty out into ponds, lakes, or rivers, or simply dissipate, with the water soaking into the soil. More advanced sewer models are closed and below the street level, but frequently carry the raw sewage directly to the nearest water course. This model is followed in many cities around the world (such as Bangkok, as described earlier). The sewer system model followed in most developed countries includes channeling the sewage to treatment plants, where it undergoes a series of physical, chemical, and biological changes before being discharged into the environment.

Although sewage treatment is enormously successful on many levels, it does not solve all urban waste problems. One of the most important, and only recently recognized, problems is that of street run-off. Most municipalities have two sewer systems: one for sewage and one for run-off from streets. *Storm sewers* are designed to accommodate large, periodic volumes of water from storms (or car washing or overwatered lawns). Two models exist for storm sewers, and each has its own water quality problems.

In one model, the storm sewers are entirely separate from sewers carrying domestic sewage. Run-off from these storm sewers undergoes no purification, but is channeled directly to receiving waters (e.g., a lake or a stream). In 1980, nearly 50% of the U.S. population was served by separate storm and sanitary sewer systems, but that figure was projected to increase to 70% by 2000 (Laws 1993). The difficulty with this model is that lawn fertilizers, domestic use of herbicides and pesticides, gas and oil leaks from cars, detergents from car washing, road salts, particles from vehicle emissions, and all other chemicals used outdoors end up in the storm sewers (Fig. 20.4). Storm sewers are now important sources of unregulated pollution, and include a full array of contaminants ranging from sediment to pathogens, pesticides, oils, and heavy metals. In fact, all 13 metals on the EPA's priority pollutant list were detected in more than 10% of urban run-off samples.

Urban run-off is also typically very high in suspended sediment (often even higher than raw sewage), chemical oxygen demand (COD) (Table 20.1), and biological oxygen demand (BOD), with nutrient concentrations high enough to cause eutrophication if not sufficiently diluted by the receiving water (Laws 1993). In many areas, lawn care contributes significantly to eutrophication of area waters. In a 1992 survey of the Minneapolis–St. Paul metropolitan area, researchers found that residents used an average of 36 lb of lawn fertilizer per household, or a total of 12,735 tons of fertilizers per year (Creason and Runge 1992). The same study reports that herbicide use on residential lawns is approximately the same, per acre, as herbicide use on area agricultural farms.

Approximately 20% of the U.S. population is served by an alternate sewer system that combines storm and sanitary sewers and channels both waste streams to waste treatment plants. Combined sewers serve the majority of European populations, including 90% of the Netherlands and 70% of France, Germany, and the United Kingdom (Marsalek 1993). Unfortunately, significant problems also exist with the combined sewer model because few facilities are designed to accommodate the large increases in water volume resulting from a

Fig. 20.4 While sanitary waste disposal is effective and comprehensive in most European cities, run-off from cars, pet waste, city dust, and other by-products of urban life are all washed directly into the rivers and canals around which the cities were built. Only recently has the importance of these cumulative, small impacts been recognized. (Photo by D. Vanderklein)

Table 20.1 Average concentrations of various components of urban run-off and raw sewage. (Adapted from Torno 1896; Laws 1993; Novotny 1994)

Variable (mg/L)	Urban Run-off			Raw Sewage	
	General Urban Areas	Developing Urban Areas	Commercial Areas	Industrial	Commercial
BOD$_5$	74	25–50	100	500–750	150–300
Suspended solids	200–4800	27,500	50–830	450–1700	100–250
Total phosphorus	0.3–4.8	23	0.1–0.9	0.9–4.1	1–3
Total nitrogen	0.2–18.0	63	1.9–11.0	1.9–14.0	2–10
Chloride	130–750	N/A	10–150	75–160	25–100
Lead	0.14–0.5	3	0.7–1.1	2.2–7.0	0.1–1.0
Copper	0.02–0.21	N/A	0.07–0.13	0.29–1.3	0.05–0.75
Zinc	0.3–1.0	N/A	0.25–0.43	3.5–12.0	0.5–2.5

storm. Surges of water after a storm frequently overload the treatment facilities and result in the release of raw sewage (and untreated storm waters) to receiving waters (Marsalek 1994). In addition to raw sewage, these combined sewer overflows (CSOs) contain very high suspended solids, BOD, COD, nitrogen, phosphorus, lead, zinc, and oil. Recent declines in top-quality rivers throughout Europe are generally attributed to continuing problems with CSOs (House 1993).

Cities with combined sewers generally experience overflows between 10 and 50 times a year. Novotny and Olem (1994) report that the average in Milwaukee, Wisconsin, is 40 times per year. While the overflows may represent only 3% to 5% of total yearly effluent release, the annual load from overflows for BOD are approximately equal to yearly BOD for secondary treatment plants, and sediment loads are 15 times higher from CSOs than secondary treatment plant effluent (Novotny and Olem 1994). In New York, for example, every time more than 0.75 inch of rain falls, 500 million gallons of mixed sewage pours into area rivers and harbors.

Waste Treatment Plants and Their Effluent

Although the issue of storm sewers is ongoing and problematic, even the effluent from waste treatment plants can be a source of water quality degradation. Because of the expense (and value judgments), most treated effluent remains high in nutrients and may, depending on the level of treatment, contain high levels of BOD, toxins, and other compounds (Table 20.2).

Most waste treatment plants are relatively small, serving towns and small cities. For example, more than 12,000 waste treatment plants in the United States (i.e., more than 75% of them) have daily volumes of less than 1 million gallons. Most of those smaller systems have limited technology, and operate by gravity flow and "natural" processes of biological treatment. The majority of U.S. sewage treatment plants use secondary treatment, but approximately

Table 20.2 Average concentrations of primary and secondary treatment plant effluents.

Variable (mg/L)	Type of Wastewater Treatment			
	Raw, Untreated	Primary	Secondary	Tertiary
COD	300	250	75	60
BOD_5	300	150	30	30
Suspended solids	300	100	30	20
Total phosphorus	10	5	1–3	0.005
Total nitrogen	50	15	5	0.010

Table 20.3 Percent of the population of various countries served by waste treatment plants.

Country	% Served	Country	% Served
Sweden[1]	100	France[2]	63.7
Denmark[1]	98	Canada[1]	62.2
Netherlands[1]	90	Norway[2]	51.0
New Zealand[3]	88	Japan[4]	36.0
Germany[5]	86.5	Italy[6]	30.0
Switzerland[5]	85	Spain[3]	29.0
United Kingdom[1]	84	Belgium[6]	22.9
Luxembourg[3]	81	Portugal[2]	12.0
Finland[1]	74	Ireland[6]	11.2
United States[4]	74	Turkey[6]	3.3
Austria[2]	67	Greece[2]	0.5

[1]Gleick 1987.
[2]Gleick 1983.
[3]Van der Leeden et al. 1990.
[4]Gleick 1984.
[5]Gleick 1985.
[6]Gleick 1980.

15% of the plants use only primary treatment while another 20% use tertiary treatment. A small percentage have no discharge at all, as they discharge to land or use an evaporation process.

These figures refer to the 74% of the U.S. population that had access to a waste treatment system. The remaining 26% of the population used on-site disposal systems such as septic tanks. Table 20.3 shows similar treatment percentages for other countries. Faulty septic tanks are another important source of water quality problems, particularly for groundwater contamination. In Minnesota, for example, an estimated 300,000 faulty septic systems discharge waste into streams and groundwater.

Managing Urban Run-off and Urban Sewage

The problem of urban run-off is growing worse. By some estimates, approximately 90% of the U.S. population is expected to live in urban or suburban settings by the year 2000 (Laws 1993); similar percentages are projected for other industrialized countries. Although education is an important component of management, it is unlikely to affect the degree of necessary improvements; for example, the lawn care survey cited earlier (Creason and Runge 1992) found that approximately 75% of Minneapolis–St. Paul residents used fertilizers or herbicides on their lawns, yet 75% also felt that not enough was being done to eliminate harmful chemicals in urban areas. The same survey also indicated that residents would continue to buy lawn care products even if prices were increased. Thus, the discrepancy between awareness and action may be particularly strong when it comes to remedying urban run-off problems.

Technical solutions to urban run-off usually revolve around the establishment of settling basins in which storm water is allowed to settle temporarily before either seeping into the ground (recharge basin) or joining stream flow. The effectiveness of settling basins depends largely on the amount of time the storm water spends in the basin. Studies reported by Laws (1993) indicate that little improvement in suspended sediment, BOD, or nutrients is achieved in settling times of less than 1 hour. After 16 hours, however, one study reported approximately 50% removal of suspended sediment and phosphorus and approximately 40% reduction in BOD. Longer settling times may remove up to 70% of suspended sediment, but have little further effect on BOD or phosphorus. Although percolation removes suspended solids and most pathogens, settlement basins that recharge groundwater run the risk that some of the chemicals in the run-off will seep into underlying aquifers. This risk is especially great in areas whose run-off is fed by industrial areas.

Some cities (e.g., Berlin and Chicago) have undertaken extremely large and expensive projects that channel storm water to underground storage tanks, where it is held until the sewage treatment plants can accommodate the flow. While effective, these projects are prohibitively expensive in almost all cases. The price for Chicago's tunnel and processing system, for example, is estimated at $4 billion.

More practical technical solutions attempt to reduce the percentage of contiguous impervious surfaces in urban areas. Run-off from suburban areas may be up to three times that of a woodlot; the more urban an area (Laws 1993), the less seepage found and the more run-off occurs. Some of the techniques being used effectively to reduce the impervious surface area include using porous pavement (a method widely used in Japan), replacing traditional curbs and gutters with grass swales, using modular pavement (concrete interspersed with sod, sand, or gravel sections), and increasing the use of gravel and crushed rock in low-use areas.

The Problem of Industrial Wastes

Globally, industrial growth is enormous. Currently, about one-fourth of total water withdrawals are used for industry production. Industrial expansion is proceeding at an extraordinarily rapid rate, and current water withdrawals are expected to double by 2005. In addition to the quantity of water withdrawn for industrial purposes, industrialization presents unique problems for waste management and potentially severe stresses on aquatic and human populations because of both the diversity and the toxicity of some of the compounds contained in industrial effluent.

While industrialization is commonly considered a modern problem, industrial pollution contributed significantly to degradation of many waterways in the late 1800s and early 1900s. In turn-of-the-century Minneapolis and St. Paul, for example, so much sawdust and bark was generated from area sawmills that the material formed deposits that obstructed river flow and fouled the water. In addition to sawmills, the Twin Cities hosted a pulp mill, coal gas works, several large breweries, packing houses, and stockyards, all of which dumped their wastes directly into the Mississippi River.

Production and Pollution in Developing Nations

Industrial wastes are also typically thought of as a problem of developed, industrialized nations. While it is generally true that developing nations have fewer industries and less industrial pollution, this picture varies greatly from nation to nation. Nicaragua, for example, uses 45% of its water for industry and Barbados 35%. Perhaps more importantly, developing nations typically have fewer controls on their industries so the pollution potential of these industries is much greater. In India, an estimated 79% of the rivers are polluted with industrial waste. One 50 km stretch of the Yamuna River, which flows through the capitol of New Delhi, receives 20 million liters of industrial effluent daily, including about 500,000 liters of "DDT wastes." Throughout India, the production of hazardous and toxic chemicals increased 2000-fold between 1950 and 1980, and most of the wastes from this production reached waterways.

The nature of the industrial threat to water quality in developing nations is also changing as the world economy expands, and as tight pollution controls in many developed nations cause companies to look for more "hospitable" environments for their industries. One example of this trend can be found in Latin America, which in 1990 had only 5% of the world total of value-added industry, but had much larger shares of pollution-intensive industries such as petroleum refining (18%) and chemical production (15%). In Mexico, industrial effluent constitutes 90% of nonagricultural water pollution.

Throughout the developing world, food processing typically represents one of the most polluting-intensive industrial activities. As many as 90% of Malaysian rivers, for example, are estimated to be incapable of sustaining fish, shellfish, crab, or shrimp life because of pollution from oil-palm and rubber processing effluent. In Kenya, where coffee is the leading cash crop, 120,000 tons of coffee is processed by more than 1200 factories that discharge effluent directly into adjacent water courses. A survey of streams and rivers during the Kenyan coffee processing season found that every river and stream studied was anaerobic. Similarly, in Latin America, an estimated 10 million tons of food processing wastes are discarded with little or no treatment.

Some industries, such as mining, are land-intensive and thus have significant landscape-level impacts. In a dramatic example from Estonia, open-cast oil shale mining has influenced an area of approximately 10,000 ha (Vallner and Sepp 1993). Within this area, forests have been cut and the area strewn with rows of spoil heaps from exploded boulders and excavated rock. Draining the mines and quarries has caused groundwater levels to fall and created a 600 km^2 pumping cone depression (see chapter 10). In addition, Vallner and Sepp (1993) detail groundwater pollution from oil leaks and underground fires and groundwater sulfate concentrations as high as 500 mg/L. The production of electricity from oil shale is also a source of severe water (and air) pollution in Estonia, with power plants emitting large quantities of arsenic, cadmium, sulfur dioxide, ammonia, and other elements that pollute soil and surface waters.

Important Industrial Pollutants

Industrial pollutants are extremely varied and may include nutrients, suspended sediments, bacteria, oxygen-demanding material, and toxins. Of these, biological oxygen demand (BOD) and suspended sediments are the most common. Laws (1993) reports suspended sediment concentrations in the effluent of Hawaiian sugar cane mills that are 40 times higher than levels in raw sewage (6900 ppm versus 200 ppm, respectively). He also notes BOD that is three times the level found in raw sewage and COD seven times that of raw sewage. Pulp and paper mills generally have suspended sediment concentrations similar to those found in raw sewage (although they may range from 0.5 to 2.5 times the concentration of raw sewage).

Because electricity invariably results from urban growth, power plants are an important part of most urban landscapes. The effluent water from power plants is typically 5 to 15 °C above the ambient temperature (Laws 1993). This increase and other sudden changes in the temperature through plant closings or production changes exceed the tolerance of sensitive species and processes (such as rearing young). In the United States, complying with EPA criteria requires power plants to mix effluent and receiving water vigorously or to cool the water significantly before release.

More unique to industrial effluent is the presence of industry-specific toxins. Of the industrial pollutants, four present the most severe and long-term impacts: organic chemicals, heavy metals, discharge of saline waters, and acidification (see chapter 17 for details of how each of these toxins affects water quality).

ORGANIC CHEMICALS. Organic chemicals are produced by a wide variety of industries, including the manufacture of synthetic detergents, fuel additives, solvents, plastics, resins and synthetic fibers, and chemicals, and from petrochemical refining, petroleum drilling, pharmaceutical manufacturing, iron and steel plants, wood pulp and paper processing, coal mining, and food processing. On a global basis, it is estimated that nearly 2 million chemical compounds have been synthesized, and an estimated 100,000 were in commercial use in the mid-1980s. About 30,000 man-made chemical compounds are used in the Great Lakes basin, for example, and more than 1000 man-made compounds have been measured in the Great Lakes ecosystem. Toxic effects including carcinogenicity, mutagenicity, teratogenicity, and other effects are associated with a wide range of organic contaminants.

HEAVY METALS. Industrial sources of heavy metal pollution include dust from smelting and metal processing; discharge of heavy metal solutions used in plating, galvanizing, and pickling; uses of metals and metal compounds in paints, plastics, batteries, and tanning; and leaching from solid waste dumps. In addition, coal, phosphate, and metal mines pollute fresh water with heavy metals. For example, heavy metals leaching from mine wastes contaminated rice fields in the Ichi River basin in Japan; these contaminants have been correlated with a high incidence of kidney failure. Heavy metal pollution from zinc and lead ore mines (and the world's largest zinc processing plant) in the Upper Silesia area of Poland have eliminated fish in the Szola River and contaminated sediments in the Vistula, Biala, and Przemsza rivers (Suschka 1993). A 1980 study showed that about 70% of anthropogenic heavy metals in the Federal Republic of Germany's Ruhr River came from industrial sources. Heavy metals can cause serious health hazards (see chapter 17).

SALINE WATER. Brine and industrial causes of salinity come from a wide variety of sources, including mining, petroleum extraction (oil wells), and salt and potash mines. Saline discharges from mines in Germany have made water from the Rhine unsuitable for greenhouse gardening in the Netherlands. Water withdrawals for mine drainage can also cause salt water intrusion into fresh water aquifers. Saline discharges from coal mines in Upper Silesia in Poland are also enormous: 950,000 cubic meters of saline water are pumped daily. Nearly two-thirds of that amount, containing 7000 metric tons of salt, is released into the tributaries of the Oder and Vistula rivers, rendering these rivers unfit for all purposes.

ACIDIFICATION. Acidification of fresh waters is a widespread phenomenon that is unique because the pollutants are carried by air, rather than water. Industry may acidify lakes and rivers through industrial wastes as well as through atmospheric deposition. Significant acidification of the Mississippi, for example, has occurred in the lower 300 km of the river, which is known as the industrial corridor. As documented by Rutherford and Walker-Bryan (1992), more than 80% of the 130 major industrial point source discharges (each averaging at least 190 tons/day) along the lower Mississippi River occur downstream from Baton Rouge. During 1987, 218,000 tons of industrial waste were discharged to the Mississippi River; more than 92% of the total was released in the approximately 150 km stretch between Baton Rouge and New Orleans. Approximately 61% of the toxic wastes discharged were acidic or acid forming.

Managing Industrial Waste

For industries to produce cleaner effluent, they must typically modify their processes. One approach is to use new equipment to recover chemicals, a strategy employed successfully by DuPont for both fungicide and paint production (Laws 1993). Other industries have switched chemicals used in production—for example, printing operations that use water-based inks generate virtually no hazardous waste, whereas the traditional solvent-based ink result in significant hazardous waste. Substituting water for chemical solvents has also been adopted successfully in other industries, such as pharmaceuticals.

In-plant recycling effort and adoption of new treatment/production systems are strategies employed in many industries. Pulp and paper mills have adopted on-site treatment facilities such as water clarifiers and bio-oxidation ponds. These on-site facilities are immensely successful in reducing suspended solid and BOD pollution. While the transition to new technologies is often expensive initially, long-term cost savings frequently result.

Dams and Channelization

Dams and channelized rivers are by no means unique to urban areas; they have been a hallmark of water management in rural and urban areas alike. Nevertheless both modifications frequently result from urbanization pressures, in the broadest context—expanding populations, demands for increased services (e.g., electricity), and the requirement of increased irrigation to feed urban markets. Consequently, these effects are covered in this chapter, with the understanding that they represent a broad interpretation of "urbanization."

Levee construction, river channelization, filling of lakes and wetlands, dredging lakes and rivers, and harbor or marina construction are among the most common physical modifications in urban areas. They are generally undertaken to make the urban environment more predictable and suitable for development but can initiate a series of unintended water quality effects. Physical modifications, therefore, cover an enormous scope of activities. Common to each, however, are changes in water flow and sedimentation rates and changes in shore and riverine buffer zones (e.g., wetland areas).

River Channels and Levees

One typical urban protection structure is coastal flood barriers. Because these structures hinder the free flow of water between the river and the coastline, they slow water flow, thereby increasing sedimentation and trapping sediment-bound chemicals behind the barrier. In a 1993 study, Ikonnikov (1993) demonstrated that a 26 km long flood prevention barrier for St. Petersburg, built in the 1980s, has led to steadily decreasing water quality in the Neva River (Gulf of Finland). The effect of the barrier was to reduce (i.e., concentrate) the size of the sedimentation basin in the Neva estuary, leading to rapid and significant sediment accumulation in the estuary. Some parts of the estuary have decreased 1 m in depth in the past 5 years. Because the Neva River sediment carries significant pollution from industrial and municipal wastes, riverbank landfills, and agricultural and cattle-breeding operations, the concentration of sediments and reduction in flow has created a serious situation.

Structural changes such as channel and levee construction have largely localized impacts. Local conditions and engineering specifications will determine whether effects are beneficial or detrimental to water quality. Typically, channelized rivers decrease habitat diversity and increase flow rates (if natural slopes are not altered; otherwise, flow rate may be slowed). Because the increased flow rate often intensifies bank scouring, channelized rivers are usually lined with stone or concrete. Consequently, there is little habitat diversity and even greater irregularity in flow regimes, with higher and quicker water flow occurring during rainfalls (Fig. 20.5). The irregular flow restricts establishment of stable macrophyte or invertebrate communities. In addition, biotic diversity is commonly lower in channelized rivers than unmodified ones. Stream temperature in channelized streams also commonly rises due to changes in bank vegetation; that is, trees and bushes are usually set back further than in unmodified channels, providing less shading (Haslam 1990). Channelized streams are typically shallower than their unmodified counterparts, in part because of urban effects on groundwater and because of fewer tributary connections to the streams (Haslam 1990).

Other structures such as coastal levees have the opposite effect. By providing new substrate and structures in "cleared" coastlines, these modifications

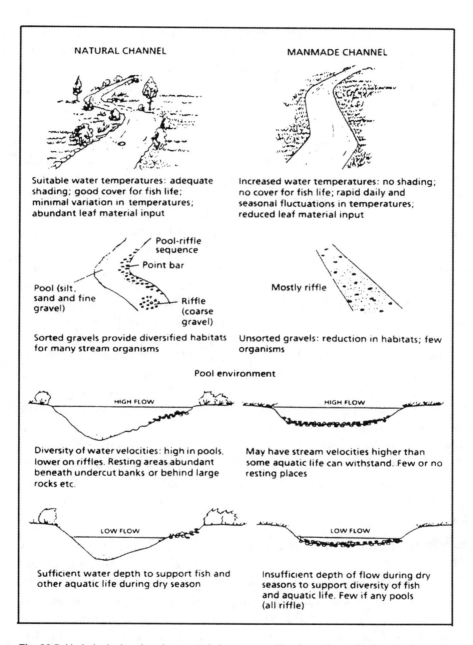

Fig. 20.5 Hydrological and environmental changes resulting from channelization create a wide range of impacts on water quality and biological communities of streams. (After Brooks 1994)

may increase habitat and consequently biotic diversity. They can provide shading and relief and act as refuge for larval and young fish. They also act as sediment traps and interrupt the flow regime.

The Phenomenon of Dams

Although they are located outside of urban areas, dams are one of the more dramatic effects of urban development. Providing water for drinking and irrigation has been a critical component of countless civilizations, from the Aztecs to the Romans to modern industrialized nations. Damming has occurred at unprecedented rates since the 1940s, however, led by the societal need to have energy (generated by hydropower) and to control unpredictable rivers. By 1980, China alone had completed 80,000 reservoirs, with a combined storage accounting for 16% of annual run-off; more than 12,000 large dams (with a height greater than 15 m) are found worldwide (Gleick 1993). Large dams are currently being completed at a rate of about 500 per year throughout the world (Gleick 1993); by the year 2000, it is estimated that more than 60% of the world's total stream flow will be regulated (Canter 1985). Dams block normal river flows, slowing renewal, altering river channels, creating lakes, and flooding terrestrial vegetation. They create a cascade of biotic effects, although these effects vary tremendously with dam size and the size of the resultant reservoir, with nature of the vegetation flooded, with area climates and soils, and with dam management practices such as water release (Fig. 20.6). Judging specific effects of a dam in one region based on measurements in another region has yielded a tenuous correlation at best. Nonetheless, some general patterns are evident in fisheries, human health, bank erosion and sediment load, temperature changes, reduced flows and velocities, variable water levels, and altered habitats. Most of these effects are interrelated.

IMPACTS ON FISH. Immediately after completion of a dam, a decline is often seen in diversity of native fishes because the anthropogenic reservoir environment differs dramatically from the original river ecosystem. Dams also interrupt many species' seasonal upstream migrations and alter the flow, dissolved oxygen, and temperatures that affect reproduction and survival. In addition, populations of native fishes generally decline because of newly erected barriers to long-distance, upstream spawning migrations. Fish ladders and special fish channels are designed into many dams, but they do not compensate entirely for changes in the original river. Other effects on fisheries come from large seasonal variations in reservoir levels and from nonseasonal drawdowns in flood control reservoirs, which can significantly affect near-shore habitats where fish feed and spawn.

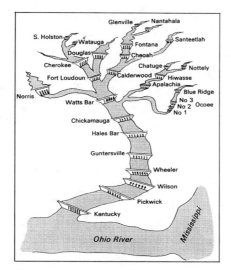

a

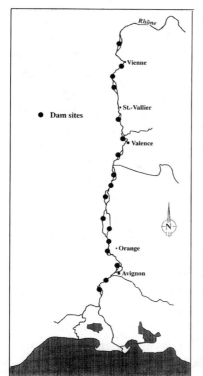

b

Fig. 20.6 Dams are a common feature of many riverine landscapes. Larger rivers are often regulated by multiple dams, with subsequent cumulative effects on water quality and biological assemblages. Typical examples of multiple dams come from (a) the Tennessee River in the United States (after Newson 1992) and (b) the Lower Rhone River, France (modified from Fruget 1992).

Once a community of reservoir-adapted fishes becomes established, fishery harvests are often excellent (although the diversity of fish may be lower than before dam construction). Yields can often increase significantly after completion of large dams because of the greater availability of nutrients from decaying vegetation and soils. These high productivity rates rarely remain stable, however (Gleick 1993). Over the long term (i.e., several decades), reservoir productivity will decline to a level consistent with, and controlled by, upstream nutrient loadings.

HEALTH EFFECTS. Because dams create large reservoirs with low recharges, they frequently create environments conducive to the development of human parasitic diseases such as schistosomiasis and malaria. Increased water availability also usually attracts more people to the water resource, resulting in an increased disease incidence. Other health effects may accumulate over time if sediment trapped in the reservoir contains heavy metals or organoclorines from upstream sources.

WATER LEVEL AND FLOW RATES. The effect of dams is felt both in the newly created reservoir and in the river below the reservoir. Within the reservoir itself, the flow of the river is interrupted and slowed, leading to longer retention time of nutrients, and hence local increases in phosphorus and reductions in organic carbons. Eutrophication of reservoirs and their rivers is a problem frequently encountered by dam managers. Below the dam, the velocity and depth of the stream is controlled by the dam's water release schedule. Typically, the water levels will be more consistent throughout the year. In addition, the flow may be lower and slower than before dam construction. Liu and Yu (1992) document annual changes in stream temperature because of the changes in minimum and maximum flow and depth. In the formerly low-flow season, the effect of the dam was to lower temperatures by 4–6 °C, while stream temperatures were lowered by the same amount in formerly high-flow seasons.

Dumnicka *et al.* (1988) document other impacts of the water release on the below-dam stream. They found (and other studies confirm) that dams generally do not change the physico-chemical nature of a stream: what flows into a reservoir typically flows out. The placement of outflow pipes is critical for downstream conditions, however. Pipes placed low in the reservoir, for example, typically release nearly anoxic water into the stream, making the stream inhospitable to biota for some distance below the dam. Other dams release surface water that is oxygen-rich, but warm and high in nutrients; eutrophic streams result. Dumnicka *et al.* summarize their findings by noting: "In clean or only slightly polluted rivers changes in physico-chemical properties of the water are responsible for the changes in the zoocenoses. In reservoirs from which only a small amount of water flows out reduction in flow is most important, while in polluted rivers it is pollution which determines the composition of the zoocenosis" [p. 183].

SEDIMENTATION. Two problems exist with sedimentation and dams. First, sediment builds up behind the dam, reducing reservoir volume and creating a sink for toxins and nutrients that can be mobilized as reservoir levels and water temperatures change. Sediment buildup also alters reservoir depth, changing the nutrient cycling and making the reservoir more susceptible to eutrophication and less able to store water as designed. Second, river sediments are important mechanisms in maintaining flood plain productivity: some level of variable flooding is typically required in natural stream ecosystems to sustain functional flood plains and high productivity (Gleick 1993). Although the dams successfully reduce the devastating effects of floods, they also reduce the beneficial side effects of floods—that is, fertile flood plain soils.

TRIBUTARY JUNCTIONS. One particularly productive and vulnerable habitat lies at the confluence of a main river channel and its side channels. Side channels often serve as refuge for plants and animals. Because they are usually shallower than the main channel, they may represent a richer feeding ground. In Nepal, where the proposed Chisapani Dam is under review, flooding the tributary junctions may lead to extinction of the endangered river dolphin. Such choices (i.e., hydropower versus protection of an endangered species) have become common concerns in both developed and developing nations.

HUMAN IMMIGRATION. Another effect of many dam projects is that the provision of irrigation or the lowered risk from flooding attracts more people to areas with sensitive environments. Increased populations in and around dams can greatly complicate waste management infrastructures, increase demand for water above predicted levels, further modify habitat, and create new demands not considered in the initial project designs.

Managing the Effects of Dams

Dams have been the traditional answer to problems of irregular water supply, but many smaller-scale and less environmentally damaging alternatives do exist in many areas. For instance, flood control is frequently aggravated by development in flood plains and channelization of rivers. Protecting flood plains and vegetation in drainage areas, establishing riparian buffer strips, and zoning flood-prone areas for parks and green belts remain viable and less expensive alternatives to dams. In addition, dam construction for water storage could be avoided in some regions if water was used more efficiently. Improved planning of nonconstruction alternatives, appropriate maintenance of infrastructural components to reduce leakage, and implementation of more realistic water rates are other effective means of improving efficiency.

Many areas in the United States are currently debating whether to dismantle certain dams. In some cases, it will cost less to dismantle aging dams than to repair them, and communities have placed greater importance on the amenity values that would be gained by recovering fast-flowing, unregulated rivers. At least four dams are being considered for demolition in Minnesota, and debate over removing large dams such as the Glines Canyon Dam in Washington state have made national news (Egan 1990).

Sensitivity to below-dam water quality problems and requirements of biota for changes in annual flow regimes have led to changes in water release schedules in some areas. Allowing flood-like releases on a scheduled basis, and restricting flow during seasons in which biota are cued to low-flow regimes (e.g., spawning), may help alleviate some downstream effects on diversity and flood plain fertility.

Monitoring and Scale Effects

As described previously, urbanization involves two principal influences: significant changes in hydrology (e.g., increased impermeable areas, more flashy streams) and increased loading of a variety of wastes. The hydrologic changes increase the velocity with which water leaves the landscape and increase the erosive power of that water. Thus, the variables most often affected by changes in hydrology and considered in monitoring include sediment load, total water volume, peak-stage height, the timing at which peak stage height is reached, and the presence of petroleum products such as oil and grease. These changes influence the ecology of receiving waters in relatively predictable ways. For example, changes in hydrology and sedimentation effects will be similar to those discussed in chapters 18 and 19 for forestry and agriculture, respectively. Oil and grease will have little ecological impact unless they are present in relatively large quantities, such as in the case of an oil spill. Oil and grease will present a threat to a downstream potable water supply before they have a significant ecological effect. As a rough estimate of spatial impacts, effects of hydrology and sediment usually will be detectable less than 10 times the distance downstream that a stream flows through an urban area (Fig. 20.7).

A second significant focus of monitoring efforts is the discharge of municipal wastes. These point sources can be highly variable in composition as well

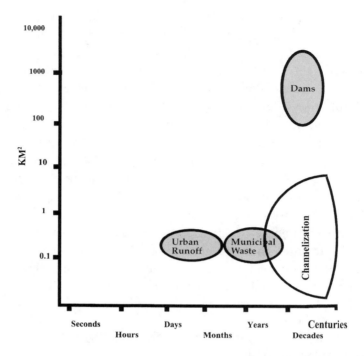

Fig. 20.7 Urban water quality concerns cross a diversity of spatial and temporal scales. Deoxygenated waters, for example, are generally midscale concerns, while dams can affect the fish communities along a broad stretch of river. Industrial pollution may be concentrated in a small area, but create a significant concern within that area.

as volume, but a few consistent patterns have been identified. Municipal effluents discharged without any treatment or with only primary treatment will result in relatively large deoxygenated zones in the receiving water body, where the benthos is dominated by animals such as *Atherix* larvae and some chironomids that are tolerant of nearly anaerobic conditions. Little to no primary productivity and very high ecosystem respiration will occur under these conditions. As outlined in chapter 17, reaeration from the atmosphere and stream turbulence will create a zone of severe impact that covers a relatively small spatial area (e.g., several tens to a few hundred meters). In contrast, the recovery zone may be 10 times the length or area of the impact zone. The zone of recovery will be characterized by low productivity, relatively high rates of respiration and decomposition, and a low diversity benthic community. Temporally, these kinds of discharges have relatively low variability. Monitoring of impacts should focus on dissolved oxygen, organic load (e.g., BOD_5), benthic communities, and (rarely) ecosystem respiration.

In most areas of the world, municipal wastes are treated via secondary treatment to an effluent standard of ≤ 30 mg/L BOD_5 and ≤ 30 mg/L suspended solids. In these cases, the zone of impact will have relatively high organic loads but responses in the benthic community will be limited to specific changes in certain populations. Usually, primary productivity will increase because of the increased nutrient loadings. Temporally, these systems will have relatively low variability. Monitoring should focus on organic loads and on nutrient dynamics. Nutrient supplements and their impacts on the biology (e.g., increased primary productivity in downstream still waters) would be expected.

Because industrial effluents vary widely in quality, quantity, treatment, and variability, it is very difficult to characterize them in a meaningful way. Generally speaking, they are discharged as point sources and usually reflect the process and product of the associated industry. As described in this chapter, industrial effluents often contain elevated metal and toxic organic compounds and elevated nutrient loadings. Actual concentrations are highly dependent upon the industry involved and the degree to which it treats its effluent.

Combined Pressures of Urbanization on the Upper Mississippi River

The nearly coincidental construction of sewers and dams along the upper Mississippi in the early part of this century was a unique occurrence. Sewer lines emptied raw sewage into the Mississippi, while dams backed up water containing those wastes. Instead of having the wastes washed downstream, residents of St. Paul and Minneapolis were confronted with increasing waste accumulation and severe water quality impacts. Eventually this situation forced construction of the first waste treatment plant on the upper Mississippi, a move far ahead of its time.

The development of the upper Mississippi area is typical of much urban development in the United States. Rapid population growth in the late 1800s forced a new structure on the neighboring cities of Minneapolis and St. Paul. Before that time, local waste management was solved by transferring the problems down river. During the last three decades of the nineteenth century, however, the population of Minneapolis doubled four times, from slightly more than 13,000 in 1870 to about 200,000 in 1900. During the same time, the population of St. Paul grew from 20,000 to 165,000 (Scarpino 1985). Along with the increase in total numbers of residents came new businesses, industries, and services that resulted in an exponential increase in waste. The most obvious "solution" to immediate problems was the Mississippi River—a cheap, convenient vehicle for carrying away garbage and sewage. The easy availability of the Mississippi made development of sanitary sewers feasible financially. Once sewers were built, the river "took care of" disposal. Other cities and towns used the same practice all along the upper Mississippi. By the late 1880s, Minneapolis alone dumped about 500 tons of garbage per day into the river (Scarpino 1985).

Scarpino (1985) notes that sewers represented a major urban reform, waterway dumping transformed a local waste problem into a regional water pollution crisis and initiated a series of chain reactions. Much of twentieth-century pollution-related history may be understood in terms of public response and institutional adjustment to unintended consequences of this sewer technology.

By the 1920s and 1930s, sizable segments of the Mississippi had become degraded by domestic sewage, especially south of Minneapolis–St. Paul and below St. Louis. Garbage shoals and disease followed: for example, St. Louis suffered a typhoid fever outbreak in the early 1900s when the Mississippi polluted its main water source (Scarpino 1985). Federal and state laws lacked authority to force river cleanup and little incentive existed for individual cities to implement expensive sewage treatment facilities. Treatment was seen as a large outlay of money for someone else's (i.e., downstream residents) benefit.

The growth of the Twin Cities brought with it increased pressure to improve navigation on the upper sections of the Mississippi. In response, the Twin Cities lock and dam was constructed and opened in 1917. The lock and dam successfully accomplished its navigation goal, but planners had failed to consider that most of the Twin Cities' new sewers emptied into the 5 mile long pool created by the new structure (Scarpino 1985). Instead of the Mississippi carrying waste away downstream, waste was now held in a reservoir flowing through the middle of the Minneapolis–St. Paul metropolitan area. A seemingly easy solution was to open the dam in the winter and flush out accumulated sewage sludge. In 1924, however, the Army Corps of Engineers contracted to sell hydroelectricity generated at the dam to a Ford Motor Company assembly plant. After that date, the dam remained closed year-round and pollution emerged as a significant local issue (Scarpino 1985). One report maintained that for nearly 50 miles below the Twin Cities, the river lacked sufficient dissolved oxygen to support any fish life; instead, the river offered little but floating sewage, odors, and disease.

Scarpino (1985) reviews the history of a second lock and dam completed in 1930 in Hastings, Minnesota, 37 miles below St. Paul. The reservoir behind Hastings' dam extended to the foot of the Twin Cities lock and dam and served to collect the remainder of the Minneapolis–St. Paul sewage outfalls, including wastes from several packing houses and stockyards. The situation came to a crisis point in July 1930, when Congress passed a Rivers and Harbors Act that authorized construction of several locks and dams between Alton, Illinois, and the

Twin Cities, which created a 9 foot deep navigation channel. Pools behind these dams threatened to become septic tanks for wastes from cities and towns along upper stretches of the river.

Aggravation of pollution problems caused by the Twin Cities dams was largely responsible for the "crisis" that led local governments to propose a combined sewage system for the Twin Cities consisting of a single treatment plant located just south of St. Paul. Assistance from the federal government during Franklin Roosevelt's administration helped the Twin Cities cover the multi-million-dollar price tag of their proposal. Construction of the combined sewage system began in 1934 and was completed in 1938. Twenty-five years later, the project was pronounced a success when game fish and healthy biota could be found in the river. After the Twin Cities committed themselves to the treatment plant construction, several smaller cities as far downstream as La Crosse, Wisconsin, signed contracts for their own facilities (Scarpino 1985). Interestingly, the same year that Minneapolis–St. Paul began the sewage treatment plant construction, St. Louis asked for permission to dump all of its garbage into the Mississippi (it had previously contracted with local hog farmers for some of its refuse disposal).

Eventually, the single outstanding difference in decision making between St. Louis and Minneapolis–St. Paul remained construction of navigation dams in the upper river. Although they did not cause the pollution problem, dams aggravated the situation and focused attention on the deteriorating quality of the river. River dumping ceased to be an acceptable alternative when sewage backed up into the metropolitan areas, and residents began to pay the social and environmental costs of pollutants that the river had formerly washed away. The Twin Cities found it possible to finance a treatment facility, overcome inadequate legislation, and smooth over differences among river-related interests when the dams again made pollution a local problem. At St. Louis, the much larger span of the river continued to flow unimpeded to the Gulf of Mexico. For St. Louis officials, the Mississippi remained a cheap, expedient means of carrying wastes out the city.

Summary

- Urbanization is a major change in land use that is occurring widely throughout the world. Although not a core natural resource management issue, urbanization results from changing land use patterns and causes significant changes in other natural resource practices.

- The hydrograph of urban streams is significantly more flashy than that of non-urban streams—that is, precipitation events cause a more rapid and larger spate (flood) event. In addition, a much greater quantity of total run-off leaves the urban landscape for a given quantity of precipitation.
- Urban waters are usually under stress from altered chemical–physical properties. For example, temperatures are usually higher, sediment loads are usually heavier, substrates are usually less diverse and smaller in median grain size, stream channels are usually geomorphologically simplified (e.g., cleaned, straightened), and water chemistry is usually altered (i.e., higher dissolved solids, more oil and grease, often higher nutrient concentrations).
- The stresses they face make urban waters more susceptible to invasion by exotic species. Within any given landscape, urban waters are often the first locus of invasion for such species as purple loosestrife and zebra mussels.
- Increasingly, urban waters are being seen as a highly valued resource rather than as a drain for urban wastes. In many programs around the world, residents, businesses, governments, and nongovernmental organizations are using waterways as the focal point for urban action groups. Both the number and diversity of such activities can be expected to increase in the future.

For Further Reading

Fruget J. Ecology of the lower Rhone after 200 years of human influence: a review. *Regulated Rivers: Research and Management* 1992;7:233–246.

Haslam S. *River pollution: an ecological perspective.* London: Bellhaven Press, 1990.

Ikonnikov V. Pollution of sediments in the Lake Ladoga catchment area and the gulf of Finland: influencing factors and processes. *Land Degradation and Rehabilitation* 1993;4:297–306.

Marsalek J, Dutka B, Tsanis I. Urban impacts on microbiological pollution of the St. Clair river in Sarnia, Ontario. *Water Science and Technology* 1994;30:177–184.

Suschka J. Effects of heavy metals from mining and industry on some rivers in Poland: an already exploded chemical time bomb? *Land Degradation and Rehabilitation* 1993;4:387–391.

21

Special Issue: Cultural Eutrophication

Overview

One of the most pervasive water quality problems around the globe is eutrophication. Eutrophication begins when nutrients are added to a lake or stream, promoting plant and algal growth. High nutrient inputs can lead to excessive productivity that deoxygenates water, reduces light infiltration, changes species composition, and impairs use of waters for drinking, navigation, and recreation. Because the mechanism of eutrophication involves influx of nutrients, nearly every land use contributes to the problem. Although attempts have been made to counteract effects of nutrient fluxes, reducing the flow of nutrients (in particular, phosphorus) into lakes and running waters remains the only effective, long-term management strategy for reducing eutrophication and enabling ecosystem recovery.

Introduction

Eutrophication is a natural process with a poor image. The term "eutrophication" refers to the gradual increase in productivity and decrease in depth that occurs over time in most lakes. The process results from the slow accumulation of nutrients (which increases primary production) and the parallel filling in of the lake with silt and decaying matter (Fig. 21.1). Cultural eutrophication is similar, except that it occurs more quickly, has less relationship to lake depth, may be more dramatic, and, most importantly, conflicts with other uses of an affected lake or river. For some uses, such as raising catfish, eutrophic conditions are desirable and encouraged. In lakes and rivers that are valued for

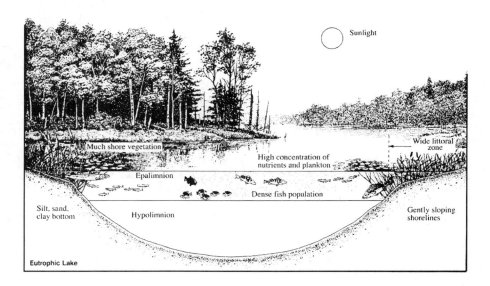

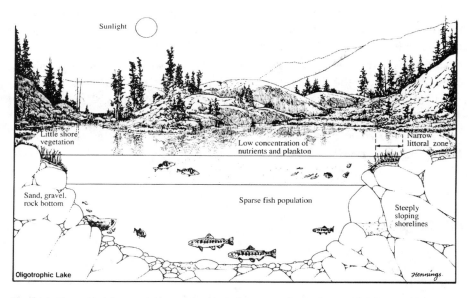

Fig. 21.1 Eutrophic lakes are characterized by being generally shallow, with high nutrients and rich vegetation growth. Deeper layers are frequently anoxic, and substrate is generally sandy or silted. Oligotrophic lakes are low in nutrients, but are generally well oxygenated throughout. (After Miller 1992)

aesthetic, recreation, and economic uses, however, eutrophication poses a serious and prevalent problem.

The eutrophication process begins with nutrient enrichment of waters, primarily with input of phosphorus and nitrogen. Because plant growth is typically nutrient-limited, the extra nutrients enhance the growth of both algae and macrophytes. At low levels, this process can lead to productive biotic communities. When nutrient input is high, however, plant growth can be excessive, leading to algal blooms and both floating and submerged mats of algae and macrophytes. As microorganisms decay these large quantities of plant material, they take up oxygen from the water. Consequently, large phytoplankton and algae growth can lead to severe depletion of water oxygen reserves. In turn, fish die, decomposition-related gases such as CO_2, CH_4, H_2S, and toxins are released, and offensive odors drive away would-be users of the affected water body.

Eutrophication is a global phenomenon, affecting between 30% and 40% of fresh water lakes worldwide (Meybeck *et al.* 1989). The impact of the problem is even greater than this enormous percentage would imply, because the affected lakes represent the majority of those in and around settled areas. Eutrophication—or more correctly, cultural eutrophication (i.e., caused by human activities)—is tightly linked to demographic growth and the nutrient inputs from sewage, fertilizers, and urban run-off. As a result, eutrophication is primarily a local problem solvable through local efforts. In many cases, changes in local and regional-scale land use practices have led to improvement of eutrophic conditions.

The cultural practices that cause eutrophication represent a wide variety of activities. A significant percentage of nitrogen and phosphorus inputs come from nonpoint sources such as agricultural fertilizers. In the industrial sector, production of agricultural products and fertilizers, pulp wastes, food processing, and distilleries contribute to eutrophication by discharging nutrient-rich wastes to water bodies. Effluent from sewage treatment plants is also typically high in nutrients, as is urban run-off that reaches receiving waters through storm sewers. In the United States, 55% of estuaries, 40% of lakes, and 37% of rivers are reported to be negatively impacted by nutrient inputs (EPA 1992).

Natural Versus Cultural Eutrophication

As described earlier, eutrophication is a natural process, and highly productive, eutrophic lakes and rivers can occur naturally. Like all ecosystems, lakes change over time as the natural processes of sedimentation, afforestation, and deforestation affect the watershed. Harper (1992) notes that most lakes and reservoirs have a finite lifespan; if undisturbed, they will fill with sediment and be replaced by terrestrial communities. This period may vary from a few years for lakes created by channel changes of rivers, to millions of years for deep, large lakes created by tectonic processes. When a lake becomes sufficiently shallow, the smaller volume of water and reduced stratification leads to improved nutrient recycling and a gradual increase in in-lake nutrient concentrations.

The progression from oligotrophy to eutrophy is not a steady and predictable path. Sediment cores reveal that many lakes have gone through several trophic cycles (Harper 1992). Many North American lakes, for example, were eutrophic after formation from glacial retreat. Many were inherently shallow and received nutrients following erosion from postglacial, unvegetated catchments. As catchments became vegetated and erosion declined, these lakes entered a long (several thousand years) and stable period of oligotrophy. Records from ancient lakes reveal even greater fluctuation in trophic state, including several reversals in productivity. For example, Lake Biwa, an ancient lake in Japan believed to be at least 4 million years old, has passed through two oligotrophic phases, two mesotrophic phases, and one eutrophic phase in the last half million years (Harper 1992). Similar patterns have been traced (through sediment cores) for lakes in areas with fluctuating climates such as the equatorial rift valley in Kenya. Many Kenyan lakes have experienced dramatic changes in water levels and subsequent changes in trophic state over the past 3000 years (Harper 1992).

The long natural cycles have generally been eclipsed by human intervention, although from 5000 years ago until approximately 500 years ago, the pace of cultural eutrophication remained relatively slow. Improvements in techniques of cultivation in the Middle Ages brought the first large-scale historical increases in eutrophication, although the eutrophication during this time was usually not severe or problematic. Many fish-rich waters important to early colonists in the United States were the result of moderate nutrient enrichment. In Asia, farmers intentionally increase yields from fish ponds by fertilizing them with human and other animal wastes. The pace of cultural eutrophication accelerated dramatically in the 1930s and in subsequent decades, however, and has now become a common problem in nearly every community (Fig. 21.2).

Demonstrating the Link

Although the linkages between nutrients and eutrophication are now well known, they are not intuitively obvious and went undetected for decades. One of the first documented cases showing that urban and agricultural run-off was related to eutrophication of lakes was developed in Madison, Wisconsin (and is reviewed in Harper 1992). The city's first sewer system was constructed in 1884.

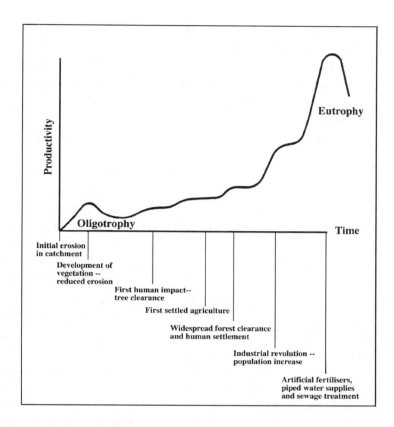

Fig. 21.2 Illustration of the increase in eutrophic conditions over time, relative to different periods of human settlement. (Redrawn from Harper 1992)

The sewers and (15 years later) effluent from a sewage treatment plant discharged into Lake Mendota. By 1918, algal nuisances in Lake Mendota had become serious enough that the city began treatment with copper sulfate (which is toxic to many algal species). Although an engineering report discounted effects of the sewage treatment plant effluent, public attention was raised when adjacent Lake Waubesa began to show similar effects only 4 years after a new sewage treatment plant was constructed in 1926. Detailed studies did not come until 1942, when it was revealed that at least 75% of inorganic nitrogen and 88% of inorganic phosphorus in Lake Mendota came from sewage effluent inflows.

Sources of Nutrient Input

Although eutrophication is a widespread phenomenon, the source of nutrient inputs is almost always local. For example, Mason (1991) notes that the brewing industry in England and Wales releases approximately 11,000 m³ each day of an effluent that is extremely rich in nitrogen and phosphorus (156 mg/L and 20 mg/L, respectively). In general, food processing industries are high in nutrient output; for example, nitrogen and phosphorus levels in sugar cane mill effluent are nearly equivalent to those found in raw sewage (Laws 1993).

Municipal sources of nutrients also create critical loads in area lakes and rivers. In 1971, for example, it was estimated that phosphate detergents were responsible for 30% to 40% of the phosphorus released into U.S. surface waters (Laws 1993). While the mandated reduction in phosphate levels of detergents did not have a significant impact on large lakes, it did have important local effects. For example, the city of Syracuse passed a law mandating that detergent phosphate concentrations not exceed 8.7%; this law was responsible for an 80% decline in phosphorus inputs to Onondaga Lake, New York (Laws 1993). Although detergent phosphates no longer represent a significant source of phosphates, most sewage treatment plants still release 70% of the nitrogen and phosphorus concentrations of raw sewage. In addition, urban run-off from separate storm sewer systems contributes nitrogen and phosphorus. The eutrophication potential of these releases is determined largely by the size, depth, and dilution capacity of the receiving water.

Agricultural fertilizers are responsible for a large proportion of increased nutrient loads to surface waters. Worldwide, fertilizer use is increasing dramatically, with nearly 40% increases in Africa and South America and a doubling of fertilizer use throughout Asia between the 1970s and 1980s (Fig. 21.3) (World Resources 1992). While fertilizer use in Europe has not increased, the region remains the highest user of all continents.

Livestock farming also contributes significantly to nutrient loads, and management of livestock manure remains a contentious issue. Manuring of agriculture fields is an important source of nitrogen leachate. In many European countries (e.g., the Netherlands) farmers must now contend with strict regulations on the timing and amount of manure they can spread on their fields. In addition to manuring, even supposedly secure storage sources such as slurry and silage liquor tanks frequently leak or burst, causing significant water quality damages. Mason (1991) reports that farm pollution incidents in England rose from 1500 per year in 1979 to more than 4000 in 1988 (representing 19% of total yearly incidents). In 1984, 23% of these incidents involved slurry stores, 28% involved livestock yards, and 20% involved silage effluent. The nutrient (and biological oxygen demand) contributions of these releases can be extremely high in local streams and lakes receiving these accidental effluent releases.

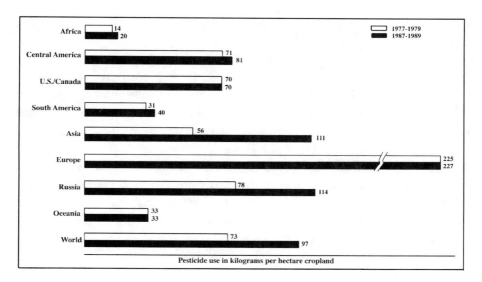

Fig. 21.3 The increase in agricultural use of fertilizers in selected countries from the period 1977–1979 to 1987–1989. Europe has, by far, the largest fertilizer use per hectare of cropland, but its use is no longer rising. In contrast, Asia's fertilizer use has doubled in 10 years and other regions show less dramatic but still alarming increases. (Data from World Resources 1992)

Susceptibility to Eutrophication

Different levels of nutrient loading may have vastly different effects, depending on geomorphic and climatological features of the affected system. These differences primarily revolve around a lake's susceptibility to changes in the oxygen regime as well as the distribution of oxygen changes (i.e., whether they are confined to the hypolimnion).

A change in the oxygen regime resulting from decay of large volumes of plant material is one of the central issues of eutrophication. Oxygen depletion leads to significant decreases in fish and zooplankton populations. This decline results in not only loss of major functions of the water resource, but also breakdowns in structure and function of affected communities. Shallow lakes are more susceptible to eutrophication because all water layers are illuminated and can support the full productive potential of the vegetation. Although the greater water volume in the hypolimnion of a deep lake slows oxygen depletion, oxygen depletion can be more long-lasting if hypolimnetic and epilimnetic layers do not mix (i.e., the lake is stratified). (As described in chapter 12, stratification occurs when the upper, illuminated, oxygen-rich layer (epilimnion), floats on denser, cooler water underneath. Primary production and nutrient uptake occur primarily in the epilimnion. When plants and animals die, organisms and their nutrients sink into the hypolimnion, where bacteria decompose the material, using up oxygen and releasing nutrients. In lakes with permanent stratification, the hypolimnion operates as a permanent, anoxic nutrient sink.)

Although the specific effects of eutrophication in water bodies depend on climatic, limnological, oceanographic, and other conditions, symptoms and manifestations of eutrophication are similar in both fresh and marine waters. Realizing that all productive waters are not necessarily the result of anthropogenic eutrophication is important, however. Naturally productive waters also exist and often sustain a high level of production. The principal water quality issues relate to the ways in which eutrophic waters conflict with other values and the degree to which cultural eutrophication leads to undesirable environmental conditions.

Implications of Eutrophication

The most common practical problems caused by eutrophication fall into five categories, four strictly related to human use and a fifth encompassing ecological qualities:

1. Drinking water
2. Recreation
3. Fisheries
4. Reservoir holding capacity
5. Reduction in biological diversity and resilience

These impacts may also be linked to other causes such as mineral turbidity, humic substances, low dissolved solids, and thermal discharges (Meybeck *et al.* 1989).

Drinking Water and Health Effects

Although rarely hazardous to humans, eutrophic waters are generally unpalatable and require extensive treatment to solve problems of taste, color, and odor. Removing algae and other plant materials also complicates and increases treatment costs because it necessitates more frequent changes in filtration, flocculation, and sedimentation. In addition, eutrophication can lead to corrosion problems in pipes and other anthropogenic structures, raising maintenance costs and increasing risks from secondary health problems such as metals leached from the corroding pipes. Other drinking water problems from eutrophication affect livestock; for example, blue–green algae can be toxic and can devastate local cattle populations (as well as some fish) (Mason 1991). In marine eutrophication, toxic algae (i.e., certain dinoflagellate species) can infect bivalves; if these molluscs are eaten by humans, they may suffer from paralytic shellfish poisoning (Meybeck *et al.* 1989).

Recreation Impacts

Recreational impairment due to eutrophication stems from navigation and dermatological problems from swimming in affected waters (Fig. 21.4). Algal and macrophyte mats can foul boat motors and oars, making boat travel complicated or even impossible. The smell of rotting vegetation also renders many lakes unattractive for any use. Health problems experienced by bathers may be due to high pH (dermatitis, conjunctivitis), swimmer's itch (schistosomiasis), and, in warm climates, bilharziasis and diarrhea (the latter condition due to ingestion of toxin-producing algae). In marine waters, occurrence of many zoeae (the larval forms of certain crustaceans) has also been linked to cases of dermatitis.

a

Fig. 21.4 (a) Culturally eutrophic lakes and rivers are low in recreational value due to floating algal mats, dermatological parasites and smell. (Photo by D. Vanderklein)

b

Fig. 21.4 (*cont.*)(b) In contrast to the algae-covered surface of eutrophic lakes and rivers, oligotrophic lakes have high visibility and support macrophytes such as water lilies. (Photo by D. Vanderklein)

Fishery Effects

While mildly eutrophic waters can create productive fisheries, highly eutrophic waters can lead to large-scale fish kills. Milder eutrophication may not affect the biomass of fish in a lake, but can cause important changes in species composition, especially in the form of increased dominance of undesirable fish stocks. For example, salmonids and coregonids are generally replaced by cyprinids in eutrophic waters (Mason 1991). Fish composition changes primarily because of reduced oxygen in the hypolimnion, which forces oxygen-sensitive species into warmer, upper layers of the lake. Cold-water fish such as salmonids may suffer from thermal stress and eventually be replaced by species tolerant of higher temperatures and reduced oxygen (e.g., cyprinids). Total fish biomass, however, may often increase or at least be maintained in eutrophic waters because of increased total food supply (Laws 1993). Secondary changes such as increased pH from greater plant growth can also modify composition and hardiness of fish.

Overall Biotic Diversity

Other biotic changes related to lake eutrophication include increases in zoo-plankton density and increases in algal, bacterial, and detrital biomass. Species composition may change as well, with effects apparent in everything from zoo-plankton feeding patterns to size-selective impact of fish predation. In addition, benthic fauna may change. Species less tolerant of reduced oxygen levels may decline and those that remain, especially species of chironomid midge larvae and oligochaete worms, may reach high densities. Uptake of nutrients by epi-phytic and planktonic algae typically increases mean biomass and productivity and alters plant seasonal patterns. Species changes then occur, leading to lower species diversity and dominance by species of diatoms, cyanobacteria, and uni-cellular green algae. Even when overall diversity remains stable, eutrophic con-ditions can cause important shifts among dominant species.

Macrophyte populations may also undergo modifications under eutrophic conditions. In special situations such as calcareous waters, eutrophic conditions may cause increases in biomass of submerged macrophytes. More commonly, however, populations of attached macrophytes are reduced because of compe-tition for light with phytoplankton or epiphytes. Diversity declines as species intolerant of low light, higher dissolved solids, or competition disappear.

Holding Capacity Impacts

Another area in which eutrophication can cause difficulties for water quality managers relates to the reduced flow and reduced holding capacity of basins. Many dry areas depend on reservoir water for public supply. In Spain, for ex-ample, fully one-third of the 800 reservoirs are highly eutrophic; serious eu-trophication in reservoirs is also reported in South America, South Africa, Australia, and Mexico (Meybeck *et al.* 1989). Through siltation and clogging of pipes and screens, eutrophic waters can be costly and problematic.

Eutrophication of Rivers

A recent report that the Minnesota River is eutrophic received front-page news coverage in the Minneapolis–St. Paul area. Eutrophication is familiar in the state due to its many lakes, but most people were shocked to hear a river called eutrophic. The Minnesota River is more than the state's namesake—it is a major tributary of the Mississippi and runs through both Minneapolis and Saint Paul. Although eutrophic lakes receive the most attention, many rivers are eutrophic. Other important world rivers that suffer from eutrophication are the Nile, the Volga and Dnieper (USSR), and the Loire (France). Commonly, affected rivers are impounded (as in the above examples) (Meybeck *et al.* 1989). Eutrophication effects in rivers are generally less acute than those observed in standing waters; some increased benthic productivity may even be viewed as beneficial in terms of the self-purification capacity of the system.

Cultural eutrophication of rivers is a more widespread phenomenon than that seen in lakes. Worldwide, rivers have doubled their content of nitrogen and phosphorus in the last 100 years, with local increases of as much as 50 times in Western Europe and North America. Biological responses to elevated nutrient levels in rivers are less dramatic, however. Few rivers run slowly enough to produce a true phytoplankton community, and those that do tend to be dominated by small single-celled species that turn the water green rather than those that concentrate in floating mats or accumulate on the shoreline. Enrichment does commonly lead to increased growth of tolerant macrophyte species or filamentous algae that may cause night-time deoxygenation. Animal species may change in response to the proportions of food available, especially as oxygen-sensitive species are eliminated. Overall, enrichment tends to cause biological effects analogous to those involved in a shift of communities naturally found in lower reaches of the river and to more upstream reaches.

Marine Eutrophication

Marine environments subject to eutrophication are usually those with limited mixing capacity. Some important examples include Oslo Fjord, Tokyo Bay, Manila Bay, Chesapeake Bay, Potomac River, Long Island Sound, Bay of Fundy, Straits of Juan de Fuca, Lake of Tunis, Malaya Straits, Seto Sea, and Java Sea (Meybeck *et al.* 1989). Some unconfined waters have also been subject to eutrophication, such as part of the North Sea, some coastlines of Spain, France, Germany, the Baltic, the northern Adriatic Sea, the Black Sea, California, Florida, regions of Indonesia, the Caribbean Islands, the mouth of the Amazon, and the Rio de la Plata (Meybeck *et al.* 1989). The potential or actual damage in these areas goes beyond an aesthetic nature to direct effects on fisheries and indirect effects on human health. On various occasions, beaches and fishing have been closed temporarily in France, Spain, Italy, and Florida because of eutrophication-related problems.

Control and Recovery

Eutrophication control involves understanding the relationship between nutrient concentration in the lake, external nutrient supply, flushing rate, and sedimentation rate. Once these linkages are established, water quality managers may estimate critical loading of water bodies—the nutrient loading level that, when exceeded, leads to eutrophication. Phosphorus binds to sediments and is lost through erosion of agriculture lands. Consequently, management practices that control erosion contribute enormously to reduced loss of phosphorus. Some reports cite phosphorus losses of as high as 60% of the applied fertilizer (Mason 1991). Some management techniques (from Meybeck *et al.* 1989) are described below.

Approaches that reduce nutrient input include the following:

- Treatment of wastewater to eliminate nitrogen and phosphorus
- Diversion of wastewater from the affected lake or river (ring canalizations)
- Channeling of effluent to sedimentation basins before release into affected waters
- Direct precipitation of nutritive substances in the effluent
- Watershed protection (reforestation; restriction of fertilizer use; restriction of industrial livestock breeding establishments; controlled fertilization/ irrigation)
- Substitution of phosphate detergents by zeolites

Corrective measures within the water body include the following:

- Physical manipulation (destratification; hypolimnetic aeration; withdrawal of hypolimnetic water; alteration of flushing regimes)
- Chemical and sedimentary manipulation (nutrient precipitation inside the water body; inactivation and removal of sediments)
- Biological manipulation (mechanical harvesting of biomass; macrophyte, algal, and fish community manipulation; application of toxic substances; herbicides, algicides, and pesticides; direct manipulation of the food chain and the biological equilibrium)

Experience with these alternatives reveals that eutrophication can be reversed. To be successful, the choice of corrective or preventive measures must be not only based on general knowledge of biological and physical dynamics of water bodies, but also tuned to specific properties of the target water body. For example, hypolimnetic aeration is appropriate only for water bodies of intermediate depth, while ring collectors can protect a lake but subsequently divert nutrients to downstream waters (Meybeck *et al.* 1989).

Recovery from eutrophication occurs slowly following reductions in phosphorus, and extends over time down through the water column into sediments. For example, because of land use changes, phosphorus concentrations in Lake Geneva increased from 10 mg/m^{-3} in 1957 to a maximum of 90 mg/m^{-3} in 1979. Slow decreases from that high yielded concentrations of 55 mg/m^{-3} in 1990 (Lang and Reymond 1992). Following this reduction, phytoplankton biomass, chlorophyll concentration, and planktonic primary production all declined. Last to respond was the oligochaete community, indicating that reversal of eutrophication finally extended to the sediments.

Strategies and Legislation

Because eutrophication generally stems from nonpoint sources, the process is difficult to legislate and control. Controlling nonpoint sources means restricting the land use practices of private property holders; it is, consequently, a contentious issue in the United States. Powerful and sweeping legislation to control nonpoint pollution that had Congressional support one year (1994) lost that backing the next year following major political moods swings. The proposed Clean Water Act of 1994 had included increased federal aid to assist in improvement of wastewater treatment and financial assistance for watershed planning; requirements to submit and implement plans for cleaning up lakes, rivers, and streams (especially in identifying ways in which to reduce nonpoint pollution); and new regulations governing street sweeping and other methods of reducing storm sewer pollution (including citizen education and reductions in lawn-care products). While these measures may not become national law, many are being implemented at local levels by affected communities. Typical of many communities, Madison, Wisconsin, and surrounding communities have implemented intensive public education campaigns, modified street sweeping programs, and increased enforcement of point sources of nutrient inputs into area lakes. The city of Syracuse's phosphate detergent regulations described earlier provide another example of successful local-level efforts to improve area lakes and rivers.

Meybeck *et al.* (1989) report that Sweden, Denmark, Germany, Austria, Switzerland, and Italy have legislative programs for treating eutrophic waters. In Sweden, 80% of treatment plants include a third stage for eliminating phosphorus and only 20% of water discharged into waterways receives no treatment.

Denmark, where agricultural uses contribute heavily to eutrophication, has introduced a computerized fertilizer control system. Individual farmers are responsible for reducing their use of fertilizer. If desired reductions are not reached within a stated time, fertilizers will be highly taxed. The regional government of Emilia-Romagna, Italy, has introduced legislation to combat inputs of nutrients to waterways from intensive livestock breeding. Part of the legislation imposes a maximum limit on the number of hogs per unit land area; other elements provide incentives to farmers for improved farm waste use.

Economic costs of eutrophication are enormous; the cleanup alone is extremely expensive. Between 1989 and 1995, Austria, for example, spent approximately US $740 million, or US $1.3 million per square kilometer, on eutrophication control (Meybeck *et al.* 1989). In Switzerland, eutrophication control costs for two lakes are estimated at US $2.5 million, with long-term maintenance costs of US $0.5 million per year.

Eutrophication has been most widely studied and modeled in the northern hemisphere. Managers must exercise caution, however, in applying regional models to other areas. Tropical and subtropical lakes respond to nutrient additions differently than temperate lakes. Greater turbidities, more erratic rainfall, year-round productivity, and highly erosive soils in tropical areas all modify the process of eutrophication.

Case Study: The Great Lakes

Cultural eutrophication in the Great Lakes basin has been extensively documented, mostly with respect to changing water chemistry and changes in species composition and abundance. In Lake Ontario, whole-lake total phosphorus concentrations measured in spring reached peak values of about 50 $\mu g/L^{-1}$ in 1971 (Sly 1991). In response to nutrient controls under the Great Lakes Water Quality Agreement (1972) between Canada and the United States, mean total phosphorus concentrations have since decreased by 80%. Loadings of nitrate continue to rise, however, largely because of increasing contributions from atmospheric sources.

One of the most intensively studied areas is the Bay of Quinte, an estuaine arm of Lake Ontario. As described by Sly (1991), until the 1940s phosphorus loadings to the bay increased almost coincidentally with human population increases. Following the introduction of detergent phosphates, loadings increased dramatically during the 1950s. Changes in nutrient loadings have had profound impacts on macrophytes, phytoplankton, zoobenthos, and fish species. These changes have been the principal influence behind most recent ecosystem changes in the bay.

During the late 1800s, as many as 37 species of aquatic plants were recorded in the Bay of Quinte. During the 1950s, moderate eutrophication maintained lush weed beds. The increasing nutrient levels subsequently favored algal growth, causing higher levels of turbidity and decreased light penetration. Particularly because of light attenuation, macrophytes were soon outcompeted by phytoplankton, and the extent and diversity of weed beds were severely diminished in the 1960s. As vegetation cover disappeared, resuspension of bottom sediment increased as shallow water areas lost their protection from wave action. While nutrient loading has since returned to the early 1950s level and summertime algal biomass and ambient water phosphorus concentration have been reduced, Sly (1991) reports that much of the bay appears to be locked into its current degraded state.

In addition to macrophytes, many fish species found in the Bay of Quinte declined in the 1950s and 1960s. As spawning grounds were covered by dead and rotting algae, dissolved oxygen was depleted, killing fish eggs during fall and winter. Degradation of *Cladophora* and milfoil became a serious threat to lake trout spawning shoals in other parts of the lake.

Monitoring Eutrophication Effects

The process of eutrophication is usually rather slow and location-specific. As described in this chapter, water bodies may receive increased loads of nutrients through the passage of time and through changes in land use practices. Phosphorus is of primary concern in fresh water systems in the temperate zone, while nitrogen is critical in tropical and marine systems. The chemical–physical variables most often affected in the process of eutrophication are nutrient loads entering the water body, nutrients in the water column, and in-lake properties such as water column transparency, chlorophyll-*a*, and sediment phosphorus concentrations. Water column primary productivity increases as nutrient loads increase, resulting in increased diel swings in dissolved oxygen and often in pH and alkalinity. Eutrophic lakes frequently stratify because solar energy does not penetrate the water column, significant oxygen deficits occur in the deeper water, and high densities of algae in the upper waters limit the mixing effects of wind.

The variables most often altered as lakes undergo cultural eutrophication include transparency (measured by Secchi depth), hypolimnetic dissolved oxygen, diel swings in dissolved oxygen, nutrient concentrations (usually dissolved or soluble reactive phosphorus [SRP]), and chlorophyll in the water column. As a result of those changes, macrophyte communities will decrease due to light limitation, density of benthic animals may become reduced due to anoxic conditions near the substrate, water column primary productivity will become greatly increased, and the water column P/R ratio will far exceed 1.0.

Those changes will gradually become evident over time. Eutrophication is a gradual process, primarily driven by external nutrient loadings. In even the most rapid examples of eutrophication, where external nutrients take the form of wastewater treatment, perhaps two to five decades are required before significant changes become apparent in the characteristics of an entire lake (Fig. 21.5). Recovery of a lake after it has undergone eutrophication and nutrient loads have been reduced requires two to five times as long as the original process (i.e., several times the water renewal rate). Even when aggressive nutrient lake management strategies are instituted, it will require several decades to achieve recovery.

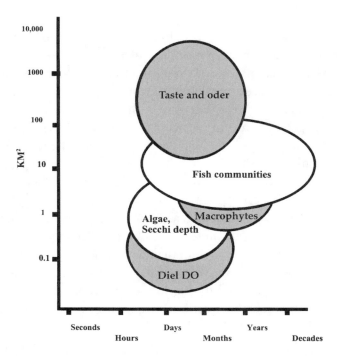

Fig. 21.5 Spatio-temporal considerations of eutrophic systems. Although many impacts, such as taste and odor, may subside within the space of a year if conditions improve, biological communities such as fish and macrophytes will take significantly longer to recover.

Spatially, nutrients are distributed by wind and water currents. In most lakes, those forces cause relatively complete mixing through the seasons and responses in the biological community will be relatively uniform around the lake. In areas where basin morphology creates discrete bays, one bay will often be strongly eutrophic while another may be mesotrophic or even oligotrophic (e.g., Lake Jackson, Florida; see chapter 12).

Monitoring of eutrophication should take into consideration that changes in the entire lake are relatively slow and variables undergoing change may be highly dynamic during short timeframes. Thus, for example, the average epilimnetic dissolved oxygen in late summer may change by only 2–3 mg/L over a 20-year period. On the other hand, diel changes in those waters may approach 20 mg/L in late summer. A sampling program should include assessments of diel changes in oxygen, annual and seasonal nutrient loads, and water column chlorophyll. Any individual lake might be sampled only two to four times per year in the program.

Summary

- Eutrophication is a natural process that is analogous to aging of a water body. Through the process of eutrophication, nutrients from the landscape are delivered to lakes, streams, coastal zones, and wetlands. In those water bodies, they contribute to accelerated plant growth and changes in the community.
- *Cultural eutrophication* is an increase in the rate of eutrophication caused by anthropogenic sources of nutrients. The fact that it is simply a changed rate of a natural process does not render this form of eutrophication less important. In many water bodies near population centers, algal productivity is very high; the water contains many tastes and odors; and wide swings occur in oxygen concentration, nearing zero at night.
- Flow-through or turnover time is a significant determinant of eutrophication effects. Thus, lakes are more susceptible than streams, and streams that enter lakes require more attention than do fast-flowing streams and rivers.
- Cultural eutrophication was seen as a major management issue in the late 1960s and societies around the world took many aggressive management actions to manipulate the process (e.g., banning phosphorus-based detergents). During the late 1970s and 1980s, eutrophication was seen as a problem under control. We now recognize that many sources of phosphorus are associated with urbanization and that population growth will continue to require careful attention to localized phosphorus management for protection of temperate fresh waters.

- In marine and tropical waters, eutrophication is accelerated by nitrogen rather than phosphorus. For example, algal growth on coral reefs threatens to destroy many tropical resorts. In those cases, nitrogen often comes from sewage or other municipal wastes.

For Further Reading

Harper D. *Eutrophication of freshwaters: principles, problems and restoration.* London: Chapman and Hall, 1992.

Land C, Reymond O. Reversal of eutrophication in Lake Geneva: evidence from the oligochaete communities. *Freshwater Biology* 1992;28:145–148.

Sly PG. The effects of land use and cultural development on the Lake Ontario ecosystem since 1750. *Hydrobiologia* 1991;213:1–75.

22

Special Issue: Acidification of Fresh Water Resources

Overview

Acidification of fresh water is a critical problem in many regions of the world. It can result from point source effluents but is most problematic and unique as nonpoint source pollution in which acidic compounds are transported by wind and rain. This transport mechanism makes acidification a large-scale problem of interregional and international proportions. Acid deposition begins with the release of sulfur dioxide and nitrogen oxide into the atmosphere, primarily through burning of fossil fuels. Chemical reactions in the atmosphere transform these chemicals into sulfuric and nitric acids, which are returned to the earth in snowfall, rain, fog, and dust. Importantly, the acidified precipitation typically falls "downwind," often by hundreds of miles.

Whether acid deposition affects fresh waters depends not only upon the presence of acidified precipitation, but also on seasonal patterns and on buffering capacity of soils. Calcareous soils such as limestone have nearly unlimited buffering capacities; lakes and rivers underlain with these soils generally will not show effects of acid precipitation. Granitic soils have virtually no buffering capacity, however, and are extremely susceptible to acidification. Biotic responses to acidification include loss of diversity through the elimination of acid-sensitive species; shortening of food chains; increased dominance of a few fish and invertebrate species; increased size and longevity of dominant species; increased growth of filamentous algae; reduced macrophyte diversity; increased water clarity (which may partially compensate for species loss and result in only minor changes in productivity); and loss of birds and mammals dependent on acid-sensitive aquatic species.

Introduction

During the 1980s, *acid rain* became a familiar term in the United States, Canada, and much of Europe. Although not all acidification of fresh waters results from acid rain per se, scientific studies and the popular press have effectively communicated to the public that burning fossil fuels releases significant and potentially devastating acidifying agents in the air. In the early 1980s, dire predictions were made about the potential effects of acidification in lakes and fresh waters; widespread sterile lakes and plummeting fish populations were common images of the consequences of acid precipitation. In some lakes and regions, these predictions have been realized and acidification is an important problem. In general, however, the natural buffering capacities of many soils have kept the geographic scale of these early predictions more limited than earliest estimates feared.

The patchwork of acid-impacted lakes evident today results from patterns of industrial emissions, patterns of geological formations that lead to acid sensitivity (or, in contrast, lend buffering capacity), and dominant wind and climate patterns that are responsible for the deposition patterns of acidified elements (Fig. 22.1). Consequently, even in areas receiving the same deposition of acidified compounds, such as the Adirondacks of New York state, nearly adjacent lakes may exhibit vastly different effects of acidification because of differences in rock structure. Similarly, although acid deposition rates are significantly higher in Eastern Europe than in Scandinavia, Scandinavian waters are severely acidified while many Eastern European waters are protected by acid-buffering soils. In addition, dominant wind patterns carry much of the acidifying agents across their boundaries.

In addition to the source and rate of acid deposition and buffering capacity of soils and bedrock, acidification effects are influenced by the timing of acid deposition. In particular, lakes and rivers that receive spring run-off from acidified snow melts may experience dramatic acid pulses during the season in which biota are most acid-sensitive.

The role of distant industrial air pollutants in causing acid rain receives the most attention by the popular press and is linked to damage of both terrestrial and aquatic environments. In fact, acid effects stem from a variety of land use sources. For example, vehicle use accounts for 30% to 50% of total nitrous oxide (NO_x) emissions in developed countries and is an important source of local acid precursors (World Resources Institute 1992). Other acid impacts may result from leaching or release of acid compounds into surface waters. For example, acid mine drainage is responsible for more than one-fourth of the acidified streams in the United States (National Research Council 1992). Other industries such as battery, iron, and copper manufacturing, textile and insecti-

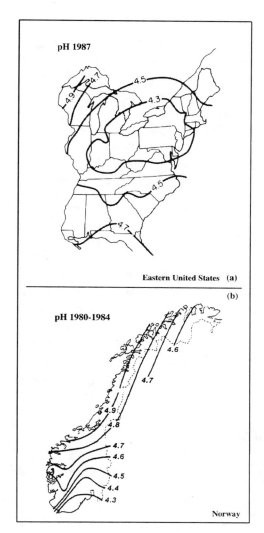

Fig. 22.1 a, b The concentration of acidified rain follows dominant wind patterns. Examples of the uneven distribution of acidified rain are given for the Eastern United States and Norway. [(a) After Laws 1993; (b) after Abrahamsen *et al.* 1994]

cide production, and even breweries produce effluents with high acid contents (Haslam 1990). These sources may create a variety of pollutant plumes in the affected waters, but the overall response of aquatic communities will be similar regardless of the sources of acid input.

Toward a Global Awareness of Acid Precipitation

Awareness of acidified precipitation is as old as the phenomenon itself. It was recognized as a local air pollution problem as early as 1662 in England. The term "acid rain" itself dates to 1872, when Robert Angus Smith, an English chemist, noted a relationship between coal burning, rainfall acidity, and damage to buildings and vegetation in and around the industrial center of Manchester, England (Laws 1993).

Despite the early identification of acid rain and decreasing fish populations in Scandinavia in the 1920s, scientific evidence for the links between industrial activity and downwind acid fog and acid rain did not accumulate until the 1950s (Laws 1993; Mason 1991). By the late 1960s, the acid rain issue had gained the attention of the public, largely through its presumed links to forest decline in Germany's Black Forest and in parts of the Applachians (World Resources Institute 1992).

The most familiar cases of acid rain impacts to fresh water resources are those in Scandinavia, the northeastern United States, and eastern Canada. In recent years, the possibility that other world regions may experience acidification problems has claimed increasing attention. Areas at risk because of rapid industrialization and acid-sensitive soils include southern China, Southeast Asia, southeast Brazil, and northern Venezuela. In addition, major acid-sensitive areas are known to exist in western Canadian provinces, the Yukon, Northwest Territories, and Labrador. In Europe, acid-sensitive areas of the Netherlands, Belgium, Denmark, Switzerland, Italy, western Germany, and Ireland have been added to the well-documented cases of Scandinavia and the United Kingdom.

The Sources and Their Expression

Sulfur dioxide (SO_2) and nitrous oxide (NO_x) are the two most important acidification precursors. While few natural sources of nitrous oxide exist, natural emissions of sulfur are significant on a global scale, and stem from both volcanoes and biogenic sources. Anthropogenic sulfur emissions increased from nearly 0 in 1850 to a point where they are predicted to reach 75 million metric tonnes by 2000; these emissions now nearly equal the annual natural sulfur emissions (Laws 1993) (Fig. 22.2). In addition to the sulfur compounds, global anthropogenic NO_x emissions, are expected to reach nearly 25 million metric tonnes by the year 2000. Global averages give a broad impression of the volume of these anthropogenic emissions, but the magnitude of their effect results from their uneven distribution around the globe. Specifically, Laws (1993) notes that most anthropogenic sulfur emissions occur on only 5% of the earth's surface. Consequently, each year nearly 100 metric tonnes of anthropogenic acidification precursors are released in a limited area and the resultant acidification becomes highly concentrated in certain regions.

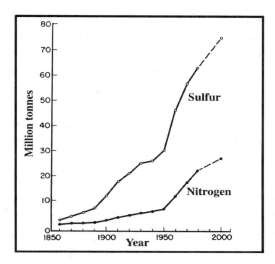

Fig. 22.2 Global anthropogenic emissions of sulfur and nitrous oxides have increased significantly since the industrial revolution. (After Laws 1993)

Rainfall saturated with carbon dioxide (CO_2) has a pH of approximately 5.6. The natural emissions of sulfur noted above are at least partially responsible for the fact that normal pH of precipitation is usually several steps below that—between pH 5 and 5.6. In areas subject to anthropogenic sulfur and nitrogen emissions, pH of rain is even lower. It has been recorded as low as 2.32 in Pennsylvania (Mason 1991), but pH between 4 and 4.5 are more common values in Europe and the northeastern United States. Because pH is measured on a log scale, a pH value of 4 is 10 times lower than a pH value of 5; a pH value of 3 would, therefore, be 100 times lower than a "normal" pH of 5. While it may be less important for fresh water acidification, it is interesting to note that acidified fog and cloud water are, on average, 10 times more acidic than acid rain. As a result, high-elevation forests may be exposed to extremely low pH conditions for large portions of the year (World Resources Institute 1992).

SO_2 and NO_x Emissions

As mentioned above, anthropogenic emissions of SO_2 (sulfur dioxide) and NO_x (nitrous oxide) are generally considered the main source of acid deposition precursors. Anthropogenic sources of SO_2 are largely stationary sources such as coal-burning power plants (70% of the total) and oil refineries (14% of the total) (Laws 1993). Figures for NO_x are less exact, but the same stationary sources are estimated to account for approximately 35% of the anthropogenic emissions of NO_x and another 30% to 50% are contributed by vehicle emissions (World Resources Institute 1992). In areas of high vehicle use (and low emissions controls) such as Sao Paulo, Brazil, and Hong Kong, cars and trucks can contribute 75% to 90% of nitrogen oxide pollution (World Resources Institute 1992). In some European countries, including the Netherlands, Denmark, the United Kingdom, and northern Italy, emissions of ammonia (NH_3) from agricultural sources have also been identified as important contributors to soil and fresh water acidification. For policy and remediation efforts, one of the important differences between these sources is their ability to carry to different regions. For example, vehicle and ammonia sources of acidifying agents are generally responsible for local problems. High smokestacks of power plants and industries (some as high as 300 m) are the primary agents of cross-border and cross-regional pollution conflict.

Once in the atmosphere, SO_2 and NO_x emissions are oxidized, creating sulfuric and nitric acid and/or sulfate and nitrate aerosols. The acids are returned to earth's surface by wet deposition (rain, mist, fog, or snowfall), while the aerosols return via dry deposition (particles and gases). Dry deposition is not, strictly speaking, acidic, but may later recombine with surface water to form sulfuric and nitric acids (Laws 1993). Consequently, although they follow different pathways, both dry and wet deposition result in the same acidic compounds (sulfuric acid and nitric acids). They differ primarily in how far they will carry in the atmosphere. Dry deposition, for example, is more important in short-range transport of acidifying agents; in the eastern United States, it may account for nearly 1.5 times the sulfuric acid wet deposition (Laws 1993). In contrast, wet deposition is a more important factor in transporting acidifying agents further from source areas. In Canada, for example, wet deposition accounts for nearly 70% of sulfuric acid deposition. The transport range of these pollutants is enormous, varying from 10 to 1000 km from the source region. For instance, approximately 17% of the acid precipitation that reaches Norway derives from Great Britain, and 20% of that falling on Sweden comes from Eastern Europe (Mason 1991).

Geographic and Political Patterns in Acid Deposition and Acidification

The greater awareness of acid rain in the latter half of this century is no surprise. In most parts of Europe and North America, SO_2 emissions approximately doubled between 1950 and 1970. Since the early 1970s, however, public outcry and new regulations have been effective in causing a 15% to 40% decline of SO_2 emissions in many developed nations. Paces (1994) traces a correlation in developed market economies between increasing GNP (dollars per capita) and decreasing SO_2 emissions (Fig. 22.3).

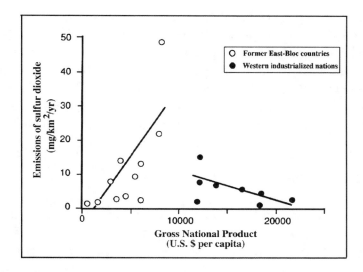

Fig. 22.3 Two distinct trends are evident in the relationship between acid-causing emissions and the gross national product of diverse nations. One model, represented by western industrialized countries, demonstrates that as GNP increases, acid-causing emissions decrease. In contrast, many former East Bloc countries demonstrate a reverse trend. See text for discussion. (After Paces 1994)

In the rapidly industrializing countries of Asia and South America, SO_2 emissions continue to increase. Globally, SO_2 emissions were highest in the (formerly) centrally planned economies of Eastern and Central Europe. Interestingly, Paces' (1994) political and economic analysis of these trends demonstrates that acidifying emission increased with annual GNP in these centrally planned economies rather than decreased as in the developed, market-economy model.

While emissions of NO_x also increased in Europe and North America between 1950 and 1970, these emissions have not yet begun to decrease. Instead, in most developed nations they have either stabilized or continued to increase, reflecting increases in automobile densities and a less widespread adoption of controls.

Soil Buffering Capacity

Emission of acidifying compounds, their conversion to acids, and their transport over large distances are the beginning of the process of fresh water acidification. The manner in which acidification is expressed in each body of water is largely dependent upon geology and soil chemistry. Certain soils (such as limestone) weather quickly and have an overall base charge that buffers the effects of acid rain and acid deposition. Lakes and streams with well-buffered soils and bedrock may, therefore, exhibit little or no effects of increased acid input. In contrast, lakes and streams with highly granitic bedrock (or other slow-weathering metamorphic rock) and thin overlying soils are highly susceptible to the effects of increased acid input. Besides soil buffering capacity, reactions at sediment surfaces provide some resistance to acidification. Sediment buffering efficiency is highest in shallower, slowly flushing lakes.

Trace Metals, Minerals, and Episodic Impacts

In addition to lowering pH, acidification creates important chemical changes in fresh waters by changing the mobility of certain trace metals. This process can significantly elevate levels of metals such as aluminum, cadmium, manganese, iron, arsenic, and mercury. Of these metals, aluminum has received the most attention, with raised levels of this substance most closely linked with toxic effects on fish. Because inorganic aluminum increases in direct proportion to decreased pH, it is often difficult to separate the effects of these two stresses. The cause of fish death often reveals the primary stress, however. In many studies, raised concentrations of inorganic aluminum have been shown to be the primary causes of fish toxicity in Scandinavian waters; in most of North America, studies have shown that pH alone is a sufficient predictor of toxic effects (Gunn and Belzile 1994).

Increased pH also generally decreases calcium concentrations in the water. Because calcium is critical to maintain the permeability of fish membranes—in particular gill surfaces—low calcium concentrations have been linked with the death of many species (Rosseland and Staurnes 1994). Low calcium concentrations have also been responsible for shifts in the macrophyte populations in Scandinavian waters to a prevalence of *Sphagnum* spp. (Brakke *et al.* 1994).

Another important toxicant that is inversely related to pH is mercury. Low pH facilitates the conversion of mercury from its unavailable forms to its easily retained methyl forms (Brakke *et al.* 1994). Although direct toxicity of mercury to fish is poorly documented (see chapter 17), this substance is toxic to fish consumers (from birds to humans). Consequently, increased mercury in the food chain of acidified lakes and streams may emerge as an important human health concern from acidification.

Acidification and its associated increase in heavy metals may be either long-term or episodic. The episodic cases generally occur in marginally affected areas and the impacts are most apparent during periods of rapid run-off (e.g., during snow melt and heavy rains). In fact, seasonally low pH may have as great an impact on biota as absolute reduction of pH. Autumn storms and spring snow melt create especially high sulfuric and nitric acid fluxes, and increase levels of hydrogen ions and aluminum. Acid surges of this kind may have disastrous effects on benthic flora, fauna, and fish populations.

Effects on Biotic Communities

Where acidified precipitation and acid-sensitive soils coincide, effects of acidification can be dramatic. In the United States alone, it is estimated that more than 50% of the species in some taxonomic groups have been eliminated from lakes in heavily impacted areas of the Northeast. On the other hand, estimated species losses for Maine, upper Michigan, northeastern Minnesota, and the remainder of the upper Great Lakes region are small.

The greatest damage to aquatic communities appears when species are eliminated that fill a unique role in the lake (or river) food web. If only one species fills a certain food or productivity niche, then the loss of that species is especially significant for the aquatic community (i.e., the the keystone species concept, discussed in chapter 15). Such a loss has been observed in acidified lakes in Canada with respect to highly acid-sensitive benthic crustaceans; their elimination destabilizes the food web by removing important food resources from predatory fishes (Schindler 1990).

Aquatic biota are subject to many impacts from acidification of fresh waters. Some important effects are described below.

Diversity

Acidified waters experience an overall decline in species diversity, coincident with increased dominance by a few species (Fig. 22.4). A related effect is a decline in redundancy. Whether effects occur directly through toxicity of acid-sensitive species or indirectly via food web changes, all components of the aquatic community decrease in diversity in acidified lakes (Schindler 1994). In one study, Schindler (1990) catalogued a 30% decline in total number of species when an experimentally acidified lake was decreased from approximately pH 6.5 to pH 5.0.

Community Structure

While species at all levels of the food chain are affected by acidification, Schindler (1990) suggests that those in the middle of the food chain are affected most dramatically. The consequence is a shortened food chain and altered

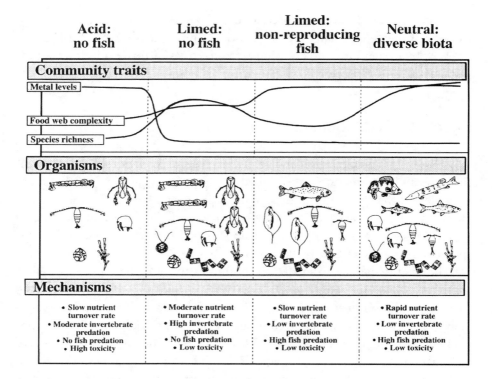

Fig. 22.4 Changes in aquatic community structure with acidification and treatments of lime to buffer acid impacts. Acidified lakes have high metal concentrations and low species richness and diversity, including no fish. The addition of lime decreases metal concentrations and increases species richness and diversity. As recovery proceeds, fish return, conditions improve for successful spawning, and nutrient cycling and diversity return to pre-acidified conditions. (After Stenson *et al.* 1993)

predation and prey pressures for both top-level and bottom-level species. Appelberg *et al.* (1993) suggest that community structure is significantly altered by the compounded effects of changes in algal and macrophyte communities, changes in invertebrate communities (due to the modified vegetation habitat), and the loss of fish predators. As a result, the invertebrates become more closely regulated by abiotic factors than by predation. Increases in dragonflies and water boatmen, for example, are particularly pronounced where populations of predatory fish have decreased. In turn, habitat and prey availability may increase for certain species of diving birds and other top-level predators.

Other far-reaching changes in community structure include the apparent acid sensitivity of many shredder species (Appelberg 1993) and the decreased rate of fragmentation of allochthonous input. The crustaceans responsible for much of the shredding activity—bivalves and snails—are especially sensitive to pH. Above pH 6.3, nearly all species are present; below 4.7, none is found (Mason 1991). Similarly, crayfish and certain species of fresh water shrimp are particularly acid-sensitive (Schindler 1990). Some declines in crustacea populations may be attributed to the decrease in calcium and sodium availability that corresponds with decreasing pH (Mason 1991). Stream biota appear to be more sensitive to acidification than those of lakes, perhaps because of increased sensitivity to acid pulses or the extreme decreases in pH from those pulses.

Phytoplankton and Macrophyte Growth

Acidification favors change in algal population composition, in particular promoting sometimes extensive growth of filamentous algae or dense mats of benthic algae. Overall algal diversity declines in acidified waters, with the relatively few acid-tolerant species being dominant. Decreased predation pressure and increased water clarity appear to compensate for loss of species diversity, as overall phytoplankton biomass typically remains unchanged in acidified waters.

Studies summarized in Schindler (1994) document a series of changes in the littoral zone as lakes acidify. In the early stages of acidification, rock-attached filamentous algae tend to predominate. As acidification proceeds, these species give way to fast-growing populations of unattached, filamentous algae (such as *Mougeotia*, *Zygnema*, and *Zygogonium*) that are responsible for large summer algal blooms.

Acidification appears to have a less predictable effect on macrophyte growth. In many European lakes, sphagnum moss has invaded acidified lakes and now forms dense benthic mats (Appelburg *et al.* 1993; Mason 1991). Other species have been eliminated. Haslam (1990) suggests that because acidification affects available carbon (through decreasing CO_2 availability), species that require larger pools of carbon are eliminated, while CO_2 source species (such as sphagnum) are favored.

Decomposition and Nutrient Cycling

Early studies appeared to conflict in their findings on the impacts of acidification on decomposition rates. Some studies show as much as a 20-fold decrease in decomposition rate, with a concomitant decrease in pH from approximately 5.6 to 4.6 (Haslam 1991). In contrast, Schindler (1990) found increased summer decomposition rates of the littoral area of acidified lakes and slower decomposition during the winter. One reason for the discrepancies may have related to the differing scales of the studies. It is now evident that metabolism of old carbon in sediments is unaffected by acidification, while decomposition of new material is influenced by decreases in shredder populations (Schindler 1994; Appelberg 1993). In addition, decomposer activity shifts from bacteria to fungi (Haslam 1991).

Because acidification eliminates many shredders, most acidified lakes accumulate a mat of coarse benthic debris. Slower turnover causes acidified lakes to act increasingly as nutrient sinks rather than nutrient cyclers (Appelberg 1993). In addition to slower decomposition, the carbon cycles of acidified lakes are disrupted by chemical reactions between acid elements and inorganic carbon. These interactions result in low available forms of carbon, which limits periphyton and some macrophyte growth.

The cycling of other important elements such as calcium, sodium, and trace metals is also affected by acidification. In general, trace metal concentrations (especially aluminum, as mentioned above) increase with increasing acidity, while calcium, sodium, and chloride availability is reduced. The aquatic nitrogen cycle is also disrupted by changes in both water chemistry and the microbial community. Nitrification declines at pH values lower than approximately 5.5; once inputs of nitrogen exceed phytoplankton demand, the remaining nitric acid contributes even further to the acidification process (Schindler 1994).

Fish Populations

Effects of acidification first became apparent through changes in fish populations; acidification is particularly devastating for many fish. Generally, fresh waters with a pH less than 5 support few or no fish species. Aluminum is especially highly toxic to fish, affecting transport of sodium and other elements across gill membranes (Brakke *et al.* 1994). Accumulations of manganese and mercury have also been detected in fish from acidified waters, and sublethal effects such as endocrine and hormone disruptions can weaken the fish and make them more susceptible to further stress (Brakke *et al.* 1994).

Effects depend upon species and fish age or development stage. In general, however, low pH and high aluminum concentrations are particularly toxic to early life stages and bring high egg and fry mortalities. Thus, many acidified waters contain unsustainable populations dominated by older and larger fish. Migratory fish are most vulnerable at spawning and smolting stages—stages that coincide with spring snow melt and acid pulses. Eels seem to be relatively insensitive to acidic conditions; other less sensitive species include pike, trout, minnow, and roach.

Besides acid toxicity and sublethal physiological changes, alterations in the food web can eliminate some fish populations, in effect starving large predatory fish long before acid toxicity affects them directly. Other changes in fish behavior are still speculative, but are likely related to acidification. For example, high aluminum concentrations interfere with fish olfactory sensors. Because these sensors may play important roles in the ways fish search for refugia, or, in the case of salmon, imprint on their home stream, such changes may have important long-range ramifications (Brakke *et al.* 1994).

River-dwelling fish are generally more sensitive to acid effects—in particular, acid pulses—than lake-dwelling fish (Rosseland and Staurnes 1994). In part, this finding relates to the limited escape possibilities for river fish; they cannot swim away from acid pulses or escape to less affected depths. Anadromous fish may also be more sensitive because their return to the river occurs at a time when pulses are high and smolt are particularly vulnerable.

Amphibians, Birds, and Mammals

Any species whose life cycle comes in contact with a lake may become affected by acidification. Embryonic and larval stages of amphibians, including the leopard frog, common toad, and natterjack toad, are acid-sensitive. Distributions of fish-eating birds, such as osprey, are also influenced by changes occurring in their prey populations. Similarly, birds that depend on macroinvertebrate prey experience changes in their own reproductive success: whether that success is improved or diminished depends on the feeding patterns of the bird. Certain obligate feeders that depend on crayfish and shrimp could be eliminated, while the success of nonspecific feeders might increase due to lake clarity. Mason (1991) reports that common loons and great northern divers in Canada have experienced declines due to decreased food in acid-affected lakes, while the goldeneye duck, a nonspecific bottom-feeder has improved survival.

As yet no concrete evidence suggests that fresh water acidification affects populations of fish-eating mammals. While these mammals can probably adjust to changes in species composition, fish from acidified lakes often carry elevated body burdens of mercury and cadmium, and these metals could potentially bioaccumulate in mammals, causing reproductive failures and behavioral abnormalities.

Clear Water

Clearer water is one conspicuous feature of an acidified lake—a result of changes in productivity and nutritional status due to the decline of fish density and precipitation of humic substances (dissolved organic matter). Increased penetration of light permits primary production at greater depths and may compensate for the negative effects on lake productivity and microbial activity by exposing more sediment to warmer waters (Schindler 1994).

Human Health

The ecological balance and integrity of lakes and rivers are significantly important to humans. Acidified fresh waters may also directly affect human health through elevated levels of trace metals such as mercury in water supplies and foods such as fish and shellfish (Nash 1993). In addition, acidity of water supplies may increase leaching rates of other trace metals (such as lead) from pipes and storage tanks.

Recovery

Lakes and streams will recover when acid deposition is reduced or intentionally mediated through application of basic agents such as lime. The speed and nature of recovery depend upon the water renewal rate and the buffering capacity remaining in the soil. Catchments that have accumulated large reserves of soil sulfuric acid will take much longer to recover. Recoveries have been widely studied in the pre-Cambrian shield of Canada where, by 1993, a combination of smelter closures and SO_2 controls had reduced emissions to approximately one-third of the values observed in the early 1970s (Keller *et al.* 1992). Recoveries occurred at rates more or less predictable from water renewal rates of lakes where deposition had decreased. Acidification can also deplete base cations in soils, however, and their recovery may take many years.

While more diverse biological communities will recover with decreased acidification, biotic composition of the lake itself may never return to its original condition. If improvements in acidity come before species elimination, the species will likely return or its reintroduction will prove successful (e.g., as with the case of lake trout in affected Canadian lakes) (Schindler 1991). Some groups appear resistant to recovery, in particular crustacean zooplankton communities.

Prevention and Alleviation of Acid Deposition

The acceptable limit of sulfuric acid deposition has been estimated at between 9 and 14 kg per hectare per year to protect the most sensitive aquatic ecosystems. These values fall far below the values of 20 to 50 kg/ha/yr currently being deposited in most of eastern North America and western Europe. Reductions in acidifying emissions of SO_2 and NO_x are generally considered the only effective long-term solution to fresh water acidification. Most of the industrialized nations are committed to reducing emissions of both compounds. European nations, for example, pledged a 30% reduction from 1980 levels by 1993. Presently, few efforts have been made to control NH_3 emissions because the process would require changes in agricultural practices such as reduced use of acidifying fertilizers.

Emission reduction is generally accomplished by conserving energy, using fossil fuel substitutes, and treating fossil fuels before nitrogen and sulfur compounds are released (Laws 1993). Some (though not enough) success has already been achieved in treating emissions through the mandated use of catalytic converters on cars. For example, Postel (1984) notes that catalytic converters used on U.S. automobiles have reduced NO_x emissions by nearly 80%. Beyond the use of fuel treatments for cars, minimizing emissions largely depends on reducing vehicle use, an enormous problem in areas dominated by urban sprawl and with generally poor public transportation.

Laws (1993) discusses no less than 10 processes that can treat emissions from stationary sources such as power plants. Stack gas scrubbing removes between 90% and 95% of SO_2 and SO_3 from stack gases, and the technique is used in more than 1000 plants worldwide. Other techniques have similar sulfur and nitrogen removal success. Laws (1993) also outlines two problems with putting these technologies to work: current electric generating plants are only half-way through their lifespans, and neither political pressure nor funds support efforts at reducing emissions.

Many countries, most notably Sweden, have relied on liming to mitigate adverse effects of acid deposition. Lime addition has successfully increased lake pH and alkalinity, and reduced levels of toxic aluminum. In turn, populations of acid-sensitive plant and animal species have expanded and fisheries have been restored. Because lime is depleted after several years, this answer is at best a short-term solution. Frequently pH is restored to higher levels than existed before acidification. Thus, liming can ameliorate acidification effects but does not necessarily return conditions to their pre-impact state. In Sweden alone, more than 100,000 tons of lime are applied annually to acidified waters, at a cost of $16 million in 1987.

Acid Mine Drainage: Another Important Acidification Source

Acid mine drainage is another—albeit more localized—critical source of acidified fresh water. It is responsible for more than 25% of acidified streams in the

United States, the vast majority of which are located in the coal-mining Mid-Atlantic Region (National Research Council 1992). Current estimates are that approximately 16,000 km of streams and nearly 12,000 ha of impoundments and reservoirs are seriously degraded by surface mine drainage (Mason 1991).

The severity of acid mine drainage impacts can be appreciated when it is considered that the acid fluid is between 20 and 300 times as acidic as acid rain, and that the mines may "leak" for several decades before effects begin to diminish (National Research Council 1992). Waters affected by acid mine drainage are typically very low in pH (less than pH 3.0) and consequently very low in species diversity. In one contaminated pool in Canada, 99.5% of the fauna were Chironomidae, while in other mine-affected fresh waters, one algae species (*Euglena mutabils*) constituted as much as 80% of the floral cover.

In addition to the affects of acidity, acid mine drainage is commonly high in sedimentary load from which an iron precipitate forms. This precipitate (ferric hydroxide) covers benthic algae and macrophytes (Mason 1991). Aluminum is also extremely high in low-pH acid mine drainage, as in the concentration of other heavy metals.

Improving aquatic conditions affected by acid mine drainage requires eliminating the drainage itself. For sites from which the drainage is leachate, a cap of clay and topsoil can effectively eliminate the causative percolation (Mason 1991). Programs to improve drainage from underground mines are more expensive and complicated, such as sealing the streambeds that leak into the mines with a polyurethane grout beneath the sediment (National Research Council 1992). One technique that is gaining in popularity is the construction of cattail wetlands to purify mine water. According to the National Research Council, more than 400 such wetlands have been constructed. Most other techniques center on reducing the ability of the pyrite to oxidize; they typically call for flooding deep mines or using surfactants that inhibit iron-oxidizing bacteria (National Research Council 1992).

Monitoring and Scale Effects

Acid deposition has two significant, direct biophysical impacts: (1) increased deposition of hydrogen ions lowers the pH of soil water and surface waters, and (2) heavy metals and other toxicants are deposited with the hydrogen ions. As hydrogen ions fall on the landscape, they are frequently neutralized (as indicated earlier in this chapter) by base exchange. When they are not neutralized, hydrogen ions may dissolve metals and other toxicants from soils and pass directly to lakes, streams, or groundwater. Therefore, the variables most commonly affected by acidification are pH and associated measures of acid–base balance. Alkalinity, a measure of the carbonate balance, is strongly affected by hydrogen ions. Alkalinities of less than 50 μeq/L usually imply a low buffering capacity and a high susceptibility to acidification impacts. Seepage lakes in granitic watersheds with thin soils routinely have alkalinities less than 10 μeq/L. As alkalinity and pH decrease, associated changes in dissolved metals often occur. The most significant of these metals is aluminum, which exists in a highly toxic, monomeric phase (Al^{+3}) between pH 5.3 and 4.7. As discussed earlier in the chapter, this form of aluminum is toxic to many fish and invertebrates.

Acidification is a regional-scale phenomenon. Acid deposition precursors are emitted into the atmosphere and travel as far as several hundred kilometers before deposition. Therefore, precipitation chemistry will usually be similar within a region (i.e., an airshed) of a few thousand square kilometers (Fig. 22.5). Impacts of that chemistry will not be uniform throughout that region and thus water quality monitoring of acid deposition must be stratified. The principal stratifying variables are soil type, lake chemistry, and hydrology; high alkalinities in soil type or lake waters indicate a low probability of acidification impact, while longer residence times imply a higher probability of impact.

Acid pulses after a rain storm or during snow melt are of relatively shorter duration but are more intense. Lowered pH and elevated heavy metals may occur for a few hours to a few days after a hydrologic event, causing all of the ecological impact. For example, the entire year-class of a fish species can be eliminated by 2 hours of low-pH rainfall that happens to occur when the larvae are very sensitive. Detection of such events can prove very challenging. Automated samplers can be installed to record pH continuously and to collect water samples when certain conditions exist (e.g., when stage height increases by a certain amount or to a certain level). Alternatively, grab samples can be collected during a hydrologic event or can be routinely collected on a water body in the hopes of detecting rare events. Anomalous development of biological samples may also indicate post facto that an event occurred.

Summary

- Acid deposition is a nonpoint source impact that begins with emissions from the landscape and results in deposition *to* the landscape and then damaging impacts to terrestrial and aquatic ecosystems. Acid *deposition* is a broader term than acid *rain*, implying that deposition can be wet, dry, or gaseous.
- The typical phases of an acid deposition sequence are as follows:

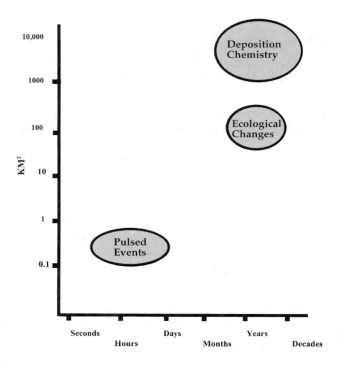

Fig. 22.5 Spatio-temporal effects of acidification of water quality show a large distinction between the dramatic impacts of pulse events and the more insidious long-term and large-scale impacts of waters acidified through atmospheric deposition.

1. *Emission* of acid deposition precursors. These substances are emitted from many kinds of sources, such as industry, automobiles, homes, or volcanoes. They are typically rich in oxides of nitrogen and phosphorus, and they are readily converted to acidic compounds through chemical–physical processes in the atmosphere.
2. *Transport* (often called long-range transport of air pollutants, or LR-TAP) involves atmospheric processes like large-scale wind systems, interaction with photochemical processes, and numerous chemical changes. Many of the substances associated with air pollution are catalytic rather than being toxic themselves, but their effect is dramatic.
3. *Deposition* of hydronium ions (H^+), substances that will release hydronium ions after reaching the earth's surface, and substances that seem to be incidentally associated with acidifying substances (e.g., metals). Deposition involves wet fall (e.g., rain, snow, fog), dry fall, and gaseous materials such as sulfur dioxide.

4. *Effects* may occur when acid deposition lowers the pH of an environment (e.g., a soil solution, a stream, or a lake), when low-pH solutions increase the mobility of metals such as aluminum or mercury, or when ecosystems and their elements are stressed by multiple impacts such as disease plus acid plus metals. We are learning that the latter case is most likely.

• Effects of acidification in water bodies are specific to species and trophic level. In general, higher-trophic-level organisms and "marginal" processes are affected most readily. For example, decomposition rates of nitrogen-rich leaf species are not affected until the pH is significantly lowered (e.g., to pH 4.5). Decomposition of more "marginal" leaf species that are low in nitrogen and already difficult to process are affected at much higher pH levels (e.g., 5.2).

• Mitigation of acid deposition is possible through very-large-scale efforts such as liming an entire watershed or a lake. Mitigation efforts have been criticized as "morally incorrect" because they suggest that emission and deposition are defensible uses of resources. In fact, each society must make its own decisions about use of resources and the impacts of those uses. In that regard, however, acid deposition has different implications than issues like eutrophication, because it is a regional (not local) issue.

For Further Reading

Appelberg M, Henrikson B, Henrikson L, Svedang M. Biotic interactions within the littoral community of Swedish forest lakes during acidification. *Ambio* 1993;22:290–297.

National Research Council. *Restoration of aquatic ecosystems: science, technology and public policy.* Washington, D.C.: National Academy Press, 1992.

Postel S. *Air pollution, acid rain and the future of forests.* Worldwatch paper 58. Washington, D.C.: Worldwatch Institute, 1984: 54 pp.

Schindler D. Experimental perturbations of whole lakes as tests of hypotheses concerning ecosystem structure and function. *Oikos* 1990;57:25–41.

23

Special Issue: Global Change: A Proactive Management Challenge

Overview

Managers must continually make decisions in the face of scientific uncertainty and lack of agreement on society's values. One of the biggest challenges faced by managers is the need to anticipate potential changes in the resource base to manage proactively, rather than reactively. Global climate change is one of the more thoroughly discussed of these issues. Most scientists believe that human dependence on carbon-based energy products has resulted in increasing concentrations of atmospheric CO_2 and the accumulation of several other gases known as "greenhouse" gases. Together these molecules increase the radiation of heat back to the earth, thereby trapping heat in the earth's atmosphere.

If predictions are correct, the cumulative effect of these changes will be an average global temperature increase between 2 and 6 °C over the next century. Because temperature is a powerful regulator of forces on earth, this increase will lead to a wide array of other changes, including (but not limited to) changes in wind and precipitation patterns, thermal expansion resulting in a rising sea level, possible melting of polar ice caps (again adding to rising sea level), increased transpiration and water demand, and shifting of global ecotypes. As these alterations occur, water quality will play an important role in determining the habitability of changed—and changing—environments. At its most general, water quality will probably decline in areas of increased aridity, while it may improve in areas with increasing precipitation. Along coastal zones, the balance of saline and fresh waters will move inland, but may be significantly altered by human attempts to prevent this movement. In both fresh and salt waters alike, temperature-linked properties such as dissolved oxygen and nutrient concentrations will change, subsequently modifying community structure.

Despite these general indications and the sure knowledge that change will occur, the exact nature of the forthcoming global changes cannot be predicted with current limitations in our knowledge and computing power. That scientific uncertainty, in concert with uncertain social values, is creating resistance at regional and national scales to developing proactive management strategies.

Introduction

The principal responsibility of natural resource managers is to make management decisions, despite any scientific uncertainty and value discrepancies. One of the most difficult situations for a manager is when strong scientific indications of potential stresses on the resource base exist, but no immediate crisis is occurring to galvanize action, stimulate public opinion, or direct management. This scenario is precisely the situation faced by today's managers regarding the issues of global climate change and other global changes in nutrient and chemical cycles. The scientific evidence is significant and irrefutable that global changes have occurred, and the implications of many of the changes are dramatic. Little consensus exists within the scientific community as to the exact nature of the impending changes, however. At the same time, significant resistance from the political and social venues has thwarted efforts to make changes to resolve an uncertain threat. Internationally, significant pressure has encouraged national governments to be *perceived* as taking action. The decision maker is left in an awkward position: because the predicted changes have a global cause and global effect, local ameliorative actions will be only a stopgap measure. Yet, advocating global solutions puts the manager in an often taboo political and social arena. This chapter outlines the manager's dilemma by reviewing what is certain and what is uncertain about global climate change, and describing how the uncertainty plays a dominant role in shaping decision-making options.

The Climate Change Debate

Most scientists agree that global climate change has occurred over the past century and will continue to occur in the coming century. These scientists point to the correlation of increasing CO_2 and other compounds in the atmosphere with an observed 0.5 °C increase in global temperatures since the Industrial Revolution. They postulate that the warming will continue and even accelerate in direct proportion to increasing "greenhouse" gases. In contrast, the relatively few dissenting scientists claim that recent apparent warming trends can easily be explained by normal variations in climate patterns that fluctuate widely over time. These scientists warn against the popularization of issues before irrefutable scientific evidence is gathered (see, for example, Brookes 1989). Such widely diverging interpretations of the same data reflect not only the degree of uncertainty pertaining to the prediction of global climates but also the fact that scientists often interpret their data through value-laden filters.

Although scientists disagree about its significance, definitive evidence exists that the proportion of greenhouse gases has increased in the atmosphere. Concentrations of CO_2 have been measured accurately since 1957, and concentrations before that date are detectable in air bubbles trapped in polar ice. These measurements catalog an increase from approximately 280 ppm before the Industrial Revolution to 360 ppm in the early 1990s (Vitousek 1994; Waterstone 1993); CO_2 continues to rise at an annual rate of 0.3% to 0.5% (Environment Canada 1992). Other important "greenhouse" gases such as methane, chlorofluorocarbons (CFCs), and nitrous oxide are also increasing in the atmosphere at alarming rates. For example, nitrous oxide is increasing by 0.7% per year (Fig. 23.1), methane concentrations are increasing by 1.0% annually, and estimates for the annual increase of CFCs range from 4% to 11% (Smith 1990). Many consider the extremely close correlations between industrial productivity and the increase of these gases to be proof that these gases are increasing as a result of human activity, in particular the burning of fossil fuels (Fig. 23.2).

The greenhouse gases play a significant and natural role in regulation of atmospheric temperatures. Natural background levels of these gases enable the atmosphere to trap heat leaving the earth's surface and reradiate it back to the earth. Without this background level, the average temperature of the earth would be 0 °F (-18 °C). Scientists postulate that, as the gases accumulate in the atmosphere, the proportion of heat trapped and reradiated to the surface will also increase. Predictions of global climate change are based on the assumption that increases in trapped heat will lead to significant temperature and climate changes on the earth's surface.

Probable Effects

To date, climate change issues that have captured public attention relate primarily to potential increases in sea level and effects on agriculture. More recent scientific and policy concern has been focused on potential changes in water resources (Ausubel 1991). Because water availability, use, and quality are all climate-sensitive, water quantity, water quality, and fisheries will all experience

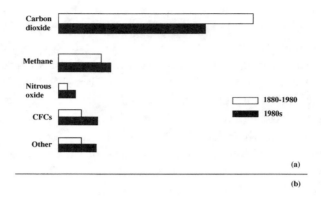

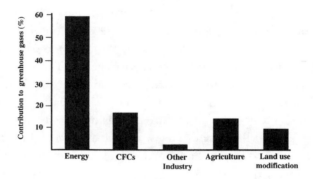

Fig. 23.1 (a) Although the amount of CO_2 in the atmosphere is increasing, the production of other gases such as methane, nitrous oxide, and CFCs has begun to make these gases more important, proportionally, in the composition of the atmosphere. Because the greenhouse effects of these secondary gases are larger than that of CO_2, their increase may accelerate predicted effects. (Data from Lashof and Tirpak 1989). (b) The contribution of human activities to global warming, illustrating that energy use and production remains the most important source of increased greenhouse gases. (Data from Lashof and Tirpak 1989)

profound shifts if global climate change predictions prove accurate. Even potable water will be severely affected. The rise in temperature will cause related increases in human water use as well as increases in evaporation and transpiration by plants. Thus, water tables in some areas could potentially be lower in the next century and pollutants in some waters could become more concentrated.

It is also projected that precipitation will increase, and, in particular, that large storms will become more frequent. These predictions have important implications for water quality management. If precipitation increases in an area, and the precipitation is primarily not from large storms, it could increase the

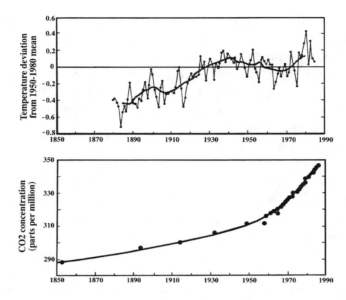

Fig. 23.2 Temperature readings taken since the mid-1800s indicate significant yearly fluctuations of temperatures, but an overall mean annual increase of approximately 0.5°C. The close correlation between the temperature increases and increases in atmospheric carbon dioxide are an important factor in global warming predictions. (Data from Houghton and Woodwell 1989)

recharge rate and thus moderate effects of increased usage while simultaneously increasing dilution in aquifers. In contrast, in areas where large storms account for an important percentage of rainfall, most water will end up as stream and urban run-off rather than aquifer recharge through soil filtration. In this scenario, precipitation may do little to offset lowered water tables in most areas, and in karstic areas it might actually increase the frequency of pollutant run-off that moves directly to the aquifer.

These examples highlight some of the complexities involved with anticipating future conditions given the current uncertainty in predictions. They also indicate that, although scientists have not come to an agreement on questions of the magnitude, direction, and location of anticipated changes, the likelihood of change itself seems certain.

The Basis for Predictions: What Is Certain

No debate exists over the fact that land use changes in the past two centuries have transformed the global landscape radically. It is also beyond contention that the proportion of atmospheric gases has changed during this same period. While some would contest the cause-and-effect relationship between these two points, global climate change predictions are based on the realization that the increase in important atmospheric gases is related to changes in land use patterns and industrialization.

As briefly described above, concentrations of carbon dioxide, ozone, nitrous oxide, chlorofluorocarbons (CFCs), and methane are known to play important roles in reflecting and absorbing heat from the earth's surface. In essence, these compounds create a chemical screen that traps heat radiating from the earth's surface and reflects it back to the earth. The most common analogy is that the lower atmosphere acts as a window, enabling some energy to pass through to space while absorbing the rest and reemitting it as heat. Just as color tinting changes the heat-trapping properties of glass used in greenhouses, greenhouse gases change the heat-trapping properties of the atmosphere.

Air bubbles trapped in polar ice reveal that historical warming trends are always reflected by increased atmospheric carbon dioxide, while atmospheric carbon dioxide is lower during ice ages. This observation, together with known chemical and heat reflectance properties of carbon dioxide, forms the basis for many of the predictions that current anthropogenic increases in CO_2 will translate into increased global temperatures. Predictions of global climate change are not restricted to observations of increased CO_2, however. As described above, gases such as CFCs are increasing dramatically; these gases have extremely high heat-reflective properties. Increases in nitrous oxide are also important elements in the development of global climate change predictions. Enormous anthropogenic production of nitrogen has significantly altered both biological nitrogen cycling and the proportion of nitrous oxide in the atmosphere. The significance of these changes appears not only in the addition of greenhouse gases, but also in the modification of the function and stress response of biological systems (Vitousek 1994).

Uncertainty in Predictions

There are many ways to understand and predict potential changes following global alterations in nutrient and chemical concentrations. For example, historical climate change patterns have been invoked to give some indication of expected patterns under a warmer climate. Historical accounts indicate that relatively minor changes in temperature have dramatic repercussions: during the last ice age, when most of North America was glaciated, average temperature was only approximately 5°C lower than today, and ice covered two-thirds of North America (Ausubel 1991). Because of the presence of noise in these historical tracings, however, they are insufficient as a predictive tool.

Another approach to predictions uses knowledge of chemical cycles to predict likely influences on biological form and function. It is known, for example, that some plants respond and acclimatize to increases in carbon dioxide or nitrogen, while other plants do not. These observations form the basis for predictions of important and large-scale shifts in plant community structure. In many cases, however, uncertainty surrounding important cycles clouds predictive ability. For example, increased carbon dioxide may decrease the concentration of oceanic calcium carbonate, decreasing the growth potential of coral reefs. This theory is based on the idea that coral reefs are sinks for atmospheric carbon dioxide, and thus are sensitive to changes in atmospheric CO_2 levels. Other research suggests that these systems are actually sources of CO_2 because the calcification process lowers the pH of water and releases CO_2 (Smith and Buddemeier 1992).

The most widely cited predictive tools in the climate change debate are computer models called global circulation models (GCMs). GCMs attempt to simulate future climate patterns by incorporating a wide range of known processes relating to climate patterns and biological responses to them. Because GCMs are more comprehensive than either historical tracings or nutrient cycle extrapolations, they are considered the best tool available, even though the predictions of each GCM are based on a different suite of assumptions. Although GCMs agree on many levels, areas of disagreement have led to considerable public distrust over the predictions. Several authors suggest that this variation has led people to focus on narrower-scope but more dramatic predictions such as potential sea level rise while ignoring more critical issues of water quality, water availability, and decreases in species diversity (Vitousek 1994; Waterstone 1993; Ausubel 1991).

GCMs AND THEIR PREDICTIONS. While all GCMs predict global warming, they differ in detail and come to very different conclusions about the appearance of the earth at the end of the 21st century. Limitations of knowledge and computing power, along with differences in underlying assumptions about climate behavior, lead to critical predictions that diverge among models. Even within a given model, changing assumptions can lead to vastly different predictions. One model's predictions of a 5.2°C increase in average global temperature became an increase of 3.2°C, 2.7°C, and finally 1.9°C as assumptions about cloud cover were changed (Smith 1991). These seemingly small differences have enormous consequences for policy and management, changing our predictions from "devastation" at 5.2°C to "moderate shifts in resource use" at 1.9°C.

If cloud cover increases due to greater evaporation and transpiration, clouds will reflect incoming solar energy, having a significant impact on keeping temperature increases lower. Some even claim that heavier cloud cover could negate any increases in global temperature (Brookes 1989). If cloud cover does not increase significantly, even extreme estimates of 6°C increases by the year 2100 could prove low. Accounting for the height of cloud formation also affects predictions on whether temperatures on the earth surface will be raised or lowered; high clouds result in surface warming, while low clouds result in cooling. Similar uncertainties surround the modeling of the behavior of the climate over the open ocean. Although ocean temperature increases will certainly lag behind air temperature increases, how much of a buffer the ocean will offer and how wind patterns and gas exchange will influence these changes remain largely matters of speculation.

The last major difficulty with model scenarios is that their ability to make regional and smaller-scale predictions remains greatly restricted. Uncertainty increases as the scale becomes finer. It may be possible, for example, to predict that Asia will experience more frequent and more numerous monsoons under predicted altered climate patterns. At a finer scale, it is not possible to predict the subtle changes in amount, duration, timing, intensity, and distribution of these rains so that managers in different countries, and in regions within those countries, have adequate information for decision making. In addition to scale uncertainties, models are limited by sheer computing power. GCMs are enormously complex and even with the fastest supercomputer's computation may take weeks. To compute details for smaller grid sizes, the time necessary for a run of the models increases prohibitively.

Despite the uncertainties described above, all GCMs agree on several issues:

- Average global temperatures will increase between 1.5°C and 6°C.
- Changes will not be uniform—higher latitudes will experience greater change than middle latitudes.
- Precipitation patterns will change due to increased evaporation and convection. In fact, large elements of the hydrologic cycle will undergo significant change.
- Daily and seasonal temperature variability will likely decrease, while precipitation variability will likely increase.

Besides these very general agreements, most models predict increases in sea level as ocean water expands with increased temperature and as polar ice caps melt. They also predict an increase in severe weather (e.g., storms, winds, and lightning strikes) and greater winter precipitation. Models disagree on other critical issues. For example, some suggest that winter warming will be greater than summer warming; other models predict exactly the opposite. Similarly, while most predict increased winter precipitation, some have predicted lower winter precipitation and higher summer precipitation. From a management point of view, a critical limitation is that models also differ in their estimates of how different regions will be affected.

Water Quality and Global Change

With one exception (i.e., rising sea level) the variety of water quality effects potentially arising from global climate change do not represent a new set of concerns. Instead, climate changes may aggravate familiar concerns such as water temperature, recharge rates, salinity, pollutant concentration, nutrient concentrations, turbidity, and sedimentation (Fig. 23.3). Water quality implications of climate change are expected to be critical from three important points of view:

1. While the predicted range of temperature increases has been experienced before, the rate of change is expected to be dramatic and will be a vital force in determining managerial response and societal adaptability.
2. Climate change will accentuate effects of existing anthropogenic stresses and probably will increase the degree of ecosystem "stress fatigue."

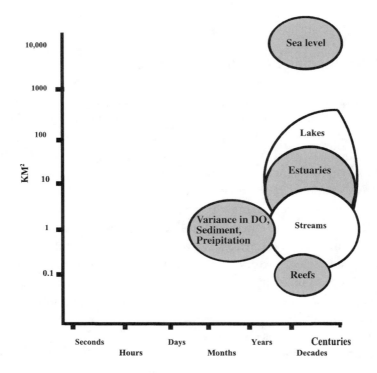

Fig. 23.3 Climate change would affect many water quality variables, such as sea level and salt water intrusion into drinking sources, changes in dissolved sediment and precipitation, and alterations in aquatic communities in streams, lakes, estuaries, and coral reefs. Unique to climate change is the clustering of nearly all of these impacts at the extended time scales of decades to centuries.

3. Changes in biological communities and water quality will change the goods and services available from local water bodies. In most cases, societies can probably adjust to these changes, but adjustment still requires advance planning and flexibility. For example, farms will remain viable but must adjust their crops and planting times. Similarly, fishing will remain viable but available fish species may change to species that are less popular today.

Adjusting to the global climate change will require a regional perspective. Demographics may change, biological communities will migrate and change through selective adaptability, and water valuation and usage may change. Overall, global change will result in both "losers" and "winners" in human societies; the greatest danger lies in attempting to preserve the status quo in the face of unyielding change. Change itself is the only certainty. The most effective management strategy will allow for maximum flexibility, maximizing local- and regional-scale abilities to respond to observed patterns on a relatively short timeframe.

Effects on Specific Ecosystems: Estuaries

Several coastal water resources are likely to be affected by climate change. Among the most obvious is the intrusion of saline water into drinking water supplies as sea level rises. This influx may create severe problems with water supply long before coastal areas become inundated. In addition, more subtle effects are likely to influence important and sensitive coastal ecosystems such as estuaries (see chapter 11 for a review of estuary ecology).

Estuarine ecology represents a balance of sea level, temperature, salinity, wind patterns, and water circulation. Other ecosystems will experience changes in one or two important influences as global change occurs, but estuaries must contend with all of these factors. Because many estuarine organisms already appear to live close to their tolerance limits (Kennedy 1990), even small changes in the balance of the estuarine environment may have large ramifications for the aquatic community.

Such large ramifications may also have significance for human populations. Estuarine and estuarine-dependent species constitute 70% of the U.S. fish catch, and estuaries serve as nursery beds for important fish and invertebrates. Estuaries also support some of the largest concentrations of human populations, and changes in the economic and ecologic base that supports these populations may lead to large-scale changes in demographics and economic vitality.

Sea Level Rise

At its simplest, a rise in sea level would flood some areas, creating new aquatic habitats. Without human interference, marshes and estuary borders would gradually move inland as sea levels advanced. While this development would tender dramatic changes for human populations, it is likely to represent little more than a shift in location for the aquatic communities (Fig. 23.4). Ausubel (1991) suggests that sea level rise will be relatively inconsequential for the next 100 years, rising only about 70 cm in that time. Predictions of a 5 m rise in sea level are common, but such increases will occur only in the most dramatic scenarios and over long time periods (Houghton and Woodwell 1989).

For estuaries, the effects of sea level rise are more likely to follow from compounded anthropogenic activities rather than from the climatic effects per se. Even small increases in sea level are likely to trigger a spate of building activity as coastal communities react to a perceived crisis by trying to "hold back the sea." While the construction of levees, banks, dikes, channels, and dams may

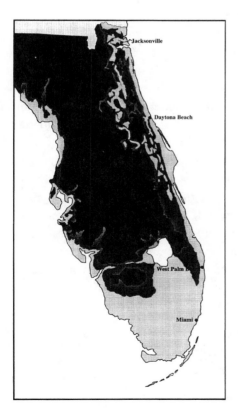

Fig. 23.4 The possible change in coastal areas of Florida if climate predictions are correct. The light gray area represents the area that would be flooded if sea level rose 4.6 m; the darker shading represents the additional area that would be flooded by a rise of 7.6 m. (Simplified from Houghton and Woodwell 1989)

temporarily protect property and familiar coastline features, they will also impede transport of sediment downriver. Kennedy suggests that the combined effect is likely to be loss of marsh habitats: "old" marshes will be submerged and creation of new ones prevented. In the most extreme case, wetlands will disappear along some coastal areas, only to be replaced by beds of submerged aquatic vegetation such as seagrasses, with their associated less diverse biological communities.

Temperature Increase

Temperature increases in shallow estuary waters will also have an enormous impact on water quality. For example, Kennedy (1990) estimates that Chesapeake Bay may become as warm as estuaries along the southeast Atlantic or Gulf of Mexico coast, that northern-affinity species such as the soft clam (a commercially important species) will be eliminated by elevated temperatures, and that warm-water organisms will expand their ranges and dominance. As with sea level rise, temperature changes may represent significant differences on a local scale, but overall may be seen as a shift of the resource base rather than elimination of it. Chesapeake Bay and other Mid-Atlantic estuaries may come to resemble semi-tropical estuaries, semi-tropical estuaries to resemble tropical ones, and high-latitude estuaries to resemble current mid-latitude estuaries. Tropical estuaries, however, could suffer large declines in diversity as tropical species now at their thermal maxima would not be able to tolerate increases.

Latitudinal shifts are the most likely result of higher estuarine temperature, but the speed with which the temperature increases occur may regulate whether species can migrate—and thus shift to higher latitudes—or whether they will be eliminated by increases that take place too quickly. If temperature increases are rapid (detectable over decades), cold-water species will quickly vacate northern estuaries as these species are already living close to their thermal maxima. The diversity of northern estuaries would then suffer while warm-water species take centuries or millennia to migrate and repopulate these estuaries (Kennedy 1990). Because humans probably will be unwilling to tolerate centuries of low diversity in northern estuaries, warm-water species will probably be introduced intentionally, causing a chain reaction of effects from exotics in fragile estuaries (see chapter 24).

Effects on the metabolic rates and behavior of biota are also likely under increased temperature regimes. For example, high temperatures raise metabolic rates, leading to greater demand for dissolved oxygen. Because warmer water holds less oxygen, increased temperatures may ultimately lead to low oxygen conditions. In addition, the northward migration of some warm-water species may be impaired by discrepancies between seasonal spawning patterns in southern and northern environments. Warm-water species frequently spawn in early spring. At high latitudes, the short spring photoperiod limits plant productivity. Kennedy (1990) theorizes that the hatching of fry may become asynchronous with the growth of plants upon which the fry feed.

Salinity

Water bodies will experience changes in salinity in three ways: (1) increased evaporation may increase salinity; (2) rising sea level will increase salinity in inundated coastal areas; and (3) precipitation increases (and an increase in severe storm events) will increase run-off of fresh water and decrease salinity. As indicated earlier, if sea level rises unabated, estuaries will largely shift inland, with the balance of salt and fresh water shifting concurrently. In the likely event that humans block sea water from its landward movement, estuarine habitats appear likely to become ever more saline. Thus, the range of saline-tolerant organisms will expand relative to the ranges of saline-intolerant species, a fact that may lead to important species shifts and introduction of new predators.

On the other hand, if predictions are correct that precipitation rates will increase under future climates, increased run-off into estuaries will become an important limiting factor. Intense run-off creates density gradients that limit transfer of oxygen from the atmosphere to deeper waters that have high biological oxygen demand. Run-off intensity has been associated with frequency and degree of low-oxygen conditions in these deep waters. Such changes will affect sessile species most adversely; mobile species, on the other hand, are likely to emigrate.

Coral Reef Ecosystems

In the tropical oceans, the predictive ability of the GCMs is highly limited; models neither agree in their predictions nor are well calibrated to current conditions (Smith and Buddemeier 1992). Because temperature increases are projected to be lower in the tropics than in temperate regions, however, changes in frequency and intensity of extreme events are probably more ecologically significant than moderate changes in temperature.

The human response to changes in specific ecosystems is likely to be an important factor controlling their ultimate conditions under global climate change. For example, in the absence of human intervention, climate change would largely promote a spatial shift of resources in estuaries. In contrast, attempts to stabilize

changing estuarine environments will dramatically alter the ability of these resources to yield services and products as well as their ecological stability.

Coral reefs are most often found off the coast of less developed nations.
Consequently, economic constraints are likely to limit intentional efforts to
modify the effects of climate change in these environments. As mentioned earlier, however, the ability of the coral reefs to adapt and shift in response to increased temperature or sea level may be severely attenuated by compounding
anthropogenic stress.

One of the most damaging stresses to coral reefs, for example, is sedimentation. Dramatic increases in sedimentation also represent one of the largest
side effects of the development and urbanization that are occurring along
most coastal zones (see chapters 11 and 20). Heavier sedimentation is also one
of the most frequent side effects of storms. Consequently, if increased storm
frequency and intensity do result from climate change, these storms would
greatly accentuate the already existing sedimentation stress (Smith and Buddemeier 1992).

Coral reefs are also very sensitive to salinity changes. Once again, a dominant effect of urbanization is increased surface run-off. In combination with
increased storm events, this run-off could lead to critical salinity depressions.

Other climate-related impacts on coral reefs, such as response to sea level
rise and temperature increases, will operate more independently of further anthropogenic modifications. Sea level rise will have a largely local effect, increasing coral community diversity and productivity in previously shallow lagoons
where salinity extremes, nutrient depletion, or other aspects of restricted circulation currently limit reef development. In contrast, deeper waters in currently sheltered communities may subject them to increased wave stress and
reduced growth. Sea level rise may also promote changes in community structure because it may selectively remove faster-growing taxa. It is unlikely to
"drown" reefs, because sea level rise (estimated at 6 mm/yr on average) is likely
to be slower than potential reef growth rates (which could reach 10 mm/yr)
(Smith and Buddemeier 1992).

A more serious concern for coral reefs may involve increases in water temperature. High water temperatures cause coral bleaching, a process in which the
symbiotic algae that inhabit corals leave their coral hosts. When bleaching persists, corals die and the reefs become susceptible to breakdown (Atwood *et al.*
1992). Bleaching represents a significant threat that has the potential to cause
widespread mortality in the case of sudden and large temperature extremes, or
selective mortality in the case of more moderate temperature increases. In the
latter case, the most temperature-sensitive corals, such as branching species,

would be eliminated (Smith and Buddemeier 1992). As with most biological systems, corals exhibit species-specific differences in heat tolerance and in phenotypic plasticity. Consequently, the long-term effects on coral reefs will depend on the speed with which temperature increases occur.

Smith and Buddemeier (1992) suggest that on large scales, reefs per se are not seriously threatened by global climate change. Instead, increased sea level is most likely to stimulate growth and have a positive impact, as increased sea surface temperatures may extend the geographic range of conditions suitable for reef development. On finer, more-local scales, however, areas of individual reefs may prove susceptible to a range of climate and development related stresses, and the loss of reef area and diversity is likely to have a significant impact on local populations. Some regions will simply be more vulnerable to change. For example, compared with other reef ecosystems, the Caribbean has low biodiversity, is highly impacted by a growing human population in its drainage basin, and is already stressed from anthropogenic effects and run-off related influences. Also, no other large-scale reef communities exist in the tropical Atlantic to serve as sources of recolonization. Of all reef systems, the Caribbean is likely to suffer the most dramatically in the coming century.

Fresh Water Ecosystems

Climate change is likely to produce significant changes in regional supply and distribution of fresh water. Like coastal systems, fresh water systems are projected to experience a shift in resources: a northward migration of ecotypes, increased water stress in some currently arid areas, and a decrease in stress resulting from increased precipitation in other regions. The "impact" of climate change will be defined by the degree to which people living in affected areas feel compelled to modify changes to preserve their lifestyle. In their review of the effects of global change on fresh water resources, Carpenter et al. (1992) suggest the following generalizations:

- Under reduced precipitation scenarios (such as those predicted for continental interiors), reductions in stream flow are likely, particularly in western U.S. basins.
- Arid basins will be more sensitive to precipitation changes than humid basins.
- Seasonal shifts in stream flow distribution will be more significant than changes in total annual run-off.

Carpenter *et al.* offer the following example of potential regional changes:

> In arid regions such as the North American Great Basin, drier conditions may decrease water supply rates below current human demand. Lakes and wetlands are expected to contract in area. Projected water quality changes are negative.
>
> In currently moist regions such as the Laurentian Great Lakes, drying will decrease stream flows, lake levels, wetland areas and water supplies. In heavily populated areas, the magnitude of change is uncertain because effects on water demand for irrigation, energy and cooling are unknown. Drier conditions may affect transportation in regions that now have abundant water. . . .
>
> Certain regions may become wetter as a result of global climate change. Increased stream flow, more frequent floods, and expansion of lakes and wetlands are likely. Reduced snow cover will reduce the need to salt roads, and water quality will benefit. [p. 121]

Those generalizations illustrate one of the most important aspects of water quality dynamics under global climate change scenarios: where conditions will become more arid, water quality will degrade through concentration of pollutants, increased temperature stress, and increased anthropogenic impacts (e.g., more use of fertilizers, insecticides, and herbicides to try to maintain lawns, gardens, and crops). Where conditions become wetter, water quality is likely to improve due to decreased inputs, increased recharge of aquifers, decreased concentrations of pollutants, and increased soil filtering capacity. Demographic patterns, such as large-scale shifts of human populations to wetter regions, and distribution of precipitation events could either accentuate these patterns or minimize them.

Several features that will respond to global climate change and are important in controlling water quality are as follows:

- *Erosion and sedimentation* rates are highest in regions having the most variable precipitation and run-off. With global change, variability will increase, so erosion, sedimentation, and turbidity are also likely to intensify.
- *Nutrient input* changes in climate will alter biomass, productivity, and species composition of riparian communities, consequently altering supplies of organic matter and nutrients to fresh waters.
- *Decreased oxygen* will result as increased water temperature increases microbial respiration, thereby reducing oxygen concentration. At low latitudes and altitudes, these changes may adversely affect eggs and larvae of fish and invertebrates.
- *Flow dynamics* will be altered. Reduced flows can concentrate pollutants, while floods decrease riparian filtering of recharge water.

- *Precipitation variability* will increase nutrient inputs to lakes, contributing to eutrophication, while increases in water renewal rates will decrease concentrations of nutrients and contaminants from point sources or sediment.

Impacts on Lakes

The impacts of climate changes on boreal lakes and indications of changes in lakes in other areas comes from the Experimental Lakes Area (ELA) of northwestern Ontario. Over 20 years, researchers have documented an increase in lake and air temperatures of 2 °C and an increase in the length of the ice-free season of three weeks. Researchers have also recorded increased average evaporation, decreased average precipitation, and subsequent decreases in water renewal rates of area lakes (Schindler *et al.* 1990).

These climate changes initiated an ecological chain reaction in which decreases in water renewal have concentrated chemical solutes, especially nitrogen, in the lakes. In addition, fire frequency has increased as vegetation changes, drought, and lightning have become more frequent. The combination of decreased water flows and increased incidence of forest fires has increased chemical concentrations and accentuated phosphorus limitation in ELA lakes and streams.

Effects of changes in light were also documented in ELA lakes. The altered hydrology has lengthened water renewal times. Decreased water flows, in turn, have decreased the transport of dissolved organic matter to the lakes and increased water clarity. The subsequent increases in penetration of solar energy combined with greater wind velocity resulting from disappearance of forest cover caused thermoclines of lakes to deepen. As a result, summer habitat was reduced for temperature-intolerant species such as lake trout and opossum shrimp that require cold, well-oxygenated hypolimnia.

Changes in the ELA ecology are clearly not related solely to increased temperature. ELA also experienced decreased precipitation and increased wind velocity. The interaction of all of these changes led to the observed effects. Whether these same effects will be expressed in other lakes depends on balances of temperature, precipitation, and landscape changes. The pattern, however, suggests that multiple effects will be the norm.

Climate Change and Streams

The two global change influences likely to cause the greatest impacts to fresh water streams and rivers are temperature and rate of flow (a function of precipitation rates). Deleterious effects are most likely to be felt by streams in the mid-temperate latitudes, which already experience summer temperatures near the thermal maximum of their biotic communities (e.g., 37 to 40 °C) (Matthews and Zimmerman 1990). In the United States, western region streams that are most susceptible to increased temperatures also have an east–west orientation, which precludes species' northward migration to cooler waters. As Matthews and Zimmerman (1990) note, if precipitation increases do not keep pace with increased evaporation in these sensitive streams, desiccation and an associated reduction in refuges may lead to dramatic changes in community composition. As many as 20 species east of the U.S. continental divide have ranges that make them vulnerable to small increases in stream heat load.

The Psychological Side of Policy

Response to temperature fluctuations has been a part of human history for centuries. Balek (1992) traces the emergence of a strong state power in Egypt to the need to organize water regulation and irrigation as climate patterns fluctuated. Lake Ontario residents adjusted their assessment of which fish species are most valued as historical climate changes affected desired fish species (Sly 1991). What may be unique to global climate change is that we have the ability to foresee the change—if not its specific nature—and to respond proactively.

Ironically, the ability to forecast may in some cases reduce the willingness of communities to make changes in their use of fertilizers or fossil fuels. Migrations to the Sunbelt of the United States and the propensity of Europeans to vacation in the warm southern regions of their continent indicate that a warmer climate may appeal to many people. Similarly, countries such as Canada may enjoy increased agricultural productivity under the global climate change and may be less concerned about adverse effects. In contrast to these examples, "losers" in the climatic change scenario may resist proactive measures because of a feeling of powerlessness. The residents of south Florida, for example, could probably be rallied to reduce fossil fuel emissions if their reductions would, by themselves, reduce the possibility that their land would be inundated. Residents, however, see no point to local measures because "saving" the Florida coast depends on global measures. In addition, many developing countries are resentful of "northern" attitudes, feeling that curbing industrial production will unfairly limit their development potential (Fig. 23.5). In a problem of global scope, however, global cooperation is essential.

Combatting the perception barrier goes beyond informing the public at large. Behavioral research suggests that water planners and other natural resource managers expect the large-scale environment to be stable; they are bi-

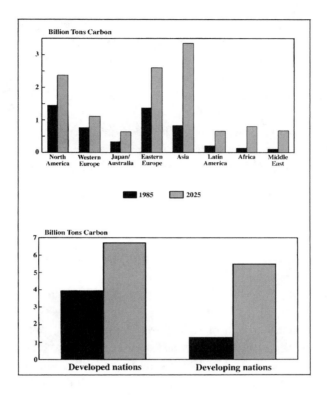

Fig. 23.5 Current and projected contributions of different nations and regions to the greenhouse effect. While many countries are expected to increase their production of atmospheric carbon between now and 2025, both developed and developing nations are expected to curb overall production by 2025. Developing nations are projected to exhibit only small reductions in overall production, while developed countries are expected to achieve larger overall reductions. (After Margenstern and Tirpak 1990)

ased against recognizing and acting on indications of change. Similarly, decision makers respond to environmental threats as they perceive them, not necessarily as they can be described, measured, or analyzed by objective criteria (Ribsame 1992). This "stability bias" is especially strong in water resource planning, where long-term conditions are assumed to emulate those observed in the relatively recent past. In fact, a survey of water resource planners in the southwestern and southeastern United States found that, while all were at least aware of global warming issues, few expected to experience actual climate change in the foreseeable future and few felt that they would be forced to change their management practices due to climate change in the next three decades (Ribsame 1992).

This stability bias is equally evident in the policies developed by most political bodies. While many countries have signed agreements limiting the emission of greenhouse gases, few have begun planning for natural resource changes under global warming scenarios. Even in less developed countries that, by all accounts, stand to be the largest losers under global climate change, the emphasis is on "exploitation" of their resources like the pattern followed by developed nations, rather than on planning for a different future. To policy makers and managers, the high degree of uncertainty in predictions represents an obstacle to planning. Yet, it should be seen more readily as the principal reason *for* planning.

Summary

- Global change is the largest environmental challenge to ever face humankind for several reasons:
  1. *Uncertainty.* We can predict the impacts of various global change scenarios using global climate models (GCMs). GCMs are capable of predicting large-scale changes in temperature, precipitation, and movement of air masses, given a series of assumptions. Unfortunately, the assumptions vary widely among the most common GCMs, as do the resulting predictions. In addition, GCMs generally fail to adequately account for oceanic dynamics (i.e., for heat and CO_2 stored in the oceans and seas)—dynamics that will lead to very large differences in resulting temperatures and hydrologic events.
  2. *Momentum.* By definition, global changes operate on a very large scale. They are driven by long-term dynamics (i.e., tens of decades). Thus, we are living with the cumulative effect of decisions made 50–100 years ago. The current global economy is based on production and consumption patterns that will result in gas concentrations that control climate 50 years hence. Some argue that decisions were made decades ago that will control future climates.
  3. *Significance.* The scale of changes that must be made to effect global change is difficult to grasp. If modern society is to change the way it affects future climates, it must conduct massive carbon sequestration programs (i.e., tree plantings) and must significantly change consumption and production to reduce the discharge of carbon dioxide to the atmosphere. The costs of those changes are global in scope.
- Water quality impacts of global change will revolve around two issues: reduced water volume and increased water temperature. Less water will be available for dilution, and assimilative capacities of many streams and

rivers will decline. At the same time, water temperature will increase in many areas. Higher temperatures will increase some assimilative capacities, but will exacerbate the effects of organic matter loadings (e.g., from sewage treatment facilities).

- Global change impacts will not be uniform throughout the globe, but will be most significant in the temperate zones and in coastal areas. In temperate areas, vegetation patterns will change, thereby changing hydrology and water quality. In coastal areas, sea levels will rise, coastal landscapes will be inundated, and erosion from storms will increase.

For Further Reading

Abu-Zeid M, Biswas A. Some major implications of climatic fluctuations on water management. In: Abu-Zeid M, Biswas A, eds. *Climatic fluctuations and water management.* Oxford/Boston: Butterworth-Heinemann, 1992;227–238.

Atwood D, Hendee J, Mendez A. An assessment of global warming stress on Caribbean coral reef ecosystems. *Bull Marine Sci* 1992;51:118–130.

Carpenter S, Fisher S, Grimm N, Kitchell J. Global change and freshwater ecosystems. *Ann Rev Ecol Systematics* 1992;23:119–139.

Gleick P. Climate change and international politics: problems facing developing countries. *Ambio* 1989;18:333–339.

Vitousek P. Beyond global warming: ecology and global change. *Ecology* 1994;75: 1861–1876.

Waterstone M. Adrift on a sea of platitudes: why we will not resolve the greenhouse issue. *Environ Mgt* 1993;17:141–152.

24

Special Issue: Exotics: A Special Biological Pollutant

Overview

Introduction of nonindigenous species is a unique problem in natural resource management, although certainly not a new one. Exotics have followed human migrations and travels throughout history. When introduced into a new environment, exotic species frequently die out, proving unable to grow and reproduce because of environmental conditions or susceptibility to local predators or pathogens. On rare occasions, the new species may find a foothold in the new environment, growing and reproducing without creating significant impacts on the resident community. The most troublesome outcome—and the subject of most exotics management—comes when a species establishes itself and displaces, eliminates, or otherwise threatens indigenous species and communities.

Exotics may be introduced intentionally or accidentally, but their impacts must be judged on two levels simultaneously: what the exotic can do *for* humans, and what the exotic can do *to* water quality, native biota, and habitat. As biological pollutants, exotics pose a unique threat because once these introductions become established, it is not possible to "turn off the tap." Even the most persistent chemical pollutants do not reproduce, grow, and disperse as do exotics. Exotics' management has proved difficult to regulate and legislate; most agencies rely on general protocols, permit systems, and species-specific laws to address an ever-changing, severe problem. Only rarely will a management agency have a proactive exotics management strategy in place to guide its reaction to the next invader.

Introduction

Exotics and introduced species have been a part of human-induced changes to the environment for millennia. Fossil records indicate that humans introduced mammals and plants into local environments more than 10,000 years ago. Aquatic exotics are traceable at least to the Romans, when the common carp was introduced from China to Europe and later from the Danube to ponds in Greece and Italy (Welcomme 1992). The inevitable escapees colonized adjacent rivers and lakes and are now effectively an indigenous population.

From the earliest civilizations through approximately 1500 A.D., exotics were introduced regularly but in moderate numbers. The rate of introductions increased steadily between 1500 A.D. and World War II, as humans removed or traveled across "natural" barriers that had previously served to isolate populations. Many introductions were accidental, but as many were purposeful: settlers and travelers sought to preserve the comforts of home by importing fruits, vegetables, and familiar animals (including fish). The age of engineering increased the rate of introductions even more as canals opened transport lanes for aquatic biota and dams created lakes ripe for new fisheries.

Since World War II, the number of introduced species has increased sharply, in large part due to the growing number of accidental introductions. In the Great Lakes alone, more than a dozen exotics became successfully established in the 1980s after being introduced from European commercial ships releasing unregulated ballast water. In Canada, nearly two-thirds of all established exotic fish were released either accidentally or illegally (Balon and Bruton 1986). Other analyses place equal or even greater responsibility for the rapid spread of exotics on fishery managers (Bain 1993; Wingate 1991). Minnesota's Department of Natural Resources has introduced fully 15 exotic species to area waters and, until the mid-1980s, freely transferred indigenous fish such as walleye and muskellenge without regard to differences in genetic strain (Wingate 1991).

Despite the history of exotics introductions and the surety that exotics will be introduced repeatedly, exotics management has changed little over time. Even the few policies and strategies in place to reduce the spread of exotics are subject to waivers and special permits if public pressure is great enough. Even when exotics exert disastrous ecological effects, the reaction and response follows a similar course everywhere: an initial short burst of intense public and scientific debate, followed by distress, and then followed equally quickly by acceptance of the new situation. Reactive, species-specific legislation may even be introduced, but the problem is rarely, if ever, seen in a larger context.

Exotics and Water Quality

Aquatic exotics come in many forms. Among the best documented are fish and aquatic plants, although molluscs and aquatic insects are also of widespread concern. Generally, the term *exotic* refers to a species of foreign origin, while an *introduced* species refers to any species from outside its local range. In management, the "origin" distinction is often less important than the ramifications of the introduction on resident populations. In fact, many more cases involve native species transplanted, either intentionally or unintentionally, to new watersheds than introductions of actual exotics. In either case, within the field of water quality management, the problem of exotics/introduced species presents a unique challenge.

Historically, water quality has been concerned with chemical and physical changes in water resources. In this framework, exotics management was not usually considered a part of water quality. In a broader view, exotics become a water quality concern because their introduction always represents a potential loss of aquatic services. Exotics do not often affect water chemistry per se, and only occasionally alter the physical habitat. Yet, they can exact a significant toll on the biological integrity of a system, altering food webs, diversity, and dominant species. In essence, exotics constitute biological pollutants that may affect community dynamics and/or chemical and physical dynamics of a water body. Their establishment may be facilitated either through degraded conditions that increase the likelihood of successful invasion, or through the predisposition by one successful exotic to other successful invasions. Management challenges are usually the same whether exotics cause water quality degradation or gain a foothold because of existing degraded conditions.

Because of the range of causes, effects, and sources of exotics, effective management hinges on the ability to predict changes. Although effective exotics management must be proactive, nearly all exotics management practiced today is reactive. The three primary challenges of exotics management are: (1) anticipating their appearance; (2) monitoring their progress; and (3) forecasting ramifications of an introduction. Reactive strategies face an enormous challenge—once established, it is usually nearly impossible (and impractical) to eliminate the intruder. Management then becomes a process of continual outlays of energy and money to keep an exotic population under control.

Exotic Fish

Because exotic fish have been introduced into new systems for centuries, and because little baseline data are available, substantial disagreement exists over the number and distribution of introduced fish species in the world. One 1992 review identified 163 species that had been introduced into 120 countries (Welcomme 1992). Regional estimates suggest that the global number is far higher. For example, Ross (1991) claims that at least 42 exotic fish species have become established in North America alone in the last 300 years. According to another report, 126 species of exotics have been collected from open waters of the United States (Courtenay *et al.* 1984). For the western United States, Williams and Jennings (1991) report 80 established exotic species and nearly 70 additional collected exotic species among 13 states. Assuming that the regional numbers are accurate, the number of global introductions is likely to be far higher than the estimates of several hundred. The specific figure is less important than the agreed-upon fact that the rate of introductions has risen sharply since the 1960s and is likely to continue to rise.

Despite discrepancies in actual numbers, most reports agree with respect to general trends of exotics establishment. A useful illustration of these trends is available from Welcomme's 1992 study of global introductions. His analysis shows, for example, that nearly 60% of 163 introductions were unsuccessful in establishing themselves in the new locality. This percentage is presumed to be low because many unsuccessful attempts fail to be reported. In particular, Welcomme notes that more than 1000 species are shipped worldwide for hobby aquaria and that many of these eventually find their way into local waterways.

Of the 40% of the 163 introductions that did become established:

- Twenty-seven species (15%) have been widely classified as pests. These introductions are either small species or large ones that stunt readily, reproduce rapidly, and have no specific value to humans. The 27 species represent a small portion of the total, but in their areas of influence, their introduction has had far-reaching effects.
- Ten percent have been generally satisfactory to humans in that they have either fulfilled their intended function or given rise to important fisheries outside the original scope of their introduction. At the same time, no significant side effects have yet been detected from these new species.
- Twelve percent have had mixed results; they have been successful in some areas of the world but not in others, or have achieved their original purpose but have also had serious negative effects on local ecosystems. This group includes some of the most widely introduced species. Some, such as common carp and rainbow trout, already have virtually global distributions.

The percentage of successful establishments rises sharply among incidences of introduced (i.e., non-native, nonexotic) species because many of these introductions occur over smaller geographic areas. Nonetheless, exact numbers on successful establishment remain elusive: introductions are commonly not documented nor are adequate baseline data known. Estimates of successful establishment range from approximately 35% to 80% (Ross 1991). Because establishment generally proves greater in areas that humans have altered or that initially exhibit low diversity, many predict that the percentage of successful establishments will increase in the future as more ecosystems are altered and anthropogenic stresses contribute to lowered diversities.

THE PURPOSE OF "TRANSFERS." Introduction of an exotic fish species can occur for a variety of reasons, ranging from "being unable to care for hobby aquarium fish" to "wanting specific species for fishing to biocontrol of already introduced exotics." While release of aquarium fish constitutes the largest percentage of total introductions, few of these species become established because of incompatible habitat requirements. Successful establishments are more common from sport introductions, escapees from aquaculture, and intentional introductions for biocontrol purposes.

- *Sport/recreation.* Nearly all the most popular sport fish species have been introduced into new areas beyond their native range. Throughout North America, for example, the brown, rainbow, and brook trouts have been introduced or transferred, as have many species of salmon, large and small mouth bass, and many species of sunfish.
- *Aquaculture.* Worldwide, a surprisingly small group of fish species is used for aquaculture. Thus, considerable transfer of a few species has taken place to countries or continents from which they were previously absent. For example, Welcomme (1992) notes that European aquaculture in the 1960s and 1970s was founded primarily on rainbow trout, a species that first appeared in Europe around 1880. Attempts to introduce aquaculture into the tropics have relied on a relatively small number of *Tilapia* species. Most recently, New Zealand began raising channel catfish, which are native to North America (Townsend and Winterburn 1992).
- *Ecological manipulation.* Some species have been introduced to "make ecosystems more efficient" or to fulfill a "missing" role. In an infamous example, the Nile perch was introduced into Lake Victoria in 1957 to increase efficiency of the lake's food chain (see the case study later in this chapter). The lake had an abundance of species that native predators and local fisheries could not exploit. By introducing the perch, administrators hoped to "convert" the indigenous population into a marketable product.

    For many years the "vacant niche" theory—which postulated unfilled niches in marine environments—justified this sort of introduction. This idea held that if the introduced species filled a vacant niche, the new species would be assimilated with no effect on community structure and function. This theory has since been widely discredited as too simplistic and operating at too narrow a spatial and temporal scale (see, for example, Herbold and Moyle 1986)

- *Biocontrol.* Another common reason for introduction of aquatic exotics is biocontrol, or introducing one species to control another. Examples of this strategy can be seen in the widespread introduction of grass carp to eliminate nuisance weeds such as hydrilla (Bain 1993) and similarly large-scale introductions of mosquito fish in tropical countries to control mosquito populations (Welcomme 1992). In such cases, the nuisance species itself is commonly an exotic and biocontrol with another exotic is judged the most effective and least disruptive approach to management.
- *Ornament.* Although now truly cosmopolitan, goldfish are native only to China. First introduced into North America in the 1600s, they have been widely introduced as ornamentals and just as widely accidentally released. Many other tropical fish, such as oscars, green swordtails, some gobies, guppies, tiger barbs, mollies, and platyfish, have been widely collected from North American waters, but not all exist as viable breeding populations (Welcomme 1992).
- *Accident.* Accidental introductions can occur from releases of ballast water, escape from aquaculture or research facilities, release of aquarium fish, or inadvertently, as in the escape of bait fish. While no exact statistics exist, accidental release of fish has expanded exponentially the numbers of introductions in the last 20 years. Management (i.e., regulation and control) is exceptionally weak in this area.

ESTABLISHMENT OF AN INTRODUCED POPULATION.   As with most other subjects related to exotic fish, the effects of introduced species on resident communities are widely debated. Frequently, even in cases of dramatic invasions, both the long-term severity of the introduction and the "reality" of a perceived crisis cause disagreement. Nonetheless, some trends are apparent in both establishment of exotics and effects of exotics on resident communities. The following characteristics are generally required for an invading exotic to become successfully established:

- *Temperature range matches* are the first requirement for successful establishment of species. Tropical species cannot survive in northern climates iced over in winter; likewise, cool-water species generally do not survive in warm tropical waters.

- *Decreased diversity* in the resident community increases the introduced species' chances for success. Low diversity can stem from many causes, including naturally harsh environments such as in many U.S. western rivers and streams, geographic isolation such as in New Zealand, or anthropogenic stress. Regardless of the cause, low diversity results in minimal resource competition or predation and favorable conditions for invaders. In the United States, those states with low diversity of aquatic biota have four times as many documented introductions as states with high diversity; they also have a higher rate of success in establishment of new species (Ross 1991). Similarly, New Zealand has only 27 native fish species but more than 20 successfully established exotics (Townsend and Winterbourn 1991).

- *Disease* and *parasitism* in resident communities are usually less severe threats for exotics than for coevolved local species. Thus, introduced species are often at a competitive advantage in a new environment.

- *The initial size of the introduced population* must be large enough to enable reproduction.

EFFECTS OF INTRODUCED SPECIES ON RESIDENT COMMUNITIES. The crux of the problem with exotic fish lies in the ways they may alter the ecological balance of the lakes and streams they occupy. It is perhaps equally important that awareness of these effects is not new. Throughout the 1980s, numerous studies recounted the history of invading species in many regions of the globe, but a literature search for effects of exotics reveals an even larger base of documentation stretching back over a period of 40 years. Although these studies have different foci, they demonstrate a most remarkable agreement on the general effects of introduced exotics. The most common effects (discussed below) influence diversity and trophic patterns, the introduction of diseases, and habitat alterations.

- *Decreased diversity and trophic alterations* are among the most obvious and frequent results of problematic introductions. Successful exotics (especially predacious fish) often reduce or eliminate populations of native fish species. For example, in New Zealand, introduced brown trout and native river *Galaxis* occur in disjunct habitats where the absence of trout best predicts the presence and abundance of *Galaxis* (Townsend and Crowl 1991). Predation and competition frequently explain the extreme shifts in food webs in many affected waters (Fig. 24.1). While some native populations shift their habitats and resource use in response to exotics, the invading exotic more typically competes with endemic species for food or eats the indigenous fish. As with most sources of water quality stress, the result is a shortened food web, decreased overall diversity, and dominance by a few

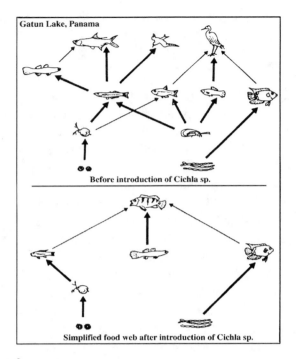

a

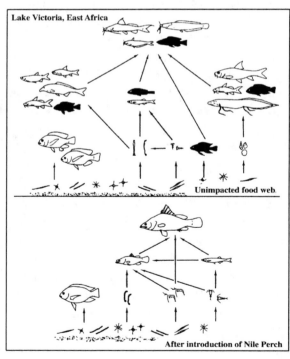

b

Fig. 24.1 Successfully established exotics typically caused adjustments in the food web. Some of these adjustments are dramatic, and have led to significant reductions in species diversity and shortening of the food webs. (a) Illustrates these impacts for Gatun Lake, Panama, where an introduction of *Cichla* sp. eliminated at least half of the members of the unimpacted food web, including organisms ranging from insects to shoreline birds. (After Welcomme 1992). (b) Illustrates similar impacts from the introduction of the Nile perch into Lake Victoria, which led to dramatically shortened food webs and extinction of indigenous species. (After Witte *et al.* 1992)

species. Although many different species have been noted in the global depletion of endemic species, the most common culprits are rainbow and brown trout; the list of waters they have affected is truly global and includes streams and lakes in Europe, Asia, Africa, South America, and New Zealand. Ross's (1991) review of exotics introductions documents that 77% of the cases reported in the literature involved a reduction or elimination of native species after (successful) introduction of an exotic.

Beyond changes in dominant species, food web alterations may have other important long-range effects. For example, in Lake Victoria, the introduced Nile perch eliminated native species that feed on algae. As those species disappeared, algal blooms increased, as did associated deoxygenation of waters (Witte 1992). While other examples of this magnitude are rare, it illustrates the importance of undertaking predictive analyses that will point out similar sensitive and critical components of ecosystem function *before* exotics are introduced.

- *Decreased size* of both native species and introduced species has been noted in numerous instances. Generally, stunting is assumed to result from early population explosions of introduced species and subsequent resource crowding. Welcomme (1992) notes that *Tilapia mossambica,* a species popular for aquaculture in the 1950s, became virtually useless commercially because of its widespread tendency to exhibit stunted growth.

- *Disease introduction.* While exotics are often immune to local diseases and parasites, they are often carriers of diseases that affect resident communities. Among the most widely documented cases is the introduction of furunculosis by North American rainbow trout to European brown trout. Welcomme cites a 1970 study that listed 48 species whose successful establishment resulted in part from disease transfer to native species. Disease introduction now represents a major concern of management agencies and provides the rationale behind some of the few protocols followed for controlling fish introductions (Wingate 1991).

- *Habitat alteration* is not the most common effect from exotics, but it has been widely discussed in relation to the globally dispersed common carp. The carp is a voracious bottom feeder that alters habitat by mechanically uprooting vegetation. Its presence increases turbidity and shading, reduces growth of macrophytes, and routinely proves detrimental to clear-water fish. Welcomme (1992) also suggests that by turning over benthic sediments, carp ingest organisms from the phosphate-rich substratum and later excrete phosphate in a more soluble form. Both the increased availability of phosphate and the reduction in macrophyte populations enable the increased growth of phytoplankton and contribute to eutrophic conditions.

Habitat alteration may represent a difficult management trade-off in some cases. For example, the grass carp is widely used to control noxious exotic weeds such as *Hydrophilla*. The carp is tolerant of a wide range of environmental conditions. According to Bain (1993), grass carp have migrated as far as 1700 km from their point of introduction, making their spread throughout the United States more rapid than any other introduced fish. This species has been implicated in the disruption of spawning grounds required by other fish species through its rooting habits. Despite their controversial side, the carp continue to be proposed as a biocontrol for weeds, perhaps indicating that current value systems place more importance on weed control than other elements of biological integrity. Bain (1992) outlines efforts to introduce sterile carp in an effort to effect vegetation control while eliminating the danger of further successful establishments.

MANAGEMENT OF EXOTIC FISH. The problem with management of any exotic species is that, once the species is established, effective control may not be possible. Even if all individuals of an introduced species could be caught and destroyed, permanent changes in the resident community may well prevent the return of previous conditions, including provision of goods and services that humans value from the ecosystem. Extirpations have been successful, but the necessary effort is usually large-scale and expensive, and ignores effects of other exotic species. In addition, effective control may require introduction of yet another exotic with its unknown attendant effects, or use of chemical or mechanical means with their unknown effects.

Another important issue in management planning is the lag effect between introduction, establishment, and ecosystem effects. For example, 25 years passed between introduction of the Nile perch and the time it was established in sufficient numbers to create the vast array of ecosystem-wide impacts since observed. This long lag time for biological processes creates difficult administrative, financial, and conceptual challenges for water quality managers.

Ultimately, the only sure way to manage exotics is to prevent their introduction. While this goal has been the expressed aim of past and present legislation and protocols, these processes are clearly ineffective today and are not likely to be sufficient. Other management tools include the following:

- Careful monitoring and early detection
- Habitat alteration to reduce breeding sites or create stresses unfavorable to the exotic
- Biocontrol (i.e., introducing predatory species, host-specific diseases and pathogens, or sterile populations)
- Public education (an aid in monitoring as well as reducing chances of spread or further accidental introductions)
- Chemical control
- Increased fishing limits, creation of short-term markets, and other attempts to intentionally overharvest a problem exotic

Exotic Aquatic Insects

Small and mobile, aquatic insect invaders are probably much more common than have been reported in the literature. In general, exotic aquatic insects are studied and recorded only when they act as disease vectors. Consequently, the majority of aquatic insect invaders that have been analyzed are mosquitoes and other nematoceran Dipterans.

As an example, de Moor (1992) traces the spread of malaria-vector mosquito *Anopheles gambiae* throughout Brazil in the 1930s. *A. gambiae* was first introduced in 1929. During the subsequent decade, the mosquitoe's range expanded and minor outbreaks of malaria became increasingly common. In 1938 and 1939, a malaria epidemic was responsible for deaths of 20,000 people. Yet, by 1940, just one year later, *A. gambiae* was successfully eradicated from South America. The eradication efforts were remarkable not only for their speed, but also for the scale of the operations and massive investment of resources that went into them. This example demonstrates that efforts to eradicate exotics most likely will be mobilized only when a direct threat to human health emerges. Luckily, the mosquito bred in open water where eradication efforts could be maximally successful. De Moor notes that other introduced malarial vector genera such as *Aedes* and *Culex* have spread more effectively because they breed readily in small water bodies—for example, in water storage vessels—and are difficult to eliminate.

When their introduction is not accidental, aquatic insects are often introduced for weed control. Coleoptera, for example, often prove to be effective biocontrol agents. Besides their feeding habits, these species show rapid population growth and colonizer attributes, but are commonly not efficient dispersers.

WHAT INFLUENCES ESTABLISHMENT OF INSECT INVADERS? Ships provide one ready mode of transport for mosquito species that reproduce in small catchments. For example, port authorities have monitored the accidental introduction of *A. aegypti* in various containers aboard boats, such as drums, plastic buckets, water tanks, life boats, and tubs.

Air travel is increasing the range of invading insects as well. Large pools of water near many airports provide breeding sites for diverse species. Inspections of planes in South Africa have revealed many "highjacking" insects; in fact, as many as 1% of planes arriving in South Africa carry arthropods (de Moor 1992).

Consistent with the invasion success of all exotics, regions most susceptible to insect invasions are those with degraded, disturbed, or modified landscapes, areas with low species diversity, and areas that are compatible with biologic tolerances of the invading insect. Because urban and urban fringe environments frequently exhibit low diversity and degraded environments, in addition to increased transportation routes, they often prove highly susceptible to invading aquatic insects.

Species characteristics increasing the chances of successful establishment include the following:

- Lack of predators or parasites (i.e., sufficient distance between source and invaded environment).
- Rapid breeding: year-around breeding capability, high fecundity, and rapid dispersal.
- Polymorphism and phenotypic plasticity that facilitate adaptability in a new environment.
- Appropriate feeding habits: generalists, in many cases; specialists, where food is abundant.

MANAGEMENT OPTIONS. Full-scale eradication of exotic aquatic insects is an enormous undertaking. As mentioned earlier, it requires extraordinary financial commitment along with extremely large crews working on the ground. In the case of malarial epidemics, benefits have outweighed costs, but this rare case has been aided by specificity of the insect and lethality of the disease. Usually, management is more appropriately concerned with controlling populations at some low level. In this context, the manager has an array of tools, including the following:

- Biocontrol (e.g., bird or reptile introduction to aid in population control or development of indigenous parasites or diseases)
- Insecticides (depending on the species and insecticide, differential susceptibility may be a benefit or hindrance to managers)

- Habitat elimination or moderation (e.g., eliminating uncovered water vessels to control mosquito populations)

Aquatic Plants

Twenty-five to thirty percent of known aquatic plants are considered true endemics, with distributions limited to a single lake, river, or portion of a river system (Ashton and Mitchell 1989). Other aquatic plants exhibit irregular distributions that may reflect historical transfer of propagules, either by humans or by waterfowl. Importantly, aquatic flora everywhere exist in a continual state of flux because of natural mechanisms of dispersal, invasion, colonization, and competition as well as physical and edaphic factors.

Besides natural forms, aquatic plants can be accidentally dispersed by the introduction of seeds and fruits along with irrigated crops, casual transport by boats, vehicles, and agricultural machinery, and escape from ornamental ponds and botanical gardens. Intentional introductions have resulted from the human fascination with exotic organisms, the desire to open new markets, and the desire to surround ourselves with familiar objects. In fact, most initial introductions of aquatic plants to new continents have been deliberate in that the introduced species was perceived to have some special human value. Accidental means of spreading within continents then becomes the dominant mechanism of dispersal.

Provided an adequate bioclimatic match exists between the non-native and native environments, reproductive capacity is a critical determinant of successful establishment. Plants capable of vegetative reproduction are especially well suited as invaders because a single viable plant is sufficient to start a new colony. Flexible habitat requirements, tolerance of environmental stress, and ability to reach a large population size all assist in later stages of an invasion.

As with other exotics, disturbed environments are particularly vulnerable to plant establishment, although introduced aquatic plants also have proved extremely successful in invading undisturbed ecosystems. In this case, the competitive advantage of freedom from native-range pests and diseases may significantly contribute to success. Many aquatic plants, however, become established without becoming invasive, as they lose the ability to set seed in their new environment.

Recent attempts to eradicate an invasive exotic from Lake Okeechobee, Florida, illustrate ways in which human values influence our perception of exotics. In the 1940s, the U.S. Army Corps of Engineers planted thousands of exotic *Melaleuca* trees (*Melaleuca quinquenervia*) on the rim canal along the southern levees of Lake Okeechobee to control erosion. The trees did their job well but have increased their range to the point that they now threaten several sanctuaries leased and managed by the Audubon Society. Concerns over erosion have given way to concerns over preserving natural habitats. Consequently, the Audubon Society, U.S. Army Corps of Engineers, Florida Department of Natural Resources, local schools, water management districts, and chemical companies are cooperating on a large-scale *Melaleuca* control (eradication) program.

Other nuisance plants such as Eurasian water milfoil have been introduced accidentally throughout the United States and continue to expand their ranges as plant parts "hitchhike" on boat trailers, anchor chains, and rudders.

MANAGEMENT OPTIONS FOR AQUATIC PLANTS. Six primary options are available for management of exotic aquatic plants:

- Manual removal (a labor-intensive, low-efficiency technique)
- Mechanical control (high rates of removal but inefficient for widespread invasions; in some cases, such as milfoil, mechanical harvesting increases the plants' rate of growth and brings only one or two months of control)
- Chemical control (while commonly used, herbicides are expensive and often nonspecific; their use is increasingly being questioned because of the indirect effects they have in the ecosystem)
- Biological control (using host-specific natural enemies, these control measures are self-sustaining when successful; identifying, collecting, and screening appropriate organisms is, however, expensive and time-consuming, often requiring decades)
- Ecological manipulation (e.g., lowering water levels has been successful in areas where water is abundant and the means exist to remove dead plant material)
- Direct use of the invasive plant for economic benefit. (By creating a market for an invasive plant (e.g., as stock feed), the plant may be harvested and thereby controlled. This option is suitable only if the harvesting and marketing activity do not interfere with the integrity or alternative uses of the water body, or more precisely if the interference does not outweigh the negatives of the exotic.)

Exotics Policy

It is a telling fact that a 1989 book entitled *Biological Invasions: A Global Perspective,* contained no reference to the legal and policy framework for exotics management. Among the hundreds of articles devoted to the problem of exotics, only a handful mention legal and policy frameworks within which exotics are managed.

An extended example of U.S. aquatic exotics legislation illustrates the reactive policy approach. The Black Bass Act of 1935 prohibits import of black bass and certain other fish, but the law has rarely been enforced. The Lacey Act prohibits import of Clariidae or uncertified (against disease) eggs of Salmonidae fish. The act further states that release into the wild of any imported animal or its progeny is prohibited except by state wildlife conservation agencies or with the written permission of one of these agencies. It also strengthens state exotics laws by making their violation a federal offense. The other pertinent U.S. federal law is the Endangered Species Act of 1973, which includes provisions prohibiting import of exotics. The rationale behind this act was to prevent depletion of endangered species in their native range, although permits may be obtained for scientific purposes or for propagation. Clearly, all of these laws contain significant loopholes through which exotics continue to pour.

In most U.S. states, exotics management usually falls under the authority of conservation departments that have regulatory authority over wildlife and fisheries. As Wingate's (1991) analysis of exotics policy notes, Maryland is the only state to pass a formal law restricting exotics introductions. Thirty-nine states responded to Wingate's survey; of these, 67% control exotics through internal policy or statute. Indeed, most introductions reported by Wingate were intentional ones carried out by the state management agency, and management agencies frequently issue permits in response to pressures from anglers and businesses. In other words, controls employed to date have been highly discretionary.

Environmental impact statements (EIS) are required by most countries for introduction of a new species. These documents are neither binding nor conclusive, however. In New Zealand, for example, enthusiastic proponents wishing to introduce channel catfish for aquaculture ignored a negative EIS finding (Townsend and Winterbourn 1992). Proponents even managed to convince the government to reject its own protocol calling for testing of proposed introduced species. Channel catfish eggs were successfully imported and held in temporary quarantine until a final decision was reached. In this specific case, an independent review team concurred with the EIS negative assessment and the eggs were destroyed. Nonetheless, this example illustrates that the requirements for EISs and protocols are not always adequate safeguards. In this case, accidental release from quarantine could have been disastrous.

Permits, port of entry restrictions, and other strategies may enable agencies to trace accidental or clandestine introduction of exotics, but they will not prevent them. Because of the regulatory difficulties surrounding the issue of exotics, some observers consider exotics to constitute the most serious threat to ecosystem integrity. Clearly a need exists to improve our predictive abilities and to establish a proactive approach to exotics management. Until we do adopt a more forward-thinking view, we will continue to operate in a circle of often futile, certainly expensive reactive and ineffective management.

Monitoring and Assessment

As illustrated throughout this chapter, introduction of an exotic species represents a change in the aquatic community, rather than a change in water chemistry or physical properties. Every native aquatic community boasts a variety of species and an evolved interaction among those species. Introduction of a new species alters those interactions. In some cases, competition for space or for nutrients will increase. In others, predation may increase. The general reaction will involve a change in species interactions; the principal kinds of variables affected will be population dynamics of individual species or dynamics in the ways that individual species interact with their environment. For example, growth rate, energy budget, or nutrient dynamics of a species may shift in the presence of an exotic. Recalling the information from our discussion of ecosystem responses to stress, we would expect communities with exotic invaders to have more leaky nutrient cycles, less efficient energy budgets, and reduced growth rates of the larger species. Available resources will shift to the exotic, which will often attain very large population densities. As a result, populations of other species will be deprived of resources. Native species will, therefore, often develop stunted growth.

Spatial Scale

Exotic species introductions are usually limited to contiguous water bodies or bodies that are linked by some distribution mechanism—that is, lakes in a chain, reaches in a river, or bays in a large lake. Colonization of new sites depends on some mechanism such as transport by boats, waterfowl, or humans. Once colonized, populations in new sites spread (often quite rapidly) until some factor limits their distribution. Therefore, monitoring plans for the effects of exotics must be designed to sample just beyond the edge of the current range of the exotic and within the range of the conditions where an invasion can be predicted (Fig. 24.2).

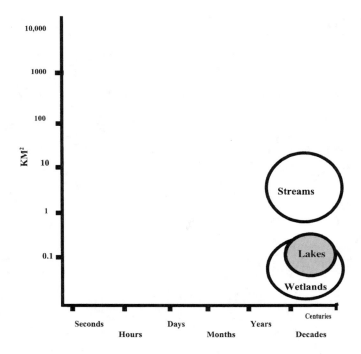

Fig. 24.2 Spatial and temporal impacts of exotic species will change for each affected species. In planning for exotic species, their tenacity is important; although lakes, streams, and wetlands each cover different spatial scales, the effects of exotic species upon each resource may be effectively permanent because of the degree to which a successful invader changes the biological assemblages of native communities.

Temporal Scale

Species' distributions are limited by their biotic and abiotic environment. A species will live and grow within the bounds of its range and the conditions it can tolerate. Once introduced into a new environment, a species will usually expand very quickly (e.g., within 1–5 years) and in many cases may remain in that site permanently. In some instances, populations have colonized a new environment rapidly, then undergone a natural decline or died back. Causes for such natural declines are not well understood. Monitoring designs for exotics must include sampling at the time of most rapid change (e.g., spring) and at times of highest probability of detection (e.g., late summer peak biomass). Sampling must occur annually at those seasons, as long as invasions are probable.

Case Study: Zebra Mussels in the Great Lakes and Beyond

The zebra mussel, *Dreissena polymorpha,* is millions of years old and originated in saline water basins that include the present Aral, Caspian, Asoz, and Black Seas. In the mouths of the rivers that flowed into those seas (e.g., the Volga, Don, Dnieper, and Danube), a stable relationship existed between the mussel and other organisms. Approximately 200 years ago, in connection with canal construction and increased water basin eutrophication, *Dreissena* began to increase its range into north European rivers that emptied into the Baltic Sea. Approximately 40–50 years ago, it began moving northward along the Volga, Don, and Dnieper rivers when large regions were flooded to construct a series of reservoirs for hydroelectric stations (Ludyanskiy *et al.* 1993).

Dreissena has an enormous dispersal capability, but in its native habitat, populations are held in check by predatory fish, most of which have several rows of pharyngeal teeth that can crush mussel shells. In North America, introduction of this mollusc has created a larger problem because of the lack of natural predators there. Consequently, large populations of *Dreissena* are expected to develop and continue to filter plankton from the water column, thereby limiting food resources of larvae of other native fish.

The Zebra Mussel in the United States

The zebra mussel was first discovered in Lake St. Clair, a northern arm of Lake Erie, in 1988. The mussel is assumed to have been introduced from the ballast water of a European ship. Within that same year, water supply facilities drawing water from the lake experienced a 50% decline in efficiency because mussels attached to insides of intake pipes. Now established in the United States, the zebra mussel is virtually uncontrollable (Fig. 24.3).

The dimensions of the zebra mussel problem are enormous. Less than three years after being introduced, mussels in the western basin of Lake Erie had reached an average density of more than $20,000/m^2$ and a maximum density of $700,000/m^2$ (i.e., 10 to 300 times its normal density in Europe). Only one year after introduction, zebra mussel biomass represented 70% of the total macroinvertebrate community biomass in northeastern Lake Erie, and the mussel's rate of production doubled in the subsequent year (Nalepa and Schloesser 1993).

Mitigation

The most immediate motivation for controlling zebra mussels is their effect on water supply intakes (e.g., power plants). Economic losses have been enormous

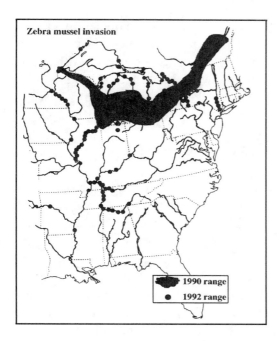

Fig. 24.3 The zebra mussel has spread extremely rapidly from its Great Lakes point of origin in the United States throughout the adjacent rivers, canals, and lakes to points as distant as Montana and Louisiana. (Simplified and redrawn from Ludyanskiy *et al.* 1993)

and are likely to represent a continuing problem. Many approaches are being investigated to control fouling effects of zebra mussels on Great Lakes power plants. Heat was initially regarded as one possibility because research indicates that a temperature of 30°C is lethal to the mussel. Although this temperature is several degrees warmer than the typical maximum experienced in the lakes, it was thought that heat could be used along intake pipes to kill fouling mussels. Unlike many other mussels, however, the zebra mussel shell stays attached to the pipe (or any substrate) even after it is killed. Consequently, heat may kill the mussel but replacing or periodically scraping the pipes is still required to allow adequate intake flow.

Another management approach uses sound waves as a control measure for the mussel. Acoustic excitation of pipes can be used to prevent establishment of mussels (or kill colonies) without affecting other biota and without introducing chemicals. Researchers also report success with treatment by an extract of the African soapberry Endod. Chlorine, permanganate, and peroxide plus iron have also proved successful in control of zebra mussels, although these chemicals have other water quality ramifications (Ludyanskiy *et al.* 1992). Many plants and facilities have turned to monitoring and early treatment as a more effective approach than mitigating against large infestations at a later date.

Over the long term, the *Dreissena* population in North American water bodies will be limited by either amounts of available food or a factor such as parasites, bacteria, or disease. In the former case, a stable population will become evident 10 to 15 years after introduction; in the latter case, the population will fluctuate broadly over periodic intervals. In either event, native fish populations will probably decline and remain at low levels due to food limitations of their larvae.

Case Study: The Nile Perch Versus Lake Victoria

The Nile perch (*Lates*) was introduced into Lake Victoria, the largest tropical lake in the world, to increase fisheries; in particular, this species was expected to feed on the less marketable endemic fish and generate a marketable product. The establishment was "successful," but the degree of its success was unanticipated. Nearly two-thirds of the endemic haplochromine species are now extinct, most other indigenous fish have declined, the Lake Victoria food web has become vastly simplified, and just three fish species have attained dominance in the lake. In addition, some people have blamed increased deforestation on the perch because processing the oily fish requires smoking rather than sun-drying, which has promoted uncontrolled cutting of local forests.

The Nile perch was not solely responsible for all of these changes, however. Kudhongania *et al.* (1992) discuss how overfishing, using successively smaller nets, and the increasing use of beach seines throughout the 1960s and 1970s resulted in damaging effects on some native stocks. *Lates* was introduced in the late 1950s, but it took more than 10 years to become fully established and 20 years until its effects caused damage. In the intervening years, fishing pressure exerted considerable stress on the lake's ecosystem. *Lates* and four other exotic species were introduced at a time when the lake was already suffering from declining fish stocks.

Other authors such as Witte *et al.* (1992) have shown that the Nile perch has played a central role in the disappearance of the haplochromine species and resultant changes in Lake Victoria's food web and ecosystem (including blooms of blue–green algae and deoxygenation of the hypolimnion that resulted from the lower densities of haplochromines). While these studies acknowledge that overfishing was occurring (e.g., within 10 years, mean trawl catches of native fish in one bay decreased from 1300 individuals per 10 minutes to zero), they note that after the Nile perch introduction haplochromines also disappeared from areas without fishing pressure. Furthermore, decreases followed irregular spatial distribution of upsurges in Nile perch populations throughout Lake Victoria. In each case, the population boom of the Nile perch corresponded with a rapid decline of haplochromines.

Despite extensive studies conducted on Lake Victoria, no viable options exist for control of the Nile perch. *Lates* is established, and the food web is permanently changed. Lake Victoria will remain an example of the need for careful control and forethought, but the lake cannot be restored to its previous state.

The Philosophical Questions of Exotics Management

Exotics management aims to remove unnatural and destabilizing elements from coevolved biotic communities. What is "unnatural" and what is "destabilizing"? The range of questions that must be addressed in exotics management is enormous. Answers to the sort of questions posed below have led people to radically different conclusions. Some believe that all exotics should be eliminated at any cost. Others suggest that, even when damaging to resident communities, some exotics might have value in providing new services. Debates may be long and few answers exist, but until these questions are addressed, values, goals, and strategies cannot be refined. Some of the intriguing questions raised in the field of exotics management cover issues of temporal and spatial scales in our judgments about effects of an introduced species. At the heart of the questions lies the fact that ecosystems are dynamic, continually losing and gaining species. What level of "background" change can be allowed, and how do we determine that background level? How do we distinguish modern-day introductions from this natural process? Some specific questions central to this discussion include the following:

- From how far away must a species be dispersed to be considered non-native?
- Is an invader less acceptable in a tightly coevolved community than a loose one?
- Invasions occur naturally—are anthropogenic invasions less natural?
- How much change in a resident population connotes destabilization?
- What elements of natural communities should be preserved (e.g., specific species, specific functional relationships, number of individuals)?
- What period of time must pass before a newcomer is regarded as naturalized? Is today's exotic tomorrow's naturalized species?
- Should an exotic be allowed to persist because its spread is occurring by natural means?
- Should a native species, historically extirpated, be reintroduced by humans?
- Ecosystems are constantly changing—how do we distinguish between "natural" levels of change or natural directions of change versus modified ones?

- Eliminating an exotic implies a desire to "restore" an ecosystem. To what level should restoration occur: 20 years previous? 50 years? before European settlement? Have changes in the intervening years made restoration feasible?
- What level of resource is tolerable to maintain a "restored" ecosystem?

Summary

- An *exotic* species is simply a species out of place. Species evolve in relation to the environment and the other species around them. When they are introduced into a new environment, they no longer face the selective pressures with which they evolve, but they become subject to other pressures. For those reasons, the vast majority of introduced species die out very quickly after introduction.
- A few species are termed "successful" from a human perspective. That is, they provide some increased human value to the ecosystem and the negative effects are less than the positive, *from a human perspective.* Examples might include fish species that create a major new fishery and have few or no significant negative impacts that threaten human use of the resource. Empty niches do not exist; when a species is introduced, it always displaces another species or members of a life stage of some species. The benefits to human decision makers may outweigh the liabilities, however.
- A very few exotic species are highly successful in the new environment, outcompeting native species and often creating massive changes in the ecosystem. For example, the common carp, which was introduced from Asia to North America, changes water clarity, alters fish and plant species composition, and significantly degrades the ecosystem, from a human point of view.
- One of the most egregious aspects of exotic species management is that the most effective management is prevention. Once an exotic species has been introduced and thriving, we can take numerous actions (at significant expense) to control its impacts. Eradication is essentially never an option.
- The greatest weakness of exotic species management in water quality programs today is that they are almost exclusively reactive. When a species has been found in a new system, we pass legislation, make policies, conduct research, and try to determine effective control strategies. In few—if any—cases has a society acted proactively to develop a strategy in preparation for the next aquatic invader.

For Further Reading

Balon E, Bruton M. Introduction of alien species or why scientific advice is not heeded. *Environmental Biology of Fishes* 1986;16:225–230.

De Moor F. Factors influencing the establishment of aquatic insect invaders. *Trans Royal Soc South Africa* 1992;48:141–158.

Drake J, ed. *Biological invasions: a global perspective.* SCOPE. New York: John Wiley & Sons Ltd, 1989.

Kudhongania A, Twongo T, Ogutu-Ohwayo R. Impact of the Nile perch on the fisheries of Lakes Victoria and Kyoga. *Hydrobiologia* 1992;232:1–10.

Nalepa T, Schloesser D, eds. *Zebra mussels: biology, impacts and control.* Chelsea, MI: Lewis Publishers, 1993.

Townsend C, Winterbourn M. Assessment of the environmental risk posed by an exotic fish: the proposed introduction of channel catfish *(Ictalurus punctatus)* to New Zealand. *Conservation Biology* 1992;26:273–282.

25

Cultural Dimensions of Water Quality Policy

Overview

Because water quality policy reflects both social values and scientific understanding, policy formation is a dynamic, fluctuating process. Every country's water quality policy reflects its unique social history and natural resource conditions, yielding as many variations of water quality policy as there are countries. These differences emerge through each country's experiences with at least seven socioeconomic and biophysical variables: religious and philosophical traditions; development status (and economic base); relative aridity of the climate and consequent value of water; population density and demographics; political structures and political will; historical influences; and international influences. Post hoc analyses show that each country's policy is the logical outcome of its biophysical and socioeconomic history. Nevertheless, these variables do not necessarily constitute a predictive formula for determining what a country's policy will be—too much depends on subtle interactions between the component influences.

This chapter explores, by example, how different countries experiencing vastly different economic, philosophical, and geographic conditions have arrived at their unique policy determinations. Even countries with similar geophysical status or political structure have arrived at widely disparate water quality policies, standards, and approaches to enforcement. For example, Singapore's strongly centralized government has achieved notable success in water quality management; in contrast, strong central leadership in Central and Eastern European countries has led to serious environmental degradation. The policies of other countries may be driven by a single dominant fact or priority, such as Canada's abundant water or Mexico's focus on rapid industrialization.

This chapter also underscores a critical point—that borrowing standards is common, but only rarely are the standards appropriate or feasible for the adopting nation.

Introduction: What Determines Water Quality Policy?

Water quality policy reflects a broad array of societal and resource conditions, although development status and resource availability are the most often cited determinants of policy directions. In reality, the picture is more complex. Countries with similar development status and similar resource characteristics remain likely to create distinctly different water quality policies. The neighboring countries of Ghana and Cameroon provide a case in point. Similar in biophysical environment and development status, these two countries exhibit different priorities and administrative structures—variations largely traceable to their separate colonial experiences. Modern-day leadership is also vitally important. The neighboring countries of Singapore and Malaysia exhibited very similar conditions 20 years ago, but a strong vision and leadership in Singapore have led to dramatic variations in environmental conditions and administration between these two countries. Similarly, different cultural and religious traditions affect the ways in which policy is expressed and may influence whether decisions are made via consensus, majority rule, or centralized control. In addition, domestic and economic pressures, as well as expectations of and advice from other nations, lead to different policies and standards (Table 25.1).

Previous chapters explored the uses and values placed on water resources, the biophysical and ecological characteristics of the different available resources, and the impact of human activities on those resources. Chapters 25–27 explore the variety of ways in which people and nations have responded to the need to effect water quality controls. In other words, how do different nations choose appropriate standards, enforcement procedures, and laws? This chapter does not attempt to provide a predictive framework for policy determination. Each variable that factors into policy judgments carries a different weight at different times and interacts dynamically with the other variables. The lack of predictive ability should not be seen as a limitation, however, but rather a strength. Too often in the past, countries have adopted the standards and approaches of other nations, to the detriment of their own water quality. The more closely water quality laws reflect the unique and evolving conditions in each country, the more likely they are to prove enforceable and effective.

Seven historical, sociological, biophysical, and political variables that influence policy are discussed below, followed by examples of different countries and how their policies reflect their own unique blend of these variables.

Table 25.1 Many countries share similar concepts in their application of water quality standards. This table compares similarities and differences between the effluent standards of several Asian countries. (After Mino 1994)

Country	Approach to Development of Water Quality Standards
Japan	Uniform standards developed for the country; more strict standards set by prefecture. Local standards often developed in collaboration with industry and stakeholder groups.
Korea	Water bodies classified on the basis of use; effluent standards based on class of receiving water.
Pakistan	Standards exist for only some industrialized areas and major cities.
Sri Lanka	Specific standards for each class of industry; based on industrial technology.
India	Each state sets its own standards for ambient and effluent waters.
Taiwan	Specific standards for industrial and residential wastes, negotiated between authorities and management.
Thailand	Standards specific to industrial class and/or size of municipal waste facility (population base).
Philippines	Standards for industrial wastes differentiated from municipal wastes; effluent limits based on receiving system classification.

Religious and Philosophical Frameworks

A critical, but elusive influence on the values that guide policy decisions stems from religious and philosophical traditions. Much of the focus on the importance of religion in an environmental context was inspired by a pivotal 1967 article by Lynn White that linked the environmental degradation of the industrialized west to a Judeo-Christian tradition of "dominance" over the earth. Since that time, many world religions have been held more closely accountable for their environmental ethics (Palmer 1990). In the final analysis, no religious tradition—no matter how strong its inherent environmental ethic—has prevented the onslaught of environmental degradation brought on by the more immediate stresses of population growth, industrialization, and urbanization. The fact that environmental ethics have not prevented environmental degradation does not imply they are powerless to play a role in environmental improvement, however. A strong movement in many world religions calls for a renewed focus on environmental ethics, including the establishment, and teaching, of the environmental ethics couched in their texts and belief systems (see, for example, Callicott 1994).

Globally, profound differences exist in the world views taught by different religions and philosophies; the environmental ethic that stems from each of these world views stresses a different aspect of people's relationship to the environment. Perhaps more importantly, to date, the social ethics of different world views define appropriate systems of social control. Thus, the policies generated by these varied traditions express different pathways for reaching similar goals. This book cannot hope to review the extensive influence of religions on national "character" and policy formation. The very brief review that follows provides only a glimpse of the ways in which major world religions create different world views—some focusing on this life, others on an afterlife; some focusing on collective action, others on individual morality; some taking an anthropocentric view of the world, others an ecocentric view.

Examples of the anthropocentric world view come from the world's primary western religions: Christianity, Judaism, and Islam. Each shares a similar creation story, in which God created the earth and then created humans, but humans are distinguished from the other animals by free will and by being "in God's image." Thus, these religions have an inherent anthropocentric focus toward the world. Even with this similarity, however, these three major religions all carry distinctively different traditions. Christian ethics contain considerable ambiguity and lack of specific rules with respect to human conduct toward nature, although Christian theologians are redefining the term "dominion" to fit the needs of modern society. Islam, in contrast, exerts more structures and proscriptions upon its followers and more directly expresses the world view that nature was created to serve human needs. Nonetheless, Muslims still live to serve God; because nature is God's creation, modern environmental interpretations of Islamic texts do find room for a stewardship ethic. Judaism, although stemming from common early scriptures, is unique among these three religions because it does not stress an afterlife as a motivation for good deeds, nor does it stress individual salvation (as do Christianity and Islam). Instead, Judaism is a collective religion that emphasizes the survival of the group over time. Because it suggests that there is no formal afterlife, Judaism invokes a strong ethic of responsible choices in this lifetime (Callicott 1994).

Hinduism and Buddhism are the two major representatives of a school of thought that views people as having no special place above other animals and plants. Hinduism is a religion of intriguing contrast for environmental ethics. On one hand, its world view of reincarnation posits a strong connection of humans to other animals and plants. In fact, through reincarnations, everything animate is essentially one being. Consequently, humans are equal to all other creatures; their souls are, in essence, one. The world view of reincarnation also creates a certain fatalism because the cycles of the world are not seen as being under an individual's control. Additionally, Hinduism is distinctly otherworldly in focus; it suggests that all facets of this lifetime are illusory and distracting to the soul. Hence, Callicott (1994) suggests that Hinduism teaches a mixed environmental ethic—that is, one strongly respecting the individual life of other plants and animals but lacking a formal sense of responsibility for caring for those lives (Fig. 25.1).

Buddhism (an ancient offshoot of Hinduism) has much in common with Hinduism, and yet much that is unique. One important difference is Buddha's teaching that the cycle of reincarnation can be broken by extinguishing all desire. Inherent in the Buddhist philosophy is a deep commitment to the importance of nature and the place of all things on this earth. Buddhism is perhaps unique, however, in its concentration on process and affirmation of change. The source of conflicting environmental ethics in Buddhism lies in the philosophy of "non-action." Different interpretations of this way of life lead to radically different lifestyles and policy choices.

Other world religions offer just as many points of departure for environmental ethics, and just as many inconsistencies and dichotomies as the five religions discussed here.

Fig. 25.1 The perception of water is also governed, in part, by cultural and religious traditions. In Hinduism, for example, water is considered religiously pure. Especially sacred waters such as the Ganges are important points of pilgrimage for ritual baths. Other uses of the same water—for washing clothes, disposing of waste, cremations, and fishing—occur simultaneously. Some suggest that the religious vision of water as "pure" may interfere with the perception of it as polluted. (Photo by D. Vanderklein)

In Callicott's (1994) extensive review of the environmental ethics of the world's religions and philosophies, he stresses that the world view of different parts of the globe was crafted more by early philosophers than by some religious traditions. Consequently, he argues that the Greek philosophers, more than the Bible and Judeo-Christian religious ethics, framed the modern western world views that have enabled widespread environmental degradation. These philosophers pursued a mechanistic rationale for nature, proposing that nature could be reduced to a sum of its parts. Because nature has no "organic integrity," it could be altered with no ill consequences. The human soul, on the other hand, bestowed a certain divination and superiority over nature.

Callicott draws a parallel to the intellectual traditions of Tao and Confucianism in Chinese thought, arguing that these traditions largely framed the environmental ethic of Chinese society until China was opened to the west. While much has changed in Chinese society since the Cultural Revolution, industrialization, and burgeoning population pressures, the foundation of Chinese thought of mind and body as a harmonized whole, as humans as part of nature, and as nature as relational and processive, remain powerful forces in Chinese society.

Historical Influences

As the above discussion illustrates, as a society changes, its successive philosophies, eras, and incidents change the value systems and priorities within that society. Thus, the origins of values become obscured under layers of influences. In the Chinese example, Confucianism and Tao, layered over Buddhism, created societal values far different than those found in Europe or India, or even those of other Far Eastern regions. Much later, the transformations stemming from the end of the dynasties, Communist takeover, and the Cultural Revolution changed millennia of traditions during a single generation. Modern value systems in China reflect their ancient roots, but more immediately reflect the potent and dramatic transformations of this century.

Similar historic influences define the path taken by Japan with respect to environmental management. In Japan, the Buddhist tradition evolved from its Indian roots to give rise to Zen, a deeply meditative style of Buddhism. While Zen is solidly rooted in the contemplation and appreciation of nature, analysts quoted in Callicott's treatise point out that "nature" in the Zen tradition is distinct from "nature/wilderness" of the western romantics. According to these interpretations, the love of nature that dominates Japanese society involves a more metaphysical concept of nature, focusing on a single place or an aesthetically significant object or species. Thus, Callicott (1994) argues, the Japanese could simultaneously love and respect nature and beauty while "not seeing" the devastating pollution overtaking other sites. Again, this interpretation is based purely on religious and philosophical traditions. In fact, Japanese society was also transformed in this century, most dramatically by its participation in, and rebuilding after, World War II.

Colonial histories also created unique layers of traditions and world views in an enormous number of the world's emerging countries. In fact, the identity and boundaries of most countries in Africa originated with colonial rule. With its imposition of language, legal, educational, and administrative systems, the introduction of an international economy, and exposure to European cultures, the colonial experience was deeply transforming and its long-term influence pervasive. The colonial experience differs from the philosophical evolutions of thought outlined above in that colonialism put entire foreign systems in place without consideration for local conditions, traditions, and relationships. After achieving independence, the countries were so transformed that no indigenous substitutes could be found for the colonial-based administrative and economic conditions. Consequently, many colonial patterns endure today.

Historical experiences may incorporate everything from experience with ecological crises (e.g., Japan's experiences with mercury and chromium poisoning) to ethnic and religious conflicts (as in much of the Middle East). Equally important are the visions of influential individuals. In the United States, for example, much of the impetus for and direction of the 1972 Clean Water Act derived from the influence of two committed senators. In centralized governments, the vision and priorities of the leaders may lead to unique environmental conditions. In each case, modern policies reflect a country's historical influence with individuals and incidents.

Development Status

Development status influences the ability of a nation to devote financial resources to water quality issues (or, in fact, to any basic services). In economically depressed nations, neither the gross national product (GNP) nor the tax base exists to fund major programs in sanitation or to provide potable water to more than a small percentage of the population. In addition, there is often a push to promote industrialization, at any cost, to jump-start failing economies and raise low living standards. Thus, both the political will and the financial resources are frequently lacking for appropriate scientific explorations, enforcement, and monitoring mechanisms.

Nonetheless, some developing countries have successfully prioritized water quality and other environmental issues. In some instances, this urgency is related to the economic base of the country. For example, the lobster industry is an important economic base for Cuba. Its importance is reflected in the careful and sustainable management regulations that govern the industry. In other developing countries, the primary economic base is related more to silviculture, as with the rubber industry in Brazil, and the coffee industry in West Africa. Consequently, activities in these countries provoke little concern for water quality issues.

Developed countries are distinguished by high GNP and high standards of living. These economic conditions bring associated improvements in average literacy, health, security, and lifespan. Values change as people begin to focus on issues beyond basic needs. In most cases, the more educated, healthier citizenry in developed nations come to expect and demand an improved quality of life, including more stringent standards for water quality. Large variations in the standards and enforcement protocols of developed countries indicate, however, that development status is not an absolute determinant of policy priorities; instead, it is only the first tier in a multilayered process of value prioritization.

Resource Availability

Arid climates generally make people value water and water quality more intensely than water-rich climates. Thus, resource availability is frequently an important factor in how a society values and legislates its water resources. In extreme cases, relative abundance or lack of water may dominate policy priorities. For example, Israel—an extremely arid country—has instituted a "no pollution" policy; Israel simply cannot afford degradation of its limited resource. In contrast, Canada—an extremely water-rich country—is relatively lax in its pollution control (in comparison with other developed nations); Canadians find it difficult to be concerned about water or water quality because of its apparent abundance.

The connection between geography and environmental management is appealing but is not always as neatly defined as the Canada/Israel contrast would indicate. In fact, the southwestern United States is the largest per capita consumer of water in the nation, but its water policies have yet to reflect the limited regional water (water scarcity continues to be conquered by technology—it is mined or diverted from distant sources). The decision as to how arid and water-rich nations view and regulate their resources may be more intricately connected to the delineation of the country's borders than to regional conditions of aridity. Israel, for example, is a small country surrounded by neighbors that do not want to share their resources with the nation. The only viable strategy, then, is one of self-containment and meticulous care of the local resources. In contrast, arid regions of the United States can draw on enormous resources, both in terms of financial assistance for development and natural resources from which to "borrow." Thus, even given the conditions of aridity and water wealth, the influencing layers of politics and cultures pose an important control over environmental management options.

Water scarcity is also relative to population increases. It has been projected that by the middle of the next century, between 45% and 65% of the world's population will live under conditions of water stress or scarcity (Engelman and LeRoy 1993); this fact will have dramatic implications for water management. Generally speaking, areas experiencing water stress are more concerned about the mere availability of water than with its quality. At the same time, decreasing quantity and increasing populations represent a formula for dramatic and serious water quality problems. To date, few countries have recognized or faced their looming future resource problems.

Population Density and Distribution

A variation on the theme of resource availability, population pressure can stress local resources and increase health risks as poor sanitation and poor water quality increase the incidence of disease. Historically, urbanization of populations has created awareness of the need for water quality improvements. It is also more economical to provide basic services to urban populations than to their rural counterparts. Urbanization, then, may "force" a country to formulate national water quality policy directives even when implementation will be directed toward only a small percentage of the population (the urban, or the urban rich who can pay for resources).

In the next 25 years, however, water quantity and quality issues are expected to play an increasingly dominant role, specifically because of the stresses placed on the resource by increasing populations. Between 1940 and 1990, the world's population doubled while the world's water consumption quadrupled. Population levels and water availability currently maintain a delicate balance. Between 1990 and 2025, nations with water scarcities are projected to rise from 20 to approximately 35, while the affected population grows from 131 million to between 817 million and 1079 million (Engelman and LeRoy 1993). China is projected to barely miss the "water stress" benchmark by 2025, but India and almost all countries in northern, eastern, and southern Africa will fall into (or continue in) a state of water stress (Fig. 25.2).

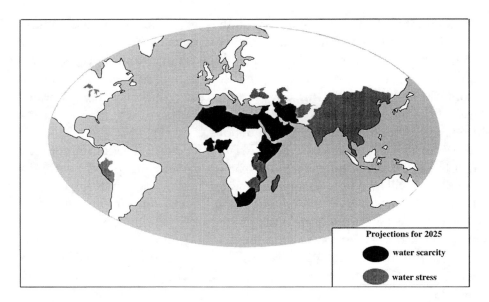

Fig. 25.2 By 2025, population increases are projected to create dramatic water shortages in many areas of the world. While all projections suggest that many countries in northern and eastern Africa will experience water stress or scarcity, projections vary according to population estimates. Following high population projections, for example, China will experience water stress by 2025; under scenarios of moderate population increases, China narrowly misses the stress benchmark. Regardless of whether entire countries such as China are included in the formal projections, significant local stress will occur (and is already occurring) throughout the globe. (Modified and redrawn from Engelman and LeRoy 1993; courtesy of Population Action International, Washington, DC)

Demographic issues can also complicate water stress and water quality management. In Mexico, for example, three-fourths of the population lives in the dry central highlands, while four-fifths of the surface water lies in the wetter coastal regions. Although the concern is not as dramatic, the development policies of the arid portions of the United States are being called into serious question as these population centers drain critical water resources of distant regions. Population issues have an enormous potential to dominate water quality discussions in the future. Even more disquieting is their potential to increase political tension both within countries and internationally.

Political Structure and Political Will

Population stress and water scarcity may set the stage for dramatic political struggles and conflicts in the future. Some hint of this possibility has already been given, for example, in Iraq's destruction of Kuwait's desalinization plants during the Gulf War. The role of these conflicts cannot be underestimated, but similarly cannot be accurately predicted.

More tangible is the political attitude that many countries express toward current water policies. Many countries have enacted comprehensive water quality policies without ever providing for their implementation. The dominant reasons behind this lack of action are in some cases political and in others economic, but in all cases they create a scenario where comprehensive laws exist "on the books" but are never enforced or implemented. Interpretations of this strategy have ranged from attempts at political domination and power to attempts to comply with international recommendations and standards despite an insufficient infrastructure to implement them. In either case, the strategy is widely used, and its causes probably reflect diverse influences.

Another influence of political structure and political will that creates intriguing differences between countries' water quality policies is the approach taken to enforcement. In some cultures, for example, the U.S. model of command-and-control enforcement would conflict with the more subtle social controls of social standing. As the discussion of Thailand given later in this chapter demonstrates, enforcement in some cultures is less combative and more focused on consensus. In fact, acceptance of command-and-control legislation may be unique to the United States. Many European countries are more comfortable with incentive programs and pollution taxes. Even within the United States, however, command-and-control approaches are being questioned as the pendulum of social values swings back toward private ownership rights and free market pressures. Consequently, the United States is exploring pollution trading permits and other concepts that emphasize personal flexibility within an overall goal of improved water and air conditions.

The Role of Public Input

National policy ownership is another area in which political structure influences policy. Although the public is an integral part of societal decision making in democratic societies, this relationship does not hold true throughout the world. Even in societies with a strong democratic tradition, administration of water use and water quality is often considered an area where public involvement in decision making is unwarranted. Most countries' water rights are vested in the government, and thus the government decides who uses water and how—a fact that does not necessarily imply national-level or participatory decision making.

In centrally controlled societies, environmental management decisions are made without public input. Agency representatives develop or adopt overall goals and long-term directions, assess environmental condition with their staffs, and then impose environmental policies to further those goals. In these cases, management goals may represent societal goals, articulated by one or a few people, or may represent the goals of just a few people.

In countries such as the United States that have a stronger democratic tradition and a common-law approach to water resources, public involvement is the spoken and sometimes acted-upon norm. The public can be involved in adopting new standards in two general ways.

In the first strategy, government agencies widely solicit public opinion, although the final decisions still rest with the managing agency. The pollution control or management agency reviews relevant history and literature and establishes proposed standards. The agency then solicits input through public hearings, legal notices, leaflets, newspapers, and other public-outreach media. The agency is usually charged with reassessing its goals and/or standards based on public input and revising its plans to reflect public concerns.

A second approach to soliciting public input is through a task force or committee. In this case, interested parties (called stakeholders) draft the initial material. This group usually includes government representatives, the public, and representatives from effluent producers (e.g., industry, municipalities, farmers). It reviews relevant information on processes causing water quality impacts and general or expected waste stream characteristics, on the expected impacts of those wastes in the receiving water, and on treatment costs and control alternatives. Based on this information, the group establishes standards and a timetable for implementation, all of which remains subject to approval (and veto power) of the implementing agency.

All of those approaches to incorporating political concerns into water quality management have strengths and weaknesses. Centrally controlled decision making can provide more rapid decisions with clear lines of authority. No guarantees exist that resulting regulations will be in the best interest of the populace, however. Parallels between the controlling agency's goals and society's goals in general are not always clear, nor are they always relevant to the standard-setting process.

International Influences

Apart from cultural traditions and colonial history, countries face increasing pressure to adopt environmental policies that minimize impacts to their neighbors and the global community as a whole. In some cases, countries work together closely to adopt mutually appropriate minimum standards (e.g., in the

European Community). Perhaps more frequently, countries responding to international pressure simply adopt the ready-made policies of model industrial nations (e.g., the United States). Lacking the research structure to design standards, and confronted with the time-consuming, costly (in the short term), and complex process of establishing laws, many nations find it simpler to adopt the standards and laws of other countries. Although these standards are supported by scientific evidence and meet international standards and judgments, they are often not appropriate to the resource conditions, economic level, or legislative style of the adopting nation. As a result, they are rarely implementable.

More recently, attempts have been made to institute water quality (and other environmental) policies that reflect indigenous patterns of resource use or other unique national and regional conditions. These moves toward nationalism and regionalism are simultaneously hailed for their appropriateness and regarded nervously for their generally lower standards and greater tolerance of pollution. In addition, many developing countries are trying to institute appropriate technologies that match the new standards and laws, using lower-cost, simpler technology alternatives to modern expensive technology.

Policy Development in Action: Selected Examples

The remainder of this chapter is devoted to examples of water quality policy throughout the world. Although tempting, it is neither possible nor instructive to include a country-by-country synopsis of water quality policies. Instead, the highlighted countries have been chosen because their water quality policies reflect the influence of one or more of the factors described earlier. Many more countries and regions could be added to a review of this nature; the following is intended to serve as only an introduction.

These highlights illustrate that water quality policy is a logical outcome of each society, once its social, cultural, political, demographic, economic, and resource characteristics are known. They also underscore that no policy formula exists in which a given combination of these variables yields a definite policy. It is far too simplistic to say "country X, a developing country, will have the following water quality policies," or "country Y, which is predominantly arid, will have the following policies." Instead, development status and resource availability must be understood in the context of other social, political, economic, and historical influences.

Managers, policy makers, and citizens desiring to improve water quality conditions need to understand the derivation of current policies. Even more important, however, is recognizing that there are a variety of culturally appropriate, sustainable strategies that different countries can follow. Ineffectual policies of today can be altered in culturally sensitive ways to yield effective policies for tomorrow. For convenience and to provide a familiar frame of reference for readers, the examples follow an approximate geographic grouping.

North America: Influences of Demography, Diversity, and Development Status

From a policy point of view, North America is intriguing because in its three countries—the United States, Canada, and Mexico—one can see examples of three very different approaches to water quality management. These variations reflect differing developmental statuses and perceptions of resource abundance. North America is also interesting because the concerns of managers stretch over such a wide array of water bodies and conditions. Policies must address water systems ranging from coastal areas to prairie potholes, from one-acre wetlands to the Great Lakes, and from mountain streams to the Mississippi River. In addition, the continent hosts the world's largest cities along with some of its smallest villages, as well as significant investments in agriculture, industry, and recreation. Perhaps even more importantly, wide disparities mark the socioeconomic conditions among the people of the continent that creates associated disparities in the ways people value water resources. This diversity makes water management very challenging.

CANADA: RESOURCE ABUNDANCE AND FIERCE DIVISION OF POWERS. The second largest country in the world (in terms of area), Canada has a unique demography that influences its environmental law traditions. More than 85% of the Canadian population lives within 120 km (75 miles) of the U.S. border. Canada also encompasses vast land areas with very low population density, and these areas could withstand significant economic activity without degrading water quality over large areas. The country's water history provides an interesting study of how the perception of abundance leads to a laissez-faire attitude toward resource management. Since the 1970s, Canada has responded to international concerns about environmental quality and has enacted laws similar to those found in other modern industrialized nations, but most Canadians have yet to recognize a water quality problem in their country.

Canada has about one-fourth of the world's stock of fresh water, and its per capita water use is close to the highest in the world. Nonetheless, these indications of abundance are deceptive because the water is, in most cases, not in the same location as the population. Most of Canada's great rivers flow away from the southern concentrations of population and industry, leaving some areas of heavy water demands with scarce supplies. In addition, all regions are susceptible to flood and drought, adding a temporal dimension to scarcity. Finally, many of the most serious management problems are not related to the adequacy of flows, but rather to the quality of water supplies, the disruption of natural flow regimes, and conflicts among uses and users.

Next to demographics and perceptions of resource abundance, perhaps the most salient issue in Canada's water management is the division of responsibility between provincial and federal governments. Authority over major uses is divided by the Canadian constitution between these two forms of government. Thus, issues of flow and quality are administered separately under different regulatory systems. The provinces, for example, have jurisdiction over natural issues impacting on property and civil rights, yet the federal government retains authority for regulating fisheries, navigation, and relations with foreign governments. The federal government also can limit provincial authority in several areas, including the ambiguous area of "works for the general advantage of Canada."

Another critical issue in Canada is the licensing and permit process. Some uses of water are specifically regulated, while others are not. Currently, among the "not" regulated category are noncommodity uses such as recreation, fisheries, wildlife protection, and other environmental and conservation purposes. As a result, the licensing system distorts the allocation of water among uses and users. Finally, constitutional uncertainty surrounds the legal basis for managing pollution in rivers that cross provincial boundaries, creating an exceedingly complicated legal and policy framework for water management in Canada. These arrangements fall short of the coherence that economic and environmental considerations suggest is needed for effective water management.

THE UNITED STATES: FEDERALISM AND CITIZEN INVOLVEMENT. The United States is the fourth largest country (in terms of area) in the world, and is both highly industrialized and extremely rich in a wide variety of natural resources. With a high literacy rate, a democratic tradition, and a strong commitment to "fairness," its environmental laws are defined by a tension between sweeping country-wide standards and sensitivity to states' rights and the rights of private individuals. In addition, the laws come under constant scrutiny by the populace. As a result, they have experienced significant conceptual and legal growth throughout their history.

The first national water quality control legislation in the United States, The Rivers and Harbors Act of 1899, was concerned with water quality only insofar as it interfered with the navigability of waters. Thus, discharges that obstructed river traffic became illegal, but urban run-off from streets and discharge from municipal sewage were exempted from the prohibitions. The 1899 Act focused on prohibiting specific practices based on their assumed impacts on water quality. In contrast, most U.S. water quality laws enacted between 1900 and 1972 focused on the expected results (i.e., they tried to regulate ambient water quality).

As U.S. society graduated from one era to the next, the emerging concerns and improved understanding were reflected in further legislation. For example, the first ambient water quality law in the United States (1912) provided funds for water supply research programs, seeking to discern relationships between water quality conditions and public and/or human health. Later laws introduced the protection of aesthetics and public health concerns. The importance of industrial growth was reflected in the concept that a certain degree of contamination is an inevitable by-product of industrialization. Early U.S. legislation and enforcement also reflected the tensions of federalism and the reluctance to allow the federal government to exercise significant control over local affairs. For example, in 1948 Congress passed the Federal Water Pollution Control Act, which made those responsibilities more explicit and which included much attention to conditions in streams and lakes; on the other hand, Congress never appropriated the funds to support this bill's grant program. In a similar vein, early legislation typically placed limits on end-points but allowed dischargers the freedom to choose the best method of compliance.

With the passage of PL 92-500, the Water Pollution Control Act Amendments of 1972, U.S. water quality policy changed dramatically. This law reflected the philosophy that any activity that affects water quality should be curtailed or controlled. Significantly expanding the role of the federal and state governments, it established a "command and control" policy approach that replaced the previously ineffective flexibility with strict standards and fines for noncompliance. While this law has gone through numerous changes since its inception, U.S. water quality policy continues to be dominated by modifications to the 1972 law rather than undergoing large-scale changes to the underlying strategy.

MEXICO: THE DRIVE TO INDUSTRIALIZE. Mexico has taken a different tack than that of its northern neighbors. Whereas lax regulation in Canada stems from a view of resources as unlimited, lax regulation in Mexico stems from a view that resources are expendable. As a developing country, Mexico has traditionally been more concerned with nation building than with water quality regulation. While federal water quality laws are sweeping and inclusive in form, they lack substance because they include few implementable regulations and responsibility has been subdivided among too many agencies. Effective management is hampered by the competition between agencies (for funding and authority) but is also hindered by the lack of broad public concern for environmental issues.

Mexico's environmental policies are the legacy of two factors. First, a largely feudal system, inherited from the Spanish conquest, operates to allow a powerful elite to control the vast majority of land and resources. Second, a strong central government regulates from the top down, with little input from local managers or populace. This pseudo-feudal system has not led to the development of a strong middle class—one of the groups that has largely carried the environmental movements in the developed world. Instead, Mexico's middle class remains small, politically weak, and strictly urban. Consequently, environmentalists have had little success in galvanizing public opinion.

Besides its Spanish legacy and the attendant social and philosophical views of the world, Mexico is characterized by a struggling economy and massive inflation. The combination is not conducive to the sort of environmental awareness that furthered pollution control regulation in the United States and Western Europe. (The environmental efforts of the United States would have taken a different shape if the environmental movement had arisen during the depression of the 1930s.) In other words, Mexican society may have concluded, at least for the short term, that it is better to have factories operating, employing people, and generating pollution than to have no factories at all. In the words of one government scientist, "Moreover, the limited resources at the command of a less developed country make it more difficult to address ecological problems in any form."

Additionally, the environmental questions and critiques in Mexico (i.e., the definition of Mexico's goals) are notably different than those in its neighboring countries to the north. In Mexico, defense of the environment is largely a class issue, so environmentalists have tied their efforts to a broader critique of Mexico's dependency and domination by an internal elite. This critique owes much to the belief that Mexico's environmental catastrophe is an integral aspect of economic dependency on the United States and other industrialized nations. As a consequence of this political framework, Mexico has adopted a series of "safe" environmental priorities that do not significantly affect industrialization and economy building. Acid rain, desertification, loss of species, contamination of the oceans, and atmospheric changes have been cited as Mexico's top environmental priorities.

Industrialized Asia: Traditions of Nonlitigious Cultures

HONG KONG: UNIMPEDED CAPITALISM AND GOVERNMENT BY CONSENSUS. In the last 20 years, Hong Kong's population has grown by 30%, and its GNP has grown by 300%. Accompanying those dramatic increases in population and productivity has been a concomitant increase in waste and pollution. Holmes (1992) comments, however, that the population issues leave little room for, or interest in, larger issues such as pollution. Part of Hong Kong's remarkable growth is attributed to the government's strategy of nonintervention in trade and industrial matters, which has enabled the territory's industry to develop and respond quickly to fast-changing market conditions. The same government strategy has not proved effective at ensuring pollution controls. The consensus-run government has not been able to respond rapidly to new problems and concerns such as pollution; effective pollution controls are barely 5 years old.

Hong Kong's pollution problems are complicated by the density and diversity of industries in the territory. In addition, the hands-off process of development has led to piecemeal growth with an overloaded and inadequate sewerage system. In fact, approximately half of all sewage and industrial effluent reaches the sea via the storm drainage system with no treatment. Hong Kong's first water quality legislation (1980) was based on a policy to stop water pollution from worsening, but did not establish approaches to prevent pollution problems until they arose. Consequently, pollution continued to increase. At the end of the 1980s, the government released a new policy document outlining priority areas: the quality of inshore waters, public sewerage, waste water treatment and disposal for public sewerage systems, and the enactment and enforcement of legislation that safeguarded the community. Although 1990 amendments to the Water Pollution Control Ordinance strengthened its enforcement capability, continued reliance on advising and persuading polluters to stop polluting has led to few real improvements in water quality. Wing-Hung Lo (1995) notes that, despite the 10-year "save the environment" program launched in 1989, "the dominant approach of policy and law enforcement through consultation has rendered impossible strict enforcement of environmental rules and regulations as local economic growth enjoys a priority over environmental protection" [p. 331].

SINGAPORE: THE INFLUENCE OF STRONG LEADERSHIP. In Southeast Asia, Singapore has progressed the farthest in environmental management and remains a remarkable example of the influence that a single individual may have in the establishment of environmental management. Like Hong Kong, Singapore is a city-state characterized by high population density and intensive development. In 10 years (1974 to 1984), Singapore's gross domestic product (GDP) rose from S$8.4 billion to S$18.3 billion, with a tripling of per capita income. Electricity consumption also tripled in the same period, industrial output increased tenfold, tourism increased fivefold, and construction vacillated between 30,000 new units per year and 80,000 units per year.

In contrast to Hong Kong, Singapore achieved its development through a strong and directed government effort. In parallel to the economic changes, the intensive government commitment to literacy and population growth achieved a near-100% literacy rate and a declining gross reproductive rate. The commitment of the one-tier government to development and quality of life is remarkable when viewed on an international scale, and is one reason that pollution is effectively controlled. Consistent with this development strategy, few private bodies have participated in promoting environmental conservation; instead, citizens generally cooperate with government efforts. Few legal cases are ever charged between private parties, and never between private parties and public authorities.

The republic has an umbrella environmental agency, with a full set of legal provisions in areas ranging from solid waste to noise pollution. Considerable investment has gone into providing physical treatment and other facilities, technical and management personnel, and matching financial resources. Environmental concern has focused on questions of public health, water pollution, and environmental enhancement. Policies are based on physical planning, economic efficiency, and a "polluter pays" principle that is supported by effective administrative and enforcement measures.

Glimpses from Developing Asia

Throughout Asia, more countries can properly be termed "developing" than "developed." While few of these nations have effective environmental policies, their development status often provides only a partial explanation for this absence.

The example of Indonesia may be one in which economic status has proved an overwhelming factor. Indonesia's progress in environmental legislation suffers from nearly every logistic barrier in existence: the country covers an enormous and fragmented area, per capita GNP is low, and both population and natural resources are unevenly distributed. The priorities and level of development of environmental legislation all indicate that Indonesia is currently more preoccupied with its economic development, population growth, and difficult administrative logistics than with water quality per se.

Other countries such as Thailand have some financial resources available and fewer logistic complications, yet they remain similarly uncommitted to environmental legislation. Despite Thailand's awareness of environmental problems, the country lacks the political commitment to effect change. In addition, despite the presence of a parliament modeled after that of England, Thais adhere to cultural traditions of negotiation rather than favoring legislation; as a result, the power of the legislative rules is limited.

Although Thailand acknowledged the importance of environmental controls early on, in the last 20 years its environmental regulation strategy has changed little. Instead, the country has maintained its tradition of a laissez-faire approach to development. In combination with an apparent lack of urgency, this tradition has led to a scenario where the laws and institutions are in place to effect more stringent controls, but neither the will, authority, nor traditions have enabled such measures to develop (Flaherty and Karnjanakesorn 1995).

Eastern Europe: Political Restructuring and Everyday Demands

The deplorable water quality of the countries in Eastern and Central Europe is legendary. Raw sewage and industrial effluents laced with heavy metals and toxic (and sometimes radioactive) chemicals are the two principal contributors to the deterioration of water quality in Central Europe's waters. In Upper Silesia in Poland, coal mine residues have rendered the Oder and Vistula rivers too polluted for consumption or industrial use.

The condition of Polish surface waters has deteriorated dramatically over the last 25 years. Class 1 water, defined as drinkable after disinfection, was present in 33% of the total length of monitored rivers in the country in 1967; by 1986, Class 1 water was found in only 4% of the country's total river length. Unclassified water, which is virtually unusable even for industrial purposes, rose from 23% of the river length in 1967 to 39% in 1986. About half of Poland's cities, including Warsaw, and 15% of its industrial facilities have no wastewater treatment systems, and about 32% of wastewater needing treatment is left untreated.

In the Czech and Slovak republics, the quality of surface waters has shown a similar downward trend, with 27% of the major river lengths classified as incapable of sustaining fish or containing inedible fish. Only about 40% of the former Czechoslovakia's wastewater is adequately treated. Sewage sludge, which was once sought as a fertilizer, is now a toxic waste in many industrial areas because of contamination by heavy metals, especially cadmium. Groundwater contamination also has increased sharply; over the past 30 years, average nitrate levels in groundwater in the built-up areas of cities and towns rose from 30 mg to 120 mg per liter.

Similar scenarios are found in Hungary, the former East Germany, and Bulgaria, where the legacies of quota-system industrial production, outdated technologies, and central planning run deep. Despite the great optimism that economic transformations would lead to rapid environmental improvement, these countries have had to face the harsh reality that their independence also brought the end of job security and food and energy subsidies. Consequently, harsh day-to-day economic issues have dominated the minds of the populace, as well as the policies of the governments since independence.

Some (temporary) improvements in water quality have been noted due to market forces. For example, Poland's 1990 stabilization program, which began to phase out energy-related budget subsidies and instated market-oriented energy prices, substantially decreased industrial output and thus resulted in emissions reductions. Although biological oxygen demand may have dropped by as much as 33%, analysts suggest that these reductions may prove only temporary because they result from recession; if this suggestion proves correct, when the economy recovers, so will production.

Interesting international issues confront the Eastern and Central European countries. Competition from the European Community and the desperate need for foreign investment have made these countries desirable end-points for hazardous waste. A Greenpeace International report found evidence of 64 trade schemes in which Poland was targeted to receive a western hazardous waste. The Polish government has stopped some waste-import schemes, however, and Hungary passed a decree in October 1990 prohibiting imports and exports based on differences in environmental standards.

South Asia: Population Explosions and Civic Strife

The issues facing India, Bangladesh, Nepal, Pakistan, and other countries of South Asia all reflect these nations' intensely individual nature. One factor that unifies and dominates policy among all of these countries is their dramatic and unsupportable population increases. A second factor that dominates all of these countries' policies—with the exception of Nepal—is tense interregional relations resulting from a deep Muslim–Hindu divide.

Population increases create tremendous pressures on water resources and water quality. In Bangladesh and Pakistan, water treatment is nonexistent; in India, some treatment exists, but more than 70% of India's waterways remain polluted (some severely so), and more than 90% of that pollution is domestic. Health issues have not yet become national priorities in these countries, nor are they able economically to keep up with the expanding infrastructure necessary to meet the exploding populations. At the turn of the century, India's population stood at 238 million. By 1991, it had increased more than threefold to 860 million, with projections that it will reach 1.16 billion in 2010. More than 40% of that population is expected to be concentrated in urban areas by 2010 (Hassan 1991). Bangladesh's population increases are even more sudden and problematic because of its small habitable land area: in 1951, population was 42 million; in 1991, it was over 116 million; and it is projected to be 177 million by

2010. Pakistan's population is expected to increase by nearly 70% by 2010. According to Hassan (1991), the resource-to-population ratio is extremely low in India, Bangladesh, Pakistan, and the Maldives, and he suggests that, for these people to survive, a general tendency will be to exploit resources to a level where the natural resource base may no longer be renewable.

Because of population pressures and the intense seasonality of rainfall (almost all rain falls in the four-month monsoon season), water issues in the region remain largely centered on quantity rather than quality—that is, supply of even some water for consumption and food production. For example, Bangladesh's location makes it remarkable susceptible to both devastating floods and droughts. Controlling water supply is a major preoccupation of both domestic and international policies in Bangladesh. It is a central and continuing source of tension with the Indian government. India and Bangladesh currently have no water-sharing agreement, despite the fact that India has a barrage on the Ganges river, just 15 km from Bangladesh, to divert water to Calcutta. Hassan (1994) notes that Bangladesh must first solve issues of land tenure (and the incentive of people to manage their natural resources) and population before natural resources management can be effective.

Other fundamental policy directives in the region are defined and subsumed by issues of ethnic clashes. Because water is viewed as both sacred and scarce, any water project has the potential to be seen as a manipulative affront to one or more groups. For example, both interstate and international relations are tenuous for India. India and Bangladesh continuously debate the issues of water distribution because Bangladesh is dependent on rivers that originate in Indian territory. Even India's plan to share the waters of the Brahmaputra river through the construction of a canal has proved a deeply divisive proposal. Bangladesh is concerned that the Indian plan will make it more—not less—secure. Assam, in which the Brahmaputra river originates, is fiercely independent and has resisted the attempts by the national government to bring the region under greater national control. Hassan (1991) attributes the rise of extremist movements in both West Bengal and Punjab to issues of water distribution and sharing—the existing tensions are catalyzed by the difficult issues and often inequitable distribution of water.

Overall Cultural Influences on Policy

These country summaries demonstrate that water quality is most often a latecomer to policy on national agendas, a feature that is counterproductive to water quality improvement. The summaries are not meant to imply, however, that water quality (or other natural resource issues) is doomed to that low level of policy influence. Historically, communities and countries have begun to prioritize water quality issues when deteriorated conditions have exerted too heavy a social and environmental cost. The challenge today is twofold, to share accumulated global knowledge about the consequences of ignoring water quality deterioration, and to end the cycle of adopting foreign standards and laws. Water quality *can* be developed proactively while honoring the cultural traditions of each country.

Each country has a set of givens that cannot be changed. Historical influences, for example, and current political systems are generally considered to be conditions within which planners must operate. Other characteristics outlined in this chapter, however, contain more flexible planning alternatives. Cuba, for instance, was mentioned as a developing country that has a sustainably managed water quality–dependent industry. The lesson from this example is that with education and integrated planning, other developing countries can also come to place a high priority on water quality issues. The values of attending to water quality should be self-evident when presented contextually.

Bangladesh is an example of a country where social conditions such as land tenure and population growth must be solved hand in hand with water quality issues. To varying degrees, the power issues and poverty that drive processes in Bangladesh are repeated many times around the globe. In each case the past inability to effect water quality planning is most often traceable to the lack of involvement of the local affected communities in their own water quality planning efforts.

Developed countries have effected improvements, but rarely proactively and rarely in a manner that recognized the social/biophysical connections that allow improvements to be sustainable. The following chapter discusses exactly this issue: the past, present, and hoped-for future of integrated water quality planning. Done well, integrated planning incorporates the myriad social conditions discussed in this chapter, rather than dismissing them as constraints to planning, and enables these conditions to guide multiple-tiered, effective, and sustainable planning efforts.

For Further Reading

Cairncross S. Domestic water supply in rural Africa. In: Rimmer D, ed. *Rural transformation in Tropical Africa.* Ohio: Ohio University Press, 1988:36–46.

Callicott JB. *Earth's insights: a survey of ecological ethics from the Mediterranean Basin to the Australian Outback.* Berkeley: University of California Press, 1994.

Calvert P. Water politics in Latin America. In: Thomas C, Howlett D, eds. *Resource politics: freshwater and regional relations.* Philadelphia: Open U. Press, 1993.

Chitale MA. Development of India's river basins. *Intl J Water Resources* 1992;8:30–44.

Mino T. Comparison of legislation for water pollution control in Asian countries. *Water Sci Tech* 1994;28:251–255.

26

Paradigms in Motion: Integrated Approaches to Water Quality Policy

Overview

To this point, three fundamental concepts of water quality management have been presented and interwoven throughout previous discussions. Chapters 1 and 2 outlined assets and failings of historical patterns of unidimensional, linear planning in water quality management and presented the emerging paradigmatic shift toward hierarchical, multidimensional planning. The failures of much of unidimensional planning became apparent, implicitly and explicitly, in discussions of land use impacts on water quality (chapters 17–24). A second recurring thread traces the concurrent paradigmatic shift in the science of water quality management. The emerging scientific paradigm views patterns and processes at multiple scales, recognizing the predictive value of large-scale processes (regional patterns and ecosystem-wide responses to stress), while acknowledging the uniqueness of each system at small scales (see, in particular, chapters 14–16; also chapters 7–13). Lastly, the importance of attitudes and values has been stressed repeatedly, as they form the core of all water quality decisions (see, especially, chapters 3 and 25).

These three concepts provide little assistance to water quality decision-making efforts if they are not integrated into a working model. Consequently, this chapter explores how the emerging paradigms are expressed under different cultural frameworks and analyzes the ingredients necessary for effective, integrated management. In particular, integrated management is presented as an outcome of an emerging philosophy—"ecosystem management"—that sees humans as part of natural systems and views natural resource decisions as critical to the systems' (and our) survival. Achieving integrated management, however, depends on the comprehensiveness of planning, astute goal setting, and resolution of issues such as boundaries, jurisdictions, and agency responsibility. Fundamentally, it relies on commitment by agencies and stakeholders.

Introduction: Trapped Between Paradigms

Water quality management is currently trapped between two paradigm shifts. Meeting multiple demands requires an expanded view of the relevant spatial and temporal scales and a comprehensive approach to water quality management. At the same time, people are demanding local ownership of decisions, expecting that decision making will actively involve stakeholders and that those with the most to gain or lose will have the strongest voice in decisions. Thus, the focus of management decisions is expanding to include multiple dimensions and scales, even as the locus of control is contracting and becoming more localized.

Concurrent with these social and administrative changes, the *science* that drives our understanding of how watersheds function is currently undergoing a scalar paradigmatic shift from point- and population-specific questions to ecosystem-level, landscape-scale, and cumulative effects questions. To understand and model cumulative impacts or longer-term dynamics, data must be collected appropriately and models must explicitly address scalar issues. Recent efforts in watershed science have made significant progress in advancing this larger-scale understanding and have demonstrated that this area of science is undergoing a paradigm shift to explicitly include issues of temporal and spatial scale in our models and to develop models and understanding that explicitly address cumulative effects, larger spatial scales, and longer timeframes. Although this emerging paradigm is widely recognized, however, its implications and implementation remain poorly understood.

The Intersection of Paradigms

Integrated water quality management involves more than the recognition of multiple values discussed above. Integrated management occurs at the intersection of the social goals of multiple-use management and the scientific understandings of biophysical and ecological patterns at different scales. Thirty years ago, the practice of water quality (or water resource) management consisted of attempts to achieve a series of population-level goals (e.g., productivity of a fish population). Acknowledging past failures and the inadequacy of this framework for multiple demands and scientific development, societies around the world have gradually rephrased their goals. Currently, the ideal management paradigm is an integrated framework that recognizes the linkages of biophysical resources within a hydrologic unit and acknowledges that attempts to achieve multiple goals within a single hydrologic unit require attention to those linkages. As a result of this paradigm, water quality management has grown into a science that integrates biophysical and socioeconomic disciplines—and into a science that strives to understand and optimize societal goals within an integrated framework on the landscape.

While the integrated paradigm remains strong, difficulties implementing the ideal framework have caused an intra-paradigm shift toward local-scale administrative models. Around the globe, national governments are shifting water management responsibilities to states, provinces, and regions. State governments are shifting decision-making authority to local communities. Private land owners and managers are demanding—and receiving—greater control over decisions as to how they manage specific, fine-scale elements of the landscape (i.e., plots of land, streamside buffer zones, wetlands). In this new paradigm, local-level goals, often dominated by "short-term economic gain," are translated directly into management practices. Strategies for water quality management then represent a summation of, and are constrained by, those lower-level tactics.

As a result of these two shifting and conflicting paradigms, managers attempting to understand and implement *integrated water quality management* face a serious dilemma. **Scientists** are developing tools, models, and understanding that explicitly address issues of increasingly coarser spatial and temporal scale and are recognizing that impacts can be understood only by incorporating explicit attention to ranges of temporal and spatial scale. **Managers** and **decision makers**, on the other hand, are being forced to make decisions on an increasingly more local basis. Effective implementation of integrated water quality management principles will require models and understanding that resolve that dilemma; in particular, these models must incorporate knowledge about longer temporal scales, greater spatial scales, and effects at those scales on more localized decision making.

Paradigms in an Evolutionary Context

Colby (1990) recognized this process of shifting paradigms but chose to phrase their development more as an evolution than a conflict. Colby's taxonomy of environmental management paradigms follows many of the attitudinal concepts presented in chapter 1, but applies them specifically to management.

Consequently, his framework provides a working model for understanding the ways in which societal views toward water quality (and the environment) are translated into specific management approaches; it also illuminates how these approaches progress as society itself evolves toward different values and goals. The evolutionary framework is not meant to imply deterministic progression of values or management schemes. As a general trend, however, Colby's evolution of environmental paradigms provides a key to understanding the present developments in, and growth of, water quality management. The model also provides an introduction valuable to the values and constraints that control the effective implementation of water quality policies.

The Two Management Extremes

Colby (1990) suggests that environmental management can be classified into five paradigms that are management counterparts of the attitudinal continuum discussed in chapter 3 (Fig. 26.1). The first of those paradigms he calls *frontier economics*. In this view, nature, the environment, or water all represent an infinite supply of resources available for human benefit—that is, they are available for human use and consumption and as a sink for wastes. In this paradigm, no environmental management is necessary because the role of management is to stimulate and protect human economic values. Because the environment's absorptive capacity is regarded as infinite, no constraint occurs and thus no management is needed.

This highly anthropocentric view is typical of both decentralized, capitalist societies and centrally planned, Marxist societies (Colby 1990). The extractive strategies implemented by early colonists of the United States and the exploitive practices of China and Central and Eastern Europe following World War II are typical of this approach. Such an approach is often taken today by developing countries that suggest that they cannot afford to constrain growth to protect environmental quality.

As a striking contrast to frontier economics, Colby presents the *deep ecology* view. This view is based on principles of ecocentrism and places natural phenomena and nonhuman species above human values. Representing the most "naturalistic" end of the attitudinal spectrum (chapter 3), management according to a deep ecology philosophy requires a regional autonomy such that geographic regions become internally balanced with regard to flows of matter, energy, and information. It requires a no-growth approach to economics and a reduction of human population density below the current levels. Implementation of deep ecology is widely regarded as an unrealistic goal for water quality or resource management, except in special cases. Today it serves more as a force

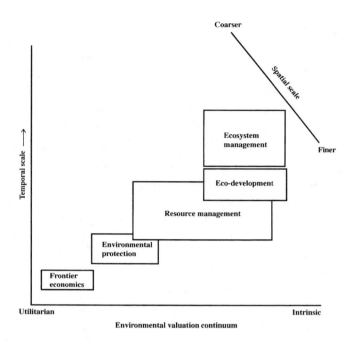

Fig. 26.1 Environmental management paradigms differ in their consideration of environmental values along an extrinsic–intrinsic continuum as well as their consideration of spatial and temporal impacts. Frontier economics, for example, is a utilitarian management approach that is concerned with small temporal and spatial scales. In contrast, ecosystem management is a larger-scale approach that places significant weight on the intrinsic values of environmental resources.

to move the center of society toward a conservation ethic than as an implementable philosophy.

Midpoint Models

Between these extremes lie a range of management approaches that derive from points on the attitudinal continuum. Similar in origin to Regier and Bronson's (1992) exploit-biased "utilists" or Meffe and Carroll's (1994) "stewardship" values, Colby (1990) presents a third management paradigm called *environmental protection*. In this damage control approach to water quality management, standards are based on command-and-control policies and end-of-pipe treatments. The environment (or the quality of the water) is regarded as an external factor in economic models and thus is used to constrain economic growth. Environmental protection relies, therefore, on use of an assimilative capacity to define an optimal level of pollution that can balance economic growth and environmental quality.

Countries or regions using this approach to water quality management usually develop distinct water quality subsets of environmental protection agencies and environmental laws. They employ a medium-specific approach to management (i.e., air laws, water laws). The philosophy is based strongly on privatization and local ownership as vehicles for managing the environment. It holds that, although the absorptive capacity of the environment is not unlimited, the growth of technology will continually increase our ability to use that absorptive capacity. Thus, this high-growth model values resources for their human benefits and relies on increases in technical knowledge to allow economic growth. Much of Central and Eastern Europe is undergoing rapid decentralization and privatization of resources since the changes of government that occurred in 1988–1991. The approach to water quality management in these regions is consistent with this model.

A fourth model, *resource management* (Colby 1990), integrates social justice values into environmental values tending toward (but not quite reaching) biocentrism. Rather than focus on the use of the environment for goods and services per se, resource management models incorporate the equity of those environmental uses into their management programs. The model represents essentially an outgrowth of the "green" and local control movements that have sprung up around the world (e.g., the Chipco movement of India). It suggests that other models do not adequately incorporate the economic losses to the poor into decision making. Consequently, this model takes a risk assessment approach to decision making in which human welfare is the dependent variable and all ecosystem functions and ecosystem stocks are assigned value. This approach relies on national-level decision making and economic constraints that would occur from any damage to humans, or reduction in human benefit.

This decision-making approach was supported by many of the environmental groups actively protesting the United Nations Conference on Environment and Development (UNCED) in Brazil in 1992. At that conference, the nations of the world discussed new approaches to valuation and decision making. Many environmental groups protested the meeting, arguing that national-level agendas that were based in economics (e.g., environmental protection philosophies) failed to account adequately for the values and the needs of the poor, the under-represented, and the disadvantaged. These groups argued instead for a more integrated risk assessment model that would explicitly value the special interests of this very large percentage of the world's population.

Colby's fifth management paradigm, called *eco-development*, would be implemented by a society that is more "biocentric" (Meffe and Carroll 1990) or "integrist" (Regier and Bronson 1992) in its environmental values. Colby's model suggests that decision making at this level explicitly recognizes a synergy existing between human functions and ecosystem functions. In paradigms such as environmental protection, an emerging emphasis focuses on "polluter pays" practices to try to allocate costs back to the generator of the problem. In the eco-development paradigm, the emphasis is on the idea that "pollution prevention pays." For example, proponents argue for bioprocess approaches to management in which waste treatment actually results in economic advantage. This paradigm is a regional philosophy, similar in that sense to deep ecology. It is conceptually and operationally based on an energy budget model in which flows are quantified and risks to human populations are judged. It differs from resource management in that uncertainty estimates are explicitly included and stakeholders are explicitly involved in the decision making. The Great Lakes ecosystem approach taken by the U.S.-Canadian International Joint Commission is typical of this model (see chapter 27).

Beyond Colby: The Integrist Paradigm

Colby's evolution of management paradigms stops a critical paradigm short of describing a viable model for current water quality needs and understandings. Consequently, a sixth paradigm may be proposed, the *ecosystem management* model. This view is becoming widely discussed and many attempts are under way to implement it. As with many emerging paradigms, one of the most difficult aspects of ecosystem management is its definition. To some views, ecosystem management represents an opportunity to implement deep ecology. To others, it provides an opportunity to integrate social and biophysical values into management practices.

Ecosystem management is a new and exciting paradigm in the world of conservation, environmental management, and natural resource management (Booth *et al.* 1995). It suggests that people are part of "natural ecosystems" and that decisions made consciously and carefully can result in the greatest probability of long-term survival of those ecosystems. Because ecosystem management is such a new paradigm, it is changing rapidly as it evolves in response to new information and greater understanding.

Individuals at the local level, managing relatively small parcels of land, will typically manage in a risk-averse fashion, choosing the management alternative with the greatest probability of "some" success and smallest associated risk. Historically, local decision making has been highly opportunistic; in many cases, the individual had much to gain and all risk was absorbed by the greater society. In such an opportunistic model, people will follow a path deemed likely to provide economic reward in a relatively short timeframe. The process of privatization shifts decision making away from the collective and toward the individual. Private ownership carries responsibility for long-term management and its associated risks. People are likely to express greater resistance to laws and regulations because those influences constrain newfound freedom. At the same time, people recognize the increased risk associated with private ownership. Implementing an ecosystem management approach and its participatory approach to decision making under such a scenario is, and will continue to be, a major challenge to managers and to the participating public.

Varied Interpretations of Ecosystem Management

The most broadly held definition of ecosystem management posits that ecosystem management is the integration of the biophysical and social sciences. As such, it involves participatory decision making, interagency collaboration, and incorporation of human valuation. This approach recognizes that human societies have the power to control the biophysical processes of any given ecosystem, and that decisions are made by human societies on the basis of value systems. By necessity, management must address human valuation and the ways in which management decisions are taken. The central goal of this philosophy remains sustainability of the ecosystem sensu latus (i.e., the biophysical elements as well as the humans in the system).

Toward Common Definitions

It is critical that water quality management be conducted in an *integrated* fashion. The word and intent of "integration" can be interpreted in many ways, however. Narrower interpretations of the concept would suggest that management is integrated if it considers simultaneously the quality and quantity of surface water and groundwater resources (Mitchell 1990). Implementing this approach involves attention to water supply, waste treatment, and resulting ambient water quality. A broader approach implies "integration of water, land, and their interactions"—that is, a concern with erosion, landscape, and riparian zone management, nonpoint source control, and management of riparian wetlands. In the broadest interpretation of integration, the term is used to imply "simultaneous consideration of the biophysical (i.e., air, water, land) and the social aspects of water quality decision making." As stressed throughout this text, water quality decisions stem from the social context and value statements; thus, the broadest of these terms would seem to be the only one that could yield truly effective, balanced, sustainable water quality decisions.

Water quality management occurs at different scales and progresses through varied planning stages, however. Consequently, the degree of comprehensive-

ness required varies at each scale and each stage (Mitchell 1990). Truly comprehensive management requires including all or a wide range of influential
variables. While this breadth makes the approach more thorough and more involving, it also adds layers of complexity that delay decisions . . . often to the
point where they are no longer useful. Mitchell (1990) suggests that three levels
of planning exist in water resource management, and that the appropriate scale
(i.e., local versus integrative) varies among those levels. He suggests that *nor-
mative* planning discusses what ought to be done. It is a comprehensive exercise, but generalized and best conducted at a coarse geographic scale and for a
relatively long planning horizon. Planning at the *strategic* level discusses what
can be done and is constrained in its list of variables to those that can reasonably be controlled. The strategic level is the one most often followed in developing policies and practices (as discussed below). Finally, *operational* planning
presents what will be done. As the implementation step of the planning process,
it is consequently the most narrowly defined.

Barriers to Implementation

The conceptual steps required to implement effective policy contain several
barriers to effective implementation, including the following:

- *Goal setting.* A common and seemingly logical strategy for interagency collaboration is for each group to establish its own goals, objectives, and agendas, often based in legal statutes. For example, the Health Department
  establishes goals for water bodies based on its statutory authority, usually
  focusing on maintenance of public health. The Environmental Protection
  Agency establishes goals for ambient water quality, while the Fisheries Department establishes goals for populations of fish species. The agencies
  then meet and attempt to resolve differences among competing or conflicting goals. That essential last step involves mediation and is rarely successful. A much more effective strategy is to meet as a group of interested
  parties and attempt to establish common goals. Such goals are usually hierarchical, where overall environmental quality is expressed in broad terms
  and detailed goals for each agency are developed within the overall structure. In this way, the interests of each agency are addressed, but the approach is synthetic rather than divisionary.
- *Boundary issues.* Commonly, agency managers focus on "typical" conditions rather than exceptions. In addition, monitoring is conducted at sites
  that are considered representative of the overall condition. A more effective
  strategy is to focus management and monitoring programs specifically at
  the boundaries of ecosystems; places where conditions change abruptly will

be more reflective of change and often will be more responsive to management action.

- *Agency responsibilities.* The world's management agencies have developed patterns of separate Environmental Protection Agencies, with discrete and parallel medium-specific groups. Many agencies, therefore, contain divisions of Air, Water, Solid Waste, Hazardous Waste, Fisheries, and Wildlife. That strategy has proved to be ineffective. The alternative that will lead us toward more effective environmental and water quality management will be integrated agencies whose charge is holistic resource management. In such a case, water quality becomes a subset of the resources under consideration, and can then be managed at the ecosystem level within an entire landscape.

Toward Implementation

Mitchell (1990) proposed an analytical framework with which to explain the role of different influences in development of water quality policy. In the context of this book, it is important to recognize that while policies are developed by country- and state-level governments, they are actually implemented by individuals working at various bureaucratic levels. A large and poorly developed step often bridges development and implementation of policy. Many different variables, including political will, economics, and resource limitation, may impede implementation of policies that have been developed and passed by governing bodies. Policies that are legally in force but are not implemented have no impact on biophysical water quality.

Mitchell stresses that the six elements of his framework are nonlinear—that is, no intended ordering in importance or in time characterizes the six. Nonetheless, all six are strong influences on development and implementation of integrated water quality policy. Mitchell's six elements are as follows:

- *Context.* As stressed throughout this text, to understand or influence water quality policy we must describe the *natural environment,* its capabilities, and its limits, as well as the *ideologies* of the society within which the policy is being developed and *existing economic, legal, and financial constraints.* Together, these factors form the context for a water quality decision.
- *Legitimation.* Established policies must be accepted by the society in which they are being developed, a process that is influenced by the objectives of the agency involved. Further influences include the responsibilities and authority of the agency. Because each agency has a purview of interest (e.g., public health, ambient water quality), unresolved conflicts may emerge among the interested parties. For that reason, another variable of impor-

tance is "rules for intervention" used by a higher authority when inter-agency negotiations prove unsuccessful. Resolution of differences among agencies is controlled by the effectiveness of administrative decision making within a department or organization, by the strength of legal statutes, and by the political commitment to resolve the problems.

- *Functions.* Implementing integrated water quality management requires that a wide variety of functions be accomplished. Those functions include generic services of value to several agencies in common (e.g., data collection, monitoring, analysis) and specific actions limited by agency responsibility. It is critical that functions be assigned to the correct level of responsibility (generally the lowest practical level). Thus, decisions should be allocated to the most local scale possible, given the sphere of influence involved in the decision (e.g., a local point source or a watershed-scale, nonpoint source program). Furthermore, assignment of functions to individual agencies must be reviewed periodically. Agency responsibilities change significantly over time, with statutes, and with political whim. Those changes will mandate changes in assigned functions.
- *Structures.* A structure is a group of people linked by a common thread (e.g., the staff of the Health Department linked by the statute to protect public health). Larger-scale structures (e.g., national-level authorities) have the advantage of economies of scale, central control, and improved coordination. In contrast, local-scale structures (e.g., city- or regional-level agencies) are close to the resource and close to the stakeholders. There is no single, correct scale. The variables of most importance in this distinction are accountability and flexibility.
- *Process and mechanisms.* Integration of water quality management requires that many individuals from different interests and with different types of responsibility coordinate their actions to accomplish specific tasks. In attempts to implement integration, the value of using creativity in selecting the mechanisms or tools cannot be overestimated. A wide variety of committees, task forces, and commissions can be used to accomplish integration. The key elements of such integration in today's water quality decision making are coordination and public or stakeholder involvement.
- *Organizational culture.* The background, attitudes, and skills inherent in an agency will dictate, to a large degree, its effectiveness in achieving integration (i.e., the agency culture will influence the way that the group approaches development of water quality policy issues). What is unwritten (i.e., the subtle rules of engagement followed by people within an agency) is often more important than what is written (e.g., more important than any Memoranda of Understanding).

The elements of Mitchell's analytical framework are all important. Ultimately, however, integration will be achieved if and when the individuals involved feel it is in their best interest, and in the best interest of their agency and their constituency, to do so. Therefore, the critical variables in achieving integrated water quality policy are the skills and views of the specific individuals involved, their abilities in negotiations and bargaining, and their commitment to the process of integration. Integration may occur at any of various administrative levels—international, national, regional, or local. For example, England's unique regional structure of integrated management is outlined in chapter 16's discussion of river basin management. England's approach removed water management responsibility from local government and gave it to a regional authority administered by an executive board (most of whom are local authority appointees). The structure also emphasizes making the regional authorities self-financing, which has led to their authority to institute actual-cost pricing. The six case studies that follow discuss the advantages and constraints of integrated management at the more common international, national, and local levels.

Integrated Management at the International Level

Water resources both transcend and define political boundaries. In either situation, they become issues for international management. The Great Lakes, for example, are a "logical" dividing line between the United States and Canada, just as the Congo River forms a substantial section of the Zaire–Congo border. Other vital waterways such as the Danube, Rhine, Nile, and Mekong rivers all flow through many nations on their way to their deltas. Africa has 54 international rivers and lakes, 11 of which drain territories of four or more states. The area drained by these international waters is more than half of the total area of the African continent (Gleick 1993).

Logically, the transboundary nature of many rivers and very large lakes and seas has resulted in many multinational attempts to manage waters. The Danube case study (chapter 27) illustrates one important feature of these attempts: they are nearly always cooperative ventures that do not have power of law. Consequently, these programs are usually poorly integrated into the specific agendas of member nations. Their success depends on the genuine goodwill and commitment of the participating nations. Generally these agreements are of two distinct types: water resource management bodies and political bodies.

Transnational water resource management bodies deal with a specific resource (e.g., Mekong Commission, Nile Basin Commission). In this case, the stated goal is to control and regulate water quality (or water resource) impacts within one country or territory to protect water quality uses in another country or territory. Thus, an upstream–downstream relationship is always involved, although it may be expressed as "two sides of the lake" (e.g., in the case of the Great Lakes basin).

The second type of multinational organization is political in nature and its water resource responsibilities are more ancillary. These organizations develop common criteria and standards that their members may either use for reference or adopt. Examples include the World Health Organization, the European Community, and United Nations programs such as UNDP and UNEP. Integrated management in two multinational organizations are reviewed below.

Case Study: The European Community

The European Community (EC) offers an excellent example of how environmental legislation is developed under the umbrella of a supra-agency. The EC is more than 30 years old, but in the last 5 years the Community's nature has changed dramatically, particularly with regard to environmental management.

The primary aim of the original EC treaty was to establish a common market for European Member States. Within this framework, environmental management and coordination of environmental programs was not within the mandate of the EC. Throughout the 1970s, however, the Community began to alter its interpretation of "economic expansion," opening the way for EC directives on the environment (Jans 1993). Despite the shift, not until 1987, when the Single Europe Act was enacted, did environmental policy first receive an explicit place in EC policy. The EC's environmental objectives, which must still operate within a broadly defined mandate of promoting free trade, are as follows:

- To preserve, protect, and improve environmental quality.
- To contribute toward protecting human health.
- To ensure prudent and rational use of natural resources.

Organizationally, the EC represents a compromise authority. The European Commission serves as coordinator, ensuring that the goals of Member States are addressed and met through national legislation and international agreements. A European Parliament and Court of Justice assist in policy making and redress of issues. The chain between Community legislation and practical action is one link longer than is the case with national governments. Importantly, the EC has no environmental inspectorate with authority to monitor or take action against polluting discharges in Member States. Monitoring and enforcement are left to each nation, although the EC may request information from states on ways that directives have been implemented. If information is not forthcoming or illustrates noncompliance, the EC may refer the matter to the Court of Justice (Jans 1993).

HARMONIZATION OF LAWS. The overriding importance of free trade is apparent in most EC environmental directives. For example, pollution taxation is allowable within the EC treaty only insofar as environmentally friendly products from other countries receive the same fiscal treatment as do domestic products. Guaranteeing "the same treatment" has important implications for each country's policies. Jans (1993) describes a case in which Denmark's strict environmental laws relating to bottle packaging effectively closed the Danish market to foreign drink manufacturers. Consequently, Denmark was brought before the European Court, where it argued for the overriding importance of protection against littering. In a precedent-setting decision, the court ruled that environmental protection constitutes an interest that could justify restrictions on movement of goods.

These cases illustrate the role of the EC in areas where national environmental laws cover different issues than EC directives (e.g., are not harmonized). These pieces of legislation are distinguished from harmonized environmental laws in which EC environmental directives serve to regulate specific national laws. Two types of directives are found in this arena: those with minimum harmonization and those with full harmonization. In fully harmonized laws, Member States cannot pursue a more stringent environmental policy than that permitted by an EC directive (nor can they allow a more lenient one). Areas such as transport of dangerous waste are fully harmonized (Jans 1993). In contrast, water quality regulations generally have minimum harmonization. In these cases, the EC establishes emission or effluent standards and expresses them as directives, but Member States are free to apply their own more (but not less) stringent standards.

RECONCILING NATIONAL LAWS AND STYLES. Integration of environmental policies is difficult and often contentious because the EC establishes directives for 12 different nations with 12 different standards of living, histories, economies, and priorities. One example of the difficulties in integration can be seen in EC directives for bathing water quality. Couched in general terms, the EC established coliform bacteria standards for areas with "a large number of bathers" (Bennett 1993). It left the interpretation of "a large number" to each Member State. In Luxembourg, a landlocked and very small nation, 39 bathing waters were identified. The United Kingdom, with 10,000 km of coastline, faced far greater economic consequences from generous definition of affected areas. Initially, the United Kingdom recognized only 27 such areas. Not until the EC threatened legal procedures did Great Britain eventually identify more than 300 additional "bathing water" sites.

Another case described by Bennett (1993) describes difficulties experienced over implementation of standards themselves. For example, early attempts to control ambient water quality were based on emission limits. The United Kingdom resisted this approach, arguing that the objective of water pollution control was water quality itself, and hence ambient quality objectives would be more appropriate. It also argued that uniform emission standards would be unnecessarily strict because many U.K. plants discharge into estuaries with high assimilative capacities. In the end, standards were issued for both emission and quality objectives, and Member States were given the option of applying either standard. Interestingly, the United Kingdom applied quality standards in four choralkali electrolysis plants. In a fifth plant, located on a low-flow canal, emission standards were applied. This option provided maximum flexibility to industry with minimal water quality protection.

For some countries (e.g., Portugal), membership in the EC has represented a major step toward increasing the speed of development (Borrego 1993). Portugal has experienced, however, and in the short term will continue to experience, difficulties with EC directives relating to water quality control, regulation of fuels, bathing waters, and ground and surface water pollution. For other countries (e.g., Denmark, the Netherlands, Germany), EC environmental regulations have been less comprehensive or stringent than currently existing national legislation (Bennett 1993).

MAKING POLICY WORK. One difficulty experienced by EC policy makers is use of "directives" for environmental regulation. Directives leave the means of compliance to the discretion of each member nation—an approach that has been both criticized and praised. Critics argue that flexibility in choosing how to comply leaves too much latitude for deviation from Europe-wide standards. At the same time, it has been argued that this flexibility is precisely the EC's strength because variations in administrative practice among states can be better accommodated.

Because compliance is accomplished within each country's own political system, many approaches to implementation are possible. In the Netherlands, for example, policy making is highly integrated. After extensive activity involving advice from interested parties, broad principles are established and are followed by strict rules of law. Consequently, implementation of laws in the Netherlands is time-consuming and protracted, but also is part of a national vision and generally has a high compliance rate. Like Germany and Denmark, the Netherlands already has comprehensive environmental legislation in effect.

For these nations, compliance requires amending already-adequate laws and sometimes altering organizational structures that are infamous for their resistance to change.

In contrast, countries such as Belgium often simply reproduce EC directives in their national legislation. Official compliance is nearly immediate but national environmental policy remains fragmented. In addition, federally structured or strongly decentralized nations such as Belgium tend to delegate environmental protection responsibilities entirely to regional authorities. In this case, the probability of effective compliance depends on a broad group of local and regional authorities (Bennett 1993).

Great Britain's tradition of regional authorities leads to yet another approach to compliance. Generally, the United Kingdom drafts EC directives into legislation, but leaves the language in general terms and allows responsible authorities to decide how the directives will be implemented. While pro forma compliance is easily accomplished within this structure, the diffusion of authority complicates the task of assuring compliance.

WHO PAYS? The issue of environmental costs and damages is one area that continues to promote fragmentation of the national laws joined by EC directives. Although the EC has adopted the "polluter pays" principle, a distinction has been drawn between private and collective situations. In private situations, this principle is firmly implemented. In collective situations, standards are less clear. In cases involving river use and pollution, countries pay their own purification measures but are not responsible for damages to other countries. For example, the Rhine receives contaminants as it traverses Switzerland, Germany, and France before entering the Netherlands. The polluted river causes problems and presents costs to the Netherlands, but polluting countries do not pay compensation. (In fact, Dutch authorities pay France to persuade the French to take preventive measures.) While complicated issues of sovereignty and free trade currently demand that damage costs remain outside of the domain of EC policies, transboundary water issues can never be fully integrated until upstream and downstream users share costs more equitably.

Case Study: NAFTA

Just as the EC has brought changes to European environmental management, so has the North American Free Trade Agreement (NAFTA) brought about substantial changes in the approach taken to environmental issues between its three members of Canada, the United States, and Mexico. Notably, the largest changes involve the relationship between the United States and Mexico. Traditionally wary of each other's claims on cross-boundary waters, the two countries have had water resource agreements since 1848. Few, if any, provisions in these agreements related to water quality. Instead, the secretive "old regime" was concerned with uses for irrigation and hydropower, decisions were made by a limited number of participants, and the agreements were deeply concerned with territorial claims on water (Mumme 1995).

These characteristics reflected the traditionally agricultural uses of border waters as well as the dependence of border communities on water for economic vitality. Demographics of the border region have changed dramatically since the 1950s. Border area populations, for example, have grown by nearly 400%, while the area of irrigated agriculture along the border increased by 10% (Mumme 1995). The clear implication is that urbanization and its attendant multiple-use values have begun to supplant rural agricultural values.

Urbanization also brought increasing environmental degradation and growing environmental awareness. The combination led to significant public pressure to adapt the existing boundary water management structures to include issues of sewage and sanitation. In 1983, the two countries negotiated the U.S.–Mexico Border Environmental Cooperation Agreement (La Paz agreement) (Mumme 1995). Although the intent and scope of the La Paz agreement was dramatic, its implementation remained remote from the general public; it had few or no educational components and had little force of law or persuasion (e.g., no ability to impose sanctions on polluting parties). Consequently, it attracted significant public criticism.

Much in the way that Colby (1990) described the evolution of management approaches, successive steps in U.S.–Mexico water resources management became progressively more participatory and implied increasingly greater inherent valuation values of water resources. The most significant shift in border water relations "piggybacked" on the tail of the large-scale, landmark NAFTA pact. NAFTA's mission is virtually identical to that of the EC: protection of free trade in member states. Its environmental protection functions also stem from issues of economic equity. The future may provide more examples where the "free trade" framework provides an important umbrella structure for legitimization of transboundary environmental issues.

Under NAFTA, three new institutions have been added to the roster of border commissions. The first—the North American Commission on Environmental Cooperation (NACEC)—is similar to the IJC (chapter 27); it has a quasi-judiciary role and monitors environmental management by the three member countries. In contrast to previous border regulatory structures, the NACEC is open and responsive to public concerns and can impose sanctions if water problems are affected by trade. Two other commissions include the

North American Development Bank, which funds environmental infrastructure improvement efforts along the border, and its associated Border Environmental Cooperation Commission, which identifies and authorizes projects for the bank.

According to Mumme (1995), two important outcomes of this process are the broadening of the range of water quality management activities along the border (increased scale of focus), and a significant improvement in the accountability of the water management agencies to local border residents (decreased locus of control). It is far too early to predict whether the new structures will prove successful. The increased opportunity for public participation, the sensitivity to changing values for border area water, and the specific mandate for cooperation among the agencies conform to the general principles of integrated management, however. As always, the success of integrated efforts will depend on personal and political commitments to water quality management.

Integration at the National Level

Few, if any, nations have implemented effective, integrated water resources or water quality management programs at the national level. Muckleston (1990) analyzes the difficulty with such implementation in terms of different perceptions of land and water that, in turn, force nations to designate separate agencies for each medium. While the fragmentation of land–water concerns within a nation partially explains the lack of integration, the difficulties of achieving consensus on values with public participation at a national level also serve as critical constraints to national-level integrated water quality management. Nearly every country continues to rely on sectoral management, mandates for interagency cooperation, and, most importantly, delegation of specific management authorities to state and local authorities.

Many nations have, however, moved toward the goal of integrated management as a general principle. The following case studies evaluate the success of these efforts to establish general national principles in two West African countries, Nigeria and Ghana. Although both cases focus on developing countries, their selection does not imply that national-level integration is more successful in developed countries. Mitchell's (1990) analysis of developed countries such as the United States, Canada, New Zealand, England, and Japan indicates that each of those countries also has failed to achieve integrated management at the national level. Some success has been achieved at the local level, as will be seen in the final case studies from localities in Japan and the Netherlands.

Case Study: Nigeria

Nigeria, the largest country in West Africa, has a widely varied landscape, ranging from a humid forest zone along the coast to semi-arid savanna in the north. Between these locations, and comprising approximately three-fifths of the land area, lies the humid to sub-humid savanna. Within these zones rainfall ranges from more than 3500 mm per year in the south to less than 500 mm in the north. Rainfall has important ramifications for development in Nigeria; it is highly seasonal, and frequently results in extended drought conditions in the north (Salau 1990).

A defining characteristic of Nigeria's water resource and water quality management is its drive toward rapid development and its recognition of the vital role that water availability plays in this development. Nigeria is a rural, agricultural country, with less than 10% of its land given over to nonagricultural uses such as towns and roads (Chokor 1993). On a local scale, the country is contending with increased urbanization, population growth, and industrialization. Meeting these new needs as well as traditional agricultural needs requires a steady supply of usable water. In fact, Nigerian investment in water supply schemes has risen by hundreds of percentage points since the 1950s (Salau 1990).

Nigeria's principal industry, oil production, accounts for approximately 80% of federal revenues, which are raised through taxes and other payments. The singular importance of this industry explains both the government's willingness to establish industrial water supply as a primary objective and the fact that oil companies do not have direct liability for damage to the environment (e.g., spills). This lack of accountability may be surprising in light of the fact that more than 2 million barrels of oil were spilled between 1976 and 1986 alone (Chokor 1994).

Nigeria has recognized the value of, and need for, integrated water resources management since the 1970s, when it adopted a river basin development strategy. As early as 1979, Nigeria had designated 11 river basin regions whose autonomous development authorities were charged with "comprehensive development of surface and ground water for multipurpose use" (Salau 1990). On the other hand, other mandates included the creation of irrigation projects, the construction of dams, and the facilitation of domestic supplies. In practice, the river basin development authorities followed those "development" mandates to the exclusion of the multipurpose mandates. Salau (1990) reports, for example, that the authorities have "directed their efforts mainly towards dam construction for water supply and irrigation projects."

The constraints to integrated management in Nigeria are many. Most important may be the attitude that economic vitality rests on "development" and that development is constrained by the seasonal and uneven rainfall. Thus, to achieve multiple crops per year throughout the country and "utilize the land fully," it is considered critical to harness the country's water resources. There is no question among Nigerian decision makers that double-cropping and expanded agriculture will, in fact, remedy the country's economic plight; similarly, little serious debate has emerged about the environmental and social trade-offs of large-scale agriculture.

In addition, the political system in Nigeria shows no real commitment to the process of integrated management. Efficient water resources management has suffered from political posturing. For example, the "river basin" boundaries correspond more closely to political boundaries than to the natural physiographic ones. Consequently, even sincere attempts to manage a basin are hampered by the lack of authority over contributing areas of the watershed. Basin authorities and state agencies also have duplicative and overlapping functions. Without a clear line of authority and a willingness to share responsibilities, management has become secondary to political authority. Lastly, political power is expressed through funding (i.e., in the way the federal government sends funds to state governments) of diverse (opposing) political parties. These problems are aggravated by the lack of adequate baseline data on which to base management decisions.

In 1988, the import of toxic chemicals to Nigeria in an Italian ship raised an enormous and hostile response by the public. According to Chokor (1993), the incident was an important catalyst in the creation of Nigeria's first professional environmental organization, the federal Environmental Protection Agency. As its first goal, the EPA is responsible for the creation of quality standards, policy formulation and implementation; its second and third goals are environmental education and public enlightenment and a strategy of dialogue and consultation, respectively (Chokor 1993). Although still in its infancy, the agency has already conducted national seminars and conferences aimed at gaining a consensus view on Nigerian environmental values and their management.

The success or failure of the new federal agency in implementing integrated management will depend on whether its creation represented a sincere commitment or a political maneuver. Already plagued by poor funding and duplication with existing agencies, its success will come slowly and remains unsure (Chokor 1993). In this regard, Nigeria is typical of many developing countries. Authorities recognize the value and need for integration and—often at the urging or with the involvement of development agencies such as World Bank and FAO—have made efforts to structure their management accordingly. Unfortunately, the structures are relatively helpless without actual commitment to the process, adequate understanding of its complexity, or involvement of the affected communities.

Case Study: Ghana

Similar to its West African neighbor Nigeria, Ghana is a developing country struggling with rapid population increases, even more rapid urbanization, and degradation of forest and water resources. Also similar to Nigeria, Ghana recently created (in 1994) a federal Environmental Protection Agency charged with coordinating pollution control efforts and formulating new and streamlined legislation. The EPA's task of implementing integrated management, however, is formidable. Interagency coordination is an articulated goal of the new EPA, but its accomplishment lies far in the future. In practice, the agency is concentrating its efforts on collecting baseline data on country resources to develop adequate and appropriate standards and enforcement protocols. The EPA views itself as catalytic, mobilizing change through interaction rather than coercion. Although the agency has chosen an innovative approach, the obstacles it faces are significant.

Ghana is primarily an agricultural nation, but one that has stressed economic development since before its 1957 independence from England. The focus on economic development has been only marginally successful, but has come at enormous environmental cost. For example, previously sustainable agricultural practices are now practiced unsustainably as they encroach on forested lands and reduce fallow periods. Growth has also occurred in industrial, mining, and export agriculture sectors, with water quality suffering in each instance. Increasing emphasis on export agriculture is creating nutrient subsidies and sedimentation from road construction. In addition, sedimentation and contamination with heavy metals (such as arsenic and mercury used for gold extraction) are associated with mining practices. Because minimal requirements are imposed for the treatment or disposal of water pollutants, effluents from industries such as food processing, textiles, chemicals, plastics, and rubber are of major concern. Within Accra alone, for example, food processing industries discharge more than 80% of the total volume of wastewater and more than 90% of the organic load of all industries. The annual population growth of 3% causes many of these impacts and exacerbates others.

In southern Ghana, urbanization is an enormous problem requiring monumental challenges of integration. While water quality is generally adequate at its source, infrastructural shortcomings such as leaky pipes and inadequate storage tanks create an enormous health hazard at the tap—for example, approximately

90% of the in-home water is contaminated with fecal coliform bacteria. One cause of the poor sanitary access and lack of potable water is due to the sector-alized and ad hoc process of infrastructure development in Ghana. Develop-ment in the form of new houses or settlements is haphazard and not governed by zoning or infrastructure development. Sewers, electricity, and transportation services are provided only after a community becomes established and lobbies for provision of services; such services are subject to routine conflict between providers of sewer systems, supply lines, and power installation. Land use deci-sions elsewhere in the country occur in a similarly random fashion, while exist-ing water quality laws and regulations are often overlapping, uncoordinated, and unimplemented. Clearly, the new EPA faces daunting challenges.

Integration at the Local and Project Scale

Management of the landscape is, at its heart, a local-scale decision; the ways in which one or a few individuals choose to act control local-scale water quality and en toto a series of those decisions control coarser-scale water quality con-dition. Therefore, water quality policy is always a local scale issue, although many policies at other levels have an impact at the local level (Glasbergen 1995). Consequently, although integration may not (yet) be achieved effectively at an international or national scale, as the above examples indicate, it is still possible to achieve integration at the local scale.

Case Study: Japan

Japan's water resources management is an intriguing juxtaposition of contrasts. By some measures, Japanese management is extremely fragmented and unco-ordinated; responsibility is shared by four ministries, two national agencies, and the local counterparts of each organization. In addition, Shirai (1990) notes that Japan lacks strong, centralized legislation or agencies to implement a com-prehensive approach. As noted in chapter 25, however, Japan has a strong cul-tural tradition of consensus and a tradition of collective, nonconfrontational conflict resolution. It has also experienced dramatic urbanization pressures that have fueled the drive to resolve development pressures. Consequently, Japan has developed very strong local institutions that influence local-scale environmen-tal management. Although Japan had no national-level mandate for integrated management or a lead agency to integrate water management until 1987, the principle of basin-level integrated management has been used successfully in different locales since 1950. Several local planning efforts have epitomized the inclusive, participatory, ecosystem- and science-based principles of integrated water quality management.

One example of Japan's local approach can be seen in the Yahagi River basin. The basin is home to more than 1.25 million residents, although the river itself is only 120 km long. The river traverses a diverse set of land uses; highly urbanized areas, thousands of hectares of rice paddies, commercial horticulture interests, and fisheries interests are all represented within the basin (Shirai 1990). Numerous pollutants are present as well. Urban pollution has resulted from the construction of new homes; industrial pollution is generated by Toyota Motor Corporation factories and from sludge following the extraction of soil for ceramics.

Steadily worsening water quality in the 1960s, combined with the lack of national-level means to resolve local issues, led to the 1969 creation of a nongovernmental management authority, the Yahagi River Basin Water Quality Protection Association (YWPA). Because 18 user groups collaborated to form the YWPA, it easily won legitimacy. As of 1987, membership on the association had grown to 49 groups.

The YWPA has established two primary objectives. First, it ensures that standards are appropriate and are being met. Within this mandate, the YWPA established local standards for key pollutants that are significantly more restrictive than the national and prefectural standards. Second, the YWPA assesses the suitability of large-scale development projects, requiring an environmental assessment and monitoring plan for each proposed development. Monitoring is conducted according to locally established, standardized procedures. While not formally stated, the YWPA's third apparent goal is communication. For example, the YWPA established the Council on Environmental Pollution Prevention. With membership ranging from prefecture officials to municipality representatives and environmental experts, the Council meets monthly to "ensure communication among the proponent, the contractor, the municipal governments and YWPA" (Shirai 1990).

The YWPA also is involved in even broader communication planning. For example, a companion association, the Yahagi River Basin Development Research Association, is a cooperative venture of 26 basin municipalities that was established to extend developments in environmental research and work to municipal environmental departments. Another initiative established intrabasin "sister cities" in which downstream cities and towns were paired with their upstream counterparts. School children from the sister cities visited their companion schools, became involved in water quality testing, and received environmental education about their local catchment.

The successful coordination of communication, research, and planning accomplished in the Yahagi River basin may have been one of the forces contributing to a national-level policy change in Japan in 1987. Prior to that year, national development plans emphasized urban, industrial, and infrastructure development. Environmental management was discussed in the 1977 plan. The Fourth National Comprehensive Development Plan instituted in 1987 adopted integrated management of the entire water system. The success of this policy will continue to be driven by community-level interactions and will remain subject to the commitment and organization of affected municipalities.

Case Study: The Netherlands

In contrast to the situation in Japan, the Netherlands has a strong socialist tradition in which national agencies take primary responsibility for managing country affairs, from transportation to schools and the environment. Citizens of the Netherlands pay very high taxes, but in turn receive guarantees of employment, housing, education, medical care, and other services. The Dutch almost universally enjoy a very high standard of living and accept a strong role of the government in their lives. This cultural and political framework can be seen in the role of the government in effecting integrated water quality management: the term "integrated" has different connotations in the Netherlands than in Japan, the United States, Ghana, or other countries highlighted throughout this book.

The process adopted in the Netherlands is project-specific and is utilized in areas having an accepted need for comprehensive planning. Planning itself involves establishing a temporary commission charged with producing concrete results in a given timeframe. Management is initiated and coordinated by national agencies, but the agencies are charged with soliciting participation from a broad base of the affected public (Glasbergen 1992).

The process was illustrated by Glasbergen (1992) for five projects in the Netherlands. The most successful project involved alterations to an estuary that had been enclosed by flood-protection dams and converted into a fresh water system. Large-scale changes—from loss of previous fisheries, to the appearance of new opportunities on created dry land, to eutrophication risks—created the need for significant planning efforts. Government officials from three ministries, three provinces, seven municipalities, and three water boards were included in the make-up of the temporary planning commission. This planning commission was responsible for developing management recommendations cooperatively, although a period of public comment was offered after the initial proposal was developed. Glasbergen (1992) argues that the inclusion of broad representation of concerned officials on the planning commission, the solicitation of public comment, the inclusion of monitoring and progress evaluation, and a detailed administrative agreement have all combined to make the Krammer-Volkerak area a successful example of integrated water resources management unique to this country.

Glasbergen (1992) also notes that the successes in Krammer-Volkerak were not repeated throughout the country. He highlights, for example, another case in which the responsible agency did not solicit involvement on the planning commission from an appropriately broad range of concerned agencies, did not involve officials high enough up in the bureaucracy, did not involve a period of public comment, and was poorly funded. Thus, even with the Netherlands' strong central management tradition, it is not a foregone conclusion that projects will be successful. The success of integrated management still depends on the elements outlined throughout this chapter, among them inclusiveness, legitimacy, and commitment.

For Further Reading

Bennett G. Implementation of EC environmental directives: the gap between law and practice. *The Science of the Total Environment* 1993;129:19–28.

Chokor BA. Government policy and environmental protection in the developing world: the example of Nigeria. *Environ Mgt* 1993;17:15–30.

Glasbergen P. Comprehensive planning for water systems: the administration of complex policy networks. *Water Resources Development* 1992;8:45–52.

Gleick PH. *Water in crisis: a guide to the world's fresh water resources.* Oxford: Oxford University Press, 1993.

Mitchell B, ed. *Integrated water management: international experiences and perspectives.* London: Bellhaven Press, 1990.

Mumme SP. The new regime for managing US–Mexican water resources. *Environ Mgt* 1995;19:827–835.

Regier HA, Bronson EA. New perspectives on sustainable development and barriers to relevant information. *Environ Monitoring Assess* 1992;20:111–120.

27

Decision Making in Practice: Case Studies

Overview

In theory and design, water quality decision making is a carefully considered process in which scientific data are integrated with social values to achieve the optimal balance of uses and impacts. In practice, however, such an orderly decision making process rarely occurs. Previous chapters of this book have outlined the scientific and social aspects of water quality decision making that *should* be, and increasingly are, part of any decision. This final chapter offers a more candid glimpse at the personal and institutional aspects of water quality decision making, asking "what combination of cultural, personal, social, and scientific factors contribute to real-life water quality decisions?"

This chapter evaluates three different case studies. The first case study, on the Danube, presents an examination of complex, large-scale, multinational planning and the resultant reactive nature of management. Water quality planning in the Danube has been hindered by national agendas that have caused each of the 10 countries in its basin to manage the river without regard to upstream causes or downstream effects. Although multinational agreements are now in place, large numbers of biophysical studies have been conducted but few attempts have been made to link the biophysical conditions to social, economic, and land use causes.

The second case study outlines different approaches taken by several Great Lakes communities to build "remedial action plans" (RAPs) for heavily degraded areas. Although the development of RAPs is a reactive process, it exemplifies a midscale approach in water quality planning that is designed to be ecosystem-level in scale and integrated in approach. The communities

highlighted in this chapter demonstrate that the RAP effort has produced uneven results, and that success depends on regional characteristics such as
power relationships in the community, economic interests, and the composition and training of the advisory committees.

The third case study recounts the efforts of a local community in Saranac
Lake, New York, to identify potential polluters and to ensure that its lake does
not become heavily degraded. Spurred in large part by one summer's dramatic
algal bloom, the lake committee has remained aggressive and has organized efforts to identify important polluters and demand changes in practices affecting
the lake. The Saranac Lake community, however, has been confronted with institutional roadblocks. The upper Saranac Lake story is unique in that community members have carried the full force of the efforts to effect change,
without the assistance of area agencies that should be allied with them.

Each case study presents significantly different political, social, biophysical,
and water quality problems. All of them share common underpinnings of decision making, however. In each case, community and political pressures have
been engaged in a push and pull of values, scientific interpretation, and economic pressures. The resolution of each situation is unique to the conditions
and personalities of the key players, but the same pattern is played out daily in
thousands of communities worldwide.

Introduction: The Pendulum of Values

As with all natural resource issues, water quality decisions are never permanent.
Because they reflect changes in scientific understanding, social values, and political and economic forces, water quality decisions, rules, and laws change continuously. Few places demonstrate this truth so elegantly as the United States.
Previous chapters have traced the evolution of U.S. water quality laws from concerns over navigational obstructions in the 1890s, to concerns over public
health in the 1940s, and to an emerging concern over the broader issues of
aquatic health and sustainability of aquatic ecosystems in the 1970s and 1980s.
The U.S. water quality story, however, is continually unfolding. While many in
the United States considered the protections in the 1970s and 1980s water quality legislation as sacred and permanent, reauthorizations continue to introduce
changes in the focus and direction of these programs. In 1995, for example,
Congressional and Senate proposals included such dramatic departures from
past laws as banning the promulgation of new water quality standards for storm
water discharge or for specific compounds such as arsenic. Exemptions for specific industrial facilities have also been included. The proposals reflect politi-

cians' beliefs that the public of 1995 is more concerned about economic vitality than about the previous goals of aquatic health and sustainability.

While dramatic—and to many a disturbing step backward in good water quality management—these changes provide an important illustration that water quality laws evolve continually and may swing on broad pendulums. An important part of such vacillations may be the dialogue created and the opportunity to review and affirm societal values and goals. For example, when first proposed, the changes cited above elicited little public comment or concern. As they received wider publicity, citizen groups began to relay their concern to the Congress and President. The public debate spurred a retraction of many early proposals and a moderation of "anti-environmental" legislation. The final resolution of the 1990s approach to water quality regulation will be driven by the continued response of local communities, public health advocates, environmental advocates, and the international community.

Similar dialogue over values and the political, social, and scientific forces that shape them can be seen everywhere from international-scale decision making to local-scale issues. The following case studies highlight the scientific considerations, social constraints, and political conditions affecting water quality decision making in three representative examples: international, reactive decision making in the Danube basin; midscale ecosystem-level, integrated (but still reactive) decision making in the Great Lakes; and local-scale, proactive efforts in one lake community.

The Not-So-Blue Danube: Failure to Manage a Multinational Waterway

The Danube Basin

The Danube is one of the world's most romantic rivers, rich in culture, tradition, history, and biophysical diversity. Poets and composers, emperors and kings, have sailed its waters and sung its praises. The Danube basin now includes at least 14 countries and is home to more than 81 million people (Fig. 27.1). The river itself is used for navigation, communication, transport, power generation, recreation, drinking water, wildlife, fisheries, and irrigation. It is also the second largest river in Europe, stretching for a distance of 2780 km and draining a catchment of more than 817,000 km^2. Its delta—the largest wetland in Europe—is built from the 6500 m^3 of water that the Danube drains every day into the Black Sea (Rodda 1994) (Fig. 27.2).

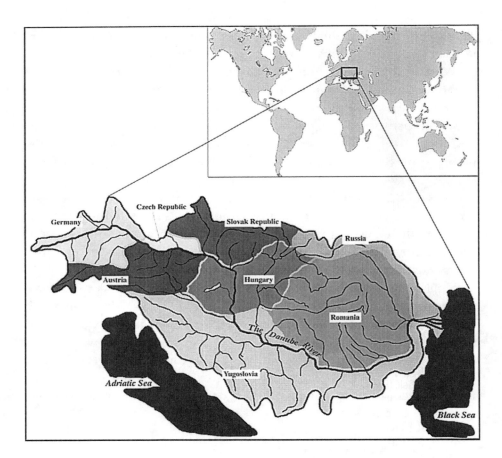

Fig. 27.1 Management of the Danube River is complicated by its many international borders and the ineffective international agreements to manage the river as a whole.

The Danube and its basin represent a major example of opportunities lost, visions held in abeyance, and political differences constraining management strategies. The Danube flows from eastern Germany into Austria, crosses the Austrian Plain, enters Slovakia, Hungary, and the republics of the former Yugoslavia, and finally creates the border between Romania and Bulgaria. That transboundary element has constrained management, limited effectiveness of water-related policies, and led to dramatic differences between expressed goals and biophysical reality.

The Danube basin contains a wide range of industrial development, mining, agriculture, forestry, and urban development. Many of the major cities of Central and Eastern Europe are either riparian (e.g., Budapest, Bratislava) or are situated on tributaries of the Danube. Those cities have followed a development

Fig. 27.2 The Danube remains a picturesque river steeped in its romantic history. Its increasing pollution threatens the livelihood of the area residents, however, as well as the ecological integrity of the river itself. (Photo by J. Perry)

path that placed great stress on the quality of the water in the river. Similarly, little if any effective cooperation or collaboration has emerged among the 14 basin or 10 riparian countries in management of the river.

Water Quality of the Danube

The Danube is one of the most studied water bodies in the world, although the studies have not been conducted in such a way as to allow spatial or temporal tracking of effects and conditions. Planning monitoring and routine data collection by the riparian countries are virtually unknown, despite the hundreds of biophysical studies of the Danube and its water quality. Therefore, we have little quantitative, controlled information on the river's water quality. Despite the limited data, continuing degradation of the river has been clearly identified. For example, nitrogen load has increased sixfold and phosphorus load has increased fourfold since the 1950s (Platt 1995). Severe localized degradation occurs due to the discharge of raw sewage and untreated industrial wastes from major cities. Budapest, for instance, releases approximately 1 million m³/day of wastewater, 90% of which is raw sewage (Benedek *et al.* 1994). In addition, much of Budapest's sewerage system, like many older cities along the basin, is more than 100 years old and not adequate to meet the increased and increasingly complex wastewater needs of today.

During the last 50 years, urbanization and industrialization have increased significantly in the Danube basin; the basin population has grown nearly 20% since the mid-1980s (Platt 1995). In addition to the population pressure per se, development in the basin has concentrated heavily along the river. Because countries of the former Eastern Bloc viewed economic motives as the most important driving variables, industries were constructed near sources of raw materials or near major transportation routes (i.e., often near the river). Industries drew people for labor and support, creating industrial population centers. As indicated above, neither the industries nor the municipalities had adequate waste treatment; in many cases, they had none at all. As a result, nitrate and BOD_5 have steadily increased, as have heavy metals in the sediments and the biota, industrial chemicals in the water column, and the high loads of suspended sediment from agricultural nonpoint sources (Miloradov 1990; Yevtushenko *et al.* 1990).

In addition to the sheer volume of complex effluents discharged into the river, other water quality problems derived from the dams and channelization that accompanied development in the basin. These hydrological changes have altered the water chemistry and biota of the river. Most of the main river channel is eutrophic and supports phytoplankton and zooplankton assemblages characteristic of eutrophic waters. Species of molluscs that were endemic, characteristic of the river, and unique to its waters are now extinct. The macro benthos are dominated by pollution-tolerant forms of Diptera (e.g., Chironomidae) and by Oligocheata. At least 10% of the native fish fauna are either extinct or threatened (Pujin 1990).

Historical Management

A critical factor in the Danube's deterioration is the separate water management schemes employed by each country through which it passes. Navigation has been the one area that has benefited from international cooperation. As early as 1856, the Danube basin countries established joint navigation commissions. Currently, the Danube Commission, which was established in 1948, oversees the hydraulic engineering and water resource management issues related to navigation (Meybeck *et al.* 1989). Not until 1991 was a multinational commission established to set expectations for the coordination and improvement of the river's water quality (Benedek *et al.* 1994).

Historically, management and legislation were implemented differently by countries in "the west" (e.g., Austria) than those in "the east." Austrian water law evolved through more than a century of legislation designed to allow maximal human use of the water with relatively controlled negative impacts. That is, the central tenets of Austrian water quality law between 1859 and 1969 were sustaining human uses and utilization of all existing assimilative capacity (Fleckseder 1990). In the late 1960s and early 1970s, western European nations actively developed waste treatment plants and restructured their water quality laws. Austria followed suit, albeit more slowly (Fleckseder 1990). By the late 1980s, Austrian legislation and treatment practices had come to reflect those of western Europe. The same cannot be said for the other riparian countries. Essentially all municipal and industrial facilities in the former Eastern Bloc lacked waste treatment until the early 1990s, and many remain without treatment facilities today. Without adequate facilities, raw wastes are discharged into streams and lakes. The results have included numerous fish kills, high rates of eutrophication, highly contaminated sediments, and extremely poor water quality. Despite those conditions, the water remains the primary source for area residents' drinking, transportation, fishing, and numerous other uses.

From the late 1880s through the early 1990s, Danube basin countries managed the river through country-specific policies, legislation, management plans, and even goals. Upstream causes and downstream effects were effectively ignored, rendering laws and policies fragmented and ineffective. Biophysical water quality was highly impaired and many uses were effectively precluded. Although upstream–downstream factors were not incorporated into legislation, they were acknowledged in separate programs. For example, the International Association of the Danube has existed since 1975. While its mandate is to coordinate research and education on Danubian water quality, little effective cooperation has developed, primarily due to a lack of political commitment and financial resources.

Present Management in the Danube Basin

Since the political events of 1989, in which the governments of many Central and Eastern European countries were overturned, numerous changes have occurred in management of the Danube. Most significantly, governments of a variety of Danube and European countries have agreed to commit large amounts of financial resources to support improved management. The Danube Task Force was created in 1991 as an agreement among several nations, the European Bank for Reconstruction and Development, and the World Bank. The resulting

"Environmental Management Programme" has produced laudable goals, including the development of an internationally coordinated program to improve the Danube's water quality, and the conservation of ecologically vulnerable areas in its basin. Part of a larger international treaty, the intent appears to be improved and effective water quality coordination among basin states. As noted below, however, the agreement is nonbinding; thus, it is likely that the immediate economic struggles of new developing states will continue to take priority over larger-scale, longer-term water quality concerns.

The Programme created by the Task Force typifies relationships that will structure water quality management worldwide in future years. For example:

- There are no legally binding agreements in the Programme. It simply is a collaboration among interested parties.
- It was stimulated by a major threat of interest to many parties (i.e., the environmental degradation of Central and Eastern Europe).
- The parties involved were willing to invest very large sums of money to address the problem.

Those three elements are seen to some degree in the IJC program in the Great Lakes (see below) and to varying degrees in other multinational river basin arrangements around the world.

The Great Lakes Areas of Concern: Experiments in Integrated, Ecosystem-level Management

Regional Action Plans

Management of the Great Lakes* has changed significantly since the early 1900s, when the first serious pollution problems became evident, but in many ways the current management approach is clearly an extension of early efforts. As outlined in chapter 1, formal management of the Great Lakes began with the establishment of the International Joint Commission (IJC) and the Boundary Waters Treaty of 1909. Formed to confront rising typhoid and cholera epidemics from sewage-laden waters, the IJC recommendations successfully ended the epidemic. Since that time, the IJC has served an effective coordinating and

---

*An overview of Great Lakes water quality history was given in chapter 1, and Lake Erie and Ontario have been highlighted in numerous chapters of section 4. For background, refer to these earlier case studies.

quasi-judiciary role, facilitating studies of Great Lakes water quality and instituting recommendations for remedial action. While the early public health efforts could be categorized as symptomatic management, the IJC's vision has continued to evolve to deal creatively with each new crisis. Notably, the IJC and Great Lakes management traditionally adhered to a crisis-response management approach, but new and experimental efforts in remediation may help move Great Lakes management away from that era of reactive management.

Until the early 1980s, Great Lakes management focused on single problems—health problems from bacterial pollution, a phosphorus problem from sewage waste and detergents, an exotics problem from introduced species, a toxicity problem of growing proportions, and a shoreline management problem as wetlands were continuously "reclaimed" (i.e., drained) and waterways altered. Although subsets of each problem had been managed successfully, as each problem has mounted and less time elapsed between identification of successive crises, it has become increasingly obvious that these efforts are truly ad hoc and that management efforts will continue to play catch-up unless a different approach is taken.

That different approach was initiated in 1978 with alterations to the 1972 Great Lakes Water Quality Agreement. The 1978 accord specifically recognized that management of the Great Lakes waters cannot be conducted separately from management of the air, water, soil, biota, and human communities of the greater basin. The 1987 revisions to the agreement strengthened the 1978 ecosystem emphasis and established a people- and region-centered approach to solving serious water quality problems in the Great Lakes. The IJC established 42 areas of concern in the Great Lakes basin ecosystem (since expanded to 43) that were degraded to a degree such that beneficial uses were impaired or the area was unable to support aquatic life. Thirty-eight of those areas have severe problems with toxic substances and have fish consumption restrictions in effect; restrictions on dredging and other signs of toxic contamination are evident in 31 areas (Hartig and Vallentyne 1989).

In establishing the areas of concern, the states and provinces adjoining the Great Lakes agreed to develop remedial action plans (RAPs) to restore beneficial uses in each affected area. Significantly different from previous sectoral approaches, RAPs mandate the involvement of local communities, citizens, and a wide range of organizations and government agencies (Hartig and Vallentyne 1989). The criteria used by the IJC to judge the appropriateness of regional action plans reflects this commitment to a broad base of involvement. For example, the eight questions used to evaluate a RAP include queries about whether the plan embodies an ecosystem approach, ones pertaining to the appropriateness of the timeframe and sustainability of the approach, and questions related to the adequacy of involvement by different stakeholders and the integration of social, technological, and ecological considerations. Thus, although the approach is defined, the specifics of implementation are left to individual communities, and different communities have had vastly different experiences with the planning and implementation of RAPs.

RAP Experiences

Many researchers have studied the RAP process, including Hartig and Law (1994), Hartig and Zarull (1992), Eiger and McAvoy (1992), Kellogg (1993), MacKenzie (1993), and others. The consensus they have reached is that the RAP *process* is enormously successful, but that RAP *implementation* may be less so; the integrated and inclusive nature of the decisions tends to narrow as communities approach implementation decisions (Kellogg 1993). Most communities have developed planning boards that have taken the mission of inclusive, integrated, ecosystem-wide planning seriously. More revealing perhaps, the different approaches among the planning communities reflect vastly different experiences with such an iterative, personal process (Fig. 27.3). The following summaries briefly describe the sites and processes of five RAPs and identify some of the key preconditions that led to wide variations in the planning and

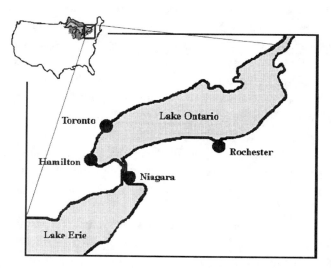

Fig. 27.3 The importance of local influences in the development of Great Lakes remedial action plans is particularly evident when comparing the close proximity of the sites discussed in this case study to their diverse approaches and solutions taken to resolving area pollution problems.

implementation process at each site. The summaries closely follow the work of Kellogg (1993) and MacKenzie (1993). Both researchers used an interview and personal presence format to learn about the motivating and constraining characters of RAP development among different sites.

HAMILTON HARBOUR. Hamilton Harbour is located on the northeast shore of Lake Ontario. The harbor itself covers a surface area of nearly 40 km², with a 37 mile shoreline. An important port, Hamilton ships more than 13 million tons of cargo annually. Development of the harbor has favored industry, with less than 5% of the shoreline being accessible to the public and more than half being occupied by industries. The city of Hamilton also hosts Canada's two largest steel companies, which together produce 60% of the country's steel. The industries, municipal outfalls from the city of 500,000, and nonpoint run-off have led to significant contamination of the harbor. Nutrients, heavy metals, PAHs, and PCBs are among the largest concerns; impaired water quality has caused reproductive failure in bird and fish populations, eutrophication, beach closings, degraded phytoplankton and zooplankton communities, and other related problems.

Since the 1960s, significant improvements have been made to Hamilton Harbour's water quality. Some have resulted from specific remedial efforts such as improvements in sewage treatment at most municipal and industrial sites, and the creation of a retention basin for combined sewer overflows. Other improvements came through displacement of the pollution—that is, the creation of a shipping canal improved circulation with greater Lake Ontario and merely dispersed the pollutants.

Hamilton Harbour's RAP process was the first undertaken in Ontario and, as such, developed differently from later processes. In particular, it occurred before the Canadian Ministry defined expectations for the composition and role of the stakeholder committees. While "advisory committees" would later be defined, Hamilton Harbour's committee was, by choice of the Ministry-hired facilitator, a technically proficient team that took a very active role in developing the RAP. The composition of the stakeholder group was also unique, as it consisted of elected officials, scientists, and other community leaders, rather than volunteers or other nonaffiliated citizens. The composition enabled the stakeholders to move forward quickly on identifying issues, because all of the stakeholders understood the technical aspects of the situation at Hamilton. The independent facilitator took a strong role in the stakeholder meetings. Because of her strong personal commitment to full consensus, she developed innovative strategies to ensure that all stakeholders agreed with the resolution reached for each issue as the process developed.

Developments within the Ministry itself eventually undermined much of the early progress made by the facilitator and "her" committee, however. Power struggles emerged midway through the 6-year process as stakeholders expressed doubts about the ability of the Ministry to implement the RAP and the Ministry attempted to revamp the structure of the stakeholder committee. The process slowed significantly because of the political turmoil, as well as health problems encountered by the RAP coordinator.

Other important issues that confused the process revolved around unclear roles and division of responsibility among the various committees involved in the RAP process. The stakeholder committee, for example, was only one of several committees working simultaneously on planning. A significant source of tension in the process arose over its role versus that of the technical writing committee. The latter typically operated in isolation, and frequently was at odds with the stakeholder group over proposed solutions to certain problems. For example, the writing team favored the option of piping Hamilton sewage directly into Lake Ontario as one solution to the local pollution problems. The stakeholder committee argued that such an approach was contrary to the "ecosystem" approach in that it was a short-term solution that merely transferred the problem elsewhere. Nonetheless, the outfall solution was contained in several versions of the writing committee's draft statements despite continuous and unyielding resistance on the part of the stakeholders.

The Hamilton Harbour approach was also unique in that it largely ignored the greater Hamilton area community. The membership of elected officials and community leaders assured a certain legitimacy in the political process, but the general populace had little to no idea that a RAP process was even under way. In fact, no efforts were made at public outreach or constituency building until the process was nearly finished. Kellogg (1993) hypothesizes that part of the rationale behind the belated efforts at public education was that the officials on the committee wanted to have final recommendations in place before they "went public." In the long run, however, the lack of public support and understanding narrowed the power of the stakeholder group to negotiate with the Ministry over the final form of the RAP.

TORONTO WATERFRONT. In contrast to Hamilton Harbour's population of 500,000, the Toronto area supports a population of 3 million residents who live in the waterfront's watershed. Only 20% of the land in the watershed is held in nonintensive production (e.g., open meadows, forest, or pasture). The remaining space is an intensively maintained urban area, with a mix of field crops, residential areas, and industrial uses. The area is also large and politically diverse:

the IJC's area of concern is affected by at least two regional municipalities and 11 local municipalities. The combined effects of differing municipalities and a large, diverse community created significant challenges for the RAP planners.

Five sewage treatment plants discharge into the Toronto watershed area of concern, serving as sources of nutrients, metals, and organic substances. In addition, discharges from storm water and combined sewer overflows represent significant sources of bacteria, nutrients, metals, and pesticides. The IJC's designation of the Toronto waterfront as an area of concern was not a surprise; problems with water quality were known and "being studied" for many years prior to the start of the RAP process. For example, the RAP process was preceded by a municipal program to reduce toxic wastes, a watershed-wide management strategy study, a Humber River water quality management plan, the Metro Toronto Waterfront Water Quality Improvement Program (known as the Crombie Commission), and an Infrastructure Rehabilitation Program for Toronto. In addition, a motivated and informed group of environmental activists had undertaken a preliminary RAP even before the official one could be mobilized. Because of these existing committees and studies, the RAP process was generally viewed as just another ineffective government program and, therefore, received little public support.

Another characteristic that reduced the effectiveness and focus of the Toronto RAP process was the lack of technical expertise on the committee. While participants were chosen through a broad-based, careful selection process, little effort was made to educate stakeholders on a technical level. Consequently, technically proficient activists in the community discredited the committee while the committee became bogged down in the essential business of educating its own members. Similar to the Hamilton stakeholder group, the Toronto committee suffered from a lack of definition over its role, vacillating between an advisory role and active participation in the development of the RAP. Inadequate funding also slowed down and discredited the progress of the Toronto stakeholder committee. Without the resources to conduct its business, the process proved slow and cumbersome and thus received poor public support.

NIAGARA RIVER. The Niagara River is a 37 mile long channel connecting Lake Ontario to Lake Erie. It is critical to the health of Lake Ontario, supplying more than 80% of the total flow into the lake. The U.S.–Canadian border runs along the river, and the river serves as a source of drinking water for more than 5 million residents on both sides of the border as well as a depository for municipal and industrial discharge, a source for boating and swimming, and a source of important tourist revenues (e.g., from Niagara Falls) (Fig. 27.4). The border

Fig. 27.4 Niagara Falls, the second largest tourist attraction in New York state, is located on the heavily polluted Niagara River. (Photo by D. Vanderklein)

also marks a distinction between classes of use: the U.S. shore is heavily used for urban residential and industrial (petrochemical) purposes, while residential, recreational, and agricultural uses dominate on the Canadian shores.

Water quality problems with the Niagara River have been documented since the 1940s, when the problems related to sewage, bacteria, oil, and nutrient pollution. Currently, the greatest water quality challenges faced by the Niagara involve toxic chemicals from the shoreline petrochemical plants. For example, aluminum, cadmium, chromium, copper, lead, mercury, arsenic, cyanide, selenium, pesticides, PCBs, and numerous oil products are all found at high levels in the river. The condition of the Niagara has degraded Lake Ontario's water, making fishing illegal in Lake Ontario and causing numerous advisories to be issued for lake fishes.

The current RAP process builds on earlier cooperative efforts to monitor and reduce toxic pollution to the river. For example, the Niagara River Toxicities Committee, formed in 1981, brought together four U.S. pollution-control agency staffs and four Canadian pollution-control agency staffs to review the river's status and make recommendations for remedial efforts. The Niagara River RAP, however, developed as two distinct planning processes: one on the Canadian side, the other on the U.S. side.

In contrast to Hamilton Harbour and Toronto, committee members in Niagara were largely self-chosen—anyone who wanted to participate could. Also in contrast to the previous examples, the Niagara group experienced significant stress from the consensus process itself. Several participants became deeply frustrated by the length of time taken to achieve consensus on different issues and felt that parliamentary-style voting would have been more efficient and more equitable. The Canadian committee was also characterized by a deep animosity toward the U.S. pollution of the Niagara River. For example, previous commissions had identified 11 toxic waste sites on the Canadian side with potential to harm water quality compared with at least 250 sites on the U.S. side. While U.S. polluters were responsible for 90% of the effluent discharged into the Niagara, the U.S. population dependent on the Niagara for drinking water was one-fifth the size of its Canadian counterpart. The adversarial relationship between the committees delayed and frustrated planning because the Canadians viewed their efforts as ineffective until the U.S. plan was complete and its intent clear.

Planning on the New York side was constrained by several preexisting conditions. For example, community relationships over the use of the river were already strained from years of public and political protests and environmental litigation. In addition, the Department of Environmental Conservation (DEC) served a dual role, operating as the lead agency in New York for the RAP process and organizing public participation, while simultaneously negotiating water quality agreements and permits with chemical industries along the Niagara. Membership of the RAP committee was chosen with astute political care in an effort not to antagonize stakeholders and to ensure more control over the process than had occurred during the RAP at nearby Buffalo. Members were nominated from a carefully selected group, including large representation from municipal and industrial polluters. The committee also functioned differently than its Canadian counterpart, with meetings that were formal, noninclusive, and conducted by majority vote.

Another problem faced by the RAP process in New York was the lack of a natural consistency along the narrow corridor that encompasses the RAP (Kellogg 1993). In contrast to other RAPs, the linear nature of the Niagara did not coincide with a contained "community." In addition, many felt that the river corridor's residents had turned a blind eye to never-ending warnings of environmental disaster; thus, it was difficult to motivate—or even reach—most community members.

In marked contrast to the Canadian RAPs, the New York process received no money to organize outreach efforts, provide for facilitators, or otherwise supply key resources for the planning and outreach process. Consequently, significant time, energy, and frustration were lost in putting forth bond issues and looking for other creative ways to fund key processes. Financial resource limitations, as well as agency reluctance to expand its jurisdiction, also precluded the committee from including an adequately large spatial scale in the RAP. Issues of land use along tributaries were considered vital by many participants, and their inability to address these concerns within the scope of the committee proved frustrating.

ROCHESTER EMBAYMENT. The Rochester embayment covers 3000 square miles. The largest single river flowing into the embayment is the Genesee, which passes through parts of Pennsylvania and 10 New York counties before passing through the city of Rochester and emptying into Lake Ontario. Land use along the river varies from urban and suburban uses in Rochester proper (including some industrial use) to rural and agricultural lands on either side of Rochester. Primary problems with the river include nutrient levels (the bay that receives water from the Genesee River is eutrophic), as well as sewage, sediment run-off, heavy metals, and some toxics such as PCBs in the sediments.

Monroe County accounts for a large portion of the Genesee's drainage area. A particularly vocal proponent of water quality planning, the Monroe County Water Authority has implemented municipal wastewater collection and treatment system improvements, detention and treatment of combined sewer systems, and regulation of industrial discharges. In addition, in the 1980s the county participated in the federally funded National Urban Runoff Program, which provided support for the county to develop policies and programs on urban run-off management, including cooperative efforts with local townships and federal agencies such as the U.S. soil and water conservation districts. Even before the RAPs began, the Monroe County water authority had participated in integrated management planning, including a basin-wide technical group working with a community-based water quality management advisory committee.

The experience of the Monroe County authority, and its role as lead agency, proved critical to the development of a smooth, minimally contentious process. As noted by Kellogg, "Monroe County, as a planning organization, not a regulatory one, was more suited for remedial action planning than the state in several ways. The county had a good history of interaction with local communities and activists and had legitimacy as an agency in the public's eye in direct contrast to the New York State's DEC, which many advisory committee members regarded with suspicion. The county's staff commitment to improving the technical expertise of the advisory committee members, including them in data gathering and educational presentations, increased the committee members' confidence about their competence and their value as technical contributors" [p. 514].

Despite the smooth process and public legitimacy achieved in the Rochester RAP, one significant contention arose over the power of the largest industry in the area, Eastman Kodak. Kodak, which is responsible for its own waste treatment, represents a significant source of pollution to the lower Genesee River. Concurrent with the RAP process, the company was seeking increases in the load limits on its waste discharge permit. RAP committee members had deep concerns over the power of Kodak in the county and on the committee, and many voiced fears that continued water quality exemptions for Kodak would undermine the entire RAP process.

Themes of the Remedial Action Plan Experiences

Each of these sites, and each of the remaining areas of concern for the Great Lakes, brings its own unique combination of influences to the planning table. From historical experiences, to the "personality" of lead agencies, to the character of the pollution problem, each community involved in the RAP process develops its own approach to the problems. Some of these distinguishing conditions, as identified by Kellogg (1993), include cultural differences, issues of technical competence by participants, economic conditions, and—interwoven throughout these conditions—issues of power and legitimacy. Three of these factors are discussed below.

CULTURAL DIFFERENCES. Canada and the United States, while both democracies, have different governmental traditions, and the federal government in each country maintains a different relationship with its citizens. Canada, for example, shows its ties to European socialist traditions. Its people *expect* the government to provide for them, and citizens in Toronto and Hamilton Harbour began each planning process with the expectation that the government could be trusted to act openly and in good faith. Canadian efforts were well funded, but citizen action groups saw themselves as taking a primarily advisory role. In contrast, citizens in the U.S. communities acted throughout the process with a certain distrust of the lead agencies. Funding was generally more limited for U.S. RAPs, and citizens took more active roles in directing the process itself.

Similar cultural differences appeared to influence the interpretation of the concept of "consensus." While Canada hired consultants to facilitate meetings and lead communication, its citizens assumed that consensus meant unanimity in decisions. Consequently, the Canadian meeting style was more informal and designed to elicit participation from everyone present. While the actual

success of this approach was more closely tied with the facilitator's active pursuit of each participant's opinion than with informality per se, the structure
stemmed from Canadian traditions that stress "harmony, cordiality and com-
monality" (Kellogg 1993).

The Monroe County efforts for the Rochester embayment also reflected a
more "Canadian-style" approach to consensus, perhaps reflecting Monroe
County's previous experiences with community meetings and basin-wide planning. In general, however, the traditions of formal majority votes in meetings
persist in most U.S. communities developing a RAP. Voting increases the efficiency of decision making, a concern to many PAC members, but does not create the commitment to the process, which has been identified as a key issue in
long-term success of community-led projects (MacKenzie 1993).

ECONOMIC CONDITIONS. The local economy has an enormous influence on the
viewpoint of citizens with respect to pollution, environmental values, and the
willingness to make environmental–economic trade-offs. In Toronto, the lack
of an industrial base in the community led to greater agreement on goal statements and steps to achieve them. In contrast, communities such as Niagara have
a long-standing division between industrial and nonindustrial interests. The
RAP process brought these parties together, but did not resolve their different
values. The more narrow the industrial base, the greater the capacity of that industry or its dependent employees to prevent issues from being discussed. This
power created significant concerns for PAC members in Rochester, who questioned the role and influence of Eastman Kodak.

PARTICIPANT SKILLS AND REPRESENTATION. Each of the cases outlined above presents different dynamics in terms of participant technical skills and the breadth
of participant representation. In some processes, the lead agency hand-selected
representatives, either in the interest of efficiency (Hamilton Harbour),
through a lack of understanding of "community representation," or in the interest of political expediency (as in Niagara). Other areas have employed an
"open door" policy. While each choice must be appropriate to the community
dynamics of the area, such choices invariably represent trade-offs. For example,
hand-choosing the committee may promote efficiency, but without concerted
effort will fail to build a broader base of support for, and cooperation in, the
goals of the process. In contrast, efforts to build a broad constituency, as in
Toronto, met with frustration and criticisms of needless delay.

Perhaps more central to the success of the RAP is the skill level of the participants themselves. Technical competency creates more confidence in the
process and its outcome— a critical motivation for stakeholders. Consequently,
the previously skilled members of the Hamilton Harbour committee and the

trained Monroe County staff both produced committees with high confidence and commitment to the process. Poor technical competency, such as in Toronto, led the committee to doubt its own role in the process and to rely heavily on the technical advice of the lead agency.

No Easy Answers

As a broad experiment being enacted simultaneously in 43 communities throughout the region, the RAP process is—not surprisingly—being watched closely. As stated earlier, the process itself has generally been lauded as enormously successful. Whether or not each committee arrives at efficient conclusions, goals, and implementation statements, the process in which broad interests come together to plan is exactly the model that has the best hope to engender public concern and action. While unique preconditions may create different experiences and remedial plans, in all cases the process is iterative and transforming for both the participants and the lead agencies. Ultimately, the effectiveness of the RAP format will remain tied to community dynamics, but with greater experience the chances for success increase for effective planning and decision making at a local level.

Saranac Lake: Local, Proactive Efforts Versus Larger Agendas

Most water quality decisions, as exemplified by the cases described above, are undertaken to resolve a crisis, or near-crisis, situation. Even those undertaken to prevent degradation to pristine lakes and streams are commonly superseded by some perceived threat to the integrity of that water body. Such is the case with the Upper Saranac Lake, which is located in the Adirondack Mountains of New York state (Fig. 27.5).

The conflict in Upper Saranac Lake has arisen over eutrophication. The largest of three lakes in the Saranac Lake chain, Upper Saranac Lake is home to approximately 550 mostly seasonal residents. The lake is naturally mesotrophic. Residents and tourists that frequent the lake are used to high water clarity, abundant fish, inlets amenable to canoeing, and little weed growth (Martin 1993). Plant growth accelerated in the 1980s until water clarity decreased and a severe algal bloom occurred in 1990 (Fig. 27.6). These changes rallied community residents to find the source of enriching nutrients and to identify ways to correct the problem.

The Upper Saranac Lake Association (USLA) is an organization of lake shore and basin residents. Active since the turn of the century, the USLA has

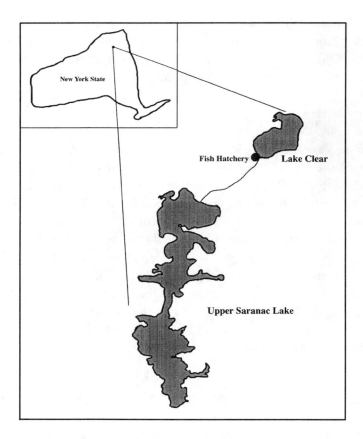

Fig. 27.5 The location of Upper Saranac Lake in New York state. Clearly evident on the map is the natural division of the lake into north and south basins, a physiographic division that aids in identifying phosphorus sources to the lake.

served as a focal point for social activities as well as information pertaining to legal and political issues that affect the zoning and management of the lake. Since the 1980s, the focus of the USLA has turned increasingly toward environmental concerns. Its recognition of increasing densities of blue–green algae in the lake and changes in fisheries initiated a series of careful and considered steps. First, the USLA established an environmental committee. Committee members realized that they knew too little of the lake's present and past condition to make recommendations. They also realized that their efforts would require the assistance of experts and, consequently, money. In response, the USLA established a foundation that allowed residents and members of the lake community to make tax-deductible contributions in support of protective efforts (Martin *et al.* 1993).

Fig. 27.6 In 1990, a severe algal bloom in Upper Saranac Lake alerted area residents to a problem in need of local action. Evident in the picture is the gradient of algal growth emanating from the mouth of the inlet from Lake Clear. (Photo by M. Martin)

Background monitoring of the lake began in 1989 through a contract with researchers at a nearby college. Although many possible sources of nutrient loading existed, the USLA and the researchers hypothesized at first that septic failures from homes and campgrounds would account for a significant portion of phosphorus inputs. They conducted dye tests for more than half of the residences and camps, but found an extremely low failure rate. Two large camps were found to have inadequate wastewater facilities, but because both operated for only a few weeks each year, they represented small sources of overall phosphorus inputs. Both camps have since upgraded their sewage systems (Martin *et al.* 1993).

Having effectively ruled out residences and septic failures as significant sources of the phosphorus input, researchers next turned to a nearby golf course and a privately run potato farm. Both, however, were already operating under nutrient management plans developed by the county Soil and Water Conservation District; those plans were being followed, phosphorus run-off was low, and neither source drained directly into the lake. The feeder streams that received run-off from these sources also contained low phosphorus concentrations.

The process of testing and elimination led the researchers to the state-run fish hatchery located at the north end of Upper Saranac Lake. Determining that the hatchery contributes a significant load of phosphorus to the lake proved relatively simple. It is the largest point source of wastewater to the whole lake, contributing 95% of all permitted wastewater flow, or 3.6 million gallons per day. This flow contains a phosphorus loading of nearly 500 g total phosphorus every day. While the concentration of 35 µg/L falls below the limit for wastewater, it is, according to Lake Manager Michael Martin, "well above the EPA eutrophic criterion." Martin suggests that "the sheer volume of the discharge, coupled with the fact that most of the phosphorus is in bioavailable form, makes the Adirondack Fish Hatchery a significant discharger within the Upper Saranac watershed."

Scientific Corroboration

Researchers have gathered several pieces of corroborating evidence to support the thesis that the hatchery is a significant source of phosphorus loading to the lake. For example, the physiography of the lake facilitates tracking phosphorus inputs. Upper Saranac Lake is naturally divided into morphologically distinct north and south basins. The north basin receives much of its inflow from Little Clear Pond outlet, the hatchery's receiving stream. The south basin receives some water from the north basin but 70% of the Upper Saranac Lake watershed drains into the south basin, primarily through Fish Creek. The north basin's water quality is consistently worse than that of the south basin, with lower visibilities, higher chlorophyll-*a* concentrations, and higher concentrations of total phosphorus and orthophosphorus. In addition, the entire hypolimnion of the north basin is anoxic for three months in the summer, while the deeper south basin experiences only a brief period of anoxia.

More corroborating evidence comes from comparing Little Clear Pond outlet, the hatchery's receiving stream, and Lake Clear outlet, a parallel stream located about one-fourth mile away. Despite the fact that the conditions in these streams were clearly identical at one time, Lake Clear outlet remains a rocky, fast-running, scenic stream. In contrast, Little Clear Pond outlet is choked with *Elodea* and duckweed, and is impassable by canoe below the fish hatchery. Organic matter has built up so deeply that evidence of the rocky bed that underlies the stream has disappeared. Although Little Clear Pond outlet contributes only 15% of the water flow into Upper Saranac Lake, it contributes 30% of the phosphorus load brought in by all three tributaries. In some summer months, it can contribute more than the other two tributaries combined.

Sediment cores provide another clue to the hatchery's contribution of phosphorus to the lake. By studying the composition of diatom assemblages present in the core layers, researchers have been able to pinpoint specific changes in fish hatchery practices, such as the switch to dry feed in the 1970s. For example, Martin (1993) reports that the relative abundance of the diatom *Fragillaria crotonensis*, a nutrient enrichment indicator species, increased in the north basin from a low of approximately 5% to approximately 18% in the early 1950s, a period corresponding to early hatchery production. In 1953, when the hatchery began full-time operations, the abundance of *F. crotonensis* rose sharply to 28% to 33%, dropping again to 19% to 23% after the hatchery switched to dry diets.

UNCERTAINTY. Despite some certain knowledge about the condition of Upper Saranac Lake and the primary source of phosphorus input, much remains unclear. For example, the three most common indicators of eutrophication (i.e., phosphorus concentrations, chlorophyll-*a* levels, and clarity) all present different pictures of the lake's long-term trends. Judging by clarity alone, the lake appears to have significantly improved since the 1990 algal bloom, even compared with baseline historical conditions. Since 1990, total phosphorus has also been lower than historical conditions. Chlorophyll-*a* has fluctuated widely, however, and remains higher than historical conditions. Blue–green algae blooms continue to occur at least twice each summer. Similar patterns have been identified in eutrophic Lake Erie (Ashworth 1986): chlorophyll-*a* concentrations remaining high, despite decreasing total phosphorus. These measures of total phosphorus are probably not instructive; both lakes appear to have high levels of bioavailable phosphorus from their years of phosphorus loading, so annual total phosphorus input is a less valuable measure. In other words, both lakes appear to have reached a critical load where wet years or point source "spills" that wash large amounts of phosphorus into the lakes cause dramatic blooms, while in dry years the lakes may appear improved and reasonably clean. The concern is that further loading will push the lake beyond a threshold such that eutrophic conditions will occur regardless of weather and loading patterns.

The Conflict According to USLA

To the USLA and researchers, it seems clear that the fish hatchery remains the primary source of phosphorus loading and, consequently, of eutrophic conditions in Upper Saranac Lake. These parties are, therefore, committed to bringing about significant reductions in phosphorus loading from the fish hatchery (including, if necessary, closing the hatchery). To date, their efforts have met with frustration. The conflict began in 1992, when the hatchery's permit came up for renewal. Based on its studies of the lake, the USLA filed an extensive comment letter documenting evidence that warranted reconsideration of the hatchery permit. The USLA requested that, in compliance with New York state law, an environmental impact statement be prepared because significant new scientific evidence was being brought to bear on the practices of an existing facility. Their comments had no effect, and the permit was issued in November 1992.

Between November 1992 and January 1993, when the statute of limitations on new permits expired, the USLA tried repeatedly to schedule meetings with the Department of Environmental Conservation (DEC) to discuss the scientific evidence warranting reconsideration of the New York DEC permit. All attempts were denied. Finally, the USLA filed suit against the DEC to halt the permitting process. After the suit was filed, several meetings between the DEC and USLA were arranged, but none ever took place. Between January and April 1993, the DEC canceled numerous meetings, as it slowly developed a proposal to settle out of court. The settlement proposal presented to the community would have allowed the hatchery to continue operations while instituting a 5-year monitoring period. According to the USLA, adequate monitoring had already been conducted; since the early 1980s, the DEC's own monitoring programs had placed Upper Saranac Lake on a "problem waters" list, although no action was taken toward remedying the conditions. The USLA wanted changes in the hatchery practices before the lake became more eutrophic.

The USLA ultimately rejected the DEC proposal, and a court date was set for August 11, 1993. The DEC requested several postponements, to which USLA agreed. When the DEC broke off all negotiations in July 1993, the USLA requested that the August 11 court date be restored. While both parties awaited a decision by the court, the DEC tried once again to submit a proposal that would enable continued operation of the hatchery without significant changes. The Franklin County Supreme Court found that the USLA and New York state studies had merit and that the DEC would have to modify practices during a repermitting process.

The settlement involved conducting a detailed 2-year lake assessment—an EPA-funded, but locally administered study. The delays and conflicts have continued, however. For example, the study was supposed to be a USLA–DEC joint venture, but the DEC had responsibility for purchasing critical equipment and awarding subcontracts for parts of the study. Neither of those steps had been accomplished a year after the court date, and 6 months after the 2-year study was slated to begin.

The DEC's Perspective

In many ways it is illogical that the DEC, charged with water quality protection within the state of New York, would be slow to take action against pollution to any lake, and certainly to one like Upper Saranac, which is a showcase lake in many ways. DEC's reluctance to change hatchery practices can be viewed from two sides, however.

From one perspective, the DEC has a larger-scale view of the fish hatchery. To residents at Upper Saranac Lake, the fish hatchery is the primary polluter of a local resource. The larger ramifications of closing down the hatchery or modifying its practices are immaterial to lake community members. To the DEC, on the other hand, the hatchery has supreme economic importance. It produces approximately 30,000 lb of young landlocked Atlantic Salmon each year—prize game fish that are stocked in Lake Ontario, the Finger Lakes, and several lakes in the Adirondacks. The hatchery has an annual operating budget of approximately $250,000 and is the basis for an estimated $19 million per year in tourist and fishing activities (Thill 1993). The DEC also just completed a $3.7 million improvement of the hatchery. (Ironically, cleaning and flushing of the old fish runs during the remodeling caused a huge load of phosphorus to enter the lake, apparently causing the dramatic algal bloom that motivated area residents to take action.) While the DEC claims that the hatchery-rearing activities are now state of the art, the wastewater treatment activities apparently are less advanced. The USLA claims that its tests found as much phosphate discharge in 1991 (after the renovations) as in 1981 (well before them). The DEC contends that more than $500,000 was allocated during the renovation to wastewater improvement and that phosphate levels have fallen since the renovation.

The second source of DEC's resistance is intra-agency conflict. Because of the importance of the fish hatchery to the DEC, significant tension has arisen within the agency that has delayed action. Water quality scientists within the DEC are loathe to become involved in the Upper Saranac Lake study, feeling that it will compromise their jobs and their chances for advancement within the organization. Some DEC employees feel that they have no incentive to advance any position not in support of the current mode of hatchery operation. Further, because environmental politics are a volatile issue in the Adirondacks, other state agencies such as the Adirondack Park Agency claim that the issues lie beyond their jurisdiction. As a result, they have refused to become involved in the dispute, even to the point of refusing to comment on the scientific merit of a monitoring design proposed by the USLA.

Conclusions

At its heart, the Upper Saranac Lake conflict is based in different perspectives of the value of the same resource. Residents want to maintain the lake in pristine condition; the hatchery management has a significant economic relationship with the lake, a long history at the site, and difficulty accepting the possibility that it may be a major contributor to lake eutrophication. Such different perspectives are not unusual, although two conditions make the Upper Saranac Lake situation unique.

The first condition is the tenacity of the local residents' group and their careful pursuit of scientific information and "appropriate channels" for action. Their process, including formation of a committee, a research plan, and a foundation, established the credibility of the USLA. It also allowed time for the committee's organization to become solidified and ensured that information gathering would proceed carefully and scientifically. Local groups are not often so organized, careful, and considered in their actions against an agency or other entrenched force.

The second condition that makes this case unique is that the fish hatchery is run by the same agency charged with administering water quality laws and permits. Had the hatchery been privately run, it is highly likely that pressure would have been brought to bear earlier and more forcefully by the DEC, and that the hatchery would have been shut down or severely modified. As it is, the USLA's efforts have achieved little or no changes in hatchery practices in the 7 years of conflict. In a contest pitting water quality against a multimillion-dollar industry, water quality is usually the last to receive full attention. In the contest of agency self-regulation, advocates of environmental caution rarely hold the most influential positions. In the meantime, monitoring will continue and the USLA is clearly unwilling to drop this issue.

Emerging Issues Facing Water Quality Managers

As these case studies—and indeed the rest of this book has shown—water quality management is a rapidly growing science; more accurately, the art of water quality management is changing rapidly. Seven central themes that differ significantly from management of the 1990s will characterize water quality management in the future:

1. *Proactive and predictive.* Management will be based on strategies and tactics that predict the ways in which ecosystems respond, rather than relying on implementing solutions to observed problems.
2. *Integrative.* Water quality management will be an integrated science focusing on the ecosystem level, in contrast to today's management of discrete populations and isolated resources such as groundwater or a river basin.
3. *Landscapes within a region.* Future water quality management will focus on the landscape and regional scale, in contrast to the more site-specific local scale that occupies today's decision making.

4. *Decade scale.* The timeframe of relevance to water quality managers will double beyond its current 2- to 3-year units. Water quality scientists and managers, as well as laypeople and decision makers, will gradually come to accept the fact that ecosystems have both resilience and momentum. Management cannot be effectively implemented or evaluated on timeframes less than several (e.g., 5–10) years.

5. *A broader science.* Of equal importance will be the growth of water quality management from an objective and analytical biophysical science to a broader science incorporating more social science tools and models (e.g., social values, decision making, qualitative data, policy support).

6. *Localization of decision making.* Globally, resource management is devolving to become a local and regional issue. National- and state-level policies will continue to evolve toward frameworks and guidance, with few specifics and great flexibility. Implementation will be conducted by, and will be highly variable among, local- and regional-level authorities such as provinces, cities, and watershed- or basin-management groups.

7. *Risk-based decision making.* The local and regional decisions of the twenty-first century will be based on discrete analyses comparing perceived costs and benefits. This process will involve formal risk analysis as well as other means of explicitly comparing alternatives.

For Further Reading

Harris HJ, Sager PE, Richman S, Harris VA, Yarbrough CJ *et al.* Coupling ecosystem science with management: a great lakes perspective from Green Bay, Lake Michigan, USA. *Environ Mgt* 1997;11:619–625.

Renner R. EPA Great Lakes guidance hits a squall. *Environ Sci Tech* 1995;29:416A–419A.

Relative to the Danube Case Study, See the Following References

Benedek *et al.* 1994; Fleckseder 1990; Meybeck *et al.* 1989; Miloradov 1990; Platt 1995; Pujin 1990; Rodda 1994; Yevtushenko *et al.* 1990.

Relative to the Great Lakes Case Study, See the Following References

Ashworth 1986; Eiger and McAvoy 1992; Hartig and Law 1994; Hartig and Vallentyne 1989; Hartig and Zarull 1992; Kellogg 1993; MacKenzie 1993; Meybeck *et al.* 1989.

Relative to the Saranac Lake Case Study, See the Following References

Ashworth 1986; Martin 1993; Martin *et al.* 1993; Thill 1993; Upper Saranac Lake Association 1993a, 1993b, fact sheets 1990–1995 and newsletters 1992–1995.

Literature Cited and References

Abrahamsen G, Stuanes AO, Tveite B, eds. *Long-term experiments with acid rain in Norwegian forest ecosystems.* New York: Springer-Verlag, 1994.

Abu-Zeid M, Biswas A. Some major implications of climatic fluctuations on water management. In: Abu-Zeid M, Biswas A, eds. *Climatic fluctuations and water management.* Oxford/Boston: Butterworth-Heinemann, 1992:227–238.

Al-Ani MY, Al-Nakib SM, Ritha NM, Nouri AM. Water quality index applied to the classification and zoning of Al-Jayish Canal, Baghdad, Iraq. *Environ Sci Hlth* 1987;A22:305–319.

Appelberg M, Henrikson B, Henrikson L, Svedang M. Biotic interactions within the littoral community of Swedish forest lakes during acidification. *Ambio* 1993;22:290–297.

Arnell N. The potential effects of climate change on water resources management in the UK. In: Abu-Zeid M, Biswas A, eds. *Climatic fluctuations and water management.* Oxford/Boston: Butterworth-Heinemann, 1992:107–116.

Ashton P, Mitchell D. Aquatic plants: patterns and modes of invasion, attributes of invading species and assessment of control programs. In: Drake J, ed. *Biological invasions: a global perspective.* SCOPE. New York: John Wiley & Sons, 1992:111–154.

Ashworth W. *The Late, Great Lakes: an environmental history.* New York: Alfred A. Knopf, 1986.

Atwood D, Hendee J, Mendez A. An assessment of global warming stress on Caribbean coral reef ecosystems. *Bull Mar Sci* 1992;51:118–130.

Austin ME. *Land resource regions and major land resource areas of the United States.* Agriculture handbook 296. Washington, DC: USDA Soil Conservation Service, 1972.

Ausubel J. A second look at the impacts of climate change. *Am Sci* 1991;79:210–221.

Bailey RG. *Ecoregions of the United States.* Map (scale 1:7,500,000). USDA Forest Service. Intermountain Region, Ogden UT.

Bain M. Assessing impacts of introduced aquatic species: grass carp in large systems. *Environ Mgt* 1993;17:211–224.

Baker D. The Lake Erie agroecosystem program: water quality assessments. *Agr Ecosyst Environ* 1993;46:197–215.

Balek J. Assessment of the historical changes of aquatic environments under climatic impacts. In: Abu-Zeid M, Biswas A, eds. *Climatic fluctuations and water management*. Oxford/Boston: Butterworth-Heinemann, 1992:91–106.

Balon E, Bruton M. Introduction of alien species or why scientific advice is not heeded. *Environ Biol Fishes* 1986;16:25–230.

Barberis JA. *International groundwater resources law*. Legislative study #40, Rome, Italy: UN FAO, 1986.

Barling RD, Moore ID. Role of buffer strips in management of waterway pollution—a review. *Environ Mgt* 1994;18:543–558.

Barton CR. Water quality and urbanization in Latin America. *Wat Intl* 1990;15:3–14.

Bartsch AF. Biological aspects of stream pollution. *Sewage Works J* 1948;20:292–302.

Bates CG, Henry AJ. Forest and stream flow experiment at Wagon Wheel Gap, Colorado. *US Monthly Weather Rev* 1928;30:1–79.

Bayley PB. Understanding large river–floodplain ecosystems. *BioScience* 1995;45:153–158.

Benedek P, Darazs A, Major V, Oszko K. The effect of Budapest on the water quality of the Danube. *Wat Sci Tech* 1994;30:147–155.

Bennett G. Implementation of EC environmental directives: the gap between law and practice. *Sci Total Environ* 1993;129:19–28.

Bhargava DS. Use of a water quality index for river classification and zoning of Ganga River. *Environ Pollut* 1983;(Ser B):51–67.

Binckley D, Brown T. Forest practices as nonpoint sources of pollution in North America. *Wat Resour Bull* 1993;29:729–740.

Birge EA, Juday C. The island lakes of Wisconsin. The dissolved gases of the water and their biological significance. *Bull Wisc Geol Nat Hist Surv* 1911;22:1–259.

Biswas AK. Sustainable water resources development: some personal thoughts. *Wat Resour Develop* 1994;10:109–116.

Blomqvist P, Bell RT, Olofsson H, Stensdotter U, Vrede K. Pelagic ecosystem responses to nutrient additions in acidified and limed lakes in Sweden. *Ambio* 1993;22:283–289.

Booth GR, Gilbert FE, Perry JA. A preliminary assessment of environmental issues in Ghana. USAID planning document. Washington, DC: Africa Bureau, USAID, 1995.

Borrego C. Water, air and soil pollution problems in Portugal. *Sci Total Environ* 1993;129:55–70.

Brakke D (rapporteur). Group report: physiological and ecological effects of acidification on aquatic biota. In Steinberg C, Wright R, eds. *Acidification of freshwater ecosystems: implications for the future*. New York: John Wiley and Sons, 1994:275–312.

Branch KM, Orions CE, Horton RL. Risk perception and agricultural drainage management in the San Juaquin Valley. *Environ Professional* 1993;15:256–273.

Brezonik PL, Eaton JG, Frost TM, Garrison PJ, Kratz TK, Mach CE, McCormick JH, Perry JA, Rose WA, Sampson CJ, Shelley BCL, Swenson WA, Webster KE. Experimental acidification of Little Rock Lake Wisconsin: chemical and biological changes over the pH range 6.1 to 4.7. *Can J Fish Aquat Sci* 1993;50:1101–1121.

Brezonik PL, Webster KE, Perry JA. Effects of acidification on benthic community structure and benthic processes in Little Rock Lake, Wisconsin. *Verh Int Theor Agnew Limnol* 1991;24:445–448.

Brinley FJ. Sewage, algae and fish. *Sewage Works J* 1943;15:78–83.

Brinson MM. Changes in the functioning of wetlands along environmental gradients. *Wetlands* 1993;13:65–74.

Brooks RP, Croonquist MJ. Wetland habitat and trophic response guilds for wildlife species in Pennsylvania. *J Penn Acad Sci* 1990;64:93–102.

Brooks W. The global warming panic. *Forbes* December 25, 1989:97–102.

Brown GW, Krygier JT. Effects of clearcutting on stream temperature. *J Soil Wat Conserv* 1970;25:11–13.

Cairncross S. Domestic water supply in rural Africa. In: Rimmer D, ed. *Rural transformation in Tropical Africa*. Columbus, Ohio: Ohio University Press, 1988:36–46.

Cairns J Jr. Lack of theoretical basis for predicting rate and pathways of recovery. *Environ Mgt* 1990;14:517–526.

Cairns J Jr., Smith JP. The statistical validity of biomonitoring data. In: Loeb SL, Spacie A, eds. *Biological monitoring of aquatic systems*. Boca Raton: Lewis Publishers, 1994:49–68.

Callicott JB. *Earth's insights: a survey of ecological ethics from the Mediterranean Basin to the Australian Outback*. Berkeley: University of California Press, 1994.

Calvert P. Water politics in Latin America. In: Thomas C, Howlett D, eds. *Resource politics: freshwater and regional relations*. Buckingham, Philadelphia: Open U. Press, 1993.

Canter L. *Environmental impact of water resources projects*. Chelsea, MI: Lewis Publishers, 1985.

Carpenter S, Fisher S, Grimm N, Kitchell J. Global change and freshwater ecosystems. *Ann Rev Ecol Syst* 1992;23:119–139.

Carpenter S, Kitchell JF. The temporal scale of variance in limnetic primary production. *Am Nat* 1987;129:417–433.

Cartwright N, Clark L, Bird P. The impact of agriculture on water quality. *Outlook on Agriculture* 1991;20(3):145-152.

Charbonneau R, Kondolf G. Land use change in California, USA: nonpoint source water quality impacts. *Environ Mgt* 1993;17:453–460.

Chitale MA. Development of India's river basins. *Intl J Wat Resour* 1992;8:30–44.

Chokor BA. Government policy and environmental protection in the developing world: the example of Nigeria. *Environ Mgt* 1993;17:15–30.

Colby ME. Environmental management in development: the evolution of paradigms. World Bank Discussion Paper 80. World Bank, 1990.

Collins SL, Glenn SM, Roberts DW. The hierarchical continuum concept. *J Veg Sci* 1993;4:149–156.

Cooper C, Lipe W. Water quality and agriculture: Mississippi experiences. *J Soil Wat Conserv* 1992;47:220–223.

Corkum L. Spatial distributional patterns of macroinvertebrates along rivers within and among biomes. *Hydrobiol* 1992;239:101–114.

Correll DL, Jordan TE, Weller DE. Cross media inputs to eastern United States watersheds and their significance to estuarine water quality. *Wat Sci Tech* 1992;26:2675–2683.

Cote RP. Marine environmental management: status and prospects. *Mar Pollut Bull* 1992;25:18–22.

Courtemach DL. Merging the science of biological monitoring with water resource management policy: criteria development. In: Davis WS, Simon TP, eds. *Biological assessment and criteria: tools for water resource planning and decision making*. Boca Raton: Lewis Publishers, 1995:315–326.

Cowardin LM, Carter V, Golet FC, La Roe ET. *Classification of wetlands and deepwater habitats of the United States*. Office of Biological Services: Fish and Wildlife Service: U.S. Department of the Interior. Washington, DC: U.S. Government Printing Office, 1979;GPO 024-010-00524-6.

Creason J, Runge C. Use of lawn chemicals in the Twin Cities. Public Report Series #7. University of Minnesota: Minnesota Water Resources Research Center, 1992.

Croonquist MJ, Brooks RP. Use of avian and mammalian guilds as indicators of cumulative impacts in riparian wetland areas. *Environ Mgt* 1991;15:701–714.

Crowell W, Troelstrup N, Queen L, Perry JA. The effects of harvesting on the growth and species composition of plant communities dominated by Eurasian Watermilfoil. *J Aquat Plant Mgt* 1994;32:56–60.

Crowl T, Schnell GD. Factors determining population density and size distribution of a freshwater snail in streams: effects of spatial scale. *Oikos* 1990;59:359–367.

Cummins KW. Trophic relations of aquatic insects. *Ann Rev Entomol* 1973;18:631–641.

Custodio E. Some aspects of groundwater pollution in Spain. *Intl Wat Environ Mgt Ann Symp* 1991;14.1–14.15.

Dana ST. *Farms, forests and erosion.* U.S. Department of Agriculture Year Book 1916:107–134.

Davis WS. Biological assessment and criteria: building on the past. In: Davis WS, Simon TP, eds. *Biological assessment and criteria: tools for water resource planning and decision making.* Boca Raton: Lewis Publishers, 1995:15–30.

De Moor F. Factors influencing the establishment of aquatic insect invaders. *Trans Roy Soc S Africa* 1992;48:141–158.

Depledge MH, Weeks JM, Martins AF, Da Cunha RT, Costa A. The Azores: exploitation and pollution of the coastal ecosystem. *Mar Pollut Bull* 1992;24:433–435.

Detenbeck NA, Johnston CA, Niemi GJ. Wetland effects on lake water quality in the Minneapolis/St. Paul metropolitan area. *Landscape Ecol* 1993;8:39–61.

Dixon RK, Perry JA. Natural resource management in rural areas of northern Pakistan. *Ambio* 1986;15:301–305.

Dixon RK, Perry JA. Upland watershed management in rural Pakistan. *J Am Wat Works Assoc* 1986;78:72–80.

Dixon RK, Perry JA, Hiol Hiol F. Carbon sequestration and conservation in forest systems: Preliminary assessments in Ghana and Cameroon. *Sci Total Environ* (in press)

Dixon RK, Perry JA, Vanderklein E, Hiol Hiol F. Vulnerability of forest resources to global climate change: case study of Cameroon and Ghana. *Climate Res* (in press)

Doudoroff P, Warren CE. Biological indices of water pollution with special references to fish populations. In: US Public Health Service. *Biological problems in water pollution.* Cincinnati, OH: Robert A. Taft Sanitary Engineering Center, 1957:143–163.

Drake J, ed. *Biological invasions: a global perspective.* SCOPE. New York: John Wiley & Sons, 1989.

Duda AM, Johnson RJ. Targeting to protect groundwater quality. *J Soil Wat Conserv* 1987;42:325–330.

Dumnicka E, Kasza H, Kownacki A, Krzyzzanek E, Kuflikowski T. Effects of regulated stream on the hydrochemistry and zoobenthos in differently polluted parts of the upper Vistula River (Southern Poland). *Hydrobiol* 1988;169:183–191.

Egan T. Dams may be razed so that salmon can pass. *New York Times* July 15, 1990:1.

Eiger N, McAvoy P. *Empowering the public: lessons and ideas for communicating in the Great Lakes Areas of Concern.* Chicago: Center for the Great Lakes, 1992.

Eleftheriadis N, Tsalikidis I, Manos B. Coastal landscape preference evaluation: a comparison among tourists in Greece. *Environ Mgt* 1990;14:475–487.

Ellis MM. Detection and measurement of stream pollution. *Bull Bur Fish* 1937;48:365–437.

Engel JR, Engel JG, eds. *Ethics of environment and development: global challenge, international response.* Tucson: University of Arizona Press, 1990.

Engel S. *Ecosystem responses to growth and control of submerged macrophytes: a literature review.* Technical bulletin no. 170. Madison, WI: Department of Natural Resources, 1990.

Engelman R, LeRoy P. *Sustaining water: population and the future of renewable water supplies.* Population and environmental program. Washington, DC: Population Action International, 1993.

Environment Canada. *Groundwater—nature's hidden treasure.* Water series #5. Cat no. En 37-81/5-1990E, 1990.

Environment Canada. *Water—vulnerable to climate change.* Freshwater Series A-9. Environment Canada, 1992.

EPA. *Ambient water quality criteria for Aldrin/dieldrin.* Washington, DC, 1980 EPA: 440-/5-80-019.

EPA. *National water quality inventory: 1992 Report to Congress.* Washington, DC: United States Environmental Protection Agency, 1992:EPA 841-94-001.

Falkenmark M, Suprapto RA. Population–landscape interactions in development: a water perspective to environmental sustainability. *Ambio* 1992;21:31–34.

Fausch KD, Karr JR, Yant PR. Regional application of an index of biotic integrity based on fish communities. *Trans Am Fish Soc* 1984;113:39–55.

Fawell JK, Miller DG. Drinking water quality and the consumer. *J Intl Wat Environ Mgt* December 6, 1992:726–732.

Feth JH. *Water facts and figures for planners and managers.* Washington, DC: U.S. Geol. Survey, 1973:Circular 601-I.

Flaherty M, Karnjanakesorn C. Marine shrimp aquaculture and natural resource degradation in Thailand. *Environ Mgt* 1995;19:27–37.

Fleckseder H. Development of water pollution control in Austria: an example of a riparian state in the drainage area of the River Danube. *Wat Sci Tech* 1990;22:219–226.

Forbes SA. The lake as a microcosm. *Bull Ill Nat Hist Surv* 1887;15:537–550.

Fox J. The problem of scale in community resource management. *Environ Mgt* 1992;16:289–297.

Frissell CA, Liss WJ, Warren CE, Hurley MD. A hierarchical framework for stream habitat classification: viewing streams in a watershed context. *Environ Mgt* 1986;10:199–214.

Fruget J. Ecology of the lower Rhone after 200 years of human influence: a review. *Regulated Rivers: Res Mgt* 1992;7:233–246.

Gallant AL, Whittier TR, Larsen DP, Omernik JM, Hughes RM. *Regionalization as a tool for managing environmental resources.* Washington, DC: U.S. Environmental Protection Agency, 1989:EPA/6–/3-89/060.

Geier TW, Perry JA, Queen L. Improving lake riparian source area management using surface and sub-surface runoff indices. *Environ Mgt* 1993;18:569–586.

George DG. Physical and chemical scales of pattern in freshwater lakes and reservoirs. *Sci Total Environ* 1993;135:1–15.

Gerlach LP. Crises are for using: the 1988 drought in Minnesota. *Environ Professional* 1993;15:274–287.

Ghetti PF, Ravera O. European perspective on biological monitoring. In: Loeb SL, Spacie A, eds. *Biological monitoring of aquatic systems.* Boca Raton: Lewis Publishers, 1994:31–46.

Glasbergen P. Comprehensive planning for water systems: the administration of complex policy networks. *Wat Resour Develop* 1992;8:45–52.

Gleick PH. Climate change and international politics: problems facing developing countries. *Ambio* 1989;18:333–339.

Gleick PH. *Water in crisis: a guide to the world's fresh water resources.* Oxford/New York: Oxford University Press, 1993.

Gopal B, Sah M. Conservation and management of rivers in India: case-study of the river Yamuna. *Environ Conserv* 1993;20:243–254.

Grayson RB, Haydon SR, Jyasuriya MDA, Finlayson BL. *J Hydrol* 1993;150:459–480.

Griffiths CM, Board NP. Approaches to the assessment and remediation of polluted land in Europe and America. In: Eden G, Heigh M, eds. *Water and environmental management in Europe and North America: a comparison of methods and practices.* New York: Ellis Horwood, 1994:11–19.

Gunn J, Belzile N. Extrapolating from toxicological findings to regional estimations of acidification damage. In: Steinberg C, Wright R, eds. *Acidification of freshwater ecosystems: implications for the future.* New York: John Wiley and Sons, 1994:217–226.

Hanson H, Lindh G. Coastal erosion: an escalating environmental threat. *Ambio* 1993;22:188–195.

Harper D. *Eutrophication of freshwaters: principles, problems and restoration.* London: Chapman and Hall, 1992.

Harris HJ, Sager PE, Richman S, Harris VA, Yarbrough CJ. Coupling ecosystem science with management: a Great Lakes perspective from Green Bay, Lake Michigan, USA. *Environ Mgt* 1987;11:619–625.

Harris RC, Skinner AC. Controlling diffuse pollution of groundwater from agriculture and industry. *J Intl Wat Environ Mgt* October 6, 1992:569–575.

Hartig JH, Law N. Institutional frameworks to direct development and implementation of Great Lakes Remedial Action Plans. *Environ Mgt* 1994;18:855–864.

Hartig JH, Vallentyne JR. Use of an ecosystem approach to restore degraded areas of the Great Lakes. *Ambio* 1989;18:423–428.

Hartig JH, Zarull MA, eds. *Under RAPs: toward grassroots ecological democracy in the Great Lakes Basin.* Ann Arbor: University of Michigan Press, 1992.

Hasan S. Natural resource management in Bangladesh. *Ambio* 1994;23:141–145.

Haslam S. *River pollution: an ecological perspective.* London: Bellhaven Press, 1990.

Hassan S. *Environmental issues and security in South Asia.* Adelphi Papers 262. International Institute for Strategic Studies. London: Brassey's Publishers, 1991:69 pp.

Hatcher BG. Coral reef primary productivity: a hierarchy of pattern and process. *Trends Ecol Evol* 1990;5:149–155.

Hellawell JM. *Biological indicators of freshwater pollution and environmental management.* London: Applied Science Publishers, 1986.

Herbald B, Moyle P. Introduced species and vacant niches. *Am Nat* 1986;128:751–760.

Hildebrand LP, Norrena EJ. Approaches and progress toward effective integrated coastal zone management. *Mar Pollut Bull* 1992;25:94–97.

Hilsenhoff WL. *Use of arthropods to evaluate water quality of streams.* Tech. Bull. 100. Madison, WI: Wisconsin Department of Natural Resources, 1977.

Hilsenhoff WL. *Using a biotic index to evaluate water quality in streams.* Tech. Bull. 132. Madison, WI: Wisconsin Department of Natural Resources, 1982.

Hilsenhoff WL. Rapid field assessment of organic pollution with a family-level biotic index. *J North Am Benth Soc* 1988;7:65–68.

Hjorth P, Thi Dan N. Environmentally sound urban water management in developing countries: a case study of Hanoi. *World Resour Develop* 1993;9:453–464.

Hofer T. Himalayan deforestation, changing river discharge and increasing floods—myth or reality. *Mountain Res Develop* 1993;13:213–233.

Hollis GE. Environmental impacts of development on wetland in arid and semi-arid lands. *Hydrol Sci J* 1990;35:411–428.

Holmes PR. Policies and principles in Hong Kong's water pollution control legislation. *Wat Sci Tech* 1992;26:1905–1914.

Hong SY. Assessment of coastal zone issues in the Republic of Korea. *Coastal Mgt* 1991;19:391–415.

Hornbeck JW, Adams MB, Corbett ES, Verry ES, Lynch JA. Long-term impacts of forest treatments on water yield: a summary for northeastern USA. *J Hydrol* 1993;150:323–344.

Hornig CE, Bayer CW, Tidwell SR, Davis JR, Kleinsasser RJ, Linam GW, Mayes KB. Development of regionally based biological criteria in Texas. In: Davis WS, Simon TP, eds. *Biological assessment and criteria: tools for water resource planning and decision making.* Ann Arbor, MI: Lewis Publishers, 1995:145–152.

Hornsby A, Buttler T, Brown R. Managing pesticides for crop production and water quality protection: practical grower guides. *Agri Ecosyst Environ* 1993;46:187–196.

Houghton R, Woodwell G. Global climatic change. *Sci Am* 1989;260:35–44.

House M, Ellis J, Herricks E, Hvitved-Jacobsen T, Seager J, Lijklema L, Aalderink H, Clifforde I. Urban drainage—impacts on receiving water quality. *Wat Sci Tech* 1993;27:117–158.

Ikonnikov V. Pollution of sediments in the Lake Ladoga catchment area and the gulf of Finland: influencing factors and processes. *Land Degrad Rehab* 1993;4:297–306.

Jacks G, Sharma VP. Nitrogen circulation and nitrate in groundwater in an agricultural catchment in southern India. *Environ Geol* 1983;5:61–64.

Jans JH. Legal grounds of European environmental policy. *Sci Total Environ* 1993;129:7–17.

Johnson BL, Richardson WB, Naimo TJ. Past, present, and future concepts in large river ecology. *BioScience* 1995;45:134–141.

Johnston CA. Cumulative impacts to wetlands. *Wetlands* 1994;14:49–55.

Kaplan E, McTernan WF. Overview of the risk assessment process in relation to groundwater contamination. *Environ Professional* 1993;15:334–340.

Karau J. The control of land-based sources of marine pollution. *Mar Pollut Bull* 1992;25:80–81.

Karr JR. Assessment of biotic integrity using fish communities. *Fisheries* 1981;6:21–27.

Karr JR. Biological integrity: a long-neglected aspect of water resource management. *Ecol Applic* 1991;1:66–84.

Karr JR. Protecting aquatic ecosystems: clean water is not enough. In: Davis WS, Simon TP, eds. *Biological assessment and criteria: tools for water resource planning and decision making.* Boca Raton: Lewis Publishers, 1995:7–14.

Kay JJ. A nonequilibrium thermodynamic framework for discussing ecosystem integrity. *Environ Mgt* 1991;15:483–495.

Keller W. Introduction and overview to aquatic acidification studies in the Sudbury, Ontario, Canada area. *Can J Fish Aquat Sci* 1992;49(supplement 1):3–7.

Kellogg WA. Ecology and community in the Great Lakes Basin: the role of stakeholder and advisory committees in environmental planning processes. Ph.D. dissertation. Ithaca, NY: Cornell University, 1993.

Kelly JR, Harwell MA. Indicators of ecosystem recovery. *Environ Mgt* 1990;14:527–545.

Kennedy V. Anticipated effects of climate change on estuarine and coastal fisheries. *Fisheries* 1990;15:15–24.

Klijn F, De Waal RW, Oude Voshaar JH. Ecoregions and ecodistricts: ecological regionalization for the Netherlands' environmental policy. *Environ Mgt* 1995;19:797–813.

Knox CE. What's going on down there: pervasive groundwater contamination prompts new cleanup approaches. *Sci News* 1988;134:362–365.

Kolasa J, Pickett ST. Ecosystem stress and health: an expansion of the conceptual basis. *J Aquat Ecosyst Hlth* 1992;1:7–13.

Kolkwitz R, Marrson M. Oekologie der pfanzlichen Saprobien. *Berichte der Deutschen Botanischen Gesellschaft* 1908;26(a):505–519.

Kotlyakov VM. The Aral Sea basin: a critical environmental zone. *Environment* 1991;33:4–9.

Kudhongania A, Twongo T, Ogutu-Ohwayo R. Impact of the Nile perch on the fisheries of Lakes Victoria and Kyoga. *Hydrobiol* 1992;232:1–10.

Kudo A, Miyahara S. Predicted restoration of the surrounding marine environment after an artificial mercury decontamination at Minimata Bay, Japan—economic values for natural and artificial processes. *Wat Sci Tech* 1993;25:141–148.

Lamb JC. *Water quality and its control.* New York: Wiley and Sons, 1985.

Land C, Reymond O. Reversal of eutrophication in Lake Geneva: evidence from the oligochaete communities. *Freshwat Biol* 1992;28:145–148.

LaPoint TW, Perry JA. Use of experimental ecosystems in regulatory decision making. *Environ Mgt* 1989;13:539–544.

Larsen DP, Dudley DR, Hughes RM. A regional approach for assessing attainable surface water quality: an Ohio case study. *J Soil Wat Conserv* 1988;43:171–176.

Lave LB, Ennever FK. Toxic substances control in the 1990s: are we poisoning ourselves with low-level exposures? *Ann Rev Public Hlth* 1990;11:69–87.

Lave LB, Males EH. At risk: the framework for regulating toxic substances. *Environ Sci Tech* 1989;23:386–391.

Laws E. *Aquatic pollution: an introductory text.* 2nd ed. New York: Wiley and Sons, 1993:611 pp.

Lawton JH. From physiology to population dynamics and communities. *Functional Ecol* 1991;5:155–161.

Lemons J, ed. *Scientific uncertainty and its implications for environmental decision making.* Blackwell Scientific (in press).

Leopold LB, Langbein WB. *A primer on water.* Washington, DC: US Geol. Survey, 1963.

Levin S. The problem of pattern and scale in ecology. *Ecology* 1992;73:1943–1967.

Lindeman RL. The trophic–dynamic aspect of ecology. *Ecology* 1942;23:399–418.

Liu J, Yu Z. Water quality changes and effects on fish populations in the Hanjiang river, China, following hydroelectric dam construction. *Regulated Rivers: Res Mgt* 1992;7:359–368.

Lloyd JW, Tellam JH, Rukin N, Lerner DN. Wetland vulnerability in East Anglia: a possible conceptual framework and generalized approach. *J Environ Mgt* 1993;37:87–102.

Lloyd R. The toxicity of zinc sulphate to rainbow trout. *Ann Appl Biol* 1960;48:84–94.

Logan T. Agricultural best management practices for water pollution control: current issues. *Agri Ecosyst Environ* 1993;46:223–231.

Ludyanskiy M, McDonald D, MacNeill D. Impact of the Zebra Mussel, a bivalve invader. *BioScience* 1993;43:533–544.

Lundin CG, Linden O. Coastal ecosystems: attempts to manage a threatened resource. *Ambio* 1993;22:468–473.

Lyklema J, van Hylckama TEA. Water, something peculiar. *Hydrol Sci Bull* 1979;24:499–504.

Lyon RM, Farrow S. An economic analysis of clean water act issues. *Wat Resour Res* 1995;31:213–223.

Machiwa JF. Anthropogenic pollution in the Dar es Salaam harbour area, Tanzania. *Mar Pollut Bull* 1992;24:562–567.

MacKenzie SH. Ecosystem management in the Great Lakes: some observations from three RAP sites. *J Great Lakes Res* 1993;19:136–144.

Martin MR. The limnological condition of Upper Saranac Lake. Final report of the 1992 monitoring program and analysis of historical trends in water quality. The Adirondack Aquatic Institute at Paul Smith's College, 1993.

Martin MR, Stager JC, DeAngelo M, Smith C. Sleuthing the slime: investigating the eutrophication of Upper Saranac Lake, NY. 13th Annual NALMS Symposium conference proceedings, 1993.

Mason C. *Biology of freshwater pollution.* 2nd ed. Essex: Longman Scientific and Technical, 1991:351 pp.

Marsalek J, Barnwell T, Geiger W, Grottker M, Huber W, Saul A, Schilling W, Torno H. Urban drainage systems: design and operation. *Wat Sci Tech* 1993;27:31–70.

Marsalek J, Dutka B, Tsanis I. Urban impacts on microbiological pollution of the St. Clair river in Sarnia, Ontario. *Wat Sci Tech* 1994;30:177–184.

Matthews GJ. International law and policy on marine environmental protection and management: trends and prospects. *Mar Pollut Bull* 1992;25:70–73.

Matthews W, Zimmerman E. Potential effects of global warming on native fishes of the Southern Great Plains and the Southwest. *Fisheries* 1990;15:26–32.

May RM. Levels of organization in ecology: In: Cherrett JM, ed. *Ecological concepts: the contribution of ecology to an understanding of the natural world.* Boston: Blackwell Scientific Publications, 1988:339–363.

Meffe GK, Carroll CR, eds. *Principles of conservation biology.* Sunderland, MA: Sinauer Associates, 1994.

Merino M. The coastal zone of Mexico. *Coastal Mgt* 1987;15:27–42.

Meybeck M. Carbon, nitrogen and phosphorus transport by world rivers. *Am J Sci* 1982;282:401–450.

Meybeck M, Chapman D, Helmer R, eds. *Global freshwater quality: a first assessment.* Oxford/Cambridge: Basil Blackwell, 1989.

Miller GT. *Environmental science: sustaining the earth.* 7th ed. California: Wadsworth Publishers, 1992.

Miller GT. *Living in the environment: principles, connections and solutions.* 8th ed. California: Wadsworth Publishers, 1994.

Miloradov M. Water resources of the Danube River Basin: sources of pollution and control and protection measures. *Wat Sci Tech* 1990;22:1–12.

Mino T. Comparison of legislation for water pollution control in Asian countries. *Wat Sci Tech* 1994;28:251–255.

Minshall GW. Stream ecosystem theory: a global perspective. *J North Am Benth Soc* 1988;7:263–288.

Mitchell B, ed. *Integrated water management: international experiences and perspectives.* London: Bellhaven Press, 1990.

Mitchell R. Intentional oil pollution of the oceans: crisis, public pressure and equipment standards. In: Haas PM, Keohane RO, Levy MA, eds. *Institutions for the earth: sources of effective international environmental protection.* Cambridge: MIT Press, 1993.

Mitsch WJ. Combining ecosystem and landscape approaches to Great Lakes wetlands. *J Great Lakes Res* 1992;18:552–570.

Mitsch WJ. *Wetlands.* 2nd ed. New York: Van Nostrand Reinhold, 1993.

Morris LA, Bush PB, Clark JS. Ecological impacts and risks associated with forest management. In Cairns J Jr, Niederlehner BR, Orvos DR, eds. *Predicting ecosystem risk.* Princeton, NJ: Princeton Scientific, 1992:153–213.

Moss B. *Ecology of fresh waters: man and medium.* 2nd ed. Boston: Blackwell Scientific Publications, 1991.

Muckleston KW. Integrated water management in the United States. In: Mitchell B, ed. *Integrated water management: international experiences and perspectives.* London: Bellhaven Press, 1990:22–44.

Muller F. Hierarchical approaches to ecosystem theory. *Ecol Modelling* 1992;63:215–242.

Mumme SP. The new regime for managing US–Mexican water resources. *Environ Mgt* 1995;19:827–835.

Naiman RJ, ed. *The freshwater imperative: a research agenda.* Washington, DC: Island Press, 1995:165 pp.

Naiman RJ, Beechie TJ, Benda LE, *et al.* Fundamental elements of ecologically healthy watersheds in the Pacific Northwest Coastal Ecoregion. In: Naiman RJ, ed. *Watershed management: balancing sustainability and environmental change.* New York: Springer Verlag, 1992:127–187.

Naiman RJ, Decamps H, Pastor J, Johnston CA. The potential importance of boundaries to fluvial ecosystems. *J North Am Benth Soc* 1988;7:289–306.

Nalepa T, Schloesser D, eds. *Zebra mussels: biology, impacts and control.* Chelsea, MI: Lewis Publishers, 1993.

Nash L. Water quality and health. In: Gleick P, ed. *Water in crisis: a guide to the world's fresh water resources.* New York: Oxford University Press, 1993:25–39.

National Public Radio. All Things Considered: Puerto Rico faces a serious water shortage. April 16, 1995.

National Research Council. *Restoration of aquatic ecosystems: science, technology and public policy.* Washington, DC: National Academy Press, 1992.

Nayak BU, Chandramohan P, Desai BN. Planning and management of the coastal zone in India—a perspective. *Coastal Mgt* 1992;20:365–375.

Neary DG, Bush PB, Michael JL. Fate, dissipation and environmental effects of pesticides in southern forests: a review of a decade of research progress. *Environ Toxicol Chem* 1993;12:411–428.

Newbold JD. Cycles and spirals of nutrients. In: Calow P, Petts GE, eds. *The rivers handbook.* Boston: Blackwell Scientific, 1994;1:379–408.

Newman RM, Perry JA. Effects of chronic chlorine exposure on litter processing in outdoor experimental streams. *Freshwat Biol* 1987;18:415–428.

Newman RM, Perry JA. The combined effects of chlorine and ammonia on litter breakdown in outdoor experimental streams. *Hydrobiol* 1989;184:69–78.

Newman TJ. *Classifications of surface water quality. Review of schemes used in EC member states.* Water Resource Center. Oxford: Heinemann Professionals Publication, 1988.

Newson M. *Water, land and development: river basin systems and their sustainable management.* London: Routledge Press, 1992.

Ngoile MAK, Horrill CJ. Coastal ecosystems, productivity and ecosystem protection: coastal ecosystem management. *Ambio* 1993;22:461–467.

Niemi GJ, Detenbeck NE, Perry JA. Comparative analysis of variables to measure recovery rates in streams. *Environ Toxicol Chem* 1992;12:1541–1547.

Nip MI, Udo de Haes HA. Ecosystem approaches to environmental quality assessment. *Environ Mgt* 1995;19:135–145.

Norton BG, Ulanowicz RE. Scale and biodiversity policy: a hierarchical approach. *Ambio* 1992;21:244–249.

Novotny V, Olem H. *Water quality: prevention, identification and management of diffuse pollution.* New York: Van Nostrand Reinhold, 1994.

Odum EP. Trends expected in stressed ecosystems. *BioScience* 1985;35:419–422.

Odum EP. Field experimental tests of ecosystem-level hypotheses. *Trends Ecol Evol* 1990;5:204–205.

Odum EP. Great ideas in ecology for the 1990s. *BioScience* 1992;42:542–545.

Office of Technology Assessment. *Protecting the nation's groundwater from contamination.* Washington, DC: US Government Printing Office, 1984:503 pp.

Oksanen L. Trophic levels and trophic dynamics: a consensus emerging? *Trends Ecol Evol* 1991;6:58–60.

Omernik JM. Aquatic ecoregions of the coterminous United States. *Ann Assoc Am Geogr* 1987;77:118–125.

Omernik JM, Rohm CM, Lillie RA, Mesner N. Usefulness of natural regions for lake management: analysis of variation among lakes in northwestern Wisconsin, USA. *Environ Mgt* 1991;15:281–293.

Omerod SJ, Rundle SD, Lloyd EC, Douglas A. The influence of riparian management on the habitat structure and macroinvertebrate communities of upland streams draining plantation forests. *J Appl Ecol* 1993;30:13–24.

Paces T. Acidic emissions and political systems. In: Steinberg C, Wright R, eds. *Acidification of freshwater ecosystems; implications for the future.* Environmental Sciences Research Report ES 14. New York: John Wiley and Sons, 1994:5–15.

Palmer M. The encounter of religion and conservation. In: Engel JR, Engel JG, eds. *Ethics of environment and development: global challenge, international response.* Tucson: University of Arizona Press, 1990.

Parker DJ, Sewell WR. Evolving water institutions in England and Wales: an assessment of two decades of experience. *Wat Resour J* 1988;28:751–785.

Parsons PA. Evolutionary rates: stress and species boundaries. *Ann Rev Ecol Systemat* 1991;22:1–18.

Patrick R. Biological measure of stream conditions. *Sewage Indust Wastes* 1950;22:929–938.

Patrick R. What are the requirements for a successful biomonitor? In: SL Loeb, Spacie A, eds. *Biological monitoring of aquatic systems*. Boca Raton: Lewis Publishers, 1994:23–39.

Paul-Wostl C. The hierarchical organization of the aquatic ecosystem: an outline of how reductionism and holism may be reconciled. *Ecol Modelling* 1993;66:81–100.

Percoda N. Requiem for the Aral Sea. *Ambio* 1991;20:109–114.

Perry JA. Water quality in the 21st century. *Limnetica* 1994;10:5–13.

Perry JA. Networking urban water supplies in West Africa. *J Am Wat Works Assoc* 1988;80:35–42.

Perry JA. Specialization in U.S. university curricula: the role of water quality in forestry schools. *Environ Professional* 1990;12:53–60.

Perry JA, Clark WH. Groundwater classification through spring chemistry: the Lower Portneuf River, Idaho. *J Idaho Acad Sci* 1990;26:55–71.

Perry JA, Dixon RK. Water resources in Thailand's resettled villages. *J Am Wat Works Assoc* 1985;77:46–57.

Perry JA, Dixon RK. An interdisciplinary systems approach to community resource management: preliminary field test in Thailand. *J Developing Areas* 1986;21:31–48.

Perry JA, Herricks EE, Schaeffer DJ. Innovative designs for environmental monitoring: are we asking the questions before the data are collected? In: Boyle TP, ed. *New approaches to monitoring of aquatic ecosystems*. ASTM, STP 1986;940:28–49.

Perry JA, Schaeffer DJ. The distribution of riverine benthos: a river dis-continuum? *Hydrobiol* 1987;148:257–268.

Perry JA, Schaeffer DJ, Kerster HK, Herricks EE. The environmental audit II: application to stream network design. *Environ Mgt* 1985;9:199–208.

Perry JA, Troelstrup NH. Whole ecosystem manipulation: a productive avenue for test system research? *Environ Toxicol Chem* 1988;7:941–951.

Perry JA, Troelstrup NH, Newsom M, Shelley B. Whole ecosystem manipulation experiments: the search for generality. *Wat Sci Tech* 1987;19:55–71.

Perry JA, Vanderklein E. Environmental problem solving in an age of electronic communication: toward an integrated or reductionist model? In: Lemmons JE, ed. *The role of scientific uncertainty in environmental problem solving*. Blackwell Scientific (in press)

Perry JA, Vanderklein E, Ek AR. Breadth versus depth in forestry education: the case for a more conceptual approach. *J Forestry* 1994;92:60.

Perry JA, Vanderklein E, Streckansky B. Environmental management training in Slovakia: teaching new ways to think. *Environ Professional* (in press)

Perry JA, Ward RC, Loftis JC. Survey of state water quality monitoring programs. *Environ Mgt* 1984;8:21–26.

Perry JA, Zeyen R, Newsom M, Ahlstrand G. X-ray microanalysis of leaf litter decomposing in lakes. *BioScience* 1989;39:260–263.

Petulla JM. Environmental values: the problem of method in environmental history. In: Bailes KE, ed. *Environmental history: critical issues in comparative perspective*. New York: University Press of America, 1985:36–63.

Phillips JD. Upstream pollution sources and coastal water quality protection in North Carolina. *Coastal Mgt* 1991;19:439–449.

Pimm SL, Lawton JH, Cohen JE. Food web patterns and their consequences. *Nature* 1991;350:669–674.

Platte AE. Dying seas. *World Watch* 1995;6:10–19.

Pointing C. *A green history of the world*. London: Sinclair Stevenson, 1991.

Postel S. *Air pollution, acid rain and the future of forests.* Worldwatch paper 58. Washington, DC: Worldwatch Institute, 1984:54 pp.

Postel S. Water and agriculture. In: Gleick P. *Water in crisis: a guide to the world's fresh water resources.* New York: Oxford University Press, 1993:56–66.

Postel S. Rivers drying up. *World Watch* 1995;8:9–19.

Power M, Power G, Dixon DG. Detection and decision-making in environmental effects monitoring. *Environ Mgt* 1995;19:629–641.

Pujin V. Changes in the composition of the Danube River Basin biocenosis resulting from anthropogenic influence. *Wat Sci Tech* 1990;22:13–30.

Purdy WC, Ritter HP. *Investigation of the pollution and sanitary conditions of the Potomac Watershed with special reference to self purification and the sanitary conditions of shellfish in the Lower Potomac River.* Washington, DC: Government Printing Office, 1916.

Rapp D, Murray-Rust D, Christiansson C, Berry L. Soil erosion and sedimentation in four catchments near Dodoma, Tanzania. *Geografiska Annaler* 1972;54A:255–318.

Rapport DJ. What constitutes ecosystem health? *Perspectives Biol Med* 1989;33:120–132.

Rapport DJ. Challenges in the detection and diagnosis of pathological change in aquatic ecosystems. *J Great Lakes Res* 1990;16:609–618.

Rapport DJ. Evaluating ecosystem health. *J Aquat Ecosyst Hlth* 1992;1:15–24.

Rebuffoni D. Interests collide over locks on vital waterway. *Minneapolis Star Tribune* Sept. 26, 1994:1A, 21A–23A.

Regier HA, Bronson EA. New perspectives on sustainable development and barriers to relevant information. *Environ Monitor Assess* 1992;20:111–120.

Ribsame W. Social constraints on adjusting water resources management to anthropogenic climate change. In: Abu-Zeid M, Biswas A, eds. *Climatic fluctuations and water management.* Oxford/Boston: Butterworth-Heinemann, 1992:210–226.

Richards C, Minshall GW. Spatial and temporal trends in stream macroinvertebrate communities: the influence of catchment disturbance. *Hydrobiol* 1992:241:173–184.

Richardson CJ. Ecological functions and human values in wetlands: a framework for assessing forestry impacts. *Wetlands* 1994;14:1–9.

Richardson RE. The bottom fauna of the Middle Illinois River, 1913–1925: its distribution, abundance, valuation and index value in the study of stream pollution. *Bull Ill Nat Hist Surv* 1928;17:387–482.

Roberts R, Lighthall D. The political economy of agriculture, groundwater quality management, and agricultural research. *Wat Resour Bull* 1991;27:437–446.

Rodda DW. The environmental programme for the Danube Basin. *Wat Sci Tech* 1994;30:135–145.

Ross S. Mechanisms structuring stream fish assemblages: are there lessons from introduced species? *Environ Biol Fishes* 1990;30:359–368.

Rosseland B, Staurnes M. Physiological mechanisms for toxic effects and resistance to acidic water: an ecophysiological and ecotoxicological approach. In: Steinberg C, Wright R, eds. *Acidification of freshwater ecosystems: implications for the future.* New York: John Wiley and Sons, 1994:227–246.

Rutherford D, Walker-Bryan B. Acidification of the lower Mississippi River. *Trans Am Fish Soc* 1992;121:369–377.

Salau AT. Integrated water management: the Nigerian experience. In: Mitchell B, ed. *Integrated water management: international experiences and perspectives.* London: Bellhaven Press, 1990:188–201.

Scarpino PV. *Great river: an environmental history of the Upper Mississippi, 1890–1950.* Columbia: University of Missouri Press, 1985.

Schindler DW. Experimental perturbations of whole lakes as tests of hypotheses concerning ecosystem structure and function. *Oikos* 1990;57:25–41.

Schindler DW. Lakes and oceans as functional wholes. In: Barnes RS, Mann KH, eds. *Fundamentals of aquatic ecology.* 2nd ed. Oxford: Blackwell Scientific Publications, 1991.

Schindler DW. Changes caused by acidification to the biodiversity: productivity and biogeochemical cycles of lakes. In: Steinberg C, Wright R, eds. *Acidification of freshwater ecosystems: implications for the future.* New York: John Wiley and Sons, 1994:153–164.

Schindler DW, Beaty K, Fee E, *et al.* Effects of climate warming on the lakes of the Central Boreal Forest. *Science* 1990;250:967–970.

Schindler DW, Frost T, Mills K, *et al.* Comparisons between experimentally—and atmospherically—acidified lakes during stress and recovery. *Proc Roy Soc Edinburgh* 1991;97B:193–226.

Shaw DG. The Exxon-Valdez oil spill: ecological and social consequences. *Environ Conserv* 1993;19:253–258.

Shiklomanov IA. World fresh water resources: In: Gleick PH, ed. *Water in crisis: a guide to the world's fresh water resources.* Oxford: Oxford University Press, 1993:13–24.

Shirai Y. Institutional arrangements for Japanese water resources. In: Mitchell B, ed. *Integrated water management: international experiences and perspectives.* London: Bellhaven Press, 1990:148–171.

Shortreed KS, Stockner JG. The impact of logging on periphyton biomass and species composition in Carnation Creek: a coastal rain forest stream on Vancouver Island, British Columbia. In: Hartman GF, ed. *Proceedings of the Carnation Creek workshop: a 10-year review.* Nanaimo, British Columbia: Pacific Biological Station, 1982:197–209.

Skaggs R, Breve M, Gilliam J. Hydrologic and water quality impacts of agricultural drainage. *Crit Rev Environ Sci Tech* 1994;24:1–32.

Slocombe DS. Environmental planning, ecosystem science and ecosystem approaches for integrating environment and development. *Environ Mgt* 1993;17:289–303.

Sly PG. The effects of land use and cultural development on the Lake Ontario ecosystem since 1750. *Hydrobiol* 1991;213:1–75.

Smith C. Riparian afforestation effects on water yields and water quality in pasture catchments. *J Freshwat Qual* 1992;21:237–245.

Smith DG, Cragg AM, Crocker GH. Water clarity criteria for bathing waters based on user perception. *J Environ Mgt* 1991;33:285–299.

Smith DG, Davies-Colley R. Perception of water clarity and colour in terms of suitability for recreational use. *J Environ Mgt* 1992;36:225–235.

Smith J. From global to regional climate change: relative knowns and unknowns about global warming. *Fisheries* 1990;15:2–6.

Smith SV, Buddemeier RW. Global change and coral reef ecosystems. *Ann Rev Ecol Syst* 1992;23:89–118.

Solley WB, Chase EB, Mann WB. *Estimated use of water in the United States in 1980.* Washington, DC: US Geol. Survey, 1983:Circular 1001.

Someya A, Kato H, Kaneko A. Report on the basic legal structure and current movement on post and coastal zone in Japan. *Coastal Mgt* 1992;20:49–72.

Southerland MT, Stribling JB. Status of biological criteria development and implementation. In: Davis WS, Simon TP, eds. *Biological assessment and criteria: tools for water resource planning and decision making.* Boca Raton: Lewis Publishers, 1995:81–107.

Sparks RE. Need for ecosystem management of large rivers and their floodplains. *BioScience* 1995;45:168–182.

Speidel DH, Agnew AF. Water. In: *The natural geochemistry of our environment.* Boulder: Westview Press, 1982.

Stanford JA, Ward JV. Management of aquatic resources in large catchments: recognizing interactions between ecosystem connectivity and environmental disturbance. In: Naiman RJ, ed. *Watershed management: balancing sustainability and environmental change.* New York: Springer Verlag, 1992:91–124.

Stenson JAE, Svensson JE, Cronberg G. Changes and interactions in the pelagic community in acidified lakes in Sweden. *Ambio* 1993;22:277–282.

Stone L, Weisburd RS. Positive feedback in aquatic ecosystems. *Trends Ecol Evol* 1992;7:263–267.

Stowe K. *Essentials of ocean science.* New York: Wiley & Sons, 1987.

Strauss SH, Bernard SM. Power and the people. *Environ Forum* 1991;8:11–15.

Streeter HW, Phelps EB. *A study of the pollution and natural purification of the Ohio River. III. Factors concerned in the phenomenon of oxidation and reaeration.* U.S. Public Health Service Bulletin 1925;146:1–75.

Suschka J. Effects of heavy metals from mining and industry on some rivers in Poland: an already exploded chemical time bomb? *Land Degrad Rehab* 1993;4:387–391.

Syme GJ, Williams KD. The psychology of drinking water quality—an exploratory study. *Wat Resour Res* 1993;29:4003–4010.

Tarzwell CM, Gaufin AR. Some important biological effects of pollution often regarded in stream surveys. *Proc 8th Indust Waste Conf* 295–316. Illinois: Purdue University Engineering Bulletin, 1953.

Tchobanoglous G, Schroeder ED. *Water quality: characteristics, modeling and modification.* N. Reading, MA: Addison-Wesley, 1987.

Thill MR. Hatchery offer nixed. *Press-Republican* May 4, 1993:13.

Thompson WT. Is irrigation detrimental to trout culture? *Trans Am Fish Soc* 1912;41:103–114.

Thornton JA, McMillan PH. Reconciling public opinion and water quality criteria in South Africa. *Wat S Africa* 1989;15:221–226.

Thurman E, Goolsby D, Meyer M, Kolpin D. Herbicides in surface waters of the midwestern United States: the effect of spring flush. *Environ Sci Tech* 1991;25:1794–1796.

Townsend C, Crowl T. Fragmented population structure in a native New Zealand fish: an effect of introduced brown trout? *Oikos* 1991;61:347–354.

Townsend C, Winterbourn M. Assessment of the environmental risk posed by an exotic fish: the proposed introduction of channel catfish (*Ictalurus punctatus*) to New Zealand. *Conserv Biol* 1992;26:273–282.

Trama FB. The acute toxicity of copper to the common bluegill (*Lepomis macrochirus* Rafinesque). *Notalae Naturae* 1954;257:13 pp.

Troelstrup N, Perry JA. Interpretation of scale dependent inferences from water quality data. *Proceedings of the 1990 Midwest Pollution Control Biologists meeting.* Chicago, Ill. April 10–13, 1990. USEPA Region V. Environmental Sciences Division, 1990.

Trojan MD, Perry JA. Analysis of hydrogeologic sensitivity in Winona County, Minnesota. *J Minn Acad Sci* 1989;54:30–36.

Turner RK. Valuation of wetland systems. In: Opschoor JB, Pearce DW, eds. *Persistent pollutants*. Netherlands: Kluwer Academic Publishers, 1991:55–63.

Tursunov AA. The Aral Sea and the ecological situation in Central Asia and Kazakhstan. Translated from *Gidrotekhnicheskoe Stroitel'stvo* 1989;6:15–19.

Upper Saranac Lake Association. Fact sheets 1990–1995; newsletters 1992–1995. New York: Saranac Lake.

————. USLA to sue state for pollution of Upper Saranac water. *Tupper Lake Free Press* January 27, 1993a.

————. DEC prepares response to USLA. *Adirondack Daily Enterprise*. March 24, 1993b.

Vallner L, Sepp K. Effects of pollution from oil shale mining in Estonia. *Land Degrad Rehab* 1993;4:381–385.

Vanek V. The interactions between lake and groundwater and their ecological significance. *Stygologia* 1987;3:1–23.

Vitousek P. Beyond global warming: ecology and global change. *Ecology* 1994;75:1861–1876.

Walbridge MR, Lockaby BG. Effects of forest management on biogeochemical functions in southern forested wetlands. *Wetlands* 1994;14:10–17.

Walker DA, Walker MD. History and pattern of disturbance in Alaskan arctic terrestrial ecosystems: a hierarchical approach to analysing landscape change. *J Appl Ecol* 1991;28:244–276.

Warren CE. *Biology and water pollution*. Philadelphia: W.B. Saunders, 1971.

Waterstone M. Adrift on a sea of platitudes: why we will not resolve the greenhouse issue. *Environ Mgt* 1993;17:141–152.

Welcomme R. International transfers of inland fish species. In: Drake J, ed. *Biological invasions: a global perspective*. SCOPE. New York: John Wiley & Sons, 1992:22–40.

Wetzel RG. *Limnology*. 2nd ed. Philadelphia: Saunders Press, 1983:767 pp.

Wiens JA. Spatial scaling in ecology. *Functional Ecol* 1989;3:385–397.

Williams J, Jennings D. Computerized data base for exotic fishes: the western United States. *Calif Fish Game* 1991;77:86–93.

Wilson CB, Walker WW Jr. Development of lake assessment methods based upon the aquatic ecoregion concept. *Lake Reservoir Mgt* 1989;5:11–22.

Windom HL. Contamination of the marine environment from land-based sources. *Mar Pollut Bull* 1992;25:32–36.

Wing-Hung Lo C. Environmental protection in Hong Kong amidst transition: is Hong Kong ready to manage its environment by Law? *Environ Mgt* 1995;19:331–344.

Wingate P. U.S. state's view and regulations on fish introductions. *Can J Fish Aquat Sci* 1991;48(suppl 1):167–170.

Winterbourn MJ, Townsend CR. Streams and rivers: one-way flow systems. In: Barnes RSK, Mann KH, eds: *Fundamentals of aquatic ecology*. 2nd ed. Oxford: Blackwell Scientific Publications, 1991:230–242.

Witte F, Golschmidt T, Goudswaard P, Ligtvoet W, van Oijen M, Wanink J. Species extinction and concomitant ecological changes in Lake Victoria. *Netherlands J Zool* 1992;42:214–232.

Wolman MG. Changing national water quality policies. *J Wat Pollut Contr Fed* 1988;60:1774–1781.

World Resources Institute. *World resources 1992–1993: a guide to the global environment.* Oxford: Oxford University Press, 1992.

World Resources Institute. *World resources 1993–1994.* World Resources Institute. Oxford: Oxford University Press, 1993.

Yap HT. Marine environmental problems: experiences of developing regions. *Mar Pollut Bull* 1992;25:37–40.

Yevtushenko NY, Bren NV, Sytnik YM. Heavy metal contents in invertebrates of the Danube River. *Wat Sci Tech* 1990;22:119–125.

Yoder CO. Policy issues and management applications for biological criteria. In: Davis WS, Simon TP, eds. *Biological assessment and criteria: tools for water resource planning and decision making.* Boca Raton: Lewis Publishers, 1995:327–344.

Index